Advances in

Geosciences

Volume 21: Solar Terrestrial (ST)

ADVANCES IN GEOSCIENCES

Editor-in-Chief: Wing-Huen Ip *(National Central University, Taiwan)*

A 5-Volume Set

Volume 1: Solid Earth (SE)
ISBN-10 981-256-985-5

Volume 2: Solar Terrestrial (ST)
ISBN-10 981-256-984-7

Volume 3: Planetary Science (PS)
ISBN-10 981-256-983-9

Volume 4: Hydrological Science (HS)
ISBN-10 981-256-982-0

Volume 5: Oceans and Atmospheres (OA)
ISBN-10 981-256-981-2

A 4-Volume Set

Volume 6: Hydrological Science (HS)
ISBN-13 978-981-270-985-1
ISBN-10 981-270-985-1

Volume 7: Planetary Science (PS)
ISBN-13 978-981-270-986-8
ISBN-10 981-270-986-X

Volume 8: Solar Terrestrial (ST)
ISBN-13 978-981-270-987-5
ISBN-10 981-270-987-8

Volume 9: Solid Earth (SE), Ocean Science (OS) & Atmospheric Science (AS)
ISBN-13 978-981-270-988-2
ISBN-10 981-270-988-6

A 6-Volume Set

Volume 10: Atmospheric Science (AS)
ISBN-13 978-981-283-611-3
ISBN-10 981-283-611-X

Volume 11: Hydrological Science (HS)
ISBN-13 978-981-283-613-7
ISBN-10 981-283-613-6

Volume 12: Ocean Science (OS)
ISBN-13 978-981-283-615-1
ISBN-10 981-283-615-2

Volume 13: Solid Earth (SE)
ISBN-13 978-981-283-617-5
ISBN-10 981-283-617-9

Volume 14: Solar Terrestrial (ST)
ISBN-13 978-981-283-619-9
ISBN-10 981-283-619-5

Volume 15: Planetary Science (PS)
ISBN-13 978-981-283-621-2
ISBN-10 981-283-621-7

A 6-Volume Set

Volume 16: Atmospheric Science (AS)
ISBN-13 978-981-283-809-4
ISBN-10 981-283-809-0

Volume 17: Hydrological Science (HS)
ISBN-13 978-981-283-811-7
ISBN-10 981-283-811-2

Volume 18: Ocean Science (OS)
ISBN-13 978-981-283-813-1
ISBN-10 981-283-813-9

Volume 19: Planetary Science (PS)
ISBN-13 978-981-283-815-5
ISBN-10 981-283-815-5

Volume 20: Solid Earth (SE)
ISBN-13 978-981-283-817-9
ISBN-10 981-283-817-1

Volume 21: Solar Terrestrial (ST)
ISBN-13 978-981-283-819-3
ISBN-10 981-283-819-8

Advances in

Geosciences

Volume 21: Solar Terrestrial (ST)

Editor-in-Chief

Wing-Huen Ip

National Central University, Taiwan

Volume Editor-in-Chief

Marc Duldig

Australian Antarctic Division, Australia

World Scientific

NEW JERSEY • LONDON • SINGAPORE • BEIJING • SHANGHAI • HONG KONG • TAIPEI • CHENNAI

Published by

World Scientific Publishing Co. Pte. Ltd.
5 Toh Tuck Link, Singapore 596224
USA office: 27 Warren Street, Suite 401-402, Hackensack, NJ 07601
UK office: 57 Shelton Street, Covent Garden, London WC2H 9HE

British Library Cataloguing-in-Publication Data
A catalogue record for this book is available from the British Library.

ADVANCES IN GEOSCIENCES
A 6-Volume Set
Volume 21: Solar Terrestrial (ST)

ISBN-13 978-981-283-808-7 (Set)
ISBN-10 981-283-808-2 (Set)
ISBN-13 978-981-283-819-3 (Vol. 21)
ISBN-10 981-283-819-8 (Vol. 21)

Typeset by Stallion Press
Email: enquiries@stallionpress.com

Printed in Singapore.

EDITORS

Editor-in-Chief: Wing-Huen Ip

Volume 16: Atmospheric Science (AS)
Editor-in-Chief: Jai Ho Oh
Editors: G. P. Singh
C. C. Wu
K.-J. Ha

Volume 17: Hydrological Science (HS)
Editor-in-Chief: Namsik Park
Editors: Ji Chen
Joong-Hoon Kim
Jinping Liu
Young-Il Moon
Sanjay Patil
Ashok Kumar Rastogi
Simon Toze

Volume 18: Ocean Science (OS)
Editor-in-Chief: Jianping Gan
Editors: Minhan Dai
Anne Mueller
Murty Vadiamani

Volume 19: Planetary Science (PS)
Editor-in-Chief: Anil Bhardwaj
Editors: Yasumasa Kasaba
Guillermo Manuel Muñoz Caro
Takashi Ito
Paul Hartogh
C. Y. Robert Wu
S. A. Haider

Volume 20: Solid Earth (SE)
Editor-in-Chief: Kenji Satake

Volume 21: Solar & Terrestrial Science (ST)
Editor-in-Chief: Marc Duldig
Editors: P. K. Manoharan
Andrew W. Yau
Q.-G. Zong

REVIEWERS

The Editors of Volume 21 (Solar & Terrestrial Science) would like to acknowledge the following referees who have helped to review the papers published in this volume:

A. Asai
Nanan Balan
Dipankar Banerjee
I. V. Chasei
Yen-Hsyang Chu
Mark Conde
Len Culhane
P. Demoulin
Marc Duldig
Bhola Dwivedi
Malcolm Dunlop
Peter Dyson
Geza Erdos
R. A. Fallows
John French
T. A. Fritz
Zenjiro Fuji
Ivan Galkin
Chaosong Huang
Boian Kirov
Masayoshi Kojima
Axel Korth
Karel Kudela
James LaBelle
Bo Li
Jun Lin
D. Maia
C. Mandrini
P. K. Manoharan
Karim Meziane
Wojciech Miloch
Kyoung Min
Yong-Jae Moon
Harm Moraal
Tomoko Nakagawa
Viviane Pierrard
Ian Richardson
Yoshifumi Saito
Brigitte Schmieder
Jan Soucek
Anming Tian
J.-N. Tu
Ilya Usoskin
Qian Wu
Andrew W. Yau
Jingxiu Wang
Y-F. Wang
Q.-G. Zong

PREFACE

The present volume set of Advances in Geosciences (ADGEO) contains papers from the Busan annual meeting in 2008 and some from the Singapore annual meeting in 2009. As Editor-in-Chief, I must apologize to the AOGS members and authors for this delay. Since 2006 we have published 20 volumes in total. This publication project has been supported by the AOGS Council, World Scientific Publication Company (WSPC), the team of hard working editors and the broad membership and participants of AOGS. As with the main purpose of the Society, ADGEO is meant to promote information exchange and to forge scientific cooperation in the area of Earth science and environmental study. As witnessed by the negotiation efforts at the United Nations Climate Change Conference in Copenhagen in December 2009, all these issues have become more and more important and vital in the Asia-Pacific region. It is not a matter of exaggeration in saying that the solution to global warming, if there is one, has to come from the emerging economies and developing countries covered by AOGS. By design, ADGEO has its fundamental role to play. In practical terms, it is actually a difficult task because of many factors involved in deciding the quality of manuscripts, editorial and review processes, publication procedure, scientific impacts, readership, policy of the AOGS Council, and last but not least, marketing from the point of view of the publisher. Any small mishap in this long chain of interactive steps could lead to a major discontinuity. We have encountered such a situation with the publication of the Busan manuscripts. It is only with the cooperation of the authors, the ADGEO editorial team, WSPC, and the AOGS Secretariat Office, that we are able to produce these volumes, albeit a long delay. With this lesson learned, we hope to consolidate the ADGEO management and editorial system so that it would become an essential publication in our understanding of Earth and space science and information tool books in the battle against climate change. Finally, I would like to take this opportunity to thank the Volume Editors-in-Chief who are the driving force in making ADGEO possible: A. Bhardwaj (Planetary Science), M. Duldig (Solar and Terrestrial Science), J.P. Gan (Ocean Science), J.H. Oh (Atmospheric

Science), N.S. Park (Hydrological Science), K. Satake and C.H. Lo (Solid Earth). They have to work very hard to ensure both the quantity and quality of the published papers in ADGEO. Of equal importance, the support from WSPC is essential and its foresight in identifying the academic and social values of Earth science and environmental study to be sustained and articulated by ADGEO is very much appreciated.

Wing-Huen Ip
Editor-in-Chief

PREFACE TO ST VOLUME

Solar terrestrial science is concerned with the interactions between the earth, the sun, and interplanetary space and the resulting effects these interactions induce at the earth such as those collectively know as space weather. This volume comprises recent research results and invited reviews on solar terrestrial research with an emphasis on work undertaken in the Asia Oceania region. Aspects of Solar Terrestrial research covered include: the solar cycle, sunspots, coronal magnetic fields, the solar wind, coronal mass ejections, the magnetosphere, the ionosphere, the mesosphere, cosmic rays and the heliosphere, and spacecraft mission instrument design. Observational measurements, modelling and theoretical developments are included and over one third of the papers are invited reviews by leading international researchers.

Marc Duldig
ST Volume Editor-in-Chief

CONTENTS

Advances in Geosciences
Vol. 21: Solar & Terrestrial Science (2008)
Ed. Marc Duldig

ACCELERATION AND HEATING OF SOLAR WIND IONS BY NONLINEAR WAVES

L. OFMAN
The Catholic University of America
and NASA Goddard Space Flight Center,
Greenbelt, Maryland, USA
Leon.Ofman@nasa.gov

Also at Department of Geophysics,
Tel Aviv University,
Tel Aviv, Israel

Recent advances in observations and modeling provide better understanding of the possible role of waves in the acceleration and heating of the solar wind. Ulysses, ACE, Helios, SOHO, TRACE, and other satellite observations found ample evidence for waves in the corona and in the solar wind. Numerical models show that nonlinear interaction between MHD waves and the plasma can provide the necessary momentum and heat input to produce the fast solar wind in coronal holes. It was also found that the heating of solar wind protons and heavy ions is more significant than of electrons. Temperature anisotropy suggests that ion-cyclotron wave heating is taking place in heavy ions. In this review article, several recent observations and numerical models of nonlinear wave driven wind are discussed.

1. Introduction

The physical mechanism that leads to heating and acceleration of the solar wind is still being intensively studied. Since the exact mechanism is not known, models of the solar wind rely on an ad-hoc modeling of the energy equation by using polytropic index below the adiabatic value,[1–3] or by including an ad-hoc heating function, and momentum input to model the acceleration and heating of the solar wind. A method of semiempirical determination of the heating function that relies on solar wind observations near the Sun and at 1AU was recently developed.[4,5] Another solar wind modeling approach relies on empirical variation of the polytropic index to match the global solar wind properties obtained from the solution of 3D MHD equations to observations.[6]

However, it is important to study possible acceleration and heating mechanism by including it from basic principles. In this review we will focus on nonlinear waves as the physical mechanism of coronal heating and solar wind acceleration. The acceleration of the solar wind by MHD waves was first suggested about four decades ago, and the heating of solar wind ions by ion-cyclotron waves was extensively studied in the past. *In-situ* observations by ACE, Ulysses, and Helios spacecraft provide ample evidence of MHD waves in the heliosphere. Slow magnetosonic waves were detected by SOHO in coronal plumes,[7–9] and observational evidence was found for Alfvén waves.[10] Interpretation of UVCS observations of O^{5+} ion emission suggests that the heavy ion temperature is anisotropic with $T_{\perp} > T_{\parallel}$. TRACE satellite observations found direct evidence of transverse waves in magnetic active region loops (first findings were reported by Refs. 11, 12). Very recent observations by Hinode satellite provide evidence that there is sufficient energy in the MHD wave spectrum in the transition region to power the solar wind.[13]

Resonant absorption of ion-cyclotron waves was one of the mechanisms suggested for the heating and acceleration of solar wind ions some time ago (for a review see, Ref. 14). However, a closer theoretical look at the ion-cyclotron heating model indicates that there are difficulties with energy transport mechanism from the corona to the solar wind in the multi-ion plasma.[15] An alternative view, is that the broad line widths detected in heavy ion emission and were interpreted as signatures for strong perpendicular heating[16,17] are due to unresolved MHD waves motions[18,19] or unresolved structures in the observational line of sight.[20]

One dimensional models that included WKB Alfvén wave pressure term to add momentum to the solar wind were studied for decades (for example, see Ref. 21). The WKB Alfvén waves were also included in 3D MHD global models.[22,23] Recently, the wave acceleration and heating of the solar wind was modeled with fully resolved waves in 2D (for a review see, Ref. 24). Nonlinear Alfvén wave acceleration and heating of the solar wind was studied with 2.5D MHD model,[25] and recently with 2.5D 3-fluid model.[26,27] In 2.5D models two spatial dimensions are considered, and three components of the magnetic field and flow are included. Acceleration and heating by shock wave trains and by nonlinear Alfvén waves was studied recently using 1.5D MHD model that start from the photospheric level,[28–30] rather than the base of the corona.

Hybrid models of solar wind plasma allow investigating the kinetics of the ions, and the resonant interaction between ion-cyclotron waves and the plasma in fully nonlinear regime, relaxing the approximations used in linear and quasi-linear theories. Ofman *et al.*[31] studied the evolution of the heavy ion anisotropy, and the conditions for marginal stability using 1D hybrid simulations for low-β coronal plasma. Gary *et al.*[32,33] investigated the evolution of anisotropy using 2D hybrid simulations for $\beta \sim 1$ conditions, and for He^{++} ions. Ofman *et al.*[34] investigated the heating and the acceleration of the solar wind plasma with heavy ions by a source f^{-1} spectrum of waves using 1D hybrid model. Liewer *et al.*,[35] and Hellinger *et al.*[36] investigated the wave heating of the solar wind in an expanding box model. The effects of drift as well as source spectrum in the multi-ion plasma were studied by Xie *et al.*[37] The observational implication of the heavy ion motion due to the effect of cyclotron waves was investigated by Ofman *et al.*[38] Recently, the effect of the source wave spectrum and velocity drift on the perpendicular heating of heavy ions was investigated using 2D hybrid code by Ofman and Viñas.[39]

2. Observational Motivation

Some of the most intriguing observations of the SOHO spacecraft were obtained with the UVCS instrument that imaged the EUV emission of oxygen ions (O^{5+}) and of hydrogen (Ly α) in solar coronal holes over several solar radii. Spectroscopic observations of the ion emission lines in the corona provide information on the velocity distribution function of the ions. By using the observed line width and intensity of the oxygen doublet (1032 Å, and 1037 Å), Mg X, and Ly α emission the range of the parallel and perpendicular velocity distribution widths, and the bulk outflow velocity of the ions were estimated.[15–17,40,41] The surprising findings of large temperature anisotropy and large outflow speed of O^{5+} ions (compared to protons) in coronal holes lead to new theoretical insights and ideas on coronal heating and acceleration mechanisms. In the above studies, the UVCS observations were interpreted as evidence for perpendicular ion heating by resonant ion-cyclotron wave spectrum. However, we note that other interpretation of these observations that does not involve ion-cyclotron waves is possible.[18,20]

The magnitude of the unresolved wave motions in the solar corona can be deduced form nonthermal line broadening of emission lines.[10] This is due to the Doppler shift $\Delta\lambda/\lambda = V/c$, induced by the motions of the plasma-carrying field lines in the observed line of sight. In the above expression λ is the wavelength of the emission line, $\Delta\lambda$ is the line width, V is the

Doppler velocity and the c is the speed of light. The effective temperature that combines kinetic and unresolved motions is given by

$$T_{\text{eff}} = T + \langle V_\phi^2 \rangle \frac{A m_p}{2k_b}, \tag{1}$$

where T is the kinetic temperature, V_ϕ is the Alfvén wave amplitude, A is the atomic mass number of the ions, m_p is the proton mass, and k_b is Boltzmann's constant.

The detailed observations of streamers with SOHO/UVCS exhibit similar preferential heating of heavy ions, although to lower temperatures than in coronal holes[42–44] (see Fig. 1). The spectroscopic streamer observations also show that the outflow speed of the solar wind is strongly dependent on the location in the streamer structure, with the plasma on closed field regions practically at rest, and the plasma on open field region exhibiting significant outflow speed on the order of 100 km/s. The waves can contribute to the observed line width through Doppler broadening, as well as by the heating and acceleration of solar wind plasma.

In situ measurements of Ulysses and Helios spacecraft in the high-speed streams have revealed a variety of nonthermal features of ion velocity distributions.[45–47] Proton distributions are often appeared as double-peaked stream with an average relative drift velocity parallel to the background magnetic field.[48,49] One of the goals of nonlinear hybrid

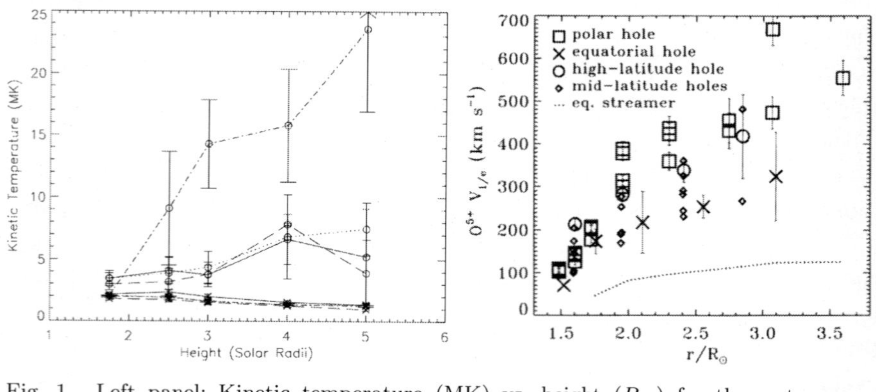

Fig. 1. Left panel: Kinetic temperature (MK) vs. height ($R_\odot$) for three streamers studied by Uzzo *et al.*:[44] Str030428 (solid lines), Str030509 (dashed lines), Str030925 (dotted lines), and for the streamer studied by Strachan *et al.*[42] (dash-dotted lines) for O^{5+} ions (circles) and protons (crosses). Adapted from Uzzo *et al.*[44] Right panel: radial dependence of $V_{1/e}$ line width velocity in polar (squares), equatorial (crosses), high-latitude (circles) and mid-latitude (small diamonds) coronal holes (left panel) for O^{5+} emission lines. Adapted from Miralles *et al.*[43]

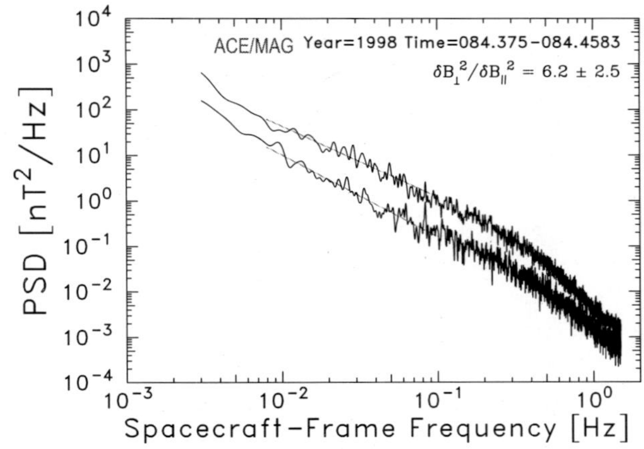

Fig. 2. Example of magnetic spectrum from the ACE database. Year/day of sample is given. The top curve represents the summed power in the two components perpendicular to the mean magnetic field, and the lower curve represents the power in the parallel component. The fit functions with power exponent -1.56 ± 0.3 (upper) and -1.8 ± 0.03 (lower) are shown. Adapted from Smith *et al.*[52]

kinetic modeling reviewed below is to examine the effects of turbulent wave spectrum on ion velocity distributions. The results can be compared to *in situ* measurements in the high-speed solar wind. Recent observations by ACE (see Fig. 2) and Wind spacecraft of the solar wind protons at 1AU indicate that the turbulent cascade rate agrees better with Kreichnan rate ($f^{-3/2}$, where f is the fluctuation frequency), rather than with Kolmogorov ($f^{-5/3}$) rate.[50,51]

3. Overview of Recent Models of Wave Driven Wind

A brief overview of several recent models of wave driven solar wind is given in this section. In particular, the results of one-dimensional and two-dimensional single fluid wave driven wind models are reviewed, as well as the results of recent 2.5D three-fluid model. The results of hybrid models of wave heating of solar wind plasma by resonant ion-cyclotron wave spectrum are presented.

3.1. *Single-fluid MHD models*

The acceleration of the solar wind by WKB wave pressure term was studied for decades in the past, using mostly one-dimensional models. The first 2.5D MHD model of fast solar wind accelerated by fully resolved nonlinear

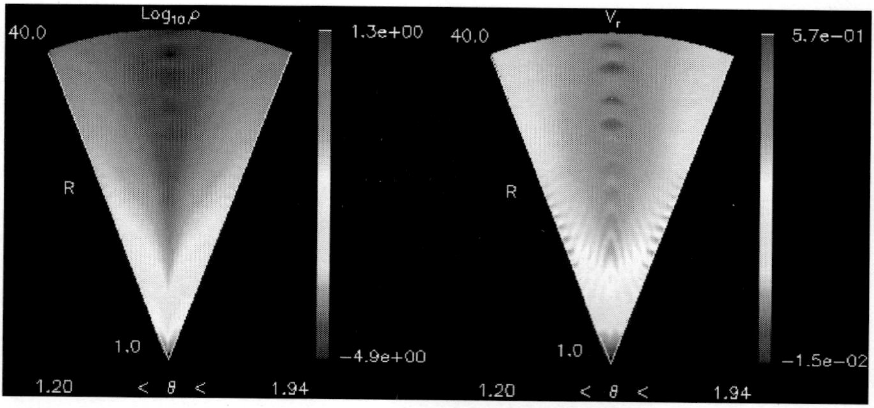

Fig. 3. Nonlinear-wave driven fast solar wind in coronal holes obtained from 2.5D MHD model developed by Ofman and Davila.[25] Left panel shows the density structure and the right panel shows the solar wind outflow speed in the region 1 to 40 $R_\odot$. Adapted form Ofman and Davila.[25]

Alfvén waves was developed by Ofman and Davila.[25] In that model the waves were driven at the base of the coronal hole, and their propagation into the coronal hole and the resulting acceleration of the solar wind were modeled self-consistently. In Fig. 3 the density structure and the fast solar wind radial outflow velocity are shown. The amplitude of the driving Alfvén wave was $v_d = 46\,\mathrm{km\,s^{-1}}$, the driving frequency was $f = 0.35\,\mathrm{mHz}$, the magnetic field in the center of the coronal hole at $r = 1R_\odot$ was $B = 7\,\mathrm{G}$, the temperature was $T = 1.4 \times 10^6\,\mathrm{K}$, the number density was $n_0 = 10^8\,\mathrm{cm^{-3}}$, and the nonradially expanding magnetic field configuration with expansion ratio $f_e = 10$ was used. Due to nonlinearity the Alfvén waves produce compressional waves in a process similar to parametric instability, as evident in v_r and ρ with highest amplitude near the center of the coronal hole.

For the purpose of comparison to one-dimensional models the radial cut through the center of the coronal hole is shown in Fig. 4. The radial (V_r) and transverse (V_ϕ) velocity components of the solar wind are shown for several wave frequencies. For reference, Parker's[53] isothermal solar wind solution is shown. The enhanced (over Parker) solar wind acceleration by the nonlinear waves is evident. It is also evident that low frequency waves provide higher momentum to the solar wind than high frequency MHD waves. This is due to increased wavelength of the wave resulting in stronger reflection from the inhomogeneous coronal hole density. Later, this model

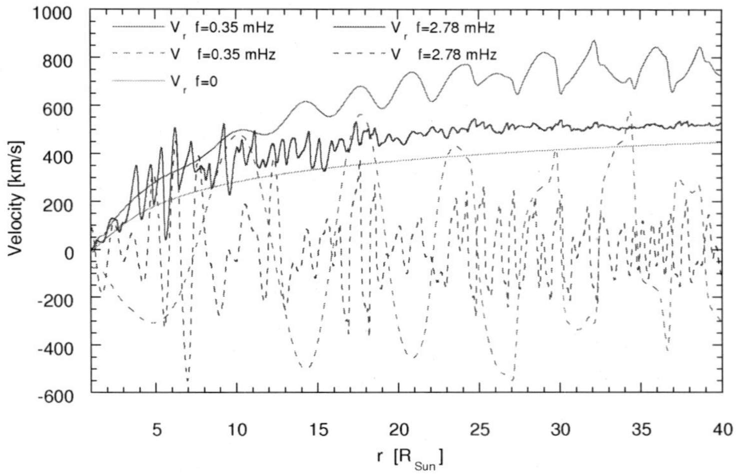

Fig. 4. A radial cut through the center of the fast wind in the coronal hole model shown in Fig. 3. The radial and transverse solar wind velocities are shown for several wave frequencies. The green curve shows the Parker's isothermal solar wind solution. Adapted form Ofman and Davila.[25]

was extended to two- and three-fluid plasma with a driving spectrum of Alfvén waves[19,26,54,55] (see below).

Recently, Suzuki and Inutsuka[29] investigate the acceleration of the fast solar wind by nonlinear Alfvén waves using one-dimensional (1.5D) MHD single fluid model. The main new feature of this model was that the Alfvén waves were initiated at the photospheric level, rather than at the base of the corona. The model allows to trace the flow of wave energy from the photosphere into the corona, taking into account the effects of reflection from large density gradients of the low frequency ($\lesssim$0.05 Hz) transverse waves. The authors performed parametric studies of the model, and investigated the effects of boundary conditions, wave amplitude, and expansion factor on the resulting solar wind. They found that the solar wind bulk properties agree with observations near 1AU.

3.2. *Multi-fluid model result*

Recently, Ofman[26] developed 2.5D three-fluid model of wave driven fast solar wind. Here, a brief review of these results is given. Ofman[26] solved the coupled three-fluid equations, that describe the dynamics of protons, and He^{++} or O^{5+} ions in a magnetized coronal hole. The electron momentum equation was solved with the approximation of negligible electron inertia.

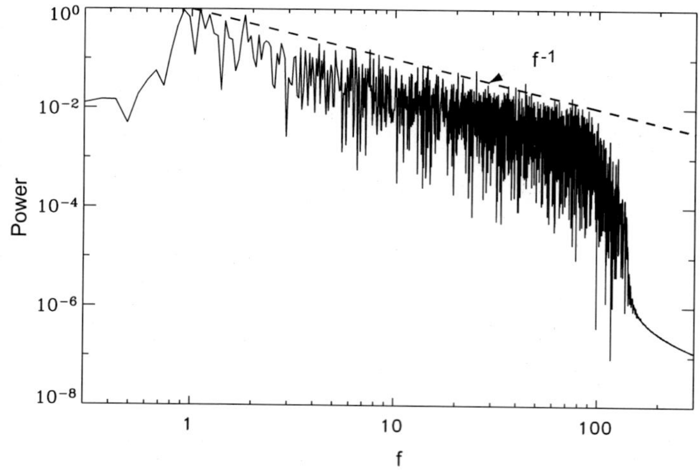

Fig. 5. The typical form of the normalized wave power spectrum used in the three-fluid and in hybrid model to drive the waves. Adapted from Ofman.[26]

An f^{-1} spectrum of Alfvén waves (see Fig. 5) was driven at the lower coronal hole boundary. A snapshot of the resulting two-dimensional solution showing the velocity components of protons and He^{++} ions is shown in Fig. 6. The Alfvénic fluctuations were applied non-uniformly with a random phase, and as a result one can see wave trains propagating into the coronal hole in the V_ϕ components. The associated compressive fluctuations in V_r are driven nonlinearly by the Alfvénic fluctuations.

The radial dependence of the solar wind velocity, density, and temperature obtained by averaging the model results over θ are shown in Fig. 7. It is evident that He^{++} ions flow faster than protons, in agreement with observations. The proton flow is driven mostly by wave acceleration and dissipation, while additional empirical heating was deposited into the He^{++} ions to obtain faster outflow. The right panel shows the corresponding densities (thick curves) and temperatures (thin curves) of electrons, protons, and He^{++} ions.

In this model, as the Alfvén waves propagate into the coronal hole, they deposit momentum into the solar wind, and dissipate via enhanced (turbulent) viscosity and resistivity. The high frequency end of the Alfvén wave nonresonant spectrum dissipates closer to the Sun, than the low frequency waves. As a result, the spectrum is modified, and becomes steeper than the source f^{-1} spectrum. The result of this process is evident in the spectrum of Alfvénic fluctuations shown at a distance of $18R_\odot$ (Fig. 8).

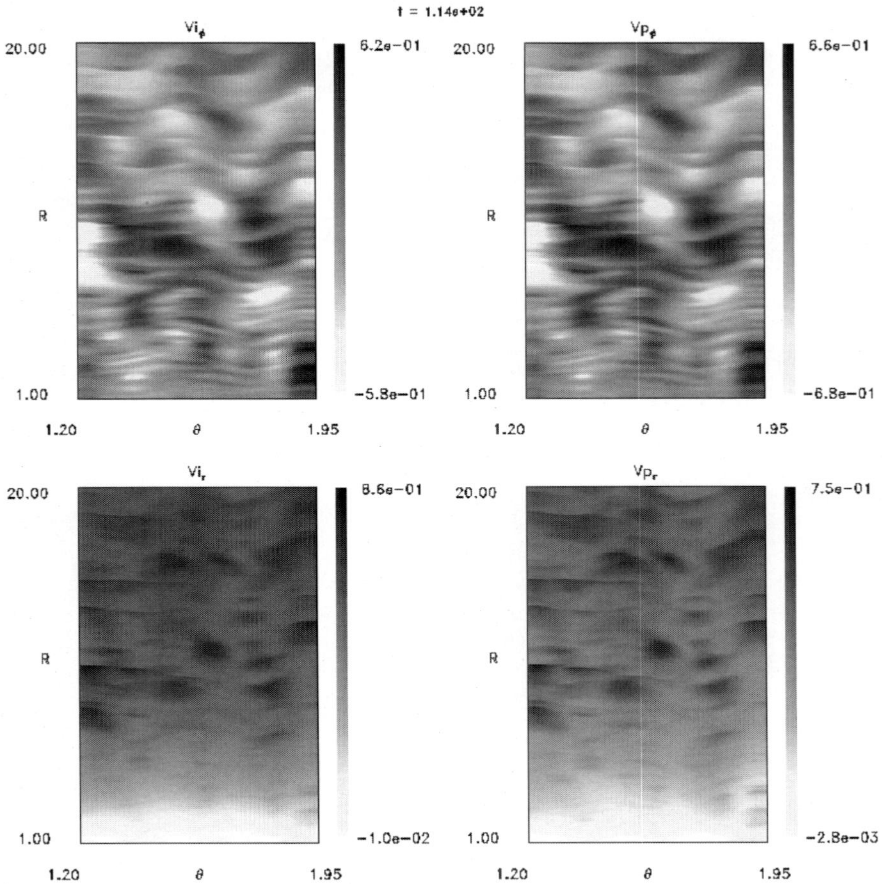

Fig. 6. A snapshot of the fast solar wind outflow driven by Alfvén waves in a model coronal hole calculated with the three-fluid model by Ofman.[26] The left panels show the He^{++} velocity components, and the right panels show the protons velocity components (upper panels: V_ϕ; lower panels: V_r). The velocities are normalized by the Alfvén speed $V_A = 1527\,\mathrm{km\,s^{-1}}$.

The steepening of the spectrum is also consistent with the dissipation of magnetosonic shocks arising from the large amplitude, nonlinear Alfvén waves.[28]

The kinetic and effective (line width) temperature of protons, and He^{++} ions calculated with the three-fluid model by Ofman[26] is shown in Fig. 9. The temperatures were averaged over the width of the coronal hole in the model. In the three-fluid Alfvén wave driven solar wind model it was found that the proton effective temperature is dominated by the contribution of

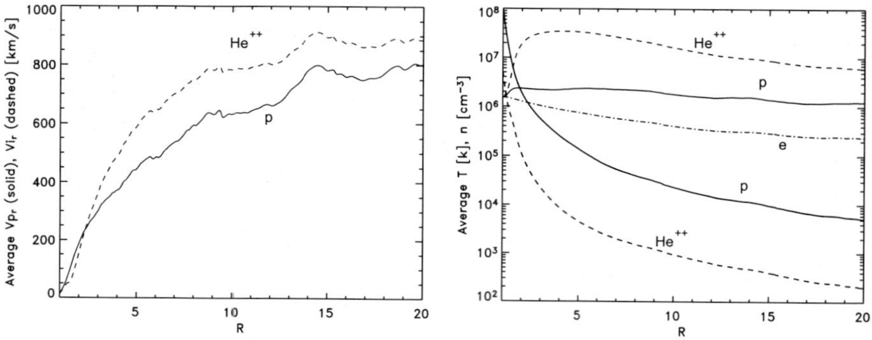

Fig. 7. The solar wind outflow obtained from the 2.5D three-fluid model with He^{++} driven by the wave spectrum. The left panel shows the outflow speed of protons (solid curve) and He^{++} (dashes). The right panel shows the corresponding densities (thick curves) and temperatures (thin curves). Additional heat was deposited into the He^{++} ions to obtain the faster than protons outflow. The above quantities were averaged over the width of the model coronal hole.[26]

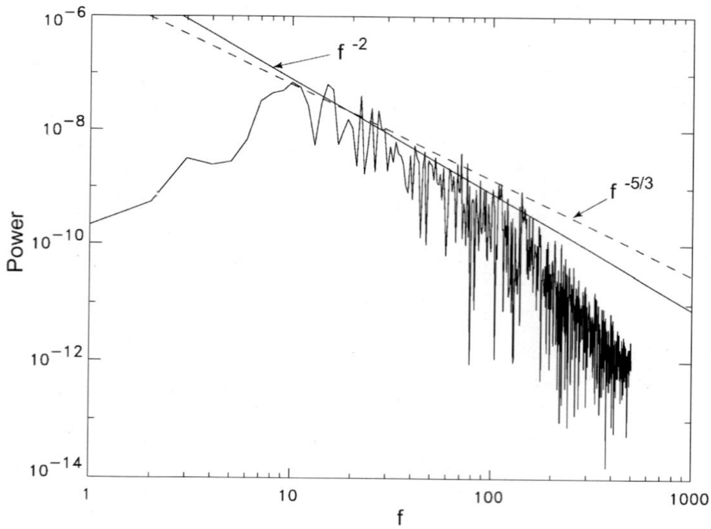

Fig. 8. The magnetic field fluctuation spectrum in stationary frame obtained at $18R_{\odot}$. The Alfvén waves with f^{-1} spectrum were launched at the coronal hole boundary ($1R_{\odot}$). Adapted form Ofman.[25]

the waves close to the Sun (the second term on the r.h.s. of Eq. 1). For the He^{++} ion component of the wind, the main contribution is kinetic (the first term on the r.h.s. of Eq. 1). The He^{++} ion heating in model is constrained

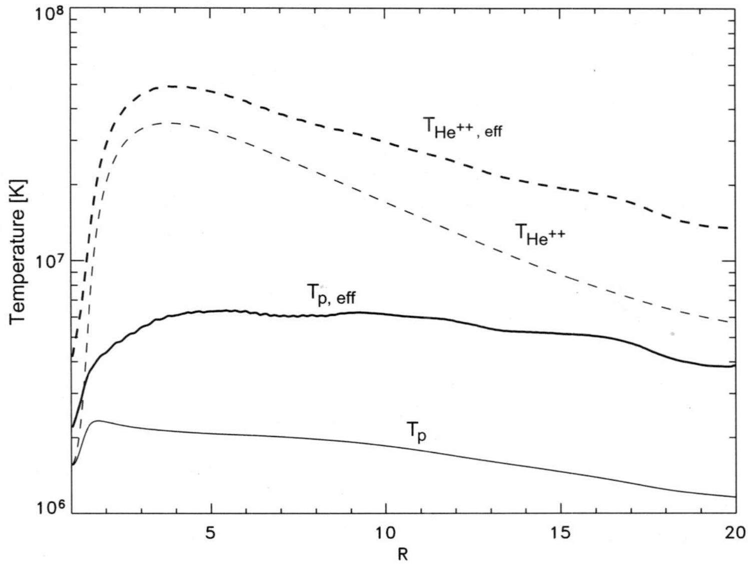

Fig. 9. The effective temperature (thick line style) and the kinetic temperature (thin line style) for protons (solid) and He^{++} ions (dashes) for wave driven fast solar wind in a coronal hole obtained with three-fluid model. Note, that the proton effective temperature is dominated by the effects of the waves, while He^{++} ion heating is mostly due to empirical heat input. Adapted form Ofman.[26]

by the requirement to match Ulysses observation of He^{++} ions flowing faster than protons by about the local Alfvén speed in the distant solar wind.

3.3. *Recent 2D hybrid model of solar wins wave heating*

Recently, Ofman and Viñas[39] studied the heating and the acceleration of protons, and heavy ions by a spectrum of waves in the solar wind, as well as the nonlinear influence of heavy ions on the wave structure using a 2D hybrid model. They considered for the first time the heating and acceleration of protons and heavy ions by a driven input spectrum of Alfvén/cyclotron waves, and by super-Alfvénic heavy ion beams in the multi-species, low-β coronal plasma in two spatial dimensions. They found that in homogeneous plasma the ion beams are more efficient in generating temperature anisotropy than the driven wave spectrum. Such super-Alfvénic beams could be generated at the sites of magnetic reconnection. The main results obtained by Ofman and Viñas[39] with the 2D hybrid model of solar wind plasma heating by a spectrum of ion-cyclotron are reviewed below.

Driven wave spectrum was applied in a small localized region in the center of the 2D domain by imposing the following circularly polarized perpendicular magnetic field fluctuations

$$B_y(t) = B_{\perp,0}F(t) \tag{2}$$

$$B_z(t) = B_{\perp,0}F(t+\pi/2) \tag{3}$$

$$F(t) = b_0 \sum_{i=1}^{N} a_i \sin(\omega_i t + \Gamma_i), \tag{4}$$

where the mode amplitude $a_i = i^{-p/2}$, frequency $\omega_i = \omega_1 + (i-1)\Delta\omega$, the frequency range $\Delta\omega = (\omega_N - \omega_1)/(N-1)$, Γ_i is the random phase, and N is the number of modes. The value of p is 1, and $N = 100$.

The frequencies are normalized to proton gyroresonant frequency (Ω_p). The value of $B_{\perp,0}$ was varied in each run. The spectrum was applied in a region of area $10\Delta x$ by $10\Delta y$ in the center of the computational domain. A limited band of frequencies in the broad-band driver of the form f^{-1} was introduced near the resonant frequency, since the rest of the spectrum does not resonate with the bulk of the particle distribution. The resulting magnetic field fluctuations of B_y, in the 2D domain produced by this driver are shown in Fig. 10. Panel (a) shows B_y at $t = 130\Omega_p^{-1}$ before the fluctuations reach the periodic boundary, and panel (b) shows the fluctuations at later time $t = 250\Omega_p^{-1}$.

The non-Maxwellian features of the ion velocity distribution were investigated with this model. The perpendicular velocity distribution of the ions is shown in Fig. 11. For the case with the driven wave spectrum (Fig. 11a) at the end of the run the perpendicular velocity distribution of O^{5+} is close to bi-Maxwellian, with centrally peaked shape of the distribution. When the initial distribution is drifting Maxwellian with the drift velocity $V_d = 1.5V_A$, where V_A is the Alfvén speed (Fig. 11b), the perpendicular velocity distribution is shell-like, with decreased phase-space density in the central part of the distribution, compared to the perimeter. When the initial drift velocity is increased to $2V_A$, the shell like structure of the phase-space density of O^{5+} ions becomes even more apparent (Fig. 11c). It is interesting to note, that the He^{++} perpendicular velocity distribution for the drifting case is nearly bi-Maxwellian, and does not exhibit the shell structure (Fig. 11d).

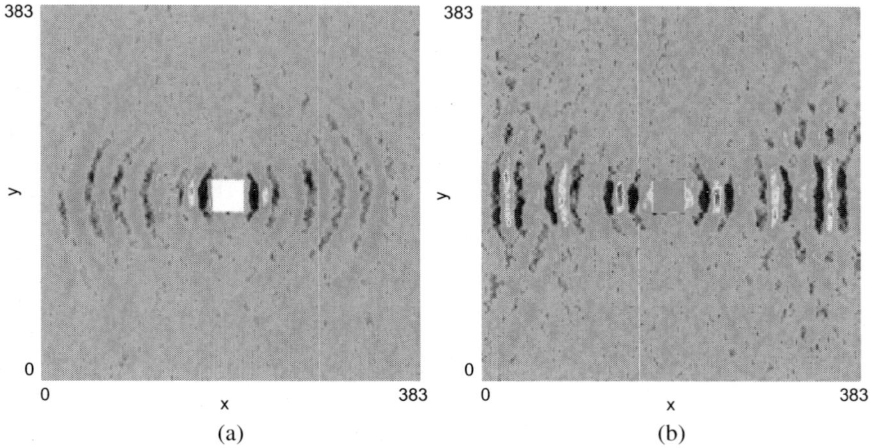

Fig. 10. The typical form of the magnetic field fluctuation, B_y, in the 2D 128×128 computational domain generated by the driver near the origin. (a) At $t = 130\Omega_p^{-1}$, and (b) at $t = 250\Omega_p^{-1}$. Adapted from Ofman and Viñas.[39]

In Fig. 12 the temporal evolution of the temperature anisotropy, and the drift velocity for proton-O^{5+} plasma for (a) driven wave spectrum, and (b) initial drifting Maxwellian distribution (i.e., O^{5+} ion beam) is shown. In both cases the O^{5+} ions undergo strong heating with significant increase of both, parallel and perpendicular temperature. In the driven wave spectrum case the amplitude of the magnetic field fluctuations $B_{\perp,0} = 0.06$, and the frequency range of the driver was $\omega_1 = 0.25$, $\omega_N = 0.4$. It was found that the O^{5+} anisotropy grows quickly (within $400\Omega_p^{-1}$) to about 4, and than saturates nonlinearly remaining in the range 4–5 throughout most of the evolution ($1500\Omega_p^{-1}$). The frequency range of the wave spectrum includes the O^{5+} ion resonant frequency at rest. The anisotropy of the protons remains close to unity throughout the run. It was found that no significant net drift has formed between the protons and O^{5+} ions.

4. 2D Hybrid Model of Wave Heating with Inhomogeneous Density

Below we present some recent results of multi-ion 2D hybrid simulations of solar wind plasma with He^{++} and O^{5+}, and inhomogeneous background ion density. The driven wave spectrum was applied in a narrow band in the center and across the computational domain (see Fig. 5). The primary simulation results show that the waves accelerate the ions away from this

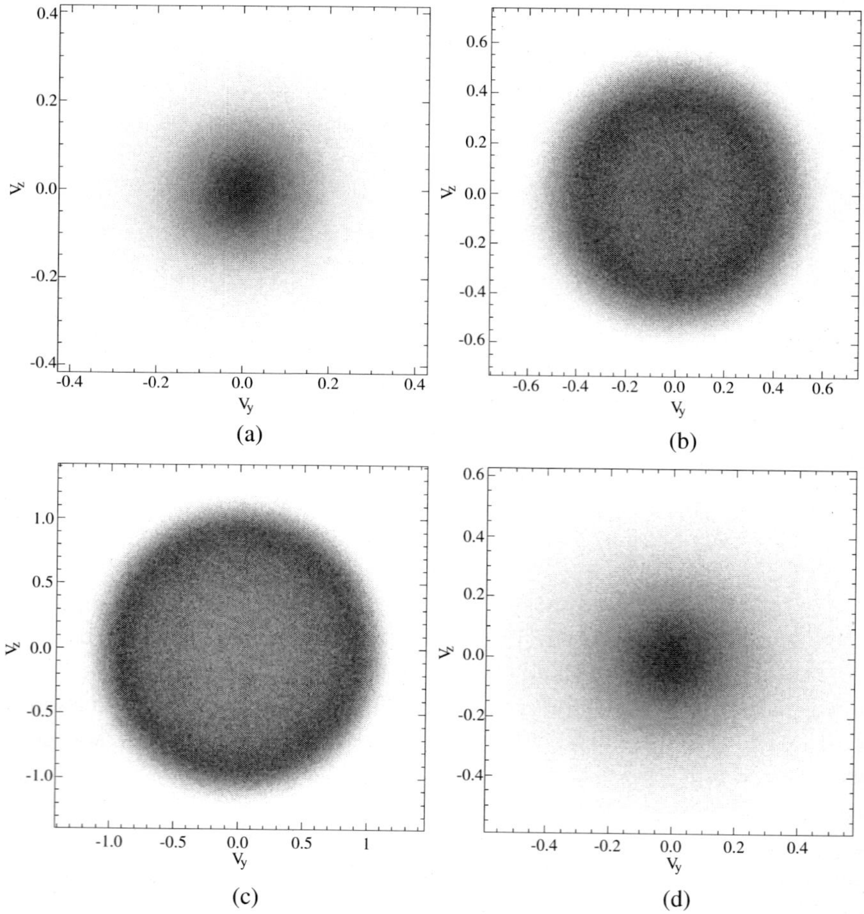

Fig. 11. The perpendicular velocity distribution of the ions. (a) O^{5+} ion perpendicular velocity distribution with driven wave spectrum at $t = 1700\Omega_p^{-1}$. (b) O^{5+} ion perpendicular velocity distribution with $V_d = 1.5V_A$, and with (c) $V_d = 2V_A$. (d) The perpendicular velocity distribution of He^{++} ions at $t = 1117.55\Omega_p^{-1}$ with initial drift of $1.5V_A$. Adapted from Ofman and Viñas.[39]

region, and produce spatial wave-like structure in the density distribution. In this model the heavy ions can resonantly interact with both inward and outward propagating waves simultaneously resulting in preferential energization of heavy ions. Thus, the heavy ion velocity is not restricted to a particular shell in phase space, contrary to quasi-linear models.[56]

Both right-handed and left-handed waves can be excited by plasma micro-instabilities and resonate with multi-ion plasma. These waves play

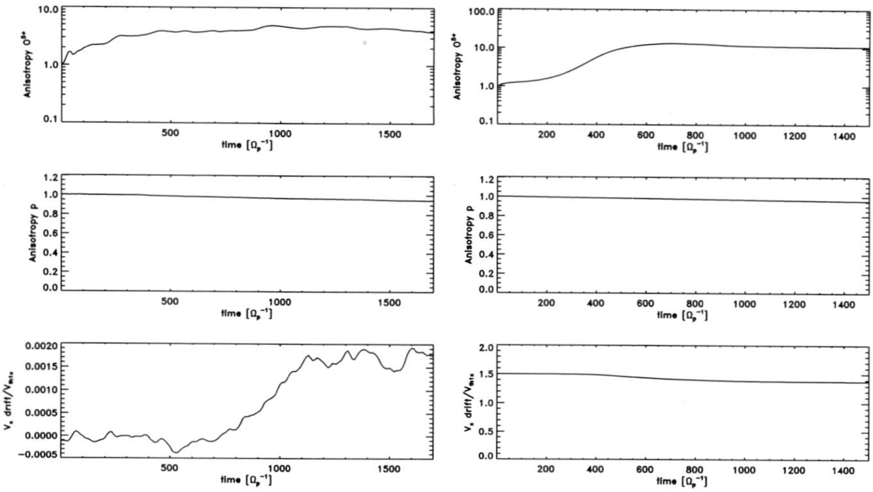

(a) (b)

Fig. 12. The temporal evolution of the temperature anisotropy, and the drift velocity for proton-O^{5+} plasma. (a) Run with driven wave spectrum with $B_{\perp,0} = 0.06$, $\omega_1 = 0.25$, $\omega_N = 0.4$, and (b) run with O^{5+} ion beam with initial p-O^{5+} drift velocity $V_d = 1.5V_A$. Adapted from Ofman and Viñas.[39]

an important role in the ion heating as fast solar wind expanding outward with increasing β and anisotropy. Strong heating of He^{++} ions are caused by the resonance of He^{++} with the growing magnetosonic wave; while resonant heating of O^{5+} with left-handed Alfvén waves occurred in a case where ion-cyclotron waves were generated by micro-instability of anisotropic protons.

Future work is needed to determine the parameter threshold for drift velocity and anisotropy instabilities by using parametric studies, and to investigate the evolving wave-particle system by including external driver, internal source (micro-instabilities), and sink (resonant absorption and damping) of ion cyclotron waves self-consistently.

The presence of the background inhomogeneity (Fig. 13a) affects the level and the efficiency of heating, due to the coupling between ion-cyclotron waves, and magnetosonic waves in the 2D density structure. Both, protons and He^{++} exhibit similar initial density structure. The evolution of the temperature anisotropy of He^{++} ions (Fig. 13b) and of protons (Fig. 13c) in the inhomogeneous magnetized plasma with an input wave spectrum is shown. The perpendicular heating of He^{++} is evident, and the anisotropy reaches a level of 4. The protons remain nearly isotropic, with some quasi-periodic fluctuations on a $200\Omega_p^{-1}$ timescale. The fluctuations appear due to

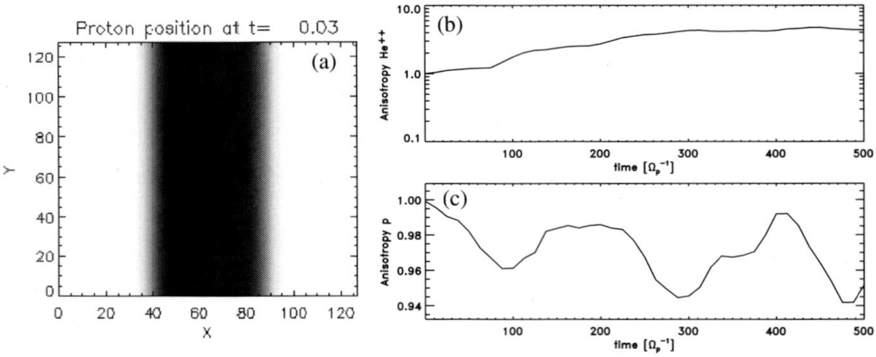

Fig. 13. (a) The initial density distribution of protons. The temporal evolution of the temperature anisotropy of (b) He^{++} ions and (c) protons (lower panel) in the inhomogeneous plasma with driven ion-cyclotron wave spectrum.

the non-homogeneous background structure, and they disappear when the run with the same parameters is repeated with homogeneous background plasma.

5. Conclusions

Several leading models of solar wind acceleration and heating by nonlinear waves are reviewed. Single fluid models show how the solar wind can be accelerated to fast wind speed of $\sim 800\,\mathrm{km\,s^{-1}}$ measured at 1AU, driven by Alfvénic fluctuations from the photosphere, or from the coronal base. It has been shown that waves can heat the solar wind plasma to corona temperatures given sufficiently enhanced effective viscosity and resistivity. However, single fluid models can not describe the kinetics of the heating process, the different ion temperatures, outflow speeds, and other properties of the solar wind constituents, as deduced from observations.

Multi-fluid models are used to model the observational signatures of heavy ion heating and acceleration. The results of multi-fluid models can be compared to spectroscopic observations by constructing line profiles and comparing to ion emission lines. The 2.5D hybrid models can successfully account for the acceleration and heating of the solar wind ions in large scale coronal structures, such as holes or streamers. Waves, in the MHD spectral range (i.e., below the gyroresonant frequency) can be included self consistently in these models, and the resulting solar wind agrees remarkably well with observations. The multi-fluid dispersion relation does contain the ion-cyclotron resonances and can be used to study ion wave velocities.[38]

However, the kinetics of the heating process can not be described in detail by the multi-fluid models.

Kinetic response of the ions to resonant and nonresonant wave heating can be modeled with hybrid codes. For example, using hybrid models it was shown that O^{5+} ions are heated more strongly than protons and He^{++} ions due to resonant interaction with driven left-hand polarized cyclotron wave spectrum, and the stability threshold of anisotropic O^{5+} was determined.[31] After short linear heating stage, the evolution becomes nonlinear, and the resulting temperature anisotropy saturates at a parameter-dependent value. It was found that the velocity distribution of the heavy ions heated by broad-band spectrum of cyclotron waves in the lower corona can be approximated well by bi-Maxwellian distribution. However, when super-Alfvénic drift is present the perpendicular velocity distribution becomes shell-like, and He^{++} as well as O^{5+} ion temperature develops strong perpendicular anisotropy. Since the hybrid models are usually limited computationally to small scales of order several hundred ion-inertia lengths, they can not be used effectively to model the global 3D solar wind from the Sun to 1AU. Therefore, the results of both, kinetic and fluid models must be combined to study the physics of the solar wind.

Acknowledgments

This work was supported by NASA grants NNX08AF85G, NNX08AV88G, and by NRL grant N00173-05-1-G027.

References

1. E. Parker, *JGR* **64** (1959) 1675.
2. J. A. Linker, Z. Mikić, D. A. Biesecker, R. J. Forsyth, S. E. Gibson, A. J. Lazarus, A. Lecinski, P. Riley, A. Szabo and B. J. Thompson, *JGR* **104** (1999) 9809.
3. P. Riley, J. A. Linker, Z. Mikić, R. Lionello, S. A. Ledvina and J. G. Luhmann, *ApJ* **653** (2006) 1510.
4. E. C. Sittler, Jr. and L. Ofman, 2D MHD model of the solar corona and solar wind: Recent results, in *Proceedings of the ILWS Workshop*, eds. N. Gopalswamy and A. Bhattacharyya 2006.
5. M. Guhathakurta, E. C. Sittler and L. Ofman, *Journal of Geophysical Research (Space Physics)* **111** (2006) 11215.
6. O. Cohen, I. V. Sokolov, I. I. Roussev, C. N. Arge, W. B. Manchester, T. I. Gombosi, R. A. Frazin, H. Park, M. D. Butala, F. Kamalabadi and M. Velli, *ApJL* **654** (2007) L163.

7. L. Ofman, M. Romoli, G. Poletto, G. Noci and J. L. Kohl, *ApJ* **491** (1997) L111.
8. C. E. Deforest and J. B. Gurman, *ApJ* **501** (1998) L217.
9. L. Ofman, M. Romoli, G. Poletto, G. Noci and J. L. Kohl, *ApJ* **529** (2000) 592.
10. D. Banerjee, L. Teriaca, J. G. Doyle and K. Wilhelm, *Astron. Astrophys.* **339** (1998) 208.
11. M. J. Aschwanden, L. Fletcher, C. J. Schrijver and D. Alexander, *ApJ* **520** (1999) 880.
12. V. M. Nakariakov, L. Ofman, E. DeLuca, B. Roberts and J. M. Davila, *Science* **285** (1999) 862.
13. B. De Pontieu, S. W. McIntosh, M. Carlsson, V. H. Hansteen, T. D. Tarbell, C. J. Schrijver, A. M. Title, R. A. Shine, S. Tsuneta, Y. Katsukawa, K. Ichimoto, Y. Suematsu, T. Shimizu and S. Nagata, *Science* **318** (2007) 1574.
14. J. V. Hollweg and P. A. Isenberg, *JGR* **107** (2002) 10.1029/2001JA000270.
15. S. R. Cranmer, G. B. Field and J. L. Kohl, *ApJ* **518** (1999) 937.
16. J. L. Kohl, G. Noci, E. Antonucci, G. Tondello, M. C. E. Huber, L. D. Gardner, P. Nicolosi, L. Strachan, S. Fineschi, J. C. Raymond, M. Romoli, D. Spadaro, A. Panasyuk, O. H. W. Siegmund, C. Benna, A. Ciaravella, S. R. Cranmer, S. Giordano, M. Karovska, R. Martin, J. Michels, A. Modigliani, G. Naletto, C. Pernechele, G. Poletto and P. L. Smith, *Sol. Phys.* **175** (1997) 613.
17. X. Li, S. R. Habbal, J. Kohl and G. Noci, *ApJ* **501** (1998) L133.
18. L. Ofman and J. M. Davila, *ApJ* **476** (1997) L51.
19. L. Ofman and J. M. Davila, *ApJ* **553** (2001) 935.
20. N.-E. Raouafi and S. K. Solanki, *Astron. Astrophys.* **445** (2006) 735.
21. J. W. Belcher, *ApJ* **168** (1971) 509.
22. Z. Mikić, J. A. Linker, D. D. Schnack, R. Lionello and A. Tarditi, *Physics of Plasmas* **6** (1999) 2217.
23. A. V. Usmanov, M. L. Goldstein, B. P. Besser and J. M. Fritzer, *JGR* **105** (2000) 12675.
24. L. Ofman, *Space Science Reviews* **120** (2005) 67.
25. L. Ofman and J. M. Davila, *JGR* **103** (1998) 23677.
26. L. Ofman, *JGR* **109** (2004) 10.1029/2003JA010221.
27. L. Ofman, Coronal Heating and Acceleration of the Solar Wind, in *Solar and Stellar Physics Through Eclipses*, eds. O. Demircan, S. O. Selam and B. Albayrak, Astronomical Society of the Pacific Conference Series, Vol. 3702007.
28. T. K. Suzuki, *MNRAS* **349** (2004) 1227.
29. T. K. Suzuki and S.-i. Inutsuka, *Journal of Geophysical Research (Space Physics)* **111** (2006) 6101.
30. T. K. Suzuki, Self-Consistent MHD Modeling of Solar Wind, in *New Solar Physics with Solar-B Mission*, eds. K. Shibata, S. Nagata and T. Sakurai, Astronomical Society of the Pacific Conference Series, Vol. 3692007.

31. L. Ofman, A. Viñas and S. P. Gary, *ApJL* **547** (2001) L175.
32. S. P. Gary, L. Yin, D. Winske and L. Ofman, *JGR* **106** (2001) 10715.
33. S. P. Gary, L. Yin, D. Winske, L. Ofman, B. E. Goldstein and M. Neugebauer, *JGR* **108** (2003) 10.1029/2002JA009654.
34. L. Ofman, S. P. Gary and A. Viñas, *JGR* **107** (2002) 10.1029/2002JA009432.
35. P. C. Liewer, M. Velli and B. E. Goldstein, *JGR* **106** (2001) 29261.
36. P. Hellinger, M. Velli, P. Trávníček, S. P. Gary, B. E. Goldstein and P. C. Liewer, *Journal of Geophysical Research (Space Physics)* **110** (2005) 12109.
37. H. Xie, L. Ofman and A. Viñas, *JGR* **109** (2004) 10.1029/2004JA010501.
38. L. Ofman, J. M. Davila, V. M. Nakariakov and A.-F. Viñas, *JGR* **110** (2005) 9102.
39. L. Ofman and A. F. Viñas, *Journal of Geophysical Research (Space Physics)* **112** (2007) 6104.
40. J. L. Kohl, G. Noci, E. Antonucci, G. Tondello, M. C. E. Huber, S. R. Cranmer, L. Strachan, A. V. Panasyuk, L. D. Gardner, M. Romoli, S. Fineschi, D. Dobrzycka, J. C. Raymond, P. Nicolosi, O. H. W. Siegmund, D. Spadaro, C. Benna, A. Ciaravella, S. Giordano, S. R. Habbal, M. Karovska, X. Li, R. Martin, J. G. Michels, A. Modigliani, G. Naletto, R. H. O'Neal, C. Pernechele, G. Poletto, P. L. Smith and R. M. Suleiman, *ApJ* **501** (1998) L127.
41. M. A. Dodero, E. Antonucci, S. Giordano and R. Martin, *Sol. Phys.* **183** (1998) 77.
42. L. Strachan, R. Suleiman, A. V. Panasyuk, D. A. Biesecker and J. L. Kohl, *ApJ* **571** (2002) 1008.
43. M. P. Miralles, S. R. Cranmer and J. L. Kohl, *Advances in Space Research* **33** (2004) 696.
44. M. Uzzo, L. Strachan and A. Vourlidas, *ApJ* **671** (2007) 912.
45. E. Marsch, *Kinetic Physics of the Solar Wind Plasma* (Physics of the Inner Heliosphere II. Particles, Waves and Turbulence, XI, 352 pp. 152 figs.. Springer-Verlag Berlin Heidelberg New York. Also Physics and Chemistry in Space, volume 21; 2, p. 45-133, 1991), pp. 45–133.
46. W. C. Feldman, B. L. Barraclough, J. L. Phillips and Y.-M. Wang, *Astron. Astrophys.* **316** (1996) 355.
47. M. Neugebauer, B. E. Goldstein, E. J. Smith and W. C. Feldman, *JGR* **101** (1996) 17047.
48. E. Marsch, R. Schwenn, H. Rosenbauer, K.-H. Muehlhaeuser, W. Pilipp and F. M. Neubauer, *JGR* **87** (1982) 52.
49. B. E. Goldstein, M. Neugebauer, L. D. Zhang and S. P. Gary, *Geophys. Res. Lett.* **27** (2000) 53.
50. J. J. Podesta, D. A. Roberts and M. L. Goldstein, *Journal of Geophysical Research (Space Physics)* **111** (2006) 10109.
51. B. J. Vasquez, C. W. Smith, K. Hamilton, B. T. MacBride and R. J. Leamon, *Journal of Geophysical Research (Space Physics)* **112** (2007) 7101.
52. C. W. Smith, B. J. Vasquez and K. Hamilton, *Journal of Geophysical Research (Space Physics)* **111** (2006) 9111.

53. E. N. Parker, *ApJS* **8** (1963) 177.
54. L. Ofman and J. M. Davila, Two-Fluid 2. 5D MHD Model of the Fast Solar Wind and the Effective Proton Temperature, in *American Institute of Physics Conference Series*, eds. S. R. Habbal, R. Esser, J. V. Hollweg and P. A. Isenberg, American Institute of Physics Conference Series, Vol. 4711999.
55. J. M. Davila and L. Ofman, *Space Science Reviews* **87** (1999) 165.
56. P. A. Isenberg, *JGR* **109** (2004) 10.1029/2002JA009449.

Advances in Geosciences
Vol. 21: Solar & Terrestrial Science (2008)
Ed. Marc Duldig

RELATION BETWEEN LOOP-SHAPED INTERPLANETARY DISTURBANCES AND THE MAGNETIC FLUX ROPE

M. TOKUMARU*, M. KOJIMA and K. FUJIKI
Solar-Terrestrial Environment Laboratory,
Nagoya University, Nagoya 464-8601, Japan
**tokumaru@stelab.nagoya-u.ac.jp*

K. MUNAKATA
Faculty of Science, Shinshu University,
Matsumoto 390-8621, Japan
kmuna00@shinshu-u.ac.jp

T. KUWABARA
Bartol Research Institute, University of Delaware,
Newark, Delaware 19716, USA
takao@bartol.udel.edu

K. MARUBASHI
Astronomy and Space Science Institute,
Yuseong-gu, Daejeon 305-348, Korea
k.marubashi@eos.ocn.ne.jp

We analyzed interplanetary scintillation (IPS) observations at 327 MHz to retrieve the 3-dimensional (3D) distribution of solar wind density enhancements associated with coronal mass ejections (CMEs) for 5 events that occurred in 2001. As a result, elongated density structures, some of which were loop-shaped, were detected from the IPS observations for four events. Comparing the geometry of these IPS density structures with that of the flux rope deduced from cosmic ray and *in situ* observations, we found that the orientation of the elongated or loop-shaped features observed by IPS were generally consistent with the axis direction of the flux rope cylinders determined by cosmic ray and *in situ* observations. This fact suggests that the magnetic flux rope may play an important role in forming the elongated features of density enhancements.

1. Introduction

A loop-shaped interplanetary (IP) disturbance was identified from ground-based interplanetary scintillation (IPS) and space-borne white-light measurements during the halo coronal mass ejection (CME) event of 2003 October 28.[1] The propagation speed of those loop-shaped disturbances was much slower than that of the associated IP shock, and the trajectory of the loop was found to be aligned along the axis direction of the cylindrical magnetic flux rope determined from muon cosmic ray and *in situ* measurements for this CME event.[2] Therefore, we considered such loop-shaped IP disturbances to represent a high-density plasma confined within the magnetic flux rope.

This result significantly differs from the typical features of interplanetary CMEs observed *in situ.* Namely, the magnetic cloud or flux rope, which corresponds to an internal part of the CME, is usually filled by a rarefied plasma, and density enhancement is ascribed in most cases to the compression region between the magnetic cloud and the CME-driven IP shock.[3] Here, we should keep in mind that these are averaged features, and that a variety of density variations within the magnetic cloud has been reported.[4] Moreover, an encounter rate of the coronal ejecta for *in situ* measurements greatly depends on the spatial extent of the ejecta as well as on the relative position of an observer in the solar wind. If we use IPS observations that enable us to cover a wide range of the solar wind in a short time, our understanding of the global features of interplanetary CMEs may be enhanced. Therefore, utilizing IPS observations, we need to investigate the relation between density enhancements and the magnetic flux rope associated with CMEs in more detail.

In this study, we analyzed our IPS observations of 5 cases to reconstruct the 3D-density feature of CME, and compared them with cosmic ray and *in situ* observations of the magnetic flux rope. The outline of this paper is as follows. In Sec. 2, we describe the CME events analyzed here. In Sec. 3 and Sec. 4, we present the 3D-density reconstruction from IPS observations of those events and a comparison between IPS density reconstructions and the 3D geometry of the magnetic flux rope determined from cosmic ray and *in situ* measurements, respectively. In Sec. 5, we address the origin of density features derived from IPS observations. In Sec. 6, we summarize the results obtained here.

Table 1. Cosmic ray cylinder CME events in 2001.

Date		Apr. 2	Apr. 10	Aug. 15	Aug. 25	Nov. 4
Flare	UT	21:51	05:26	(N/A)	16:45	16:20
	Class	X20	X2.3		X5.3	X1.0
	Imp.	(N/A)	3B		3B	3B
	Loc.	(N/A)	S23W09		S17E34	N06W18
CME	UT	22:06	05:30	23:54	16:50	16:35
	Type	PH	FH	FH	FH	FH
	V_{pos}	2505	2411	1575	1433	1810
IPS	UT[a]	4/4 3.6 h	4/11 3.1 h	8/17 2.7 h	8/27 3.2 h	11/6 1.9 h
Cosmic ray	Begin	4/4 19 h	4/11 23 h	8/17 13 h	8/27 20 h	11/6 14 h
	End	4/5 05 h	4/12 05 h	8/17 19 h	8/28 14 h	11/7 12 h
IP shock	UT	4/4 14:55	4/11 13:43	8/17 11:03	8/27 19:52	11/6 01:21

Note: FH = Full halo, PH = Partial halo, Imp. = Optical Importance, Loc. = Heliographic location.
[a]Observation mean time, V_{pos} = Plane-of-sky speed of CME in km/s derived from LASCO[6] observations.

2. Analyzed Events

We analyzed 5 CME events that occurred in 2001, and key parameters of those events are indicated in Table 1. Information of CME shown here was adopted from the SOHO LASCO CME catalog (http://cdaw.gsfc.nasa.gov/CME_list/index.html). These events were selected, since the geometry of the magnetic flux ropes associated with CMEs had already been determined from cosmic ray observations using the Global Muon Detector Network.[5] In their analysis, a cylindrical IMF model, which represents the magnetic field enhancement associated with the flux rope, was fit to the observed cosmic ray density data (detailed descriptions of the analysis of cosmic ray observations are available in Ref. 5). Cosmic ray cylinders determined from their analysis are presented in Sec. 4.

As shown in Table 1, all the analyzed CME events were halo-shaped, 4 of which were the full-halo (FH) type, and all were associated with the occurrence of ssc (i.e. the IP shock) at the Earth. Therefore, all events were basically regarded as Earth-directed, although two of them lacked the association with an optical flare, and one (i.e. the 2001 August 15 CME event) had no X-ray counterpart. IP disturbances were clearly detected for all cases from IPS observations using the 327-MHz four-station system of the Solar-Terrestrial Environment Laboratory (STEL) of Nagoya University[7] (see Table 1). The scintillation disturbance factors (g-values[8]) derived from STEL IPS observations showed a marked increase in solar wind density fluctuations for one or two days after the occurrence of CMEs. Abrupt increases of solar wind speed up to $> 1000\,\mathrm{km/s}$ were also observed for the 2001 April 2 and 10 events (no IPS speed data were available for the 2001 August 15 and 25 events).

In situ plasma and IMF data collected by the ACE spacecraft were available for four cases (no ACE plasma data were available for the 2001 November 4 event). The associated IP shocks were identified from ACE plasma data for the four cases. In addition, clear signatures of the magnetic flux rope; i.e. an enhancement of magnetic field intensity and a systematic rotation of the field vector, were observed for three out of four cases (i.e. except for the 2001 August 25 event). A cylindrical flux rope model was fit to the ACE data for these three cases, and the results are presented in Sec. 4.

3. 3D Density Reconstruction from IPS Observations

We determined the 3D distribution of solar wind density enhancements associated with a CME by analyzing g-value data obtained from IPS observations with the model fitting method. While this method was basically the same as the one used for earlier studies,[1,9] we briefly describe our method below, and then present our results.

The g-value represents a variation of the integration level of solar wind (electron) density fluctuations along the line-of-sight, and can be calculated using the following formula:

$$g^2 = \frac{1}{K}\int_0^\infty \Delta N_e^2 w(z)dz, \tag{1}$$

if wave scattering is weak. Here, ΔN_e, z, and $w(z)$ are the solar wind density fluctuations, the distance along the line-of-sight, and the weighting function of the IPS,[10] respectively. The normalizing factor K is given by $K = \int_0^\infty \Delta N_{e0}^2 w(z)dz$, and ΔN_{e0} represents the density fluctuation level of the quiet solar wind. We assume that ΔN_{e0} distributes as R^{-2}, where R is the radial distance from the sun, and $\Delta N_e = \Delta N_{e0}(1 + f_{CME})$, where f_{CME} represents the turbulence level increase caused by a CME. In this analysis, we assume that f_{CME} has a Gaussian-form spatial distribution defined by the e^{-1} radial thickness D, the e^{-1} maximum and minimum angular extent θ_0 and $AR\cdot\theta_0$, and a peak value of the ΔN_e increase C_1. We also assume that the ΔN_e-enhanced region radially expands at a constant speed $V_S = V_{S0}\cos^{\alpha}(\theta/2)$ (where θ is the separation angle with respect to the center axis). Thus, f_{CME} is given by nine parameters; D, θ_0, AR, β, which is an inclination of the ΔN_e-enhanced region with respect to the heliographic equator, V_{S0}, α, C_1, and the heliographic location of the center axis (λ_0, ϕ_0).

We determined these free parameters by fitting g_{cal} calculated using Eq. 1 to g_{obs} obtained from STEL IPS observations. In this analysis, we obtained the best fit solution by minimizing a rms (root mean-square) deviation σ, which is given by

$$\sigma^2 = \frac{\chi^2}{N} = \frac{1}{N}\sum_{i=1}^{N}\frac{(g_{obs} - g_{cal})^2}{g_{obs}}, \tag{2}$$

where N is the number of observations. Once a minimum value of χ^2_{min} is given, the estimated error is determined by finding a range corresponding to $\chi^2_{min} + 1$ (a 68% confidence level) for a given parameter. The best-fit solutions are indicated in Table 2. The χ^2_{min} and correlation coefficient are also indicated here. As shown in Table 2, the best-fit model has a high degree of correlation between model calculations and the observations for all cases.

We note that the propagation speeds derived IPS observations are systematically lower than the plane-of-sky CME speeds measured from LASCO observations (see Table 1). This fact suggests that significant deceleration of CMEs takes place between the corona and the IP space. In this study, we focus on the two model parameters of AR and β. A small value of AR (i.e. $AR < 1$) denotes that the model has an anisotropic angular

Table 2. Best-fit parameters determined from the model fitting analysis of IPS observations.

Date	Apr. 2	Apr. 10	Aug. 15	Aug. 25	Nov. 4
V_{S0} (km/s)	1060 ± 86	1515 ± 106	844 ± 70	892 ± 21	1297 ± 11
α	2.2 ± 1.1	3.0 ± 1.6	-1.0 ± 1.4	0.6 ± 0.5	3.6 ± 0.2
ϕ_0 (deg)	S10 $\pm$ 15	N07 $\pm$ 3	N07 $\pm$ 29	N02 $\pm$ 6	S16 $\pm$ 2
λ_0 (deg)	W30 $\pm$ 10	W26 $\pm$ 6	E40 $\pm$ 32	W13 $\pm$ 9	E09 $\pm$ 1
D (AU)	0.13 ± 0.06	0.08 ± 0.05	0.13 ± 0.05	0.03 ± 0.01	0.015 ± 0.007
θ_0 (deg)	140 ± 87	90 ± 41	129 ± 61	46 ± 10	127$\pm$(N/A)
AR	0.20 ± 0.18	0.04 ± 0.02	0.47 ± 0.29	0.34 ± 0.12	~ 1
β_0 (deg)	-8 ± 17	10 ± 6	-121 ± 55	-10 ± 14	(N/A)
C_1	3.1 ± 1.2	11.2 ± 6.2	2.4 ± 0.9	10.9 ± 4.3	8.7 ± 2.2
N	23	13	20	17	23
χ^2_{min}	1.632	0.617	1.076	1.166	2.621
ρ	0.706	0.648	0.833	0.794	0.762

extent, suggesting an elongated feature. For this case, the angle β defines the orientation of the maximum angular extent. As shown in Table 2, four cases are classified into an elongated structure event with $AR < 1$, and the remaining one (2001 November 4 event) is an isotropic structure event with $AR \sim 1$. It should be noted here that the 3D reconstruction for the 2001 November 4 event may be influenced by the effect of observational geometry. That CME is thought to have reached at ~ 1 AU by the time our IPS observations had been made, but such an observational condition is not suitable for retrieving an overall view of the CME feature, since the IPS data taken at ~ 1 AU were entirely dominated by the contribution from the near-Earth solar wind plasma. Thus, the result of this event is considered to be less reliable. In Fig. 1, remote-observer views are depicted of the best

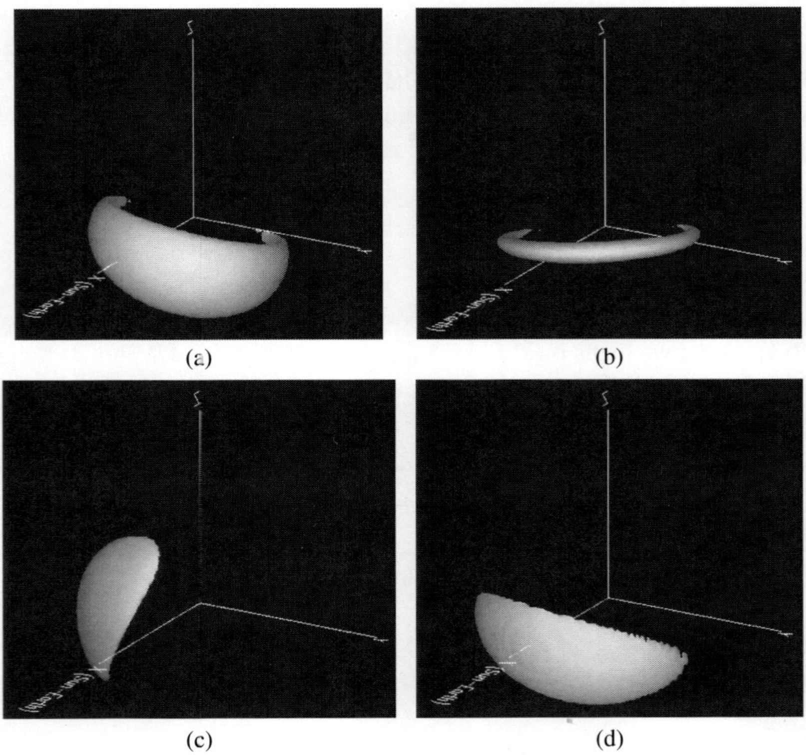

Fig. 1. Remote-observer's views of the best fit model for (a) 2001 April 2, (b) April 10, (c) August 15, and (d) August 25 CME events. Solar ecliptic coordinates are used. The XY-plane is the ecliptic, and the X-axis to the left in each image includes the Earth. The origin corresponds to the location of the Sun, and the length of a line in each axis corresponds to 1 AU; i.e. The Earth is located at the edge of the X-axis line.

fit model for $AR < 1$. Some of the models shown in the figure appear to be loop-shaped. An interesting point to note is that the best-fit models tended to elongate in an East-West direction. It has been reported that magnetic flux ropes usually have a small inclination to the ecliptic plane with their azimuthal direction close to the East-West direction,[11] which implies a close association between the IPS density structures and the magnetic flux ropes. In the next section, we compare the best-fit models determined from IPS observations with the cylindrical magnetic flux rope models deduced from cosmic ray and *in situ* observations.

4. Comparison with Magnetic Flux Rope Observations

Table 3 summarizes both the cosmic ray and *in situ* observations of the magnetic flux ropes for the CME events studied here. The latitude and longitude of the cosmic ray cylinder are shown in Table 3.[5] The geometry of the magnetic flux rope has also been determined by fitting the force-free cylindrical model to *in situ* measurements by the ACE spacecraft. The method employed here is the same as the one developed by Ref. [12]. The geometry of the force-free cylindrical models determined here is shown in Fig. 2. As mentioned in Sec. 2, the *in situ* data of the magnetic flux rope data were unavailable for 2001 August 25 and November 4 events. Owing to the large gyro-radius of the cosmic ray particles, the cosmic-ray-based method was considered to be more effective than the *in situ* method for deducing the large-scale geometry of the flux rope. The discrepancy between cosmic ray and *in situ* data may be ascribed to this fact. The latitude and longitude of the force-free cylindrical models for remaining events are demonstrated in Table 3, which reveals that the geometry of the cosmic ray cylinders is generally consistent with that of the force-free models for the 2001 April 2 and August 15 events. As for the 2001 April 10 event, the cylinder geometry

Table 3. Comparison between IPS, cosmic ray and *in situ* observations for elongated structure events.

Date		Apr. 2	Apr. 10	Aug. 15	Aug. 25
IPS	β_0 (deg)	-8 ± 17	10 ± 6	-121 ± 55	-10 ± 14
Cosmic ray	Lat.(deg)	+7	+7	+31	+10
	Lon.(deg)	56	95	226	124
In situ	Start (UT)	4/4 19 h	4/12 8 h	8/17 21 h	(N/A)
	End (UT)	4/5 9 h	4/13 6 h	8/18 13 h	
	Lat.(deg)	+19	+5	+51	
	Lon.(deg)	286	352	334	

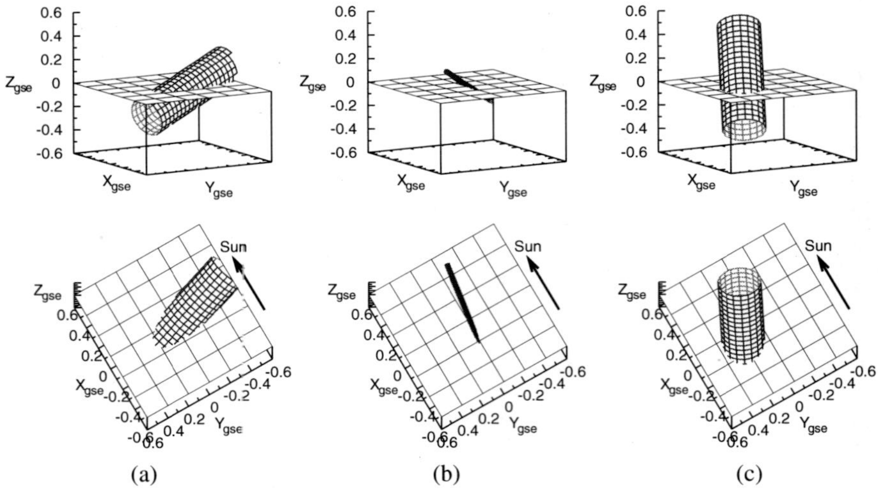

Fig. 2. Geometry of the cylindrical flux rope model determined for (a) 2001 April 2 (b) April 10, and (c) August 15 CME events.

determined from cosmic ray observations significantly differed from that from the *in situ* data. The reason for this discrepancy may be attributed to the small flux rope radius (0.02 AU) of this event (see Fig. 2). In the case of such a thin flux rope, the fitting analysis can be readily affected by fluctuations in the observed data.

A comparison of the geometry between the high-ΔN_e structure observed by IPS and the cylindrical structures of IMF by cosmic ray and *in situ* observations is demonstrated in Fig. 3 and Fig. 4 for the four cases shown in Fig. 1. The solid curve in each plot indicates the central axis of the ΔN_e model fit to IPS data, and the arrows A and B correspond to the axis directions of a cylindrical IMF structure determined from cosmic ray and ACE observations, respectively. Here, it should be noted that the cylindrical IMF structures correspond to local features of the magnetic flux rope near the Earth, whereas IPS data represent global features of ΔN_e enhancements travelling between the Sun and the Earth. In comparing between IPS and IMF structures, therefore, we should pay attention to those directions near the Sun-Earth line. As shown in these figures, the high-ΔN_e structure is closely aligned along the cylindrical IMF structure, except for the 2001 April 10 event. We may recall here that the geometry of the cosmic ray cylinder was inconsistent with that of the cylindrical flux rope model for the 2001 April 10 event. The close association among IPS,

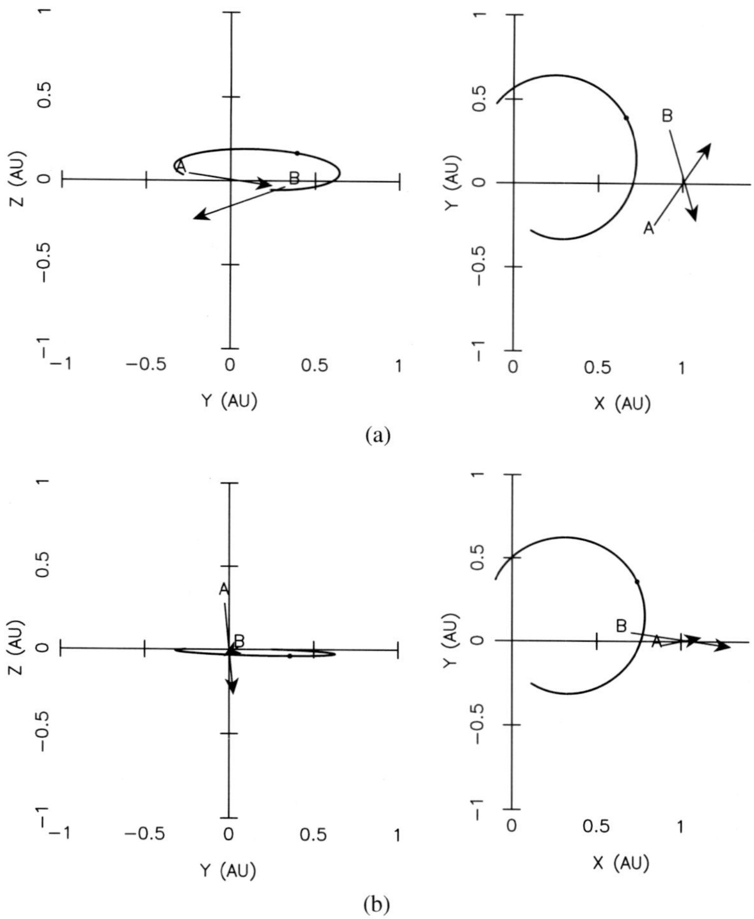

Fig. 3. The IPS ΔN_e model (solid curve), the cosmic ray cylinder (arrow A) and the magnetic flux model (arrow B), projected onto the YZ (left) and XY planes (right) in the ecliptic coordinate, for (a) 2001 April 2 and (b) April 10 events.

cosmic ray and *in situ* observations for the remaining events suggests that the magnetic flux rope plays an important role in forming the elongated high-ΔN_e structures.

5. Discussion

There are two possible explanations for the close association between the elongated high-ΔN_e structure and the magnetic flux rope. The first one is that the ΔN_e enhancement represents the compression region associated

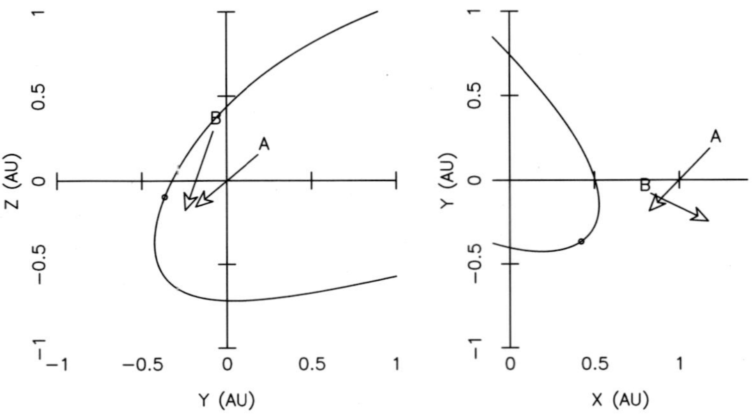

(a)

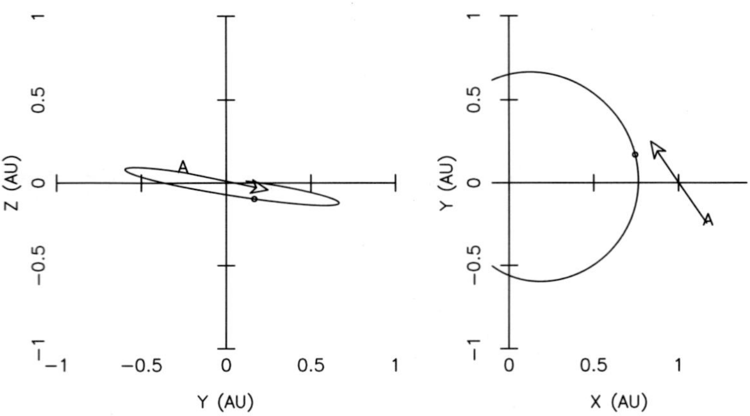

(b)

Fig. 4. The same as Fig. 3 but for (a) 2001 August 15 and (b) August 25 CME events.

with the IP shock. The magnetic flux rope is considered to have a cross section elongated in its axis direction. Therefore, a loop-like density-enhanced structure may be formed at the sheath region between the IP shock and the magnetic flux rope, which is a shock driver, since strong compression takes place there. The average value of the radial thickness D determined from IPS data is ~ 0.08 AU, which is consistent with the typical value (~ 0.14 AU) for the thickness of the shocked gas shell observed by *in situ* at 1 AU.[3] The second one is that it corresponds to the dense coronal plasma confined within the magnetic flux rope, as is the case of the 2003 October 28 CME event.[1] While little is known about the interplanetary counterpart of a CME's core owing to lack of observational data, the high-density region at the rear of the magnetic cloud on 1997 January 10-11 was attributed to the prominence material.[13] The radial extent of the high-density region was ~ 0.02 AU, and the diameter of the magnetic cloud was 0.20 AU. The radial extent of the IPS reconstruction is larger than that of the prominence material, but not too different from it. Here, we should notice that the spatial resolution of the IPS reconstruction greatly depends on the number of lines-of-sight (LOS's), and there is a possibility that the radial extent of the IPS reconstruction broadens owing to the limited number of LOS's used for our IPS observations. In this study, we lack definitive information to determine the origin of the loop-shaped high-ΔN_e structure, and it must remain an unsettled question.

6. Summary

STEL IPS observations were used to determine the 3D distribution of the solar wind density (ΔN_e) enhancements associated with CMEs for the 5 events in 2001. As a result, the elongated (i.e. loop-shaped) features of ΔN_e enhancements were revealed by this analysis in 4 cases, whereas an isotropic shell-shaped feature was determined for the remaining one. The elongated high-ΔN_e structures were compared with the magnetic flux rope geometry determined from cosmic ray and *in situ* observations, and were found to be closely aligned with the magnetic flux ropes for three cases. This fact suggests that the magnetic flux rope may play an important role in forming the high-ΔN_e structures.

Acknowledgments

This work was partly supported by the Grant-in-Aid for Scientific Research (16340147) and by the Grant-in-Aid for Creative Scientific Research (17GS0208) from the Japan Society for the Promotion of Science. The IPS observations were carried out under the solar wind program of the Solar-Terrestrial Environment Laboratory (STEL) of Nagoya University. We thank the ACE Science Center for providing the ACE data via the web site (http://www.srl.caltech.edu/ACE/level2/).

References

1. M. Tokumaru, M. Kojima, K. Fujiki, M. Yamashita and B. J. Jackson, *J. Geophys. Res.* **112** (2007) A05106, doi:10.1029/2006JA012043 .

2. T. Kuwabara, K. Munakata, S. Yasue, C. Kato, S. Akahane, M. Koyama, J. W. Bieber, P. Evanson, Z. Fujii, M. Tokumaru, M. Kojima, K. Marubashi, M. L. Duldig, J. E. Humble, M. R. Silva, N. B. Trivedi, W. D. Gonzalez and N. J. Schuch, *Geophys. Res. Lett.* **31** L19803, doi:10.1029/2004GL020803 (2004).
3. G.Borrini, J. T. Gosling, S. J. Bame and W. C. Feldman, *J. Geophys. Res.* **87** (1982) 4353.
4. R. P. Lepping, D. B. Berdichevsky, C.-C. Wu, A. Szabo, T. Narock, F. Mariani, A. J. Lazarus and A. J. Quivers, *Ann. Geophys.* **24** (2006) 215.
5. T. Kuwabara, J. W. Bieber, P. Evenson, K. Munakata, S. Yasue, C. Kato, Z. Fujii, M. L. Duldig, J. E. Humble, M. R. Silva, N. B. Trivedi, W. D. Gonzalez, L. Alisson and N. J. Schuch, *Proc. 31th International Cosmic Ray Conference (ICRC 2007)* SH.2.1 Paper 716 (2007).
6. G. E. Brueckner, R. A. Howard, M. J. Koomen, C. M. Korendyke, D. J. Michels, J. D. Moses, D. G. Socker, K. P. Dere, P. L. Lamy, A. Llebaria, M. V. Bout, R. Schwenn, G. M. Simnett, D. K. Bedford and C. J. Eyles *et al.*, *Sol. Phys.* **162** (1995) 357.
7. M. Kojima and T. Kakinuma, *Space Sci. Rev.* **53** (1990) 173.
8. G. R. Gapper, A. Hewish, A. Purvis and P. J. Duffet-Smith, *Nature* **296** (1982) 633.
9. M. Tokumaru, M. Kojima, K. Fujiki, M. Yamashita and A. Yokobe, *J. Geophys. Res.* **108** (2003) 1220, doi:10.1029/2002JA009574.
10. A. T. Young, *Astrophys J.* **168** (1971) 543.
11. V. Bothmer and R. Schwenn, *Ann. Geophys.* **16** (1998) 1.
12. K. Marubashi and R. P. Lepping, *Ann. geophys.* **25** (2007) 2453.
13. L. Burlaga, R. Fitzenreiter, R. Lepping, K. Ogilvie, A. Szabo, A. Lazarus, J. Steinberg, G. Gloecker, R. Howard, D. Michels, C. Farrugia, R. P. Lin and D. E. Larson, *J. Geophys. Res.* **103** (1998) 277.

Advances in Geosciences
Vol. 21: Solar & Terrestrial Science (2008)
Ed. Marc Duldig

SOLAR WIND AND CME STUDIES OF THE INNER HELIOSPHERE USING IPS DATA FROM STELAB, ORT, AND EISCAT

M. M. BISI* and B. V. JACKSON
Center for Astrophysics and Space Sciences,
University of California, San Diego,
9500 Gilman Drive #0424, La Jolla,
CA 92093-0424, U.S.A.
**mmbisi@ucsd.edu or Mario.Bisi@gmail.com*
http://ips.ucsd.edu/

R. A. FALLOWS and G. D. DORRIAN
Institute for Mathematical and Physical Sciences,
Aberystwyth University, Penglais Campus, Aberystwyth,
Ceredigion. SY23 3BZ Wales, UK
http://www.aber.ac.uk/maps/en/

P. K. MANOHARAN
Radio Astronomy Centre,
TATA Institute of Fundamental Research,
Udhagamandalam (Ooty) 643001, Tamilnadu, India
http://www.ncra.tifr.res.in/

J. M. CLOVER, P. P. HICK and A. BUFFINGTON
Center for Astrophysics and Space Sciences,
University of California, San Diego,
9500 Gilman Drive #0424, La Jolla,
CA 92093-0424, U.S.A.

A. R. BREEN
Institute for Mathematical and Physical Sciences,
Aberystwyth University, Penglais Campus,
Aberystwyth, Ceredigion. SY23 3BZ Wales, UK

M. TOKUMARU
Solar-Terrestrial Environment Laboratory,
Nagoya University, Furo-cho, Chikusa-ku,
Nagoya 464-8601, Japan
tokumaru@stelas.nagoya-u.ac.jp
http://stesun5.stelab.nagoya-u.ac.jp/index-e.html

Interplanetary scintillation (IPS) observations provide views of the solar wind at all heliographic latitudes from near 1 A.U. down to fields of view covered by coronagraphs. These observations can be used to study the propagation of the solar wind and solar transients out into interplanetary space, and also measure the inner-heliospheric response to co-rotating solar structures and coronal mass ejections (CMEs). We use a three-dimensional (3D) reconstruction technique that obtains perspective views from solar co-rotating plasma and outward-flowing solar wind as observed from the Earth by iteratively fitting a kinematic solar wind model to IPS data from various observing systems. Here we use the model with both Solar Terrestrial Environment Laboratory (STELab), Japan, and Ootacamund (Ooty) Radio Telescope (ORT), India, IPS observations. This 3D modeling technique permits reconstructions of the density and velocity structures of CMEs and other interplanetary transients at a relatively coarse resolution for STELab and better for Ooty; and is dependent upon the number of observations. We present 3D reconstructions of CME events around 4-8 November 2004 from Ooty IPS observations and some preliminary reconstructions of STELab IPS observations around the Whole Heliospheric Interval (WHI). We also present some preliminary results of a CME observation by both the European Incoherent SCATter (EISCAT) radar IPS observations and those made by the Solar TErrestrial RElations Observatory (STEREO) of a CME in May 2007.

1. Introduction

Interplanetary scintillation (IPS) observations of the solar wind, solar wind transients, and the inner-heliosphere have now been used for around 45 years e.g. Refs. 1–6. IPS is seen as a powerful tool to probe the interplanetary medium. It is the rapid variation in signal received by radio antennas on Earth arising from the diffraction of radio waves from a distant, compact, natural radio source due to density variations within the outwardly-propagating solar wind. IPS observations allow the solar wind speed to be inferred over all heliographic latitudes and a wide range of heliocentric distances (dependent upon source strength and observing frequency, and also the location of observable sources in the sky), e.g. Refs. [1, 7–10]. Using the scintillation-level converted to g-level as a proxy, the solar wind density can also be inferred from such IPS observations, e.g. Refs. [6, 11–13]. See Equation 1 later.

The sources of IPS data discussed in this paper were taken from observations made by three different IPS systems around the World. These are the radio arrays of the Solar Terrestrial Environment Laboratory

(STELab),[14] University of Nagoya, Japan; the Ootacamund (Ooty) Radio Telescope (ORT),[15–17] India; and the mainland radio telescopes of the European Incoherent SCATter (EISCAT)[18–20] radar in Northern Scandinavia (Norway, Sweden, and Finland). The two Asia-based systems (STELab and Ooty) operate at an observing frequency of 327 MHz where STELab typically observes 20-40+ radio sources per day, and Ooty is currently capable of observing up to 1000 radio sources per day. EISCAT (based in Europe) however, currently operates in IPS mode on a campaign basis only at frequencies of either 928.5 MHz or 1420 MHz and is capable of observing up to a dozen sources per day at best.

As fully described by Ref. [12], the scintillation-level measurements have been available from the STELab radio antenna at Kiso, Japan, from 1997 to the present, and more recently from mid-2002 from the STELab radio antenna at Fuji. The New Toyokawa site will also be used for these measurements when it is fully operational. Another STELab antenna is located at Sugadaira, making four antenna sites in total with geographic distances between sites ranging from 98 km to 207 km. Ooty has been providing scintillation-level measurements on a similar time span as that of STELab. Unfortunately, the EISCAT system presently does not provide a scintillation-level measurement. The 3D velocity reconstructions are based directly on the IPS velocity observations; for 3D density reconstructions the g-level provides a proxy for density. This requires a "conversion" from IPS scintillation level, expressed as g-level, to density. The resulting reconstructions are of an inner-heliosphere region typically ranging from 15 solar radii ($R_\odot$) out to approximately 3 astronomical units (AU). The g-level, or disturbance factor, g, is defined by Eq. (1).

$$g = m/\langle m \rangle \tag{1}$$

Here, m is the observed scintillation level, and $< m >$ is the mean level of scintillation for the source at the elongation at the time of observation. Scintillation-level measurements from STELab are available for each astronomical radio source as an intensity variation of signal strength (resulting from small-scale variations in density, ΔN_e). These data are automatically edited to remove any obvious interference. Further discussion regarding the calculation of and use of g-level as a proxy for density (and also the real-time calculation used for space-weather forecasting) can be found in Refs. [11, 12].

When two or more radio antennas are used (such as with the STELab and EISCAT systems) and the separation of the ray-paths in the plane of the sky from source to each telescope lies close to radial (the solar wind flow direction) centered at the Sun, a high degree of correlation between the patterns of scintillation recorded at the two telescopes may be observed, e.g. Ref. [21]. The time lag for which maximum cross-correlation occurs (taking into account 'plane-of-sky' assumptions) can then be used to estimate the outflow speed of the irregularities producing the scintillation, e.g. Refs. [22, 23, 10]. More sophisticated methods involving the fitting of the observed auto- and cross-correlation spectra with the results from a weak-scattering model, have also been adopted for IPS data analyses, e.g. Refs. [10, 24–27]. For a single radio antenna (such as with Ooty) the outflow speed is determined from the power spectrum of the IPS observation as described in Ref. [28].

The three-dimensional (3D) reconstructions[11] of both speed and density are obtained through the use of a purely kinematic solar wind model and a technique which uses perspective views of solar co-rotating plasma[29] and of outward-flowing solar wind[12] crossing our IPS lines of sight from Earth to the radio source. The IPS data are iteratively fitted using our model over 18 iterations to make sure the model converges to a final solution. We then compare the resulting 3D reconstructions with *in situ* measurements from both the Wind — Solar Wind Experiment (Wind|SWE)[30,31] instruments and the Advanced Composition Explorer — Solar Wind Electron, Proton and Alpha Monitor (ACE|SWEPAM).[32,33]

Section 2 discusses some low-resolution 3D reconstructions of the Whole Heliospheric Interval (WHI) using STELab IPS data. Section 3 provides a preliminary summary of Ooty IPS 3D density reconstructions of the interplanetary counterparts of Coronal Mass Ejections (CMEs) seen by the SOlar and Heliospheric Observatory — Large Angle Spectrometric COronagraph (SOHO|LASCO)[34,35] responsible for the early-November 2004 geomagnetic storms. Section 4 discusses some preliminary work on a comparison of a CME observed by both EISCAT and the Solar TErrestrial RElations Observatory Ahead (STEREO-A)[36] spacecraft's Sun Earth Connection Coronal and Heliospheric Investigation (SECCHI)[37,38] instrumentation. We give an overall summary in Sec. 5.

2. Low-Resolution 3D Reconstructions of STELab IPS Data for the WHI Period

The Whole Heliospheric Interval (WHI) covers Carrington Rotation 2068 (CR2068) and runs through the period 20 March 2008 through 16 April 2008. Figures 1 and 2 are summaries of the portion of CR2068 for which

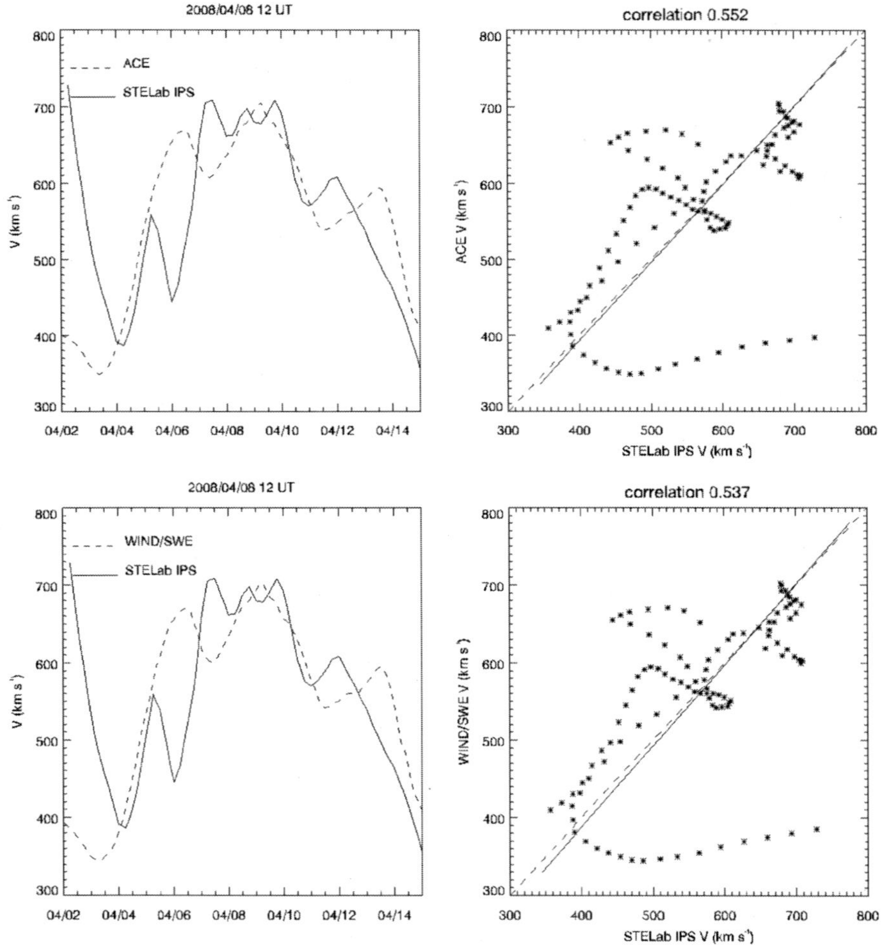

Fig. 1. 3D-velocity-reconstruction time-series extraction from STELab IPS data as compared with (top) ACE and Wind (bottom) *in situ* plasma measurements extracted from the reconstruction at the L_1 position (left-hand plots). Also shown are the correlation plots (right-hand plots) of the two data sets in each case as a measure of the accuracy of the reconstruction. In both cases, the results are very similar and give a good visual match as well as reasonable correlations.

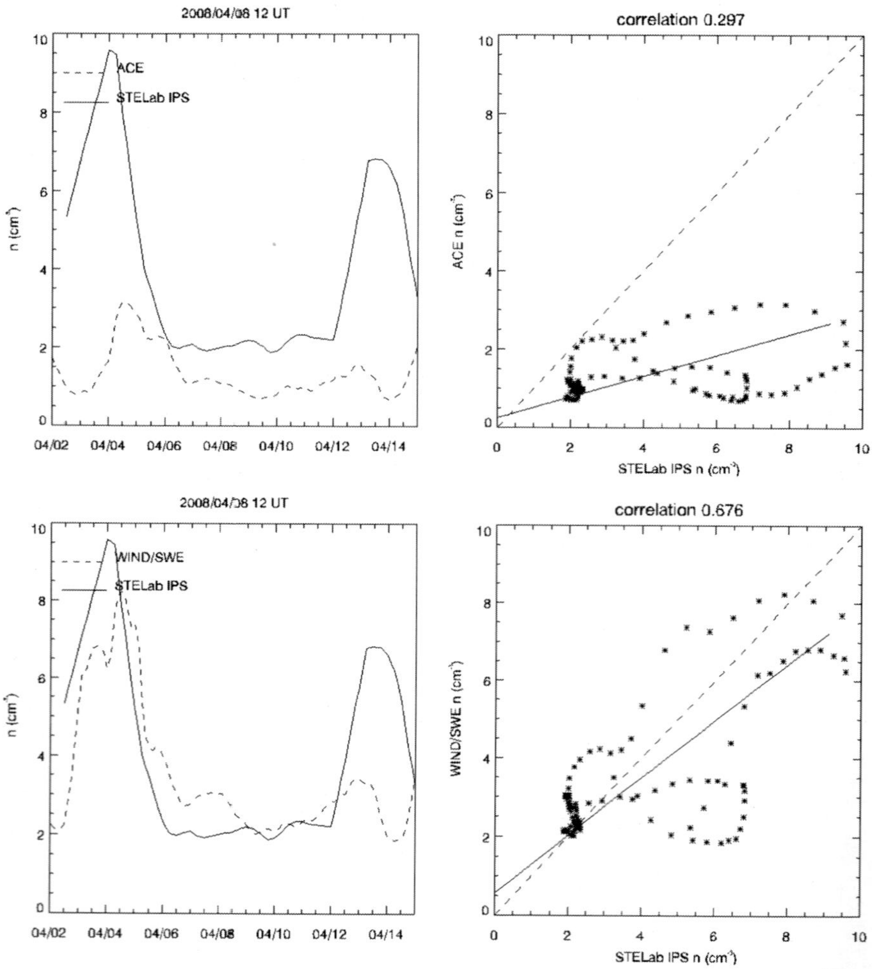

Fig. 2. 3D-density-reconstruction time-series extraction from STELab IPS data as compared with (top) ACE and Wind (bottom) *in situ* plasma measurements extracted from the reconstruction at the L_1 position (left-hand plots). Also shown are the correlation plots (right-hand plots) of the two data sets in each case as a measure of the accuracy of the reconstruction. The spacecraft data differ somewhat in their measurements of velocity, but as only Level-0 ACE data are presently available, it is likely that the Wind density measurements are more reliable; and it is these that provide both a better visual match when comparing with the extracted 3D-reconstruction time series and also the better correlation value.

there were STELab observations and these show successful reconstruction comparisons with the near-Earth *in situ* spacecraft ACE and Wind.

The velocity measurements recorded at each spacecraft are very similar when averaged on a daily cadence and result in equally-good correlations with time-series velocity extracts from the reconstructions at the position of the spacecraft. There are only relatively minor differences between the two spacecraft's daily-averaged velocity measurements. However, the density measurements differ somewhat between spacecraft. As only the ACE Level-0 data are available at present, it is likely that the Wind data are more accurate for L_1 *in situ* measurements presently and so it is this reconstruction comparison in density with the Wind spacecraft measurements that is considered as the baseline comparison in this paper. The *in situ* data have been averaged over a period of one day to match that of the 3D-reconstruction temporal cadence. The 3D reconstruction has a latitude and longitude digital resolution of 20° by 20°.

This is a preliminary analysis of the 3D reconstructions of speed and density from the STELab IPS observations during WHI. The extraction of time series at the L_1 point produce good comparison with measurements taken *in situ* by spacecraft and the goal in the future when all the spacecraft data are available is to perform full multi-point *in situ* comparisons with other spacecraft such as the STEREO spacecraft plasma measurements. However, from the analysis performed here, it seems that the heliosphere is well reconstructed at least in the direction of the Earth. The reconstructions seem to reveal a co-rotating feature around a period of three days from 04 April 2008 to 06 April 2008 (as do the *in situ* measurements). Further work on identifying this feature, by comparing these data with multi-point *in situ*, coronagraph, and other white-light data to attempt to identify transient candidates during this time, will be the subject of a forthcoming paper.

3. Summary of 3D Density Reconstructions from Early November 2004 using Ooty IPS Data

Interplanetary CMEs (ICMEs) around early-November 2004 led to a series of geomagnetic storms (as previously discussed[6,39,40] for STELab IPS and for the Solar Mass Ejection Imager (SMEI)[41,42] white-light reconstructions[6,40]). These geomagnetic storms formed part of the Coordinated Data Analysis Workshop (CDAW) master paper[43] and its corrected table.[44] Here we present results of improved 3D IPS density reconstructions of this time period using Ooty data (previously STELab low-resolution 3D reconstructions were performed).

The greater number of sources observed by Ooty allows for an increase in the resolution (and therefore the detail) for the reconstructions (in the same way that 3D reconstructions of SMEI white-light observations can be performed at much higher resolutions, e.g. Ref. 45). The density reconstructions used here from Ooty data are at a 10° by 10° latitude and longitude digital resolution, with a half-day temporal cadence.

Figure 3 shows a time-series extraction and correlation of the Ooty 3D density reconstructions at the position of, and compared with, the *in situ* spacecraft density measurements around L_1 (Wind and ACE). In both cases, the spacecraft data were smoothed over half-day periods to be commensurate with the temporal cadence of the reconstructions. The data from both spacecraft differ from each other slightly, but overall agree well. In addition, as can be seen from both the shape and position of the reconstruction-extraction time series, the Ooty density reconstructions agree well on both the timing and the magnitude of the main peak in density, and agree well with the timing of the smaller peaks. Overall, the Ooty reconstructions seem to have a better agreement with the Wind spacecraft measurements than that of the ACE spacecraft measurements since these yield a higher correlation value. A similar case was seen when comparing both STELab density and SMEI density reconstructions with the same *in situ* data.[6,40]

These are the first results from 3D tomographic reconstructions using Ooty IPS data and a more-detailed discussion of these will be carried out in a future paper.

4. Preliminary CME Observation in EISCAT IPS Data of the 16 May 2007 CME and its Comparison with STEREO-A SECCHI White-light Observations

On 16 May 2007 at 00:22:30 UT, a CME was seen off the East limb of the Sun by the STEREO-A Outer Coronagraph (COR2) instrument. Later at 13:30 UT it was seen rather well developed by the STEREO-A Heliospheric Imager (HI) 1 instrument. During this time the EISCAT system in northern Scandinavia was conducting IPS observations of radio source J0431+206 located 50.1° East and 12.1° South of the Sun-Earth line. The point of closest approach (P-Point) of the line-of-sight (LOS) to the Sun lay at around 52.2 $R_\odot$ throughout the 30-minute observation on 16 May 2007 from

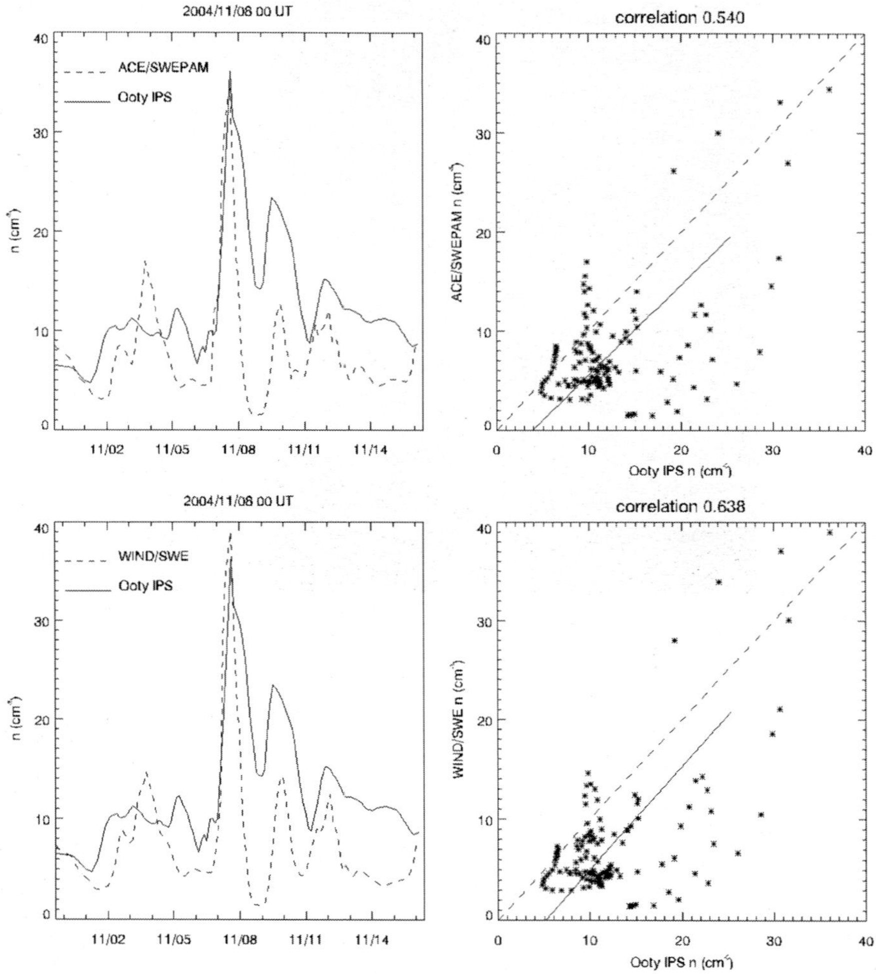

Fig. 3. Time-series comparison in density of the 3D Ooty reconstruction data extracted at the Sun-Earth L_1 point and compared with solar wind plasma measurements taken with (top) the ACE spacecraft and (bottom) the Wind spacecraft in the left-hand plots. The right-hand plots show a correlation of the two time series for each of the spacecraft. Note that the higher correlation is obtained with the Wind-spacecraft comparison.

13:30 UT to 14:00 UT. Figure 4 shows an image from the STEREO-A HI1 at 13:30 UT with the approximate apparent P-Point position of the radio source projected onto the image.

The two EISCAT antennas which gave suitable geometry to perform the two-site cross-correlation analysis were Kiruna in Sweden, and Sodankylä

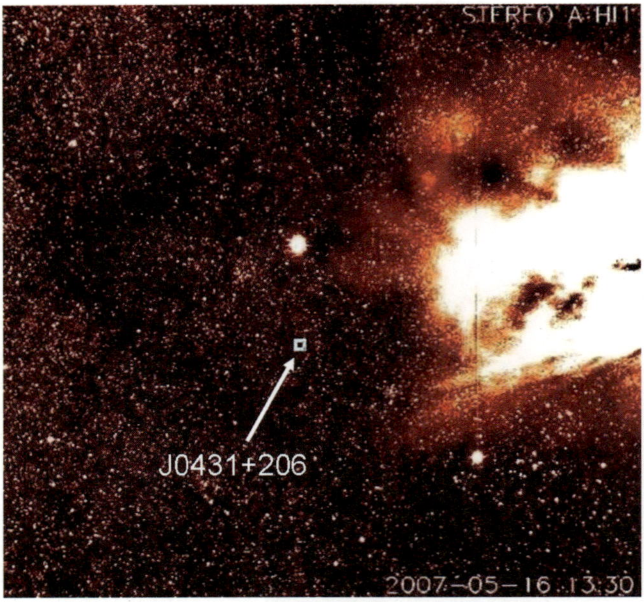

Fig. 4. A STEREO-A HI1 image frame on 16 May 2007 at 13:30 UT showing the CME expanding out from the Sun (on the right-hand-side of the image). The projection of the P-Point of the LOS for the EISCAT IPS radio source J0431+206 is approximately marked in the field of view of the imager. At this time, the angular separation between the STEREO-A spacecraft and the Earth was small (5.461°).

in Finland. The cross-correlation of the two signals along with the baseline geometry allows an approximation of the outward solar wind flow speed to be calculated from the location of the peak maximum in the cross-correlation function (CCF) of the two simultaneous IPS signals. A more detailed description of how the solar wind velocity can be directly obtained from the CCF can be found in Refs. [46, 47]. Negative lobes at, or near to, zero time lag in the CCF can occur as a result of a transient, e.g. a transient, passing through the LOS such as a CME.[48] Other determining factors include an enhanced level of scintillation.[49,50] The shape and evolution of the negative lobe associated with the passage of a transient feature along with the shape of the CCF can be used to infer the motion of various parts of a CME crossing the IPS LOS.[5] Figure 5 displays the changing CCF with time in 5-minute intervals of the Kiruna-Sodankylä observation.

The six parts of Fig. 5 are each a 5-minute segment of the 30-minute observation showing the CCF for each segment. The time intervals of the

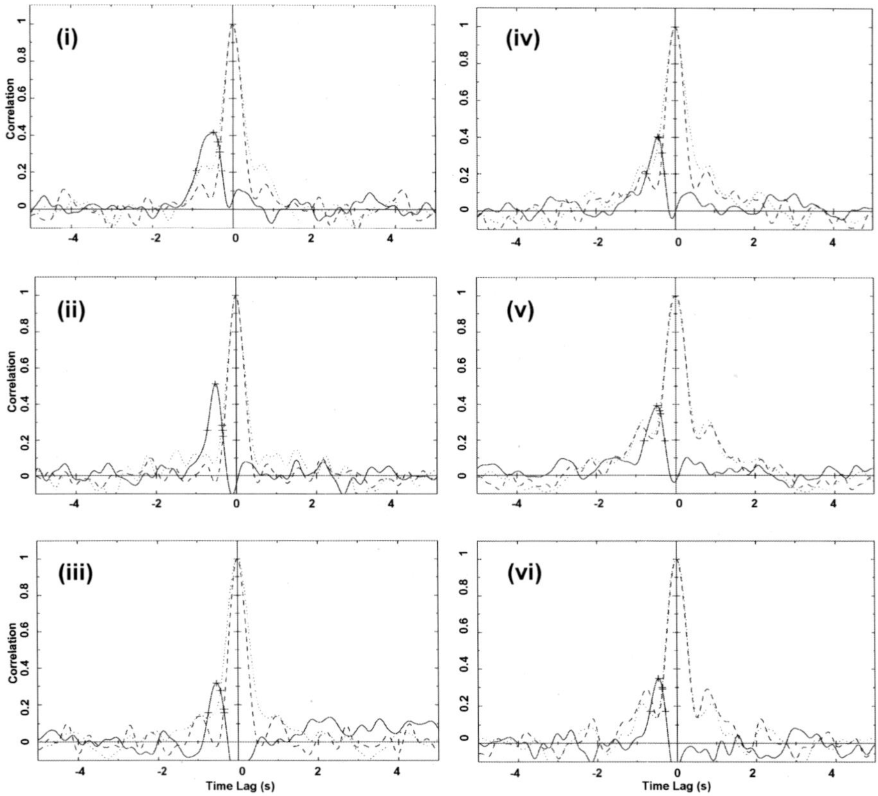

Fig. 5. The cross-correlation functions (CCFs) of the Kiruna-Sodankylä IPS observation in 5-minute intervals starting with (i) 13:30 UT to 13:35 UT up to (vi) 13:55 UT to 14:00 UT. Velocities range from 390 km s^{-1} to 530 km s^{-1} with a baseline around 240 km. The description of the observation can be found in the text.

segments are as follows: (i) 13:30 UT to 13:35 UT, (ii) 13:35 UT to 13:40 UT, (iii) 13:40 UT to 13:45 UT, (iv) 13:45 UT to 13:50 UT, (v) 13:50 UT to 13:55 UT, and (vi) 13:55 UT to 14:00 UT. Interpreting each of the CCFs with the assumption of the passage of the interplanetary counterpart of the CME as observed by the STEREO|HI white-light instrumentation, then (i) shows the onset of what may be the passage of a transient across the LOS since there are hints that a negative lobe in the CCF at zero time lag may be forming. 5-minutes later in (ii), a negative lobe around zero time lag is clearly forming which indicates the passage of a transient; likely in this case to be an ICME. It is possible that the negative lobe is caused by the geometry of the LOS with respect to the compression at the "nose" of the ICME. In (iii) we see the passage continue as a very deep negative lobe occurs. (iv) shows the negative lobe disappearing and thus signifying the passage of the ICME as coming to an end. However, (v) shows what seems to be another ICME/transient passage starting and (vi) confirms this. It should be noted though, only one CME was seen in that direction by the STEREO spacecraft. This means that (iv), (v) and (vi) are likely to be showing a different "part" of the transient than is seen in (i), (ii), and (iii) since there are hints that a negative lobe in the CCF at zero time lag may be forming. 5-minutes later in (ii), a negative lobe around zero time lag is clearly forming which indicates the passage of a transient; likely in this case to be an ICME. In (iii) we see the passage continue as a very deep negative lobe occurs. (iv) shows the negative lobe disappearing and thus signifying the passage of the CME as coming to an end. However, (v) shows what seems to be another ICME/transient passage starting and (vi) confirms this. However only one CME was seen in that direction by the STEREO spacecraft. This means that (iv), (v) and (vi) are likely to be showing a different "part" of the transient than is seen in (i), (ii), and (iii).

An argument is generally made in the IPS community that the changing and rotation of the baselines/geometry can cause a negative lobe at zero time lag to fade and return throughout an observation. However in this case, it should be duly noted that the baselines and geometry varies almost inconsequentially throughout the observation thus removing this possibility as an option for the fading and re-initiating of the negative lobe seen at zero time lag.

The only present solution for this occurrence must be that different parts of the ICME are crossing the LOS. This may be perhaps the "front" of the ICME where there are often shocks and there can be a high amount of turbulence, and perhaps the "ejecta" portion (of a CME) where the majority of the mass generally occurs (which can also have an effect on scintillation level even though the level of scintillation is usually determined by the level of turbulence in whatever material crosses your LOS). The other solution is that the first part represents the initial shock ahead of the ICME front, and the second part the front itself. Further investigation is needed to determine exactly what is being observed here, but the timings, geometry, and velocities measured by both EISCAT and STEREO for this event seem to agree. The reader is referred to Ref. [51] for a more-detailed analysis of this event and its comparison with the HI-1A white-light observations.

5. Summary

This paper provides a brief summary of the most-recent advances of using the 3D-tomographic-reconstruction technique and also of EISCAT IPS observations as compared with those of STEREO|SECCHI white-light instrumentation. These include 3D tomographic reconstructions of both STELab and Ooty IPS data, and showing the improvements to the density reconstructions of the Ooty data for the complex November 2004 period; and a good correlation between the observation of a CME by both STEREO and EISCAT.

The 3D-reconstructed heliosphere from STELab data from a latter part of the WHI period show what appears to be a co-rotating region passing across the Sun-Earth L_1 point with time. The time-series extraction from the reconstruction in both speed and density give good comparison with the ACE Level-0 and Wind data presently available. Further investigation is needed in determining the exact characteristics of this feature and also to check in greater detail for possible transient candidates from white-light corona and heliosphere observations. The 3D reconstruction can also be verified further for it's global accuracy by comparing with multi-point *in situ* observations; as is the aim of future work when the *in situ* data are more readily available in the very-near future.

The geoeffective storms during November 2004 discussed here and also in greater detail by Refs. [6, 39, 40, 43, 44] are very well reproduced in g-level-to-density 3D reconstructions using Ooty IPS data, and are a great improvement over previous[6] reconstructions using these data. The reconstructions are shown to be associated with known *in situ* signatures by comparing these data with the Wind and ACE spacecraft. Overall, the Ooty data reconstruction appears to show improved density values compared with the STELab[6,40] density values when time-series of the two are compared with *in situ* measurements. The Ooty reconstructions have the advantage of having double the spatial and temporal resolution from those of STELab since there are many more Ooty IPS observations during this time, and thus should provide a better comparison with those measured *in situ*. The Ooty IPS velocity measurements were also incorporated into these reconstructions which further improved on the density reconstructions shown previously and these improved reconstructions as well as velocity reconstructions will be discussed in a future paper.

An EISCAT IPS observation from May 2007 shows evidence of the passage of an ICME that was noted from STEREO-A HI white-light imaging on the same day. The CCFs of 5-minute blocks of the 30-minute observation showed some substantial variation, indicating that different parts of the ICME passed through the LOS during the course of the observation. Attempts will be made in a future paper to firmly link the CCF variations with the passage of different parts of the CME seen by the STEREO-A spacecraft's remote-sensing instruments.

In conclusion, for our 3D tomography we continue to follow CMEs from near the solar surface outward until they are observed *in situ* near Earth, and a study is now underway to compare in near-real-time with both STEREO spacecraft, as is already being done routinely with Wind, ACE, and Ulysses. These events, reconstructed in 3D in terms of both speed and density for WHI and density alone for November 2004, show that the heliospheric response to CMEs and co-rotating features can be large. We also see that there is good comparison with the EISCAT IPS observations with those see by STEREO|SECCHI ahead instrumentation, along with such timings of the 16 May 2007 event. We look forward to other (true-multi-point) *in situ* comparisons such in forthcoming papers for the WHI period now that the *in situ* data are becoming more widely available. As our 3D tomographic models become more sophisticated, possibly incorporating a 3D MHD solar wind model, and multi-point calibrations are realised, we expect the comparisons to improve even further from those demonstrated in Ref. [40] and in this paper, as well as a full paper on the 16 May 2007 event.

Acknowledgments

The authors (MMB, BVJ, JMC, PPH, and AB) acknowledge AFOSR grant FA9550-06-1-0107, NASA grants NNX08AJ116 and NNG05GM58G, and NSF grant ATM0331513, to the University of California at San Diego (UCSD) for support to work on these 3D tomography and STEREO-comparison analyses. The authors (RAF, GDD, and ARB) acknowledge STFC for their funding of the EISCAT-STEREO analyses carried out in this paper. The authors wish to thank the Wind|SWE and ACE|SWEPAM groups for use of the solar wind proton density and velocity measurements used in this paper for *in situ* comparisons, and also to the STEREO|SECCHI Consortium for making their data and images available on the World Wide Web. EISCAT is supported by the research councils of Norway, Sweden, Finland, Japan, China, the United Kingdom, and Germany.

References

1. A. Hewish, P. F. Scott and D. Wills, *Nature* **203** (1964) 1214.
2. M. Kojima and T. Kakinuma, *Space Science Review* **53** (1990) 173.
3. A. Canals, A. R. Breen, L. Ofman, P. J. Moran and R. A. Fallows, *Annales Geophysicae* **20** (2002) 1265.
4. M. M. Bisi, Interplanetary Scintillation Studies of the Large-Scale Structure of the Solar Wind, Ph.D. Thesis, The University of Wales, Aberystwyth 2006.
5. R. A. Jones, A. R. Breen, R. A. Fallows, A. Canals, M. M. Bisi and G. Lawrence, *J. Geophys. Res.* **112** (2007) 8107.
6. M. M. Bisi, B. V. Jackson, P. P. Hick, A. Buffington and J. M. Clover, *Advances in Geosciences* **14** (2009) 161.
7. W. A. Coles and J. K. Harmon, *J. Geophys. Res.* **83** (1978) 1413.
8. R. A. Fallows, A. R. Breen, P. J. Moran, A. Canals and P. J. S. Williams, *Advances in Space Research* **30** (2002) 437.
9. M. M. Bisi, B. V. Jackson, R. A. Fallows, A. R. Breen, P. P. Hick, G. Wannberg, P. Thomasson, C. A. Jordan and G. D. Dorrian, *Proc. SPIE* **6689** (2007) 668911.
10. R. A. Fallows, A. R. Breen and G. D. Dorrian, *Ann. Geophys.* **26** (2008) 2229.
11. P. P. Hick and B. V. Jackson, in *Telescopes and Instrumentation for Solar Astrophysics. Proc. SPIE*, eds. S. Fineschi and M. A. Gummin 2004.
12. B. V. Jackson and P. P. Hick, Astrophys. and Space Sci. Lib. Vol. 314 (Kluwer Academic Publ., Dordrecht., 2005) pp. 355–386.
13. A. R. Breen, R. A. Fallows, M. M. Bisi, R. A. Jones, B. V. Jackson, M. Kojima, G. D. Dorrian, H. R. Middleton, P. Thomasson and G. Wannberg, *ApJL* **683** (2008) L79.
14. M. Kojima and T. Kakinuma, *J. Geophys. Res.* **92** (1987) 7269.
15. G. Swarup, N. V. G. Sarma, M. N. Joshi, V. K. Kapahi, D. S. Bagri, S. V. Damle, S. Ananthakrishnan, V. Balasubramanian, Bhave and R. P. P. Sinha, *Nature* **230** (1971) 185.
16. P. K. Manoharan, M. Kojima, N. Gopalswamy, T. Kondo and Z. Smith, *ApJ* **530** (2000) 1061.
17. P. K. Manoharan, M. Tokumaru, M. Pick, P. Subramanian, F. M. Ipavich, K. Schenk, M. L. Kaiser, R. P. Lepping and A. Vourlidas, *ApJ* **559** (2001) 1180.
18. H. Rishbeth and P. J. S. Williams, *Monthly Notices of the Royal Astronomical Society* **26** (1985) 478.
19. G. Bourgois, W. A. Coles, G. Daigne, J. Silen, T. Turunen and P. J. Williams, *Astronomy and Astrophysics* **144** (1985) 452.
20. G. Wannberg, L.-G. Vanhainen, A. Westman, A. R. Breen and P. J. S. Williams, in *Proc. Union of Radio Scientists (URSI)*, 2002.
21. J. W. Armstrong and W. A. Coles, *JGR* **77** (1972) 4602.
22. A. R. Breen, R. A. Fallows, M. M. Bisi, P. Thomasson, C. A. Jordan, G. Wannberg and R. A. Jones, *J. Geophys. Res.* **111** (2006) 8104.
23. M. M. Bisi, A. R. Breen, R. A. Fallows, G. D. Dorrian, R. A. Jones, G. Wannberg, P. Thomasson and C. Jordan, *EOS Trans. AGU, Fall Meeting Supp. — Abstract SH33A-0399* **87** (2006) 52.

24. W. A. Coles, *Astrophysics and Space Science* **243**(1) (1996) 87.
25. M. Klinglesmith, The Polar Solar Wind from 2.5 to 40 Solar Radii: Results of Intensity Scintillation Measurements, Ph.D. Thesis, University of California, San Diego (UCSD) 1997.
26. R. A. Fallows, A. R. Breen, M. M. Bisi, R. A. Jones and G. Wannberg, *Geophys. Res. Lett.* **33** (2006) 11106.
27. M. M. Bisi, R. A. Fallows, A. R. Breen, S. R. Habbal and R. A. Jones, *J. Geophys. Res.* **112** (2007) A06101.
28. P. K. Manoharan and S. Ananthakrishnan, *MNRAS* **244** (1990) 691.
29. B. V. Jackson, P. P. Hick, M. Kojima and A. Yokobe, *J. Geophys. Res.* **103** (1998) 12049.
30. K. W. Ogilvie and M. D. Desch, *Advances in Space Research* **20** (1997) 559.
31. K. W. Ogilvie, D. J. Chornay, R. J. Fritzenreiter, F. Hunsaker, J. Keller, J. Lobell, G. Miller, J. D. Scudder, E. C. Sittler, Jr., R. B. Torbert, D. Bodet, G. Needell, A. J. Lazarus, J. T. Steinberg, J. H. Tappan, A. Mavretic and E. Gergin, *Space Science Reviews* **71** (1995) 55.
32. E. C. Stone, A. M. Frandsen, R. A. Mewaldt, E. R. Christian, D. Margolies, J. F. Ormes and F. Snow, *Space Science Reviews* **86** (1998) 1.
33. D. J. McComas, S. J. Bame, P. Barker, W. C. Feldman, J. L. Phillips, P. Riley and J. W. Griffee, *Space Science Reviews* **86** (1998) 563.
34. V. Domingo, B. Fleck and A. I. Poland, *Space Science Reviews* **72** (1995) 81.
35. G. E. Brueckner, R. A. Howard, M. J. Koomen, C. M. Korendyke, D. J. Michels, J. D. Moses, D. G. Socker, K. P. Dere, P. L. Lamy, A. Llebaria, M. V. Bout, R. Schwenn, G. M. Simnett, D. K. Bedford and C. J. Eyles, *Sol. Phys.* **162** (1995) 357.
36. M. L. Kaiser, *Advances in Space Research* **36** (2005) 1483.
37. R. A. Howard, J. D. Moses, D. G. Socker, K. P. Dere and J. W. Cook, *Advances in Space Research* **29** (2002) 2017.
38. R. A. Howard, J. D. Moses, A. Vourlidas, J. S. Newmark, D. G. Socker, S. P. Plunkett, C. M. Korendyke, J. W. Cook, A. Hurley, J. M. Davila, W. T. Thompson, O. C. St Cyr, E. Mentzell, K. Mehalick, J. R. Lemen, J. P. Wuelser, D. W. Duncan, T. D. Tarbell, C. J. Wolfson, A. Moore, R. A. Harrison, N. R. Waltham, J. Lang, C. J. Davis, C. J. Eyles, H. Mapson-Menard, G. M. Simnett, J. P. Halain, J. M. Defise, E. Mazy, P. Rochus, R. Mercier, M. F. Ravet, F. Delmotte, F. Auchere, J. P. Delaboudiniere, V. Bothmer, W. Deutsch, D. Wang, N. Rich, S. Cooper, V. Stephens, G. Maahs, R. Baugh, D. McMullin and T. Carter, *Space Science Reviews* **136** (2008) 67.
39. L. K. Harra, N. U. Crooker, C. H. Mandrini, L. van Driel-Gesztelyi, S. Dasso, J. Wang, H. Elliott, G. Attrill, B. V. Jackson and M. M. Bisi, *Sol. Phys.* **244** (2007) 95.
40. M. M. Bisi, B. V. Jackson, P. P. Hick, A. Buffington, D. Odstrcil and J. M. Clover, *JGR* (2008) **113**, doi:10.1029/2008JA013222.
41. C. J. Eyles, G. M. Simnett, M. P. Cooke, B. V. Jackson, A. Buffington, P. P. Hick, N. R. Waltham, J. M. King, P. A. Anderson and P. E. Holladay, *Sol. Phys.* **217** (2003) 319.

42. B. V. Jackson, A. Buffington, P. P. Hick, R. C. Altrock, S. Figueroa, P. E. Holladay, J. C. Johnston, S. W. Kahler, J. B. Mozer, S. Price, R. R. Radick, R. Sagalyn, D. Sinclair, G. M. Simnett, C. J. Eyles, M. P. Cooke, S. J. Tappin, T. Kuchar, D. Mizuno, D. F. Webb, P. A. Anderson, S. L. Keil, R. E. Gold and N. R. Waltham, *Sol. Phys.* **225** (2004) 177.
43. J. Zhang, I. G. Richardson, D. F. Webb, N. Gopalswamy, E. Huttunen, J. C. Kasper, N. V. Nitta, W. Poomvises, B. J. Thompson, C.-C. Wu, S. Yashiro and A. N. Zhukov, *JGR* **112** (2007) 10102.
44. J. Zhang, I. G. Richardson, D. F. Webb, N. Gopalswamy, E. Huttunen, J. C. Kasper, N. V. Nitta, W. Poomvises, B. J. Thompson, C.-C. Wu, S. Yashiro and A. N. Zhukov, *JGR* **112** (2007) 12103.
45. B. V. Jackson, M. M. Bisi, P. P. Hick, A. Buffington, J. M. Clover and W. Sun, *J. Geophys. Res.*, in-press (2008).
46. R. R. Grall, Remote Sensing Observations of the Solar Wind Near the Sun, Ph.D. Thesis, University of California, San Diego 1995.
47. A. R. Breen, W. A. Coles, R. Grall, U. P. Løvhaug, J. Markkanen, H. Misawa and P. J. S. Williams, *Journal of Atmospheric and Terrestrial Physics* **58** (1996) 507.
48. M. T. Klinglesmith, R. R. Grall and W. A. Coles, 933MHz IPS Velocity Measurements at EISCAT, in *Solar Wind 8*, 1996.
49. A. Canals, Interplanetary Scintillation Studies of the Solar Wind during the Rising Phase of the Solar Cycle, Ph.D. Thesis, The University of Wales, Aberystwyth 2002.
50. R. A. Jones, Interplanetary Scintillation Observations of the Inner Solar Wind, Ph.D. Thesis, The University of Wales, Aberystwyth 2007.
51. G. D. Dorrian, A. R. Breen, D. S. Brown, J. A. Davies, R. A. Fallows and A. P. Rouillard, *Geophys. Res. Lett.* **35** (2008) 24104.

Advances in Geosciences
Vol. 21: Solar & Terrestrial Science (2008)
Ed. Marc Duldig

RELATIONSHIP BETWEEN EARTH-DIRECTED SOLAR ERUPTIONS AND MAGNETIC CLOUDS AT 1AU: A BRIEF REVIEW

VASYL YURCHYSHYN
Big Bear Solar Observatory 40386 North Shore Lane,
Big Bear City, CA 92314
vayur@bbso.njit.edu

DURGESH TRIPATHI
Department of applied Mathematics and Theoretical Physics,
University of Cambridge, Wilberforce Road,
Cambridge CB30WA, UK

We review relationships between coronal mass ejections (CMEs), EIT post eruption arcades, and the coronal neutral line associated with global magnetic field and magnetic clouds near the Earth. Our previous findings indicate that the orientation of a halo CME elongation may correspond to the orientation of the underlying flux rope. Here we revisit these preliminary reports by comparing orientation angles of elongated LASCO CMEs, both full and partial halos, to the post eruption arcades. Based on 100 analysed events, it was found that the overwhelming majority of halo CMEs are elongated in the direction of the axial field of the post eruption arcades. Moreover, this conclusion also holds for partial halo CMEs as well as for events that originate further from the disk center. This suggests that the projection effect does not drastically change the appearance of full and partial halos and their images still bear reliable information about the underlying magnetic fields. We also compared orientations of the erupted fields near the Sun and in the interplanetary space and found that the local tilt of the coronal neutral line at 2.5 solar radii is well correlated with the magnetic cloud axis measured near the Earth. We suggest that the heliospheric magnetic fields significantly affect the propagating ejecta. Sometimes, the ejecta may even rotate so that its axis locally aligns itself with the heliospheric current sheet.

1. Introduction

Geomagnetic storms are major disturbances in the Earth's magnetosphere that occur when the interplanetary magnetic field turns southward and remains so for a prolonged period of time.[1–3] Reconnection between the southward-directed (relative to the ecliptic plane) component of the solar wind-carried magnetic field, B_z, and the northward-directed geomagnetic field can occur at the dayside magnetopause, resulting in the transfer of significant amounts of energy from the solar wind into the Earth's magnetosphere. During a typical storm, an enhanced ring current is created and the auroral electrojets in both hemispheres move (often dramatically) equatorward. This is evidenced by a decrease in the geomagnetic field at near-equatorial magnetic stations, and a large decrease in the *Dst* index that is currently compiled from the records from four selected stations.[4]

All regions of geospace are affected by geomagnetic disturbances, and the operations of various technological systems can be impaired or even totally disrupted. For example, during large geomagnetic disturbances the rapidly changing electric currents in the ionosphere (regions within the moving electrojets as well as at distances from them), largely driven by the precipitation of charged particles from the magnetosphere, cause rapid changes in the magnetic field at Earth's surface. These changing fields in turn can induce enhanced electric currents to flow in the Earth's crust and mantle ("telluric currents"). These currents seek the highest conductivity paths, which are often long, e.g. grounded electric grid systems and communications cables. As another example, the enhanced disturbances in the ionosphere can produce serious scintillations in trans-ionospheric radio-wave propagation, even at GHz frequencies. Scintillation effects can be especially large in the equatorial regions at times (producing convective ionospheric storms, often referred to as sporadic E-events), but can be significant across all latitudes.

The occurrence of geomagnetic storms is well associated with Earth-directed coronal mass ejections (CMEs), which appear in coronagraph images as bright halos around the occulting disk [halo CME,[5]]. CMEs are eruptions of the solar magnetic field and plasma into interplanetary space, which occur following a large-scale magnetic rearrangement in the solar atmosphere[6–9] [also see several excellent reviews in Space Science Rev., 2006, DOI: 10.1007/s11214-006-9027-8]. Studies of soft X-ray images of the solar corona indicate that eruptions are more likely to occur in large magnetic configurations that contain a bright sigmoidal structure,[10] which is thought to be related to magnetic flux ropes.[11,12] Potential magnetic field modelling[13,14] suggests that the origin of CMEs is determined by the global structure of the solar magnetic field and is related to the helmet streamers that are sensitive to the evolution of active regions and/or filaments.

Depending upon the orientation of the magnetic field in the CME-originating region and the sign of magnetic helicity (twist), the Earth-directed CME may or may not have an intense negative B_z field.[15] Hence, the origin of CMEs, the structure of their source regions and their signatures in the solar wind near the Earth are of fundamental interest in the physics of the Sun, space plasmas, and space-weather research.

This paper is organised as follows. In Section 2 we describe the relationship between CME source regions and interplanetary ejecta. Particular attention is paid to the magnetic structure of a CME, and how it corresponds to that of the associated post eruption arcades (PEAs). Section 3 compares magnetic orientations of halo CMEs, tilt of the main coronal neutral line near the eruption site and the directional parameters of the corresponding magnetic clouds. In Section 4 we briefly summarise the results and discuss their implications.

2. Relationship Between Magnetic Fields of CME Source Regions and the Corresponding Interplanetary Ejecta

Interplanetary counterparts of CMEs are called interplanetary CMEs (ICMEs) or interplanetary ejecta. The internal structure of an ICME is not trivial and may include an ICME body that is often preceded by a shock. The ICME body, in turn, may be observed as a complex ejection[16] or a magnetic cloud (MC).[17] While MCs are known to be the cause of the most severe geomagnetic storms,[18] interplanetary shocks[19] and co-rotating interaction regions[9] are also quite geoeffective, although the storm associated with them are, on average, less intense than those associated with the MCs.

For a number of years researchers have been trying to establish a relationship between CMEs, their complex progenitors, and interplanetary ejecta.[20] Since complex ejecta represent unorganized and compound magnetic structures they were not subject of these studies. Instead, effort was focused on MCs that generally exhibit a magnetically organized geometry, which is thought to correspond globally to a curved flux rope as shown in Fig. 1 [also see Figs. 1 and 2 in Refs. 21–23] and can be related to various solar parameters such as magnetic field helicity, strength and direction.[8] In particular, significant research efforts have been devoted to finding whether the direction of the magnetic field and the sense of twist

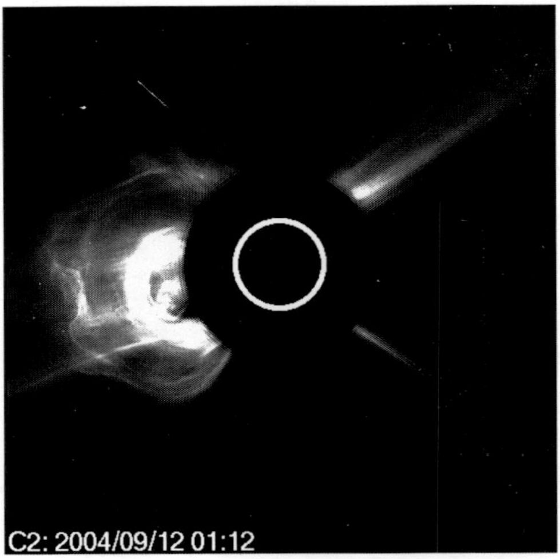

Fig. 1. LASCO C2 image of a limb CME that erupted on 12 September 2004. This image represents an event with a distinct three part structure and suggests that, at least some CMEs can be described as a flux rope structures.

found in MCs correspond to those associated with CME source regions (which include both active regions and filaments).

2.1. *Association between ejecta and post eruption arcades*

According to various case and statistical studies many interplanetary ejecta maintain nearly the same orientation and twist as the source regions they are associated with [e.g. Refs. 12, 15, 22, 24–26]. This is in agreement with helicity conservation law established for laboratory plasmas. Several statistical studies, however, produced inconclusive results.[27,28] Harra *et al.*[29] conducted a thorough case study of flaring activity and concluded that the same active region may produce interplanetary ejecta with different magnetic orientations. Therefore, the question about the magnetic connection between interplanetary ejecta and solar surface phenomena is still open. There are no well defined schemes to predict the magnetic field structure at 1AU based on solar surface measurements present at this moment.

Here, we further explore this problem by utilising magnetic field information measured at various places between the solar surface and the Earth. In particular, we compare directional angles of halo CMEs, heliospheric current sheet (HCS), MCs, and EUV PEA, observed with Extreme Ultraviolet Imaging Telescope (EIT)[30] on board Solar and Heliospheric Observatories.[31]

2.1.1. *Measurements of CME orientations from white light coronagraph images*

White light coronagraph data, such as provided by the Large Angle Spectroscopic COronagraph LASCO[32] instrument also on board the SOHO spacecraft, show that halo CMEs have various sizes and shapes. Many of them can be enveloped by an ellipse and can be fitted with a cone model.[33–36] CMEs are also found to be highly structured three-dimensional features[37,38] and they are thought to represent a flux rope.[17,38–42] In particular, Krall & St. Cyr[38] showed that statistical parameters measured for observed CMEs, such as eccentricity and the axial aspect ratio, are in agreement with those obtained for a parameterized model flux rope.

The structured CMEs appear to be magnetically organised in the axial direction, which corresponds to the axis of a large-scale twisted flux rope.[37] At the same time, modelling of erupting flux ropes[43–45] shows that modelled halo CMEs appear to be elongated in the direction of the axial field of the underlying flux rope. Therefore, it is possible the ellipse-shaped appearance of halo CMEs may be related to their originating magnetic structure.

Yurchyshyn *et al.*[45] [hereafter Paper I] studied elongation parameters of 25 halo CMEs and compared them to the corresponding parameters determined for MCs. The CME-MC events were selected from the Master Data Table compiled during a *Living With a Star* Coordinated Data Analysis Workshops[46,47] and the list published in.[48] All orientation angles were determined in the geo-centric solar ecliptic coordinate system (GSE), where y-axis is in the ecliptic plane pointing towards dusk, x-axis is directed from the Earth towards the Sun and z-axis is pointed northward. The CME orientation angle (or tilt) was determined in Paper I by fitting an ellipse to an irregularly shaped "halo" around the C3 occulting disk (Fig. 2) and measuring its tilt in the clockwise direction from the positive GSE y-axis to the ellipse semi-major axis. For each event, we measured the angles in 3-5 images depending upon their availability. The final tilt angle listed in Table 1 of Paper I was calculated as the mean of angles determined from individual frames.

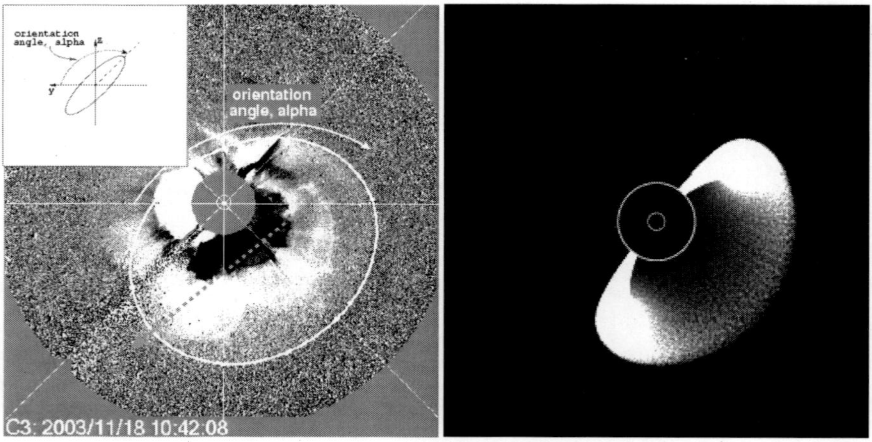

Fig. 2. Left: LASCO C3 difference image for the 2003 November 18 eruption used to measure the CME elongation. The halo CME was measured at its outer edge at intervals of 45 degrees (lines). An ellipse was then fitted to the eight halo points and its tilt angle was determined in the clockwise direction from the y-axis that points eastward to the ellipse semi-major axis (see inset). Right: A synthetic "halo" CME image produced for the same event by the erupting flux rope model. The tilt of the synthetic elongated halo represents the orientation of the simulated flux rope. Comparison with the left panel shows that the tilts of those two halos are similar thus indicating the correspondence between the observed halo elongation and the magnetic axis of a flux rope.

2.1.2. *Orientation of CME elongations versus direction of the axial field in magnetic clouds*

The main conclusion drawn from the comparison of CME elongations and MC orientations, discussed above, is that for about 64% of CME-MC events there is a good correspondence between their orientation angles. This supports the idea that the elongation of halo CMEs may indeed reflect the underlying structures of the erupting magnetic fields. If that is the case, these results also imply that the majority of interplanetary ejecta do not significantly change their orientations (less than 45 deg rotation) while traveling from the Sun to the near Earth environment. This was later confirmed by Zhao (2007, private communication). On the other hand, not-so-close-correspondence between CME and MC orientations seems to reflect the existing general confusion about the relationship between solar and MC fields: while for the majority of events magnetic fields at the Sun and 1AU seem to be in agreement, a non-negligible minority of the events display quite different behavior. We note that this correspondence rate may be affected by the fact that some CMEs in the dataset are not full halos. Therefore, their orientations may not be accurately determined.

Another feasible interpretation of these findings was discussed in Yurchyshyn *et al.*[45,49]: the main axis of a CME may rotate as it expands into the interplanetary space. It could occur as a result of interaction between the ejecta and the heliospheric magnetic field due to coronal viscosity.[50,51] The rotation may also occur because an erupting twisted flux is subject to kink or torus instability, so that its loop top could rotate as it evolves.[52–55] Lynch *et al.*[56] and Gibson and Fan[57] [also private communication] proposed that an expanding flux rope can reconnect with the surrounding fields and the footpoints of the erupting fields can be displaced, so the erupting flux rope may change its orientation.

The above possibilities will be addressed in the following sections. In order to address the above uncertainties and estimate the reliability of the above findings, a larger statistical study is needed, which would involve more magnetic parameters.

2.2. *Orientation of CME elongations and direction of the axial field in EIT post eruption arcades*

In this section we will compare orientations of Earth-directed CMEs and the corresponding PEAs. Tripathi *et al.*[58] analyzed association between 236 PEAs observed by the EIT instrument and CMEs and found a nearly one to one correspondence between them. Based on this result, PEAs can be considered as reliable tracers of CME events. For each PEA event these authors determined its heliographic position and the length along its axis, which was overlying magnetic polarity inversion lines, when traced back to the Michelson Doppler Imager (MDI)[59] synoptic charts of the photospheric magnetic field. The heliographic length of the PEAs was found to be in the range of 2 to 40 degrees, with an average of 15 degrees. No error analysis was performed in this study since the primary focus was the range of values and a trend of parameters.

The data set presented in Ref. 58 allowed us to determine the tilt (orientation) of these PEAs relative to the solar east-west line. This tilt was measured in degrees in clockwise direction starting from the east. We estimated that in most cases the PEA angles were measured with an accuracy of ± 10 deg, although in 6 events the absolute error was estimated to be 90 deg.

The list of Tripathi *et al.*[58] includes various events and not all of them are full and partial halos. For our purposes we needed to modify this list and discard all events associated with narrow and limb CMEs. Therefore, we were left with 101 PEA-CME events. For each full or partial halo CMEs from the list we determined its orientation as described above.

Since a PEA is a magnetic structure with a well defined axial field, we can take into account the PEA's axial field and assign direction to the orientation angles. The axial field and twist (helicity sign) in PEAs can be determined from solar data based on i) direct calculations of the predominant current helicity;[60–62] ii) force free field modelling[63]; iii) magnetic orientation angles[64]; and iv) visual inspection of the loop pattern seen in chromospheric and coronal images such as S-shaped sigmoids[10,12] and/or dextral and sinistral filaments.[65]

To compare these ambiguity resolved directions of PEAs with CME orientation angles, we need to also assign direction to the CME orientations. To do so we use the magnetic field information inferred from the associated PEA. The CME ambiguity resolution is based on the assumption that the direction of the axial field and twist in a flux rope CME corresponds to those of the EIT/Hα flare arcade associated with the eruption (see more details in Fig. 3). We will, therefore, assume that the axial field of the PEA makes an acute angle with that of the CME.

In Fig. 4 we plot CME directional angles versus those of PEAs. The panels represent (from left to right) disk halo CMEs, off-disk halo CMEs and partial CMEs. First note that in all three cases the data points are mainly clustered near the bisector. In Fig. 5 we show the distribution of PEA-CME difference angles plotted for all events. As it follows from the distribution plot, for the overwhelming majority of events (72%) the difference between the angles is less that 45 deg. It is worth emphasizing here that this holds true for both the partial halos and the halos that originate away from the disk center, where the projection effect is expected to play a significant role.

The data presented in Figs. 4 and 5 indicate that, on average, the elongated halos are oriented along the axis of the corresponding PEAs. In addition, the elliptical shape of a halo CME may indeed bear information on the geometry of the underlying flux rope. These results also indicate that 64% correspondence between CME and MC orientations discussed in the previous section seems not to be caused by the projection effect in CME imaging, but is rather a result of CME evolution in the interplanetary space. In the next section we will address the relationship between solar and

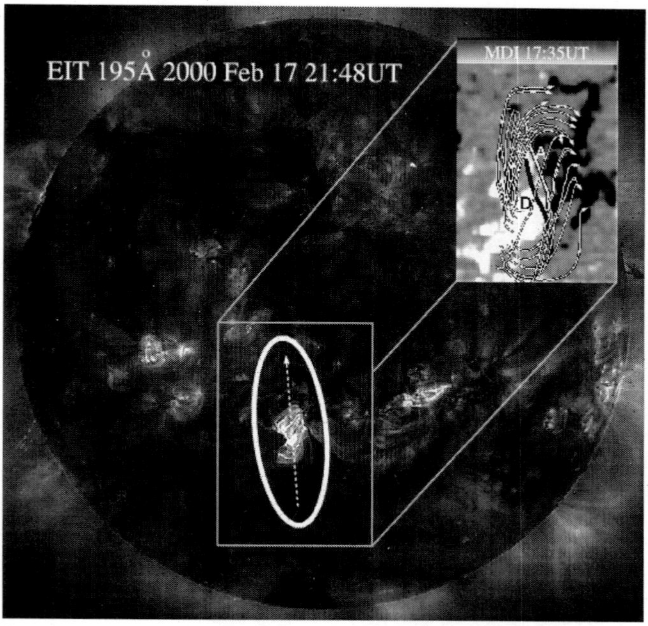

Fig. 3. EIT 195A image taken on 17 February 2000 at 2148 UT. The flare arcade associated with an eruption is inside the white ellipse that indicates the orientation of the corresponding CME. The white arrow inside the ellipse shows a northwardly directed axial field in the EIT arcade, determined from an Michelson Doppler Imager (MDI)[79] magnetogram: from positive (white) to negative (black). The "S" shaped arcade indicates a right handed (positive helicity) magnetic structure. The field direction in the corresponding CME was ascribed, based on the EIT arcade and the acute angle method.

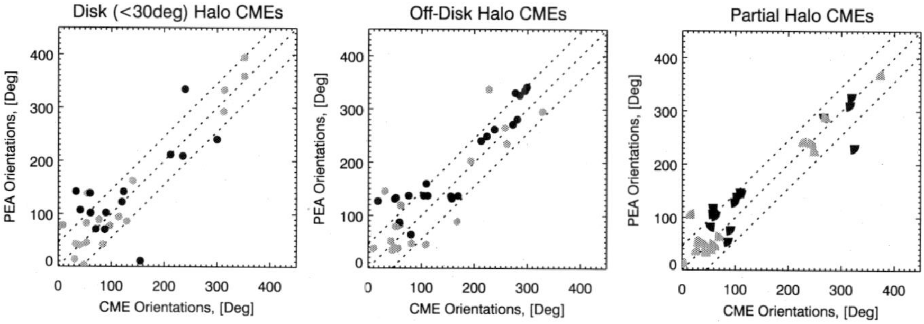

Fig. 4. CME directional angles plotted versus those of PEAs. The panels show (from left to right) data points for disk halo CMEs that originated within a 30 deg circle centered at the disk center, off-disk halo CMEs and partial CMEs. Black (gray) symbols represent events launched from southern (northern) hemisphere. The partial halo events are plotted with pie segments, whose orientation indicates the solar disk quadrant where they originated.

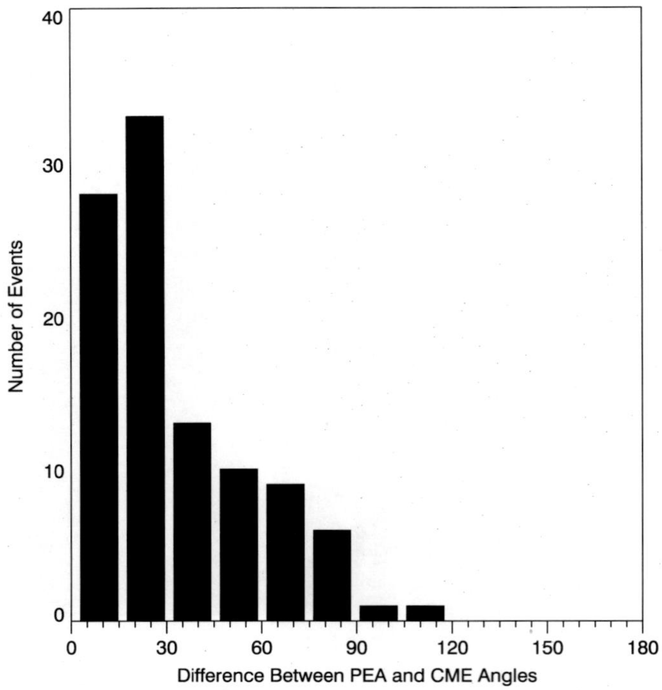

Fig. 5. Distribution of angle difference between CME and PEA orientations. For 74 out of 101 events (73%) the difference is less than 45 deg.

interplanetary magnetic fields by invoking magnetic field data at 2-3 solar radii above the solar surface.

3. Directional Parameters of CMEs, Coronal Neutral Line and Magnetic Clouds

It is generally believed that the wavy and spiral heliospheric current sheet (HCS)[66,67] interacts with solar ejecta: whenever the ejecta are moving at a faster speed than the upstream plasma there must be an upstream influence. About half of these events drive upstream shocks, testifying to the fact of their "superior" speed in general.[68,69] Zhao and Hoeksema[70] reported that the HCS may be blown out by a CME but shortly reforms near the same location.

Crooker *et al.*[71] suggested that the base of the HCS, which may be observed as a coronal neutral line (CNL), may often include multiple helmet streamers and that most CMEs might then be spatially associated with it. Subramanian *et al.*[72] estimated that about 65% of CMEs are associated with streamers. Moreover, the HCS is considered to be a conduit for CMEs[71] and the angular position (orientation) of the ejecta may, in general, be affected by its large-scale structure. When the current sheet is not aligned with the ejecta's plane, it is expected that the CME may displace the heliospheric magnetic field as it expands in the heliosphere.[73] As the magnetic field is distorted, it drapes around the ejecta and causes its deflection in longitude and/or azimuth. Since CMEs disrupt the global heliospheric magnetic field and the current sheet within it, it is natural to expect that CMEs themselves can be influenced, to some degree, by the heliospheric structures. If this is so, then the interaction between CMEs and the heliospheric magnetic field may explain the present difficulty in relating solar and interplanetary magnetic fields. However, details of this interaction are not yet studied well and the scale of the effect that the heliosphere may have on CMEs is largely unknown. Thus, the footprint of the photospheric magnetic fields associated with a CME may weaken and/or disappear due to changes in magnetic connectivity of the CME's field as suggested in Gibson and Fan,[57] and/or changes in ejecta's shapes, structures, and dynamics caused by the interaction with ambient solar wind.[74–76]

3.1. *Orientation of the coronal neutral line as measured from coronal field maps*

The background in Fig. 6 is the Wilcox Solar Observatory (WSO) coronal magnetic field map showing the polarity distribution (light and dark gray) during Carrington rotation (CR) 1968. The black solid line represents the major CNL. This map was calculated from a synoptic photospheric field map with a potential field model.[77,78] The green diamond in Fig. 6 indicates the location of the CME source region relative to the CNL, i.e., on the day when it crossed the central meridian. The averaged tilt of the neutral line near the eruption site was measured (in clockwise direction) as the tilt of a thick line segment centered on the point closest to the eruption site.

In order to compare directional angles of CMEs and MCs with the tilt of the CNL, which has 180-deg ambiguity in it, we needed to assign the direction to the CNL orientation. This was done in a way similar to that for CMEs: by requiring that the CNL directional angle makes an acute angle with the axial field of the corresponding MC. Our choice of MCs as a reference is justified as follows: i) both MCs and CNL are low-order, large-scale heliospheric structures, as opposed to the PEAs that represent

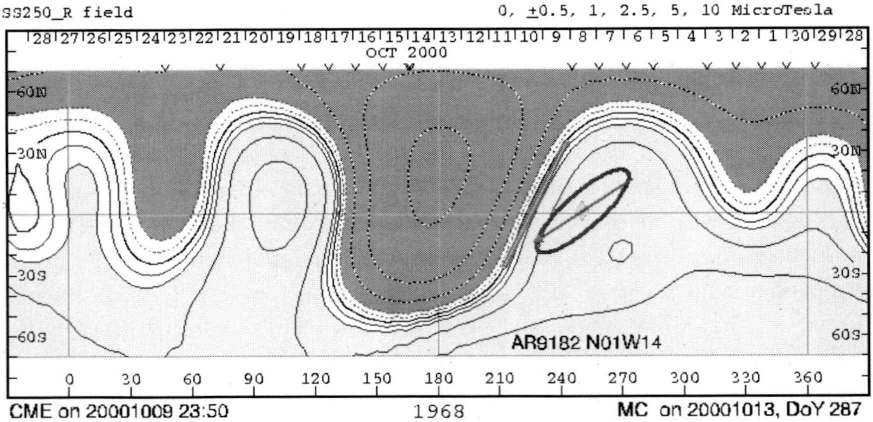

Fig. 6. Wilcox Solar Observatory coronal field map calculated for the height of 2.5 solar radii for CR 1968. Dashed red contours (dark grey) represent negative fields, blue contours (light grey) — positive fields. Thick solid line is the coronal neutral line (CNL). A red line segment shows the estimated orientation of the CNL near the site of eruption on October 13, 2000. The tilt of the CNL was measured in the CW direction from the east. The ellipse, centered at the location of the CME source (diamond) indicates the orientation of a halo CME, while the red arrow shows the axial field direction of the corresponding magnetic cloud.

high-order solar surface fields and ii) it is not necessary that each PEA is formed under the streamer belt; therefore, the PEA orientation may not always be related to the CNL.

3.2. *Magnetic cloud parameters*

In a statistical study that involves solar surface activity, coronal eruptions and interplanetary disturbances, it is crucial to select events with reliably identified association between solar surface phenomena, CMEs and MCs. As soon as MCs are involved in the selection process, the number of available and suitable events sharply decreases so that the next statistical study is based so far on only 25 events.

For each event, the MC orientation angle was obtained by averaging MC orientations produced by different MC fitting routines [see Table I in Ref. 45]. The MC orientation angle is the direction angle of the projected MC flux rope axis onto the GSE yz plane, measured in the clock-wise direction from the positive y-axis. Current methodology and techniques allowed us to determine the MC axis position with accuracy no better than 20 deg.

3.3. *Results*

In Fig. 7 (top row) we present relationships between CME, CNL and MC orientations. There are two groups of events presented with black and gray symbols. These events are separated as follows. In the upper left panel, black symbols indicate the 15 (out of 25) events that showed a good correspondence between the CMEs and MCs directions (angle difference < 45 deg). The gray symbols are events with the difference angle exceeding 45 deg. In the middle panel CME directions are compared to the CNL tilt. The black symbols here are also mainly clustered around the bisector. All

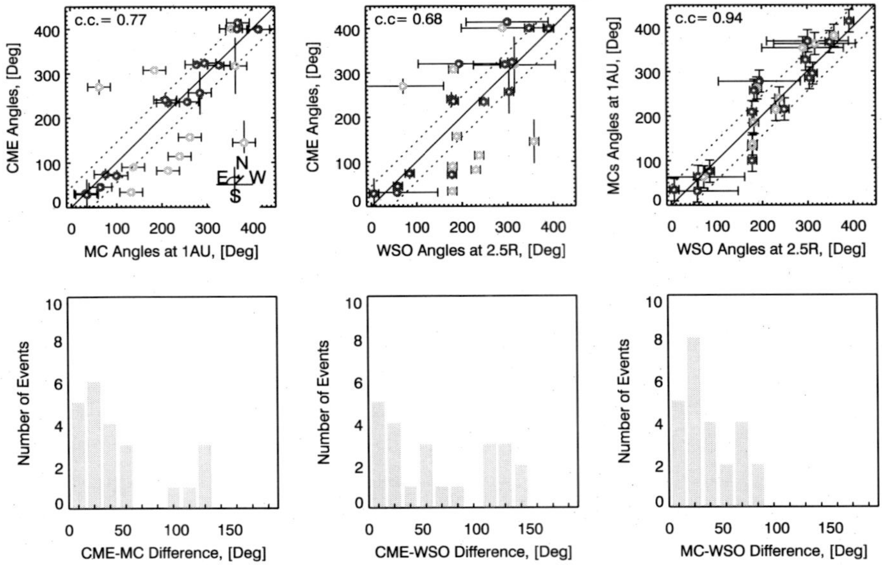

Fig. 7. Top row, left: CME orientation angles plotted versus the axial field directions in MCs. The black symbols indicate those events that showed a good correspondence (difference < 45 deg) between the two directional angles, while the gray symbols are the events that display no correspondence (difference > 45 deg). Open error bars show standard deviation of the CME measurements, while closed bars are the absolute error of the MC data. Two dashed lines indicate ± 45 degree interval around the bisector. Middle: CME orientation angles versus the orientation of the CNL near the eruption site. The black/gray symbols in this plot indicate the same events as in the left panel. Note that the majority of black symbols are also located at or near the bisector. Right: Directions of MC axial fields versus the orientation of the CNL. The black symbols indicate the same events as in the left panel. The majority of black and gray symbols in this plot are located at or near the bisector. Bottom row: Distribution of the angle difference between CME and MC (left), CME and CNL (middle) and MC and CNL (right) orientations. The bin size is 15 deg. The distribution of CNL-MC differences is very tight: for the majority of events (74%) the difference does not exceed 45 deg.

"gray" events but two are also located outside the ±45 deg range centered on the bisector. Thus, the data presented in these two panels seem to suggest that those erupting flux ropes that where initially aligned with the CNL (streamer belt) at the early stage of eruption (black dots), appear to remain so when they reach the Earth, and thus that their orientations match those of the MCs.

In upper right panel of Fig. 7, MC axis directions are plotted against the tilt of the CNL near the eruption site. This panel displays a high correlation between the parameters with the data points tightly clustered around the bisector. The lower panels in Fig. 7 show corresponding distribution of difference angle. The figure indicates that the MCs in our data set tend to be aligned along the coronal neutral line. Considering that the CNL is the base of the heliospheric current sheet, it ultimately means that the MCs tend to be embedded in the current sheet.

4. Conclusions and Discussion

We have reviewed the relationships between CMEs, EUV-PEAs and the coronal neutral lines, associated with global magnetic field and magnetic clouds near the Earth. We find that i) there is a good correlation between the directions of the axial field in PEAs and the elongations of halo CMEs; ii) there is a good correlation between the orientations of the CNL and MCs: 80% of analysed events display the difference between the CNL and MC orientations to be less than 45 deg; iii) the majority of the events that had PEA and MCs similarly oriented also had the CNL co-aligned with them; iv) those events that showed no agreement between the PEA and MC orientations, had their MCs aligned with the CNL only.

As we mentioned earlier, observations and theoretical works indicate that the coronal ejecta may evolve substantially as it expands out into the heliosphere and interacts with heliospheric and solar wind magnetic fields. CMEs might have a tendency not only to be deflected toward the heliomagnetic equator and channeled into the HCS,[70,71,79,80] but also to be locally aligned with the heliospheric current sheet. This inference, based on a detailed study of 25 events is in agreement with the earlier reports that i) MCs oriented between ±30 deg, tend to be detected more frequently[25] and ii) during solar minimum (maximum) bipolar (unipolar) MCs dominate.[79] Note that bipolar (unipolar) MCs are nearly parallel (perpendicular) to the ecliptic.

An example that illustrates the suggested CME rotation is presented in Fig. 8. We used the Potential Field Solar Surface (PFSS)[77] coronal field model results for CR 2006 obtained at Community Coordinated Model Center (CCMC). The 14 August 2003 eruption was relatively slow (~400 km/s) and associated with a weak C3.6 flare in NOAA AR 10431. The left panels in Fig. 8 show coronal field maps at 1.6 solar radii and the halo CME (ellipse), while the right panels are coronal field map at 2.5 solar radii and the corresponding MC is indicated with the cylinder. This CME was associated with the streamer belt and its elongation initially matched the local tilt of the neutral line at 1.6 solar radii. However, further out from the Sun, the neutral line changed its

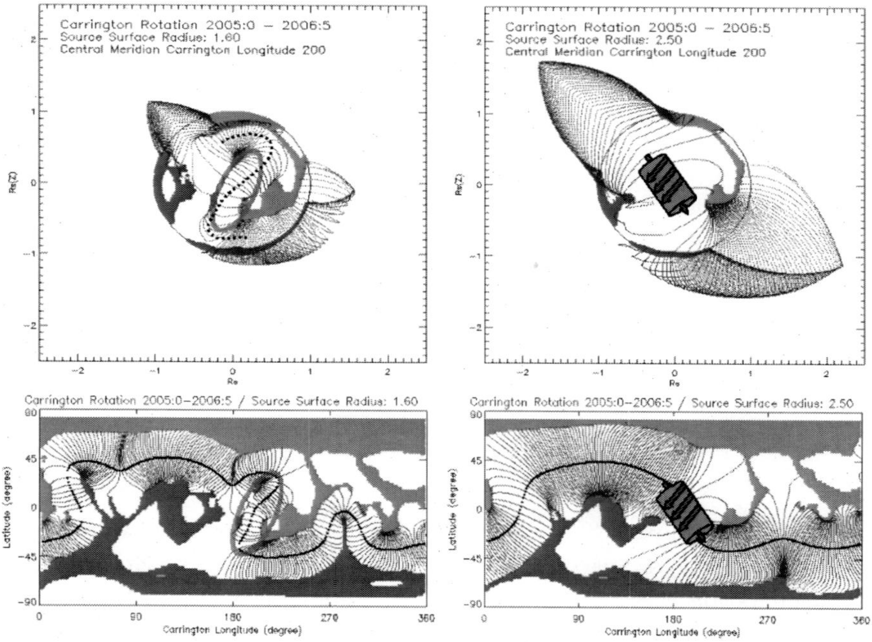

Fig. 8. Coronal field maps calculated for CR2006 with Potential Field Solar Surface (PFSS) model run at Community Coordinated Model Center (CCMC). The left panels show maps for source surface at 1.6 solar radii; the right panels show maps at 2.5 radii. The thick black contour is the coronal neutral line. The oval represents the halo CME on 14 August 2003, which was aligned with the coronal neutral line at 1.6 solar radii. Magnetic topology has changed further outward from the solar surface so that the neutral line rotated by approx. 50 deg and was co-aligned with the magnetic cloud at 1AU (cylinder, lower right panel).

orientation, which is evident from the 2.6 solar radii map. It is quite possible that the associated CME rotated too, so that the corresponding MC was well aligned with the coronal neutral line. This is in agreement with Krall *et al.*[43] who reported that the 28 October 2003 CME smoothly rotated by about 50 deg until its final orientation closely matched the MC position.

Figure 8 also gives a hint for a possible interpretation of this event. The black dotted line in panel (a) roughly indicates the location and shape of the CNL. As can be inferred from the figure, the erupting flux rope could be kinked at the beginning, so that the loop top could rotate as it expands/unkinks.[56] Gibson & Fan[57] [also private communication] proposed that an expanding flux rope can reconnect with the surrounding fields so that the footpoints of the erupting fields can be displaced. Another equal possibility is that a CME, which is not aligned with the HCS, pushes apart the magnetic field lines on both sides of the current sheet, so that the enhanced magnetic pressure and Lorentz force both work against the expanding fields, gradually deforming the loop top so that it locally aligns itself with the current sheet. The higher the speed of the ejecta, the shorter is the interaction time, so that very fast CMEs may not be affected by the current sheet at all or affected in a substantially lesser degree. Of course, nearby coronal holes may affect the ejecta, too.[76] Some CMEs originate from unipolar regions[14,76] and are, thus, associated with unipolar boundary layer. In this case the above explanation may not be valid since HCS is not present.

Acknowledgments

The CME catalog is generated and maintained by the Center for Solar Physics and Space Weather, the Catholic University of America in cooperation with the Naval Research Laboratory, and NASA. We acknowledge the usage of the list of geomagnetic storms compiled during a *Living With a Star* Coordinated Data Analysis Workshops. SOHO is a project of international cooperation between ESA and NASA. Wilcox Solar Observatory data used in this study was obtained via the web site wso.stanford.edu, courtesy of J.T. Hoeksema. The work of VY was supported under NSF grant 0536921 and NASA ACE NNG0-4GJ51G and LWS TR&T NNG0-5GN34G grants. DT acknowledges support from STFC.

References

1. C. T. Russell, R. L. McPherron and R. K. Burton, *Journal of Geophysical Research* **79** (1974) 1105–1109.
2. G. Rostoker and Falthammar, C.-G. *Journal of Geophysical Research* **72** (1967) 5853–5863.
3. B. T. Tsurutani, Y. T. Lee, W. D. Gonzalez and F. Tang, *Geophysical Research Letters* **19** (1992) 73–76.
4. M. Sugiura, http://swdcwww.kugi.kyoto-u.ac.jp/dstir (1991).
5. R. A. Howard, D. J. Michels, N. R. Sheeley, Jr. and M. J. Koomen, *Astrophysical Journal* **263** (1982) L101–L104.
6. R. Schwenn, J. C. Raymond, D. Alexander, A. Ciaravella, N. Gopalswamy, R. Howard, H. Hudson, P. Kaufmann, A. Klassen, D. Maia, G. Munoz-Martinez, M. Pick, M. Reiner, N. Srivastava, D. Tripathi, A. Vourlidas, Y. M. Wang and J. Zhang, *Space Science Reviews* **123** (2006) 127–176.
7. N. Gopalswamy, *Space Science Reviews* **124** (2006) 145–168.
8. N. U. Crooker and T. S. Horbury, *Space Science Reviews* **123** (2006) 93–109.
9. J. Zhang, K. P. Dere, R. A. Howard and V. Bothmer, *Astrophysical Journal* **582** (2003) 520–533.
10. R. C. Canfield, H. S. Hudson and D. E. McKenzie, *Geophysical Research Letters* **26** (1999) 627–630.
11. R. C. Canfield, M. D. Kazachenko, L. W. Acton, D. H. Mackay, J. Son and T. L. Freeman, *Astrophysical Journal* **671** (2007) L81–L84.
12. D. M. Rust and A. Kumar, *Astrophysical Journal* **464** (1996) L199–L202.
13. J. G. Luhmann, Y. Li, X. Zhao and S. Yashiro, *Solar Physics* **213** (2003) 367–386.
14. Zhao X.P. and D. F. Webb, *Journal of Geophysical Research (Space Physics)* **108** (2003) 1234.
15. V. B. Yurchyshyn, H. Wang, P. R. Goode, and Y. Deng, *Astrophysical Journal* **563** (2001) 381–388.
16. L. Burlaga, S. P. Plunkett and O. C. Cyr, St., *Journal of Geophysical Research* **107** (2002) 1266–1277.
17. L. Burlaga, E. Sittler, F. Mariani and R. Schwenn, *Journal of Geophysical Research* **86** (1981) 6673–6684.
18. C. C. Wu, R. P. Lepping and N. Gopalswamy, *Solar Physics* **239** (2006) 449–460.
19. N. Srivastava and P. Venkatakrishnan, *Journal of Geophysical Research (Space Physics)* **109** (2004) 10103.
20. N. Gopalswamy, *Journal of Atmospheric and Solar-Terrestrial Physics* **70** (2008) 2078–2100.
21. K. Marubashi, *Advances in Space Research* **26** (2000) 55–66.
22. V. Bothmer and R. Schwenn, *Annales Geophysicae* **16** (1998) 1–24.
23. P. Riley, C. Schatzman, H. V. Cane, I. G. Richardson, N. Gopalswamy, *Astrophysical Journal* **647** (2006) 648–653.
24. V. Yurchyshyn, Q. Hu and V. Abramenko, *Space Weather* **3** (2005) 1–13.

25. X. P. Zhao and J. T. Hoeksema, *Journal of Geophysical Research* **103** (1998) 2077.
26. K. Marubashi, *Advances in Space Research* **6** (1986) 335–338.
27. R. J. Leamon, R. C. Canfield and A. A. Pevtsov, *Journal of Geophysical Research (Space Physics)* **107** (2002) 1234–1242.
28. Y. Wang, G. Zhou, P. Ye, S. Wang and Wang, J. *Astrophysical Journal* **651** (2006) 1245–1255.
29. L. K. Harra, N. U. Crooker, C. H. Mandrini, L. van Driel-Gesztelyi, S. Dasso, J. Wang, H. Elliott, G. Attrill, B. V. Jackson and M. M. Bisi, *Solar Physics* **244** (2007) 95–114.
30. J. P. Delaboudinière, G. E. Artzner, J. Brunaud, A. H. Gabriel, J. F. Hochedez, F. Millier, X. Y. Song, B. Au, K. P. Dere, R. A. Howard, R. Kreplin, D. J. Michels, J. D. Moses, J. M. Defise, C. Jamar, P. Rochus, J. P. Chauvineau, J. P. Marioge, R. C. Catura, J. R. Lemen, L. Shing, R. A. Stern, J. B. Gurman, W. M. Neupert, A. Maucherat, F. Clette, P. Cugnon and E. L. van Dessel, *Solar Physics* **162** (1995) 291–312.
31. V. Domingo, B. Fleck and A. I. Poland, *Solar Physics* **162** (1995) 1–2.
32. G. E. Brueckner, R. A. Howard, M. J. Koomen, C. M. Korendyke, D. J. Michels, J. D. Moses, D. G. Socker, K. P. Dere, P. L. Lamy, A. Llebaria, M. V. Bout, R. Schwenn, G. M. Simnett, D. K. Bedford and C. J. Eyles, *Solar Physics* **162** (1995) 357–402.
33. X. P. Zhao, S. P. Plunkett and W. Liu, *Journal of Geophysical Research (Space Physics)* **107** (2002) 1223.
34. H. Xie, L. Ofman and G. Lawrence, *Journal of Geophysical Research (Space Physics)* **109** (2004) 03109.
35. X. P. Zhao, Determination of geometrical and kinametical properties of frontside halo coronal mass ejections (CMEs), IAU Symp. 226, Solar and Steller Mass Ejections, 2004, 42–46, (2005).
36. G. Michalek, N. Gopalswamy, A. Lara and S. Yashiro, *Space Weather* **4** (2005) S10003.
37. H. Cremades and V. Bothmer, *Astronomy and Astrophysics* **422** (2004) 307–322.
38. J. Krall and O. C. Cyr, St., *Astrophysical Journal* **652** (2006) 1740–1746.
39. J. Chen, R. A. Howard, G. E. Brueckner, R. Santoro, J. Krall, S. E. Paswaters, O. C. Cyr, St., R. Schwenn, P. Lamy and G. M. Simnett, *Astrophysical Journal* **490** (1997) L191.
40. T. G. Moran and J. M. Davila, *Science* **305** (2004) 66–71.
41. J. Krall, *Astrophysical Journal* **657** (2007) 559–566.
42. J. Chen and D. A. Garren, *Geophysical Research Letters* **20** (1993) 2319–2322.
43. J. Krall, V. B. Yurchyshyn, S. Slinker, R. M. Skoug and J. Chen, *Astrophysical Journal* **642** (2006) 541–553.
44. V. Yurchyshyn, C. Liu, V. Abramenko and J. Krall, *Solar Physics* **239** (2006) 317–335.
45. V. Yurchyshyn, Q. Hu, R. P. Lepping, B. J. Lynch and J. Krall, *Advances in Space Research* **40** (2007) 1821–1826.

46. J. Zhang, I. G. Richardson, D. F. Webb, N. Gopalswamy, E. Huttunen, J. Kasper, N. V. Nitta, W. Poomvises, B. J. Thompson, C. C. Wu, S. Yashiro and A. N. Zhukov, *Journal of Geophysical Research (Space Physics)* **112** (2007) 12103.
47. J. Zhang, I. G. Richardson, D. F. Webb, N. Gopalswamy, E. Huttunen, J. C. Kasper, N. V. Nitta, W. Poomvises, B. J. Thompson, C. C. Wu, S. Yashiro and A. N. Zhukov, *Journal of Geophysical Research (Space Physics)* **112** (2007) 10102.
48. J. Qiu and V. B. Yurchyshyn, *Astrophysical Journal* **634** (2005) L121–L124.
49. V. Yurchyshyn, *Astrophysical Journal* **675** (2008) L49–L52.
50. N. Gopalswamy, S. Yashiro, M. L. Kaiser, R. A. Howard and J. L. Bougeret, *Journal of Geophysical Research* **106** (2001) 29219–29230.
51. B. Vrsnak, D. Ruzdjak, D. Sudar and N. Gopalswamy, *Astronomy and Astrophysics* **423** (2004) 717–728.
52. T. Török, B. Kliem and V. S. Titov, *Astronomy and Astrophysics* **413** (2004) L27–L30.
53. L. M. Green, B. Kliem, T. Török, L. van Driel-Gesztelyi, and G. D. R. Attrill, *Solar Physics* **246** (2007) 365–391.
54. P. A. Isenberg and T. G. Forbes, *Astrophysical Journal* **670** (2007) 1453–1466.
55. D. R. Williams, T. Török, P. Démoulin, L. van Driel-Gesztelyi and B. Kliem, *Astrophysical Journal* **628** (2005) L163–L166.
56. B. J. Lynch, S. K. Antiochos, C. R. DeVore, J. G. Luhmann and T. H. Zurbuchen, *Astrophysical Journal* **683** (2008) 1192–1206.
57. S. E. Gibson and Y. Fan, *Journal of Geophysical Research (Space Physics)* **111** (2006) 12103.
58. D. Tripathi, V. Bothmer and H. Cremades, *Astronomy and Astrophysics* **422** (2004) 337–349.
59. P. H. Scherrer, R. S. Bogart, R. I. Bush, J. T. Hoeksema, A. G. Kosovichev, J. Schou, W. Rosenberg, L. Springer, T. D. Tarbell, A. Title, C. J. Wolfson, I. Zayer and M. D. I. E. Team, *Solar Physics* **162** (1995) 129–188.
60. V. I. Abramenko, T. Wang and V. B. Yurchishin, *Solar Physics* **168** (1996) 75–89.
61. N. Seehafer, *Solar Physics* **125** (1990) 219–232.
62. A. A. Pevtsov, R. C. Canfield and T. R. Metcalf, *Astrophysical Journal* **425** (1994) L117-L119.
63. V. I. Abramenko and V. B. Yurchishin, *Solar Physics* **168** (1996) 47–63.
64. H. Song, V. Yurchyshyn, G. Yang, C. Tan, W. Chen and H. Wang, *Solar Physics* **238** (2006) 141–165.
65. S. F. Martin, *Advances in Space Research* **32** (2003) 1883–1893.
66. M. Schulz, *Astrophyscal Space Science*, **24** (1973) 13.
67. E. J. Smith, B. T. Tsurutani and R. L. Rosenberg, *Journal of Geophysical Research* **83** (1978) 717–724.
68. T. A. Howard and S. J. Tappin, *Astronomy and Astrophysics* **440** (2005) 373–383.

69. J. T. Gosling, S. J. Bame, D. J. McComas, J. L. Phillips, E. E. Scime, V. J. Pizzo, B. E. Goldstein and A. Balogh, *Geophysical Research Letters* **21** (1994) 237–240.
70. X. Zhao and J. T. Hoeksema, *Journal of Geophysical Research* **101** (1996) 4825–4834.
71. N. U. Crooker, G. L. Siscoe, S. Shodhan, D. F. Webb, J. T. Gosling and E. J. Smith, *Journal of Geophysical Research* **98** (1993) 9371–9381.
72. P. Subramanian, K. P. Dere, N. B. Rich and R. A. Howard, *Journal of Geophysical Research* **104** (1999) 22321–22330.
73. E. J. Smith, *Journal of Geophysical Research* **106** (2001) 15819–15832.
74. J. M. Schmidt and P. J. Cargill, *Journal of Geophysical Research (Space Physics)* **108** (2003) 1023.
75. Y. Liu and K. Hayashi, *Astrophysical Journal* **640** (2006) 1135–1141.
76. Y. Liu, *Advances in Space Research* **39** (2007) 1767–1772.
77. J. T. Hoeksema, Structure and evolution of the large scale solar and heliospheric magnetic fields, PhD, Stanford University, CSSA Report No. 7, (1984).
78. J. T. Hoeksema, J. M. Wilcox and P. H. Scherrer, *Journal of Geophysical Research* **88** (1983) 9910–9918.
79. T. Mulligan, C. T. Russell and J. G. Luhmann, *Geophysical Research Letters* **25** (1998) 2959–2962.
80. S. W. Kahler, N. U. Crooker and J. T. Gosling, *Journal of Geophysical Research* **104** (1999) 9919–9924.

Advances in Geosciences
Vol. 21: Solar & Terrestrial Science (2008)
Ed. Marc Duldig

LARGE GEOMAGNETIC STORMS ASSOCIATED WITH LIMB HALO CORONAL MASS EJECTIONS*

NAT GOPALSWAMY
NASA Goddard Space Flight Center Greenbelt, MD 20771, USA
gopals@ssedmail.gsfc.nasa.gov

SEIJI YASHIRO†, HONG XIE,
SACHIKO AKIYAMA and PERTTI MÄKELÄ
The Catholic University of America, Washington DC 20064, USA

Solar cycle 23 witnessed the observation of hundreds of halo coronal mass ejections (CMEs), thanks to the high dynamic range and extended field of view of the Large Angle and Spectrometric Coronagraph (LASCO) on board the Solar and Heliospheric Observatory (SOHO) mission. More than two thirds of halo CMEs originating on the front side of the Sun have been found to be geoeffective (Dst ≤ -50 nT). The delay time between the onset of halo CMEs and the peak of ensuing geomagnetic storms has been found to depend on the solar source location (Gopalswamy *et al.*, 2007). In particular, limb halo CMEs (source longitude $> 45°$) have a 20% shorter delay time on the average. It was suggested that the geomagnetic storms due to limb halos must be due to the sheath portion of the interplanetary CMEs (ICMEs) so that the shorter delay time can be accounted for. We confirm this suggestion by examining the sheath and ejecta portions of ICMEs from Wind and ACE data that correspond to the limb halos. Detailed examination showed that three pairs of limb halos were interacting events. Geomagnetic storms following five limb halos were actually produced by other disk halos. The storms followed by four isolated limb halos and the ones associated with interacting limb halos, were all due to the sheath portions of ICMEs.

1. Introduction

Halo coronal mass ejections (CMEs) occurring on the frontside of the Sun are a potential source of geomagnetic storms because they can directly impact Earth's magnetosphere with high kinetic energy.[1,2] The

*This work is supported by NASA LWS TR&T program.
†Also at Interferometrics, VA, USA.

geoeffectiveness of halo CMEs depends on the existence of southward component of the magnetic field in the sheath and/or ejecta portions. Here we define geoeffectiveness as the ability of a CME to produce a geomagnetic storm with an intensity level measured by the Dst index at or below $-50\,\mathrm{nT}$, e.g., Ref. [3]. In a recent investigation of the geoeffectiveness of halo CMEs (Gopalswamy *et al.*,[2] herein after Paper 1), it was shown that the geoeffectiveness declines as the source region of halo CMEs has a greater central meridian distance (CMD). It was also found that halo CMEs associated with intense geomagnetic storms (Dst $\leq -100\,\mathrm{nT}$) are generally located within a longitude range of $\pm 45°$ (average longitude ~W10) whereas non-geoeffective halos (Dst $> -50\,\mathrm{nT}$) had a broad longitude distribution ($\pm 90°$). Furthermore, ~75% of disk (CMD $\leq 45°$) halos were geoeffective while only 60% of the limb ($45° <$ CMD $\leq 90°$) halos were geoeffective. The computed the delay time between the CME onset at the Sun and the peak of the geomagnetic storm was surprisingly different on the average for limb halos (56 hr) and disk halos (70 hr). Paper 1 attributed this difference to the possibility that the sheath of the interplanetary (IP) CMEs (ICMEs) developing from limb halos must have produced the geomagnetic storms (sheaths are typically ahead of ICMEs by ~ half a day[4–6]). It is also known statistically (from ICME observations) that the sheath storms are generally ahead and the cloud storms are behind the arrival of ICMEs.[7] However, detailed investigation of the IP counterparts of individual halo CMEs and the associated geomagnetic storms was not made in Paper 1. The purpose of this paper is to provide a direct confirmation that the geomagnetic storms associated with limb halos are due to sheaths in the corresponding ICMEs. To this end, we examine the IP counterparts of the limb halos reported in Paper 1 to see if the sheaths of the ICMEs from limb halos have large southward magnetic field component to make them geoeffective.

2. Data Selection

Paper 1 listed 37 limb halos ($45° <$ CMD $\leq 90°$) that were followed by Dst values at or below $-50\,\mathrm{nT}$. The listed CMEs may overlap with other sources of geomagnetic storms, such as corotating interaction regions (CIRs) formed by high speed streams from coronal holes. It is well known that CIR storms generally have a Dst index $\geq -100\,\mathrm{nT}$.[8] To eliminate the possibility that some of the weaker storms may be caused by CIRs, we consider only strongly geoeffective limb halos (Dst $\leq -100\,\mathrm{nT}$). There were 17 such limb

Table 1. List of limb halos followed by intense geomagnetic storms (1996–2005).

No	CME Date & Time	V km/s	Source Location	Dst Peak Time	DT hour	Dst (nT)	Notes
1	00/04/04 16:32	1118	N16W66	04/07 00	55.5	−288	Sh
2	00/10/24 08:26	800	S23E70	10/29 03	—	−127	CC
3	00/10/25 08:26	770	N09W63	10/29 03	90.5	−127	Sh
4	00/11/25 01:31	2519	N07E50	11/29 13	—	−119	CC
5	01/10/01 05:30	1405	S24W81	10/03 14	—	−166	CC
6	01/11/22 20:30	1443	S25W67	11/24 16	43.5	−221	INT
7	02/03/22 11:06	1750	S10W90	03/24 09	—	−100	CC
8	03/06/15 23:54	2053	S07E80	06/18 09	57.0	−141	Sh
9	04/11/09 17:26	2000	N08W51	11/10 19	—	−131	Rec
10	04/11/10 02:26	3387	N09W49	11/11 05	—	−113	Rec
11	05/01/19 08:29	2020	N15W51	01/22 06	—	−105	INT
12	05/01/20 06:54	3242	N14W61	01/22 06	47.0	−105	Sh
13	05/05/11 20:13	550	S11W51	05/15 08	—	−263	CC
14	05/08/22 01:31	1194	S11W54	08/24 11	—	−216	INT
15	05/08/22 17:30	2378	S13W65	08/24 11	41.5	−216	Sh
16	05/08/23 14:54	1929	S14W90	08/24 16	—	−160	Rec
17	05/09/09 19:48	2257	S12E67	09/11 10	38.0	−147	Sh

halos as listed in Table 1. The simple criterion for geoeffectiveness used in Paper 1 was that the halo CME must be followed by a geomagnetic storm during a 4-day interval starting one day after the CME onset. This criterion was based on the observation that it takes anywhere between 1 and 4 days for a CME to travel to Earth after the liftoff. One cannot avoid the situation that the time windows of CMEs overlap, especially during solar maximum when CMEs occur in quick succession from the same active region or from different active regions. This will result in some geomagnetic storms getting assigned to more than one CME: there may be a disk halo occurring around the time of a limb halo by chance, in which case one has to carefully decide which CME is responsible for the ensuing storm. We carefully examined all possible CMEs occurring around the time of the limb halos to determine whether it is truly geoeffective or not.

Column 2 of Table 1 gives the starting date and time (yy/mm/dd hh:mm format) of the limb halos with their sky-plane speed (V in km/s) and heliographic location of the solar source taken from Paper 1. The time of minimum Dst of the associated storms is listed in column 5 in the mm/dd hh format (the year is the same as in column 2). The delay time (DT) from the CME onset (column 2) to the time of Dst minimum (column 5) is listed in column 6. The minimum value of the Dst index is given in column 7.

Finally, some comments on the events are given in the last column (Sh — isolated sheath event; CC — chance coincidence; INT — interacting event; Rec — fluctuation in the recovery phase of a preceding storm).

3. Analysis

Figure 1 shows the out of the ecliptic component (Bz) of the IP magnetic field (IMF), the solar wind plasma temperature (T) and the Dst index. From the temperature signature we can identify the sheath (marked Sh) and the ejecta (also marked). The ejecta is of short duration because the CME is not directed along the Sun-Earth line. Note that the intense geomagnetic storm is entirely due to the Bz<0 in the sheath region. The short duration ejecta has no Bz<0, so it is not geoeffective. Examining plots like the ones in Fig. 1, we found that CMEs #3, #8, #12, #15 and #17 all produced geomagnetic storms because of their sheath portions.

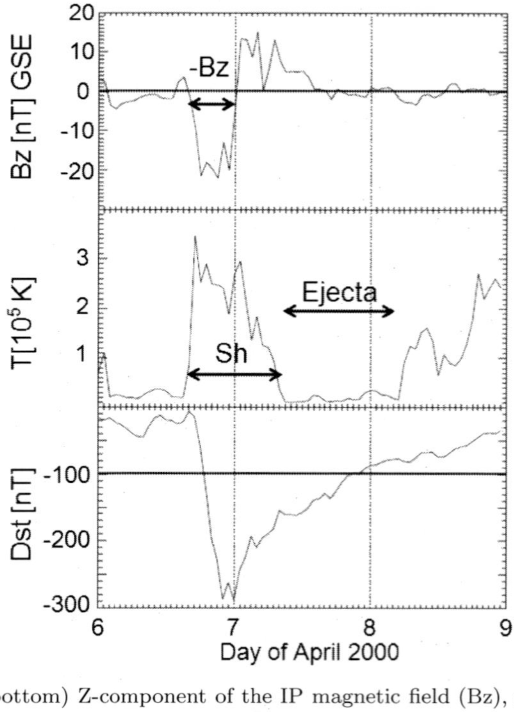

Fig. 1. (top to bottom) Z-component of the IP magnetic field (Bz), solar wind proton temperature (T) and the Dst index around the time of the April 7, 2000 storm (due to halo #1). Intervals of Bz<0 (−Bz), sheath (Sh) and the ejecta (Ejecta) are marked. Note that Bz<0 occurs only in the front part of the sheath.

The halo CMEs #2 and #3 are both candidate sources of the same storm. Looking at the solar source, we see that the eastern source is at a larger distance from the disk center. Since CMEs are deflected to the east,[9] we conclude that halo #3 is the likely candidate and regarded the association between halo #2 and the storm is by chance coincidence (CC). Halo #3 also resulted in an ejecta following the sheath. Bz<0 occurred in the sheath and partly in the ejecta, but the minimum Bz occurred in the sheath. We therefore, conclude that the storm is due to the sheath.

Halo #4 is very fast (2519 km/s), so the shock is expected to arrive in about a day. The shock actually arrives on November 26 at 11:40 UT followed by a narrow ejecta on November 27 at 12:30 UT. It is also associated with a moderate storm (~ -80 nT) due to its sheath but this is not the storm listed in Table 1. The storm listed in Table 1 is due to another CME on November 26 at 17:06 UT, which is a disk halo (N18W38). Therefore, we regard halo #4 to be a chance coincidence. Similarly halo #5 is a chance coincidence since the -166 nT storm is caused by the disk halo (N13E03) on 2001 September 29 at 11:54 UT. Halo #5 is also too close to the limb, which is unlikely to produce ejecta at Earth.

Halo #6 is followed within 3 h by another disk halo (S17W36) on 2001 November 22 at 23:30 UT. The two CMEs seem to have interacted near the Sun, so we cannot rule out the possibility that the sheath of halo #6 is swept up by the following disk halo. Therefore, we regard this as an interaction event (INT). The geomagnetic storm was caused by $Bz < 0$ in the sheath of the merged ICME at Earth.

The storm listed in the time window of halo #7 has a better candidate: the disk halo (S17W20) of 2002 March 20 at 17:54 UT, which had a shock and ejecta. The limb halo #7 did produce an IP shock that seems to pass through the ejecta from the disk halo. Therefore, we conclude that although halo #7 has an associated IP shock, it is not associated with the storm listed in its time window. The storm itself is caused by the ejecta part of the ICME associated with the disk halo.

Halo #8 is rather isolated and its association with the 2003 June 18 storm (Dst ~ -141 nT) is unambiguous. In Fig. 2, we show halo #8 (2003 June 15 CME) in the LASCO frame obtained early on June 16. The solar source is clearly near the east limb (S07E80) as evidenced by the large-scale EUV disturbance and the associated X-class flare (see soft X-ray light curve from the GOES satellite in Fig. 2). It is clear from the LASCO frame that the western flank of the disturbance has crossed the Sun-Earth line early in the event. The CME was also associated with an intense type II

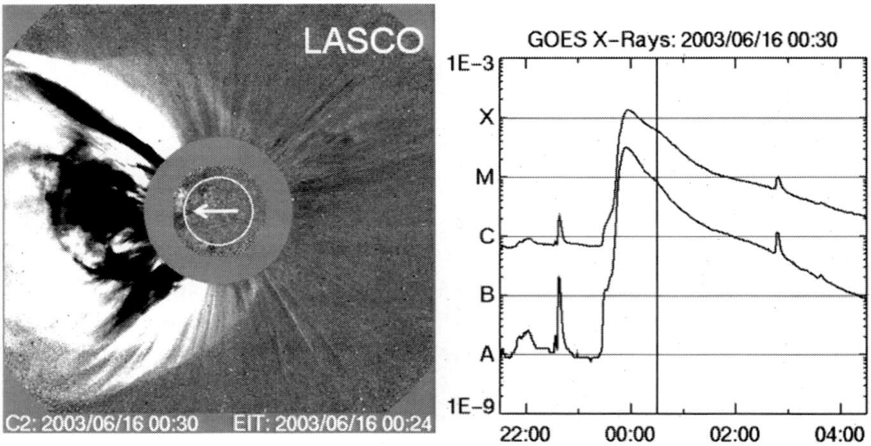

Fig. 2. (left) White-light CME (halo #8) from SOHO/LASCO with superposed EUV difference image showing the solar source (pointed by the arrow). (right) GOES light curve showing the X-class flare associated with the CME.

radio burst in the decameter-hectometric (DH) wavelengths. The DH type II bursts are indicative of CME-driven shocks in the near-Sun IP medium. There were several small CMEs (widths ranging from 13° to 40°) after the limb halo, but none of them is capable of producing a shock at 1 AU. The next significant event was a halo at the end of June 17, which was just 5 hours before the shock arrival at Earth and hence could not be the source. Halo #8 is also unique in that it is the easternmost CME to produce a major geomagnetic storm during solar cycle 23.

The solar wind plasma and magnetic signatures of halo #8 are shown in Fig. 3. The shock arrived at 04:44 UT on June 18, indicating a transit time of ~53 h. This is rather long for a 2053 km/s CME, but the Earthward speed is expected to be smaller because only the western flank of the shock seems to have arrived at Earth. The sheath that follows the shock is rather extended (more than one day). The Bz plot shows that the interval of $Bz < 0$ occurs right after the shock, in the front end of the sheath. The Dst minimum occurs just 4 hours after the shock arrival, again corresponding to the front end of the sheath. There is no indication of an ejecta after the shock, because the source is far from the disk center. Thus we conclude that this is clearly a sheath storm.

The storms listed in the time windows of halos #9 and #10 seem to be fluctuations in the recovery phase of the previous super storm (−289 nT on 2004 November 10 at 10:00 UT caused by the disk halo that left the

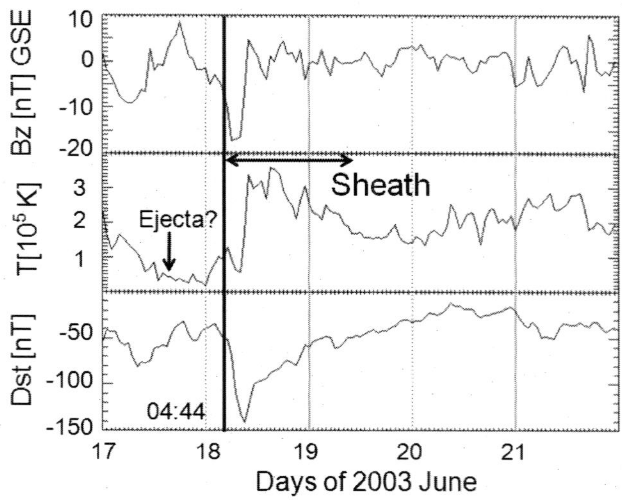

Fig. 3. Solar wind magnetic and plasma signatures of the IP disturbance associated with the 2003 June 18 geomagnetic storm (followed by halo #8). The shock maybe running into a preceding ejecta (suggested by the depressed temperature) but there is no ejecta signature following the shock sheath.

Sun on November 7 at 16:54 UT). Examination of the solar wind plasma and magnetic signatures shows that there is no shock or ejecta around the times of these two storms. There are only small negative excursions in Bz corresponding to the two Dst minima in question.

The storm on 2005 January 22 is in the time window of halos #11 and #12. Figure 4 shows the two CMEs at their first appearance in the LASCO field of view. Both appeared as non-halos in the northwest quadrant and expanded to become full halos in the LASCO/C3 field of view. The January 20 CME was visible only in a single LASCO frame because of degradation of the SOHO detectors due to impact by solar energetic particles from this CME.[10] The CME speed was estimated to be ~3242 km/s by combining the LASCO image with SOHO/EIT images that showed the eruption. Figure 5 shows the shock, the sheath, and the geomagnetic storm following the two halos. Note that Bz<0 occurs only for a short interval right after the shock at 16:48 UT on 2005 January 21. The sudden commencement in this case is extraordinarily intense with a positive excursion of ~30 nT. The two halos left the Sun within a time separation of ~23 h, so it is possible that the shocks from the halos merged to form the huge sudden commencement. The sheath shows a peculiar temperature structure (see Fig. 5), which may indicate that the sheath contains some portion of the ejecta of halo #11.

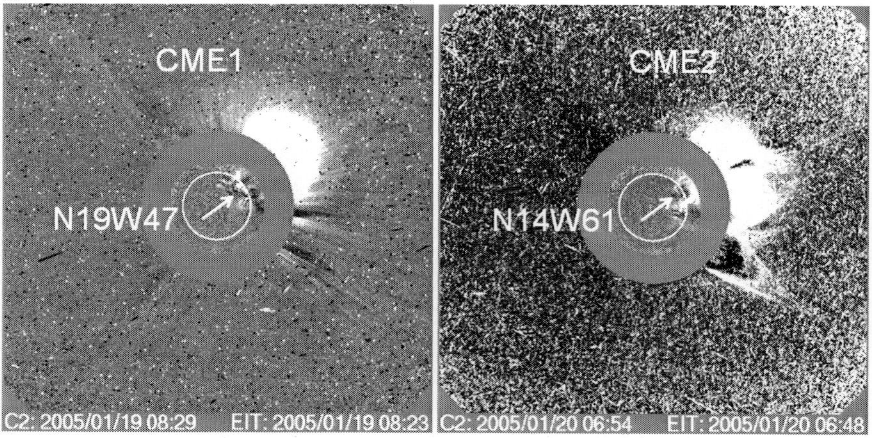

Fig. 4. Two CMEs (halos #11 and #12 in Table 1) from the same active region (AR 0720) that contributed to the geomagnetic storm on 2005 January 22. The source locations are shown on the images as EUV disturbances in the superposed SOHO/EIT difference images. CME1 and CME2 had speeds of 2030 km/s and 3242 km/s, respectively. The energetic particles from CME2 arriving at the SOHO detector severely degraded the LASCO image.

There is a slight temperature depression after the sheath region, but there is no ejecta signature in Bz and By components of the IMF. This seems to be an interaction case although one cannot rule out the fact that the first CME missed Earth.

The limb halo #13 is followed by an intense storm (Dst = $-263\,$nT), but this is chance coincidence because the storm was caused by a well-known disk halo, which occurred on 2005 May 13 in AR 0759 (N12E11) and extensively studied by many authors.[11,12]

The storm on 2005 August 24 is in the time window of the limb halos #14 and #15, both of which occurred on August 22. There is only one IP shock observed at 1 AU (on August 24 at 5:34 UT). Right after the shock, Bz becomes negative and attains a large negative value (Bz $\sim -40\,$nT). The storm is due to this Bz<0 interval in the sheath region. It is possible that a second shock is present in the sheath region followed by a mini magnetic cloud (~2 h in duration)[13] but the proton temperature remains elevated as is normally the case in shock sheaths rather than during magnetic clouds. These two CMEs originated from an active region surrounded by a coronal hole, so the interaction seems to be complicated (see Ref. [13] for more details). Halo #15 occurred ~16 h after halo #14 and must have overtaken it somewhere between the Sun and Earth because both CMEs originated

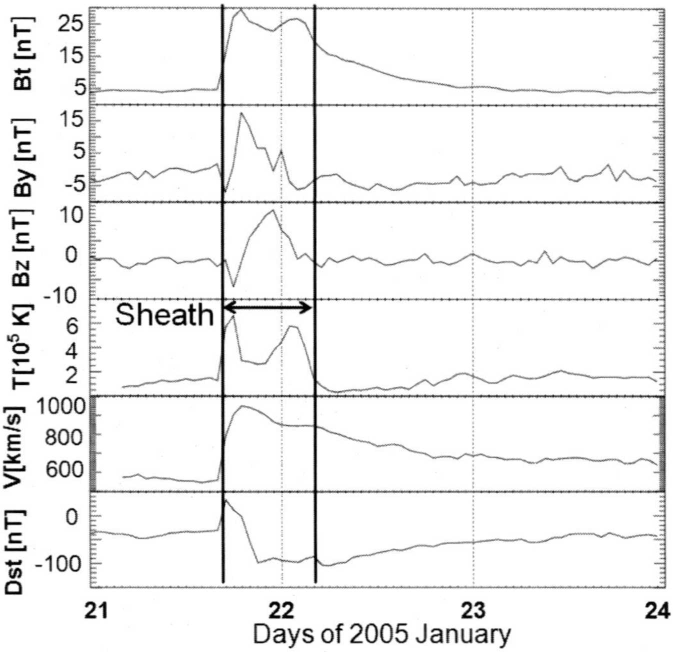

Fig. 5. Magnetic and plasma signatures following halos #11 and #12 shown as magnetic field magnitude (Bt), the By component, the Bz component, the solar wind proton temperature (T) and flow speed (V), and the Dst index. The region of enhanced temperature marked as "sheath".

from the same active region. Halo #15 is twice as fast as halo #14, so the interaction is highly likely. However, the ejecta signature is not clear at 1 AU because the proton temperature remained above the pre-shock level. As in the case of the January 2005 events, the 2005 August 22 events were also interacting and resulted in a single shock at 1 AU. Again, we compute the delay time of the storm with respect to the first-appearance time of halo #15. The storm on 2005 August 24 at 16 UT is also a fluctuation in the recovery phase of the storm associated with halos #14 and #15. Even though the fluctuation appears in the time window of halo #16, we do not see any IP signatures of this CME. Note that halo #16 originated right at the west limb.

The last halo is one of the many halos from the super active region 0808 and one of the two superfast CMEs (speed > 2000 km/s). The IP shock associated with the CME was observed at the very beginning of September 11 (00:49 UT). The Bz turns negative right after the shock, well within the high proton temperature interval, so we are certain that the storm is due to the sheath region. The magnitude of Bz is not very high (~5 nT) but the speed is extremely high, so the storm is intense. Note that this is one of the smaller storms in Table 1.

Excluding the chance-coincidence cases (5) and the three recovery-phase fluctuations (3), we get 9 limb halos that were responsible for the 7 geoeffective intervals. In every single case, the storm was caused by the sheath of the IP counterparts of the halos, thus confirming the suggestion made in Paper 1. Since there were only 7 distinct storms that can be attributed to the limb halos, we have listed only 7 delay times (from CME onset to time of minimum Dst of the storm) in Table 1. For the three pairs of interacting CMEs, we counted only the faster, overtaking CME for computing the delay time. In one case, a disk halo was overtaking a limb halo, but the time difference was very small (~3 h). Four limb halos were isolated so there is no ambiguity in the delay time. The delay ranged from 38 h to 90.5 h, with an average value of 53.3 h, not too different from the average value (56 h) reported in Paper 1 for all geoeffective limb halos (including those associated with moderate storms).

4. Discussion

We studied the geoeffectiveness of 17 limb halo CMEs by examining their IP counterparts. In particular, we examined where the Bz<0 interval occurred: within the ICME interval and/or in the sheath ahead of the CME. In all the cases, in which we can make an unambiguous association between the limb halos and IP shocks, the geoeffectiveness is caused by the sheath ahead of the ICMEs. A suggestion to this effect was made in Paper 1 without examining the IP data. In this work, we have confirmed the suggestion by examining the solar wind plasma and magnetic signatures associated with the limb halo CMEs. When the ICMEs are shock driving, the sheath provides an additional source of Bz<0. If the ejecta part is a magnetic cloud, the Bz<0 interval can occur in the front or back of the cloud for bipolar clouds, throughout the cloud interval for south-pointing high-inclination clouds, and no interval of Bz<0 for north-pointing high-inclination clouds, e.g., Ref. [5]. For limb halos, the cloud part may or may not arrive at Earth; the sheath is likely to arrive at Earth and produce a geomagnetic storm if it has a Bz<0 interval. The lack of ejecta arrival at Earth reduces the probability of limb halos producing a storm, consistent with the central-to-limb variation of geoeffectiveness of halo CMEs reported in Paper 1.

The present study also confirms the delay time between the arrival of magnetic clouds and the time of minimum Dst during storms.[5] The average delay between sheath and cloud storms can be estimated from the fact that sheath storms are typically ~3 h ahead of ICME arrival, while the cloud storms are ~11 h behind the ICME arrival.[7] Thus the sheath storms are expected to be ~14 h ahead of cloud storms. For the set of events in Table 1, we arrived at an average delay time of ~53 h, which is smaller than the value obtained for storms following disk halos by ~17 h.

One of the interesting outcomes of this study is that two of the four isolated limb halos are from close to the east limb (S07E80 for halo #8 and S12E67 for #17). This result is significant because it highlights the difficulty in forecasting geomagnetic storms based on CME observations. It is usually believed that CMEs occurring within ±30° from the disk center arrive at Earth and cause geomagnetic storms and that there is a slight western bias of the CME source regions on the Sun. Clearly CMEs originating close to the east limb also produce geomagnetic storms under extreme conditions (both the CMEs were superfast with speeds 2053 km/s and 2257 km/s).

Another surprising result is that 5 of the 9 limb halos that resulted in large geomagnetic storms were interacting with other CMEs. In one case, the limb halo (#6) interacted with a disk halo. The remaining interactions were among limb halos (#11 with #12 and #14 with #15). All the five limb halos are known producers of type II radio bursts in the IP medium (http://cdaw.gsfc.nasa.gov/CME_list/radio/waves_type2.html). Type II radio bursts are indicative of CME-driven shocks because electrons accelerated at the shock front produce Langmuir waves, which in turn produce radio emission at the local plasma frequency or its harmonic. In other words, all the five halos drove shocks in the IP medium, but at 1 AU, each pair resulted in a single shock. This may mean either the shock of the preceding CME decayed or it merged with that of the second CME in the pair.

During the study period (1996–2005), there were 75 large geomagnetic storms (Dst < -100 nT) associated with CMEs.[8] It is interesting that 7 of them (or 9.3%) are due to limb halos.

5. Conclusions

By examining the IP counterparts of limb halo CMEs using solar wind plasma and magnetic signatures, we have confirmed that the geomagnetic storms following limb halos are caused by the southward component of the

IP magnetic field contained in the ICME sheaths. Since the sheath is the first feature encountered by Earth's magnetosphere, the delay time between the onset of halo CMEs and the peak of ensuing geomagnetic storms is the smallest. The delay time is ∼20% smaller for limb halos than for disk halos reported in Paper 1. We also confirm that the overall geoeffectiveness is smaller for limb halos. This study also revealed that one of the major geomagnetic storm was caused by a halo CME originating very close to the east limb, but the CME was extremely fast. Finally, most of the large geomagnetic storms are caused by disk halos, but a significant number (∼9%) are caused by limb halos.

Acknowledgments

Data for Figures 1, 3 and 5 were obtained from NASA's OMNIweb (http://omniweb.gsfc.nasa.gov). Data for Figures 2 and 4 were obtained from the SOHO/LASCO catalog (http://cdaw.gsfc.nasa.gov) and from NOAA's GOES satellite flare listing. We acknowledge these data sources.

References

1. X. P. Zhao and D. F. Webb, *J. Geophys. Res.* **108**(A6), (2003) 1234.
2. N. Gopalswamy, S. Yashiro and S. Akiyama, *J. Geophys. Res.* **112** (2007) A06112, doi:10.1029/2006JA012149.
3. C. A. Loewe and G. W. Prölss, *J. Geophys. Res.* **102** (1997) 14209.
4. J. T. Gosling, D. N. Baker, S. J. Bame, W. C. Feldman, R. D. Zwickl, E. J. Smith, *J. Geophys. Res.* **92** (1987) 8519.
5. N. Gopalswamy, S. Akiyama, S. Yashiro, G. Michalek, R. P. Lepping, *JASTP* **70** (2008) 245.
6. R. P. Lepping, C.-C. Wu, N. Gopalswamy, D. B. Berdichevski, *Solar Phys.* **248** (2008) 125.
7. N. Gopalswamy, *JASTP* (2008) doi:10.1016/j.jastp.2008.06.010.
8. J. Zhang, *et al.*, *J. Geophys. Res.* **112** (2007) A10102.
9. J. T. Gosling, M. F. Thomsen, S. J. Bame, R. D. Zwickl, *J. Geophys. Res.* **92** (1987) 12399.
10. N. Gopalswamy, S. Yashiro, S. Akiyama, In: *Solar Influence on the Heliosphere and Earth's Environment: Recent Progress and Prospects*, ed. N. Gopalswamy and A. Bhattacharyya, Quest Publications, Mumbai, (2006) p. 79.
11. V. Yurchyshyn, C. Liu, V. Abramenko, J. Krall, *Solar Phys.* **239** 317.
12. C. Liu, J. Lee, V. Yurchyshyn, N. Deng, K. Cho, M. Karlický, H. Wang, *Astrophys. J.* **669** (2007) 1372.
13. A. Asai, K. Shibata, T. T. Ishii, M. Oka, R. Kataoka, K. Fujiki and N. Gopalswamy, *J. Geophys. Res.* **114** (2009) A00A21.

Advances in Geosciences
Vol. 21: Solar & Terrestrial Science (2008)
Ed. Marc Duldig

TWO MAJOR PROCESSES OF THE SOLAR WIND-MAGNETOSPHERE-IONOSPHERE COUPLING

XIAOYAN ZHOU
Jet Propulsion Laboratory, California Institute of Technology,
4800 Oak Grove Drive, Pasadena,
California 91109, USA
Xiaoyan.Zhou@jpl.nasa.gov

WEI SUN
Geophysical Institute, University of Alaska Fairbanks,
903 Koyukuk Drive Fairbanks,
Alaska 99775, USA

AIMIN DU
Institute of Geology and Geophysics,
Chinese Academy of Sciences, 19 Beitucheng XiLu,
Beijing 100029, China

AARON J. RIDLEY
Department of Atmospheric, Oceanic, and Space Sciences,
University of Michigan, 2455 Hayward St,
Ann Arbor, Michigan 48109, USA

Using the method of Natural Orthogonal Components (NOC) and AMIE potential data, we quantitatively separate the southward electric field (Es) and the westward electric field (Ew) in the ionospheric midnight sector auroral zone for the events in April 18, 2002 and July 15, 1997. The study results show that there are two major components existed in the high latitude ionosphere. A two-cell convection pattern in the polar cap is the fundamental pattern. A potential vortex near midnight is found to be significant during substorm expansion phases. Although the potential vortex is not a dominant pattern, its intensity can be intermittently greater than the two-cell pattern by a factor of 2 to 3. The two components indicate different processes of the energy transfer, the directly driven and unloading processes, during disturbed solar wind conditions. We find that in the midnight sector, the Joule heating related to the energy deposition by substorm expansion phases can be greater than the convection by a factor of 5–7 for the April 18, 2002 event. But during the July 15, 1997 storm, the Joule heating generated by the convection was dominant.

1. Introduction

When the dayside magnetic reconnection takes place, open field lines are swept over the polar caps at the solar wind speed. The solar wind dawn-dusk E_y maps into the polar cap ionospheres. An antisunward $\mathbf{E} \times \mathbf{B}$ drift of plasma is formed along the noon-midnight direction in the F-region ionosphere. The flows return to local noon from nightside along the dawn and dusk auroral oval, respectively. Thus, this convection creates a two-cell pattern related to the global convection electric field.[1] From 67 high-latitude ground-based magnetometers in the northern hemisphere, Kamide *et al.*[2] found the pair of potential cells that occupied the polar cap. If only the convection takes place, in nightside auroral zone the related electric field is mainly westward, E_W. But when substorm expansion phases (SEPs) occur another pair of potential cells roughly north-south aligned appears in nightside. Figure 1 (from Fig. 11 in Ref. [2]) shows these two pairs. A southward electric field, E_S, is formed in between the low latitude cells. Es drives a Hall current (which is the main portion of the westward electrojet).[3]

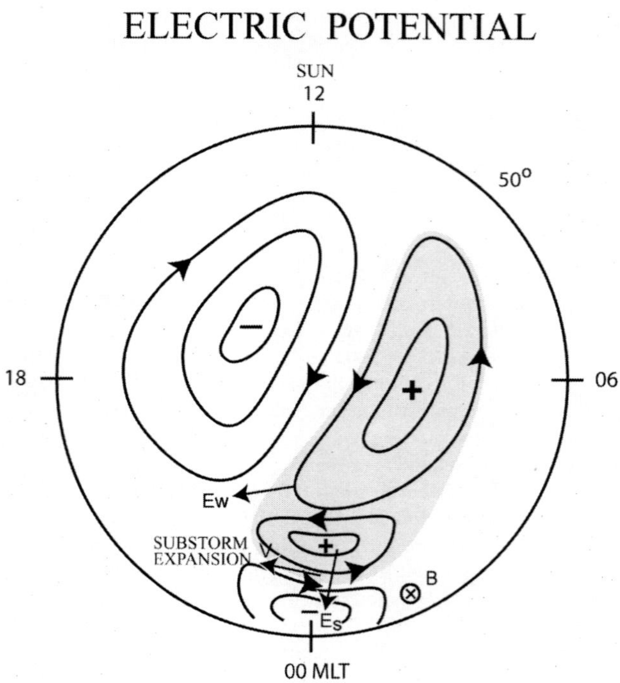

Fig. 1. Separated DD and UL components in the polar caps. In the shading area, the actual potential cells are combined and looked more complicated (from[2]).

The formation of E_S can be explained by the blockage of the northward Hall current driven by E_W due to the lack of equatorial radial current in the midnight sector of the near-Earth plasma sheet.[4,5]

The above scenario reminds us the theory that magnetospheric substorms contain two major energy transfer processes: the directly driven (DD) and the unloading (UL) processes (e.g. Refs. [2], [6–8]). During the DD process the solar wind energy is directly deposited into the high-latitude ionosphere with an appreciable time delay. A two-cell convection pattern is formed in the polar cap with auroral electrojets intensified in the dawn and dusk sectors. The convection electric field is highly correlated with the solar wind ε function that has zero value when the IMF B_z is northward.[9] As shown in Fig. 1, near midnight the electric field related to DD process is westward, E_W. In the UL process, however, the electric field related to SEP is southward, Es.

In fact, variations of the two electric field components, E_W and E_S, were observed from balloons in northern hemisphere nightside aurora zone.[10] Figure 2 shows the balloon-measured electric fields of the 19 substorms

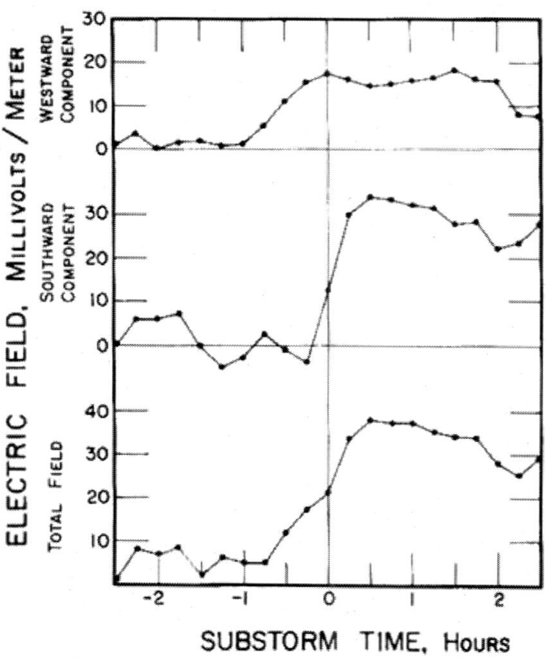

Fig. 2. Substorm electric fields in the nightside auroral ionosphere. The measurements were obtained from 19 balloon flights (adopted from Mozer *et al.*, 1971).

for 5-hour intervals near midnight between $L = 6.6$ and 8.3. The data are centered at SEP onsets, i.e. the 0 hour. The E_W component increased from ~0 to 17 mV/m during the substorm growth phase. At the SEP onset the E_S component increased about 30 mV/m. Balloon-borne electric field measurement provides vertical electric field that is associated with the ionosphere potential difference caused by global weather processes, and horizontal electric filed that is essentially equal to the average of several hundred kilometers. An electric field component was determined as the potential difference between the appropriate pairs of plates divided by their separation distance.[11] This balloon experiment showed that the nightside electric field E_S significantly increased at the SEP onset.

However, electric field measurements from balloon are very rare. The most practical and continuous monitoring of the ionospheric current/electric filed variations is from the ground-based magnetometers and radars. But such measurements only provide combined effects of the ionospheric currents. If one can separate the combined effect into simpler patterns, as described above, with clearer physics (such as those shown in Fig. 1), important information for SEP and ionospheric convection can be obtained. In addition, the current density of substorm westward electrojet as well as the corresponding energy deposition can be identified. Therefore, distinguishing E_W and E_S becomes an interesting and important issue of understanding the substorm activity and the solar wind-magnetosphere-ionosphere coupling.

It has been several decades that people tried to separate the two-cell and one-cell patterns. The first thought was to separate DP1 and DP2 equivalent currents that can be derived from the ground-based magnetograms.[1,12–14] Equivalent currents can be obtained by rotating 90° clockwise the horizontal component of the geomagnetic field and by multiplying with $2/\mu_o$, where μ_o is the permeability of the free space. Equivalent currents are not only a convenient method to study ionospheric disturbances, but also a good approximation of ionospheric currents under many conditions (e.g. Ref. [15]). Studies showed that the separated DP2 current indeed has a two-cell pattern that occupies the polar cap and the separated DP1 current appears to be a one-cell pattern near the midnight auroral oval (such as Ref. [14]). However, it is controversial about the physical meaning of equivalent currents since they are conceptual currents and lack of observational comparison.

To improve this situation, we suggested a new approach that applies the Natural Orthogonal Components (NOC) algorithm to AMIE potential observations.[16] When there is no substorm activity, the obtained convection pattern is two potential cells occupying the polar cap with positive polarity at the dawnside and the negative polarity at the duskside. When SEPs take place, an additional one potential cell appears in nightside auroral region. The two electric field components in the midnight sector, therefore, can be calculated by taking potential gradient. We also found the threshold of Es to allow a SEP onset take place is ~30–40 mV/m. This is very close to the balloon observations.

This paper presents our further studies on the two electric field components by examining SEPs during two storms one of which has quasi-periodic SEPs and energetic particle injections at the geosynchronous orbit and the other has a lack of SEP. On the basis of separated E_W and E_S, we confirmed that three out of the seven energetic particle injections are related to SEP since the enhanced E_S is about or greater than 30 mV/m. The solar wind E_y has a clear effect on the latitude location of E_W instead of its intensity.

2. Method

The Natural Orthogonal Component (NOC) algorithm is applied to solve an eigen equation

$$AX^k = \lambda_k X^k \tag{1}$$

where $\mathbf{X}^k$ is Eigen vectors (functions) and λ_k is Eigen values. The matrix $\mathbf{A}$ is constructed by a time series of electric potential $\Phi(t_i, r_j)$, time t_i $(i = 1, 2, \ldots, m)$ and location $r_j (j = 1, 2, \ldots, n)$. In the present study the time interval is 1 day. The AMIE (the Assimilative Mapping of Ionospheric Electrodynamics) potential data have time resolution of 1 min, hence $m = 1440$. $n = 576$. The potential is provided by the University of Michigan (see http://amie.engin.umich.edu/ for detailed information). The elements of $\mathbf{A}$ can be calculated by the formula

$$A_{lp} = \sum_{i=1}^{m} \phi_{il}\phi_{ip} \tag{2}$$

The DD and UL components are identified based on the potential cell patter given by the Eigen vectors. Detailed description and discussion of the NOC method and the AMIE potential can be found in.[16]

3. Analysis of Nightside Electric Fields

3.1. *The April 18, 2002 Event*

This sawtooth event has been well studied in recent years (e.g., Refs. [17–18]). Figure 3a includes the ACE solar wind observations in the top 4 panels, the geosynchronous magnetic field tilt angle ($\tan^{-1}(B_Z/B_X)$), and the one-minute resolution symmetric ring current index. The magnetic cloud started from ~0430 UT in April 18 and ended in next day at ~0220 UT. The vertical lines label the times when there was an energetic particle injection that was observed by LANL satellites.[18] Figure 3b shows the solar wind ε function on the same day and calculated total nightside electric field as well as E_W and E_S using the NOC method. E_W and E_S are calculated in the midnight sector confined between 50°–70° MLat and 23–01 MLT. The vertical lines correspond to those in Fig. 3a. Three SEPs are identified when the E_S component increases abruptly near 0300, 0600 and 0800 UT. The enhanced E_S is from ~30 to 40 mV/m. The E_S peak near 1200 UT is relative weak, but it is identified as a SEP by IMAGE UV auroral data.[17] At other times in the second half of the day, the variations

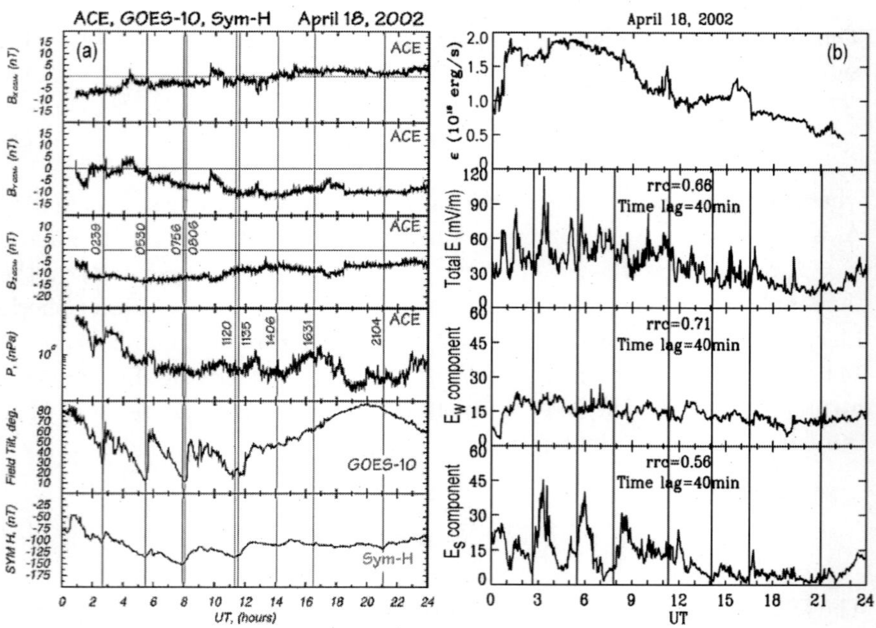

Fig. 3. The April 18, 2002 storm event. (a) Solar wind parameters and geomagnetic indices. (b) Calculated total electric field, E_S and E_W in the midnight sector.

in E_S do not show SEP occurrences. Most of the time, E_W is less than 20 mV/m and has less abrupt changes than E_S. However, the E_W does vary ~30% during SEPs as in between the vertical lines. The correlation coefficients showed in the panels are between each corresponding electric field and the ε function. E_W is the most correlated component with the solar wind energy input with a time lag of 40 min. This is consistent with the definition of the DD and UL components.[9]

Six separated potential patterns are shown in Fig. 4. A clear two-cell pattern is obtained corresponding to the first Eigen value, which indicates that the dominant pattern is the two-cell convection. Both the intensity and orientation of the pattern vary with time. From λ_2 to λ_6, a one-cell pattern becomes a major feature, but less significant than λ_1. Those one-cells with centers below 70° MLat are all in nightside. Each one-cell depicts a possible SEP onset location. How much contribution that one-cell could give is decided by the corresponding Eigen value. On average, the first seven Eigen values can provide about 80% or more of the total potential.[16]

From left to right, Fig. 5 shows the total potential, two-cell DD component and one-cell UL component at 0320 UT when the first SEP onset

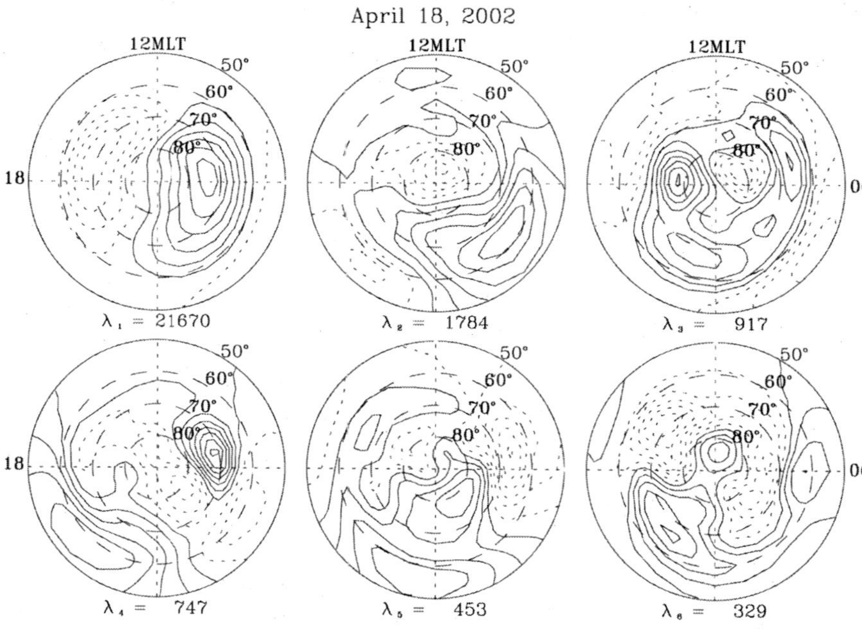

Fig. 4. Six potential patterns for the first 6 Eigen values of the April 18, 2002 event.

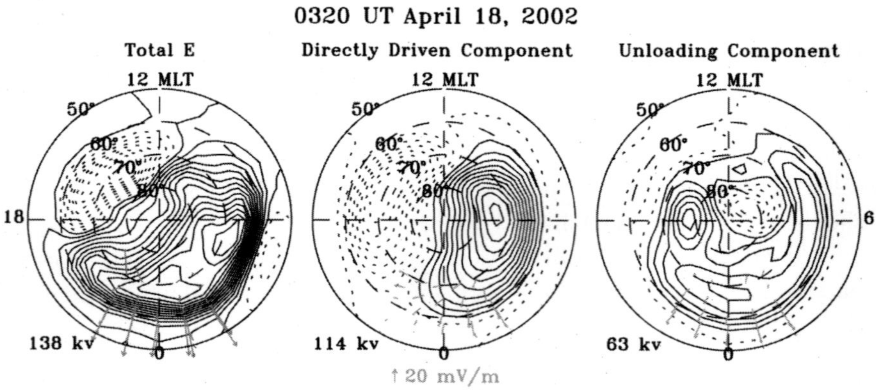

Fig. 5. Electric fields at 0320 UT during the first intense substorm expansion phase.

started. Taking gradient of the potential in the midnight sector, electric fields can be obtained. The electric field corresponding to each component is shown in grey vectors. Above 75° MLat, the convection electric field, E_W, has a clear westward orientation. But E_S in the right panel is so strong that the total E in the left panel points to south evidently. The enhancement of E_S at this SEP is ~45 mV/m that is ~9 times more than the increase of E_W as shown in Fig. 3b.

3.2. *The July 15, 1997 Event*

It has been a long-standing debate about whether geomagnetic storm can occur without SEPs. Recently, we have found some storms an be developed when there is a lack of substorms.[19,20] Using the NOC method and AMIE potential we examine the July 15, 1997 storm again. This storm (one may argue to be a SMC) was caused by a magnetic cloud. Figure 6a shows the solar wind parameters and the ground geomagnetic indices. There were no SEPs during 1300–2000 UT based on the Polar UVI imaging data. One could argue that there might be one SEP started from ~1040 UT. Except that, the AL index shows a convection bay during the magnetic cloud. Figure 6b has the same format as Fig. 3b. The sudden increase at ~1040 UT in E_S is responsible to the AL magnitude increase. But the peak E_S is ~25 mV/m that is smaller than the SEP criteria of 30–40 mV/m proposed by Sun *et al.*[16] If this E_S increase is a substorm, it only could be a small one or it may be a pseudobreakup.

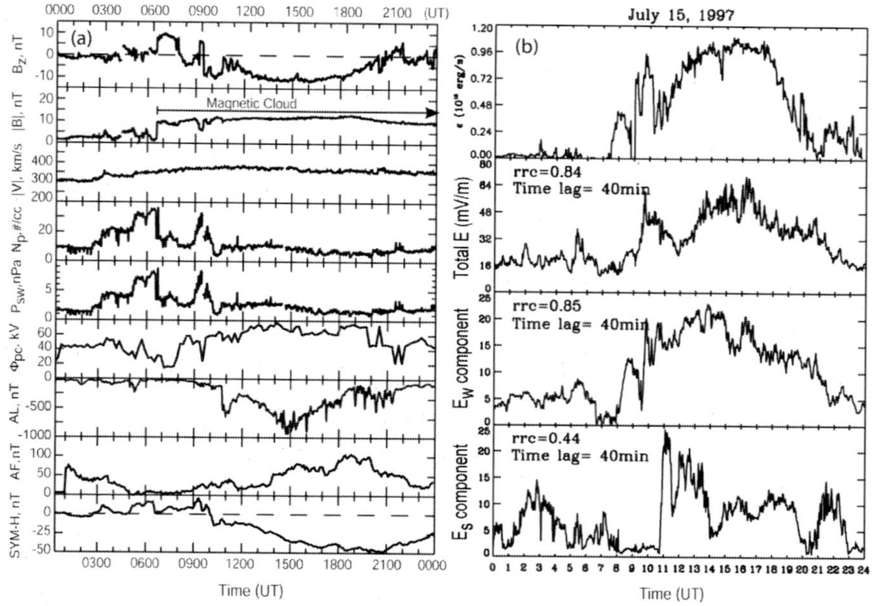

Fig. 6. The July 15, 1997 storm event. (a) Solar wind parameters and geomagnetic indices. (b) Calculated total electric field, E_S and E_W in the midnight sector.

Figure 7 shows the separated potential patterns of the first 6 Eigen values. Both λ_1 and λ_2 give a tow-cell convection pattern indicating the two-cell convection was very significant. A statistical study is needed to examine whether two Eigen values are a general signature of intense convection. We do not quite understand the pattern of Eigen 3. The one-cell pattern of λ_4 is related to the small "substorm", but the Eigen value is an order of magnitude lower than λ_1 and λ_2. In Sun *et al.*[16], we have shown that the high order Eigen values contribute less and can be ignored.

Calculated electric fields at 1100 UT are shown in Fig. 8. It is the peak time of the "substorm". So the E_S value was similar to E_W at that moment when the storm main phase just began (see Fig. 6b). Even though, the total potential (the left panel) still features a two-cell convection pattern. The convection cells were mainly constrained above 70° MLat. That implies a small polar cap. This is consistent with the Polar UVI imaging data.

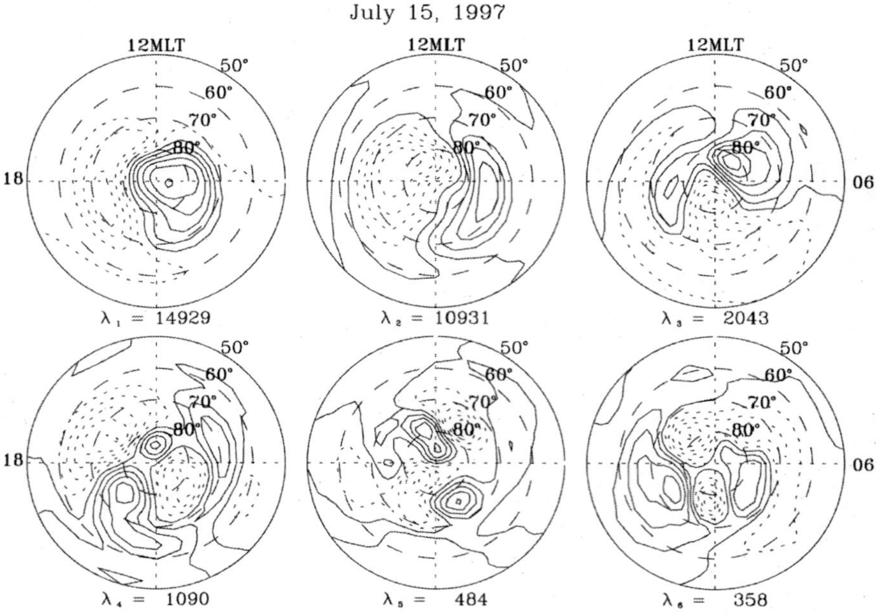

Fig. 7. Six potential patterns for the first 6 Eigen values of the July 15, 1997 event.

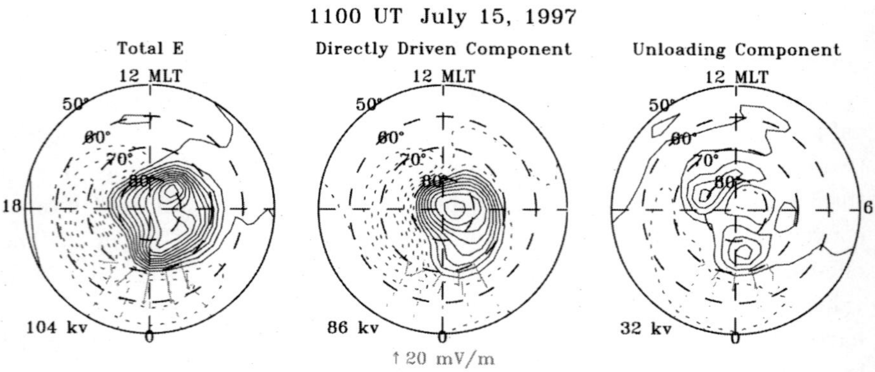

Fig. 8. Electric fields at 1100 UT July 15, 1997 during a small "substorm" or a pseudobreakup.

4. Summary and Discussion

We have shown storm-time nighside ionospheric electric fields in the auroral zone for two geomagnetic storms caused by magnetic clouds. During the April 18, 2002 event, "sawtooth" particle injections were observed at the geosynchronous orbit. Using the NOC algorithm and AMIE potential data, we confirmed that 3 out of the 7 injections were SEPs. At the SEP peaks, E_S was greater than E_W by a factor of 2–3. E_W increased ~30% in magnitude during SEPs and the maxima were less than 25 mV/m. During the July 15, 1997 storm, we did not find any substantial SEP activity except an E_S increase of ~25 mV/m at ~1040 UT, which might be related to a small "substorm" or a pseudobreakup.

Solutions of equations (1) and (2) provide Eigen functions and time variations of the potential. The former is two dimensional cell patterns (such as in Figs. 4 and 7) and the latter is the time variation of electric fields (such as in Figs. 3 and 6) that are derived from corresponding potentials. The importance of the potential pattern is given by the Eigen value. For the events that we studied so far, the two-cell convection pattern has been always weighted by λ_1. This implies that the two-cell pattern is fundamental, although its intensity varies with time and location. Comparing the bottom two panels in Fig. 3b, it is clear that when there is no substorm activity E_W is dominant, i.e., the two-cell convection pattern is dominant in pattern and in intensity. However, when SEPs occurred, E_S was double of E_W in magnitude in midnight sector. This implies that locally, the one cell pattern had a larger intensity than the fundamental convection pattern. Figure 6b shows that the two-cell convection was dominant not only in pattern, but also in intensity during the storm.

In this study, E_W is defined as the DD component and is related to the convection. Note that E_W is different from the global convection electric field that corresponds to the polar cap potential that is closely correlated with the solar wind E_y. For both events, maximum E_W is ~25 mV/m, although the solar wind E_y is larger by a factor of 2 in the April 18, 2002 magnetic cloud than the other one. Further study is needed before answering the reason. More questions are needed to be addressed. Such as how the local convection electric fields vary in the dawn and dusk sectors when there is a lack of SEP.

Energy depositions related to the processes that are described by E_W and E_S can be estimated. The Joule heating rate, U_J, is given by $U_J = \Sigma_P E^2$. Therefore, the Joule heating ratio E_S^2/E_W^2 presents the ratio of the energy deposition by convection over the energy deposition by SEP. Figure 9 shows the Joule heating ratios of the two events. For the sawtooth event, at three SEPs the ionospheric Joule heating generated by the SEPs was ~5–7 times of that by the convection. During the July 15, 1997 storm, the Joule

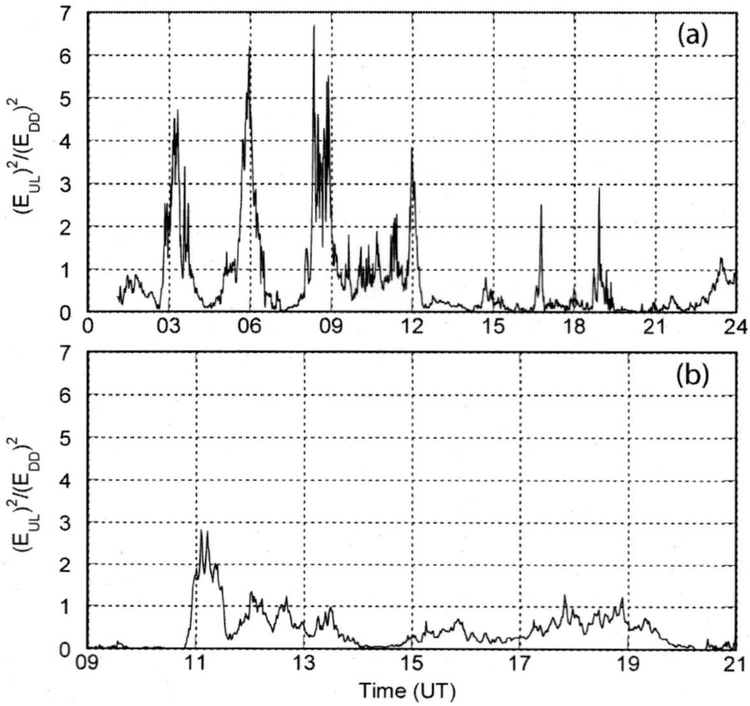

Fig. 9. Joule heating ratio of the two processes relevant to the convection and the substorm expansion phase. (a) April 18, 2002 event. (b) July 15, 1997 event.

heating related to the convection was always dominant during the storm except when there was a small substorm in ∼1100–1200 UT.

We confirm that the NOC method is a powerful tool for the substorm study. Caution should be taken when applying the NOC method and using the AMIE potential. Separated results need to be analyzed carefully case by case based on the solar wind condition and AMIE data completeness. Large data gaps in time and space, which cause anomalies in the potential distribution, may affect the potential separation as well as the calculated electric fields.

Acknowledgment

The results reported here represent one aspect of research carried out by the Jet Propulsion Laboratory, California Institute of Technology, under a contract with the National Aeronautics and Space Administration.

References

1. C. R. Clauer and Y. Kamide, *J. Geophys. Res.* **90** (1985) 1343–1354.
2. Y. Kamide, *et al.*, *J. Geophys. Res.* **99**(A10) (1994) 19451–19466.
3. Y. Kamide and W. Baumjohann, *Magnetosphere-Ionosphere Coupling*, pp. 178, Springer-Verlag New York, (1993).
4. T. Iijima, T. A. Potemra and L. J. Zanetti, *J. Geophys. Res.* **95** (1990) 991.
5. J. R. Kan, *J. Geophys. Res.* **109** A10215, doi:1029/ 2006JA012011, (2007).
6. R. L. McPherron, C. T. Russell and M. P. Aubry, *J. Geophys. Res.* **78**(16) (1973) 3131.
7. S.-I. Akasofu, What is a magnetospheric substorm?, in *Dynamics of the Magnetosphere*, p. 447, D. Reidel, Norwell, Mass., (1979).
8. G. Rostoker, S.-I. Akasofu, *et al.*, *Space Sci. Rev.* **46** (1987) 93.
9. S.-I. Akasofu, *Space Sci. Rev.* **28** (1981) 121.
10. F. S. Mozer, *J. Geophys. Res.* **76** (1971) 7595.
11. F. S. Mozer and R. H. Manka, *J. Geophys. Res.* **76** (1971) 1697.
12. T. Obayashi and A. Nishida, *Space Sci. Rev.* **8** (1968) 3.
13. A. Nishida, *Planet. Space Sci.* **19** (1971) 205.
14. W. Sun, W.-Y. Xu and S.-I. Akasofu, *J. Geophys. Res.* **103**(A6) (1998) 11695–11700.
15. J. Untiedt and W. Baumjohann, *Space Sci. Rev.* **63** (1993) 245.
16. W. Sun, X.-Y. Zhou and A. Du, *Geophys. Res. Lett.* **35** L13104, doi: 10.1029/2008GL033931, (2008).
17. A. T. Y. Lui, T. Hori, S. Ohtani, Y. Zhong, X.-Y. Zhou, M.G. Henderson, T. Nagai, T. Mukai and S. B. Mende, *J. Geophys. Res.* **109** A10215, doi:1029/2004JA010543, (2004).
18. M. G. Henderson, G. D. Reeves, R. Skoug, M. T. Thomsen, M. H. Denton, S. B. Mende, T. J. Immel, P. C. Brandt and H. J. Singer, *J. Geophys. Res.* **111**(A01S90) doi:10.1029/2005JA011111, (2006).
19. B. T. Tsurutani, X.-Y. Zhou and W. D. Gonzalez, in *Disturbances in Geospace: The Storm-Substorm Relationship*, Geophysical Monograph Series, Volume 142, edited by A. Surjalal Sharma, Yohsuke Kamide, and Gurbax S. Lakhina, Amer. Geophys. Union, Washington DC, pp. 23–36, (2003).
20. X.-Y. Zhou, B. T. Tsurutani, G. Reeves, G. Rostoker, W. Sun, J. M. Ruohoniemi, Y. Kamide, A. T. Y. Lui, G. K. Parks, W. D. Gonzalez and J. K. Arballo, *J. Geophys. Res.* **108**(A11) 1407, 10.1029/2003JA009881, (2003).

Advances in Geosciences
Vol. 21: Solar & Terrestrial Science (2008)
Ed. Marc Duldig

GLOBAL SIMULATION OF THE MAGNETOSPHERE USING NESTED GRIDS MODEL WITH INNER BOUNDARY OF $1R_E$

KEIICHIRO FUKAZAWA
National Institute of Communication and Information Technology, 4-2-1, Nukui-Kitamachi, Koganei, Tokyo 184-8795, Japan
fukazawa@nict.go.jp

The inner boundary of the real-time magnetosphere simulation performed at NICT is set at $3R_E$ from the Earth. This distance was estimated by determining the numerical stability and the computation time of the simulation using the Alfvén velocity, which is the characteristic velocity in MHD (magnetohydrodynamics) and whose value increases in the high-magnetic-field region around the Earth. Although inner boundaries do not have a considerable effect on the global configuration of the magnetosphere, they can cause problems in the simulation of the coupling between the magnetosphere and ionosphere. This is because there is a space between these two regions as a result of which we cannot connect them directly in the simulation. Thus, at present, although the simulation results of the magnetosphere can be applied to the ionosphere at high latitudes along the magnetic field lines, the results cannot be applied to the ionosphere at middle and low latitudes as the regions in the vicinity of the Earth are not considered in the magnetosphere simulation.

In order to overcome these problems, we have proposed the use of a global simulation model with a $1R_E$ inner boundary. In order to develop this model, we have used nested grids in the MHD simulation. By using nested grids, each grid can consider multiple timesteps during numerical calculations, resulting in a reduction of the calculation time. In this model, it is necessary to interpolate the data and transfer data between grids, which can result in numerical errors. However, in this model, the errors caused by interpolation are well inhibited by the use of the conservative interpolation for $\nabla \cdot \mathbf{B} = 0$ and by the use of a smoothing function.

The results of the study prove that the application of nested grids to an MHD model allows successful simulation of the magnetosphere without any numerical oscillations or abnormal increase in the plasma velocity. Further, the calculation time of the nested grids model is only slightly greater than that of the normal simulation model. Thus, nested grids are suitable for performing real-time simulations.

1. Introduction

In space weather forecasting, it is important to simulate the terrestrial environment that interacts with the solar wind before it reaches the Earth. The Advanced Composition Explorer (ACE) spacecraft is used to measure the solar wind at a Lagrangian point L1 (235R_E (1R_E: Earth radii)), which is situated upstream from the Earth. The ACE sends solar wind data back to the Earth approximately 1 hr before the solar wind actually reaches the Earth. Thus, we have an hour to simulate the magnetospheric environment. Real-time simulation of the magnetospheric environment using the ACE spacecraft data is performed at NICT (National Institute of Communication and Information Technology), and the results of the simulation are available on their web site.[1] The inner boundary in the NICT real-time simulation is set as 3R_E from the Earth; most other magnetospheric simulation models also set the inner boundary at around 3R_E. Because of the boundary, the simulation results of the magnetosphere cannot be mapped to the ionosphere at middle and low latitudes.

The reason for using an inner boundary is to reduce the calculation time. Real-time simulation calculations require both speed and stability. Numerical calculations are said to be stable when the numerical stability condition known as the CFL condition is satisfied. If Δx is the grid spacing, Δt is the time step, and c is the characteristic velocity, then the CFL condition is $c\Delta t < \Delta x$. In other words, when the velocity increases, in order to satisfy the condition, either the timestep should decrease or the grid spacing should increase.

The characteristic velocity in MHD (magnetohydrodynamics) is known as the Alfvén velocity (V_A), and it is calculated from the magnitude of the magnetic field and the plasma density ($V_A = B/\sqrt{\mu_0 \rho}$). If the magnetic-field magnitude and the plasma density are defined as $B = B_0/r^3$ and $\rho = \rho_0/r^3$, respectively, then, the Alfvén velocity becomes

$$V_A = V_{A0} r^{-\frac{3}{2}} \quad \text{where } (V_{A0} = B_0/\sqrt{\mu_0 \rho_0}). \tag{1}$$

Here, r is the distance from the Earth. Thus, the magnitude of the Alfvén velocity at a distance of 1R_E from the Earth is five times the velocity when the distance is 3R_E. In some cases, it is necessary to consider up to 25 times the value of Alfvén velocity in order to account for the diffusion effect. If the inner boundary is set at 1R_E without changing the grid spacing, then the timestep of simulation needs to be reduced drastically (less than 1/5) to enhance the Alfvén velocity. Furthermore we often use a short mesh near the Earth, thus the reduction of timestep is much more important. Although computers are expected to develop considerably in the future, as of now, the available computer resources are not sufficient to execute a simulation with such a short timestep. In addition, this problem occurs not only in real-time simulations but also in other magnetospheric simulation models.

In this paper, we have developed a simulation technique (the nested grids technique) that allows the inner boundary to be set at $1R_E$ without increasing the calculation time. The mechanism of nested grids is presented in Sec. 2. In Sec. 3, we apply the nested grids technique to a global magnetosphere model and present the test results. The study is summarized in Sec. 4.

2. Nested Grids Model

2.1. *Mechanism*

In order to allow the inner boundary to be set at $1R_E$, we have proposed the application of nested grids to the magnetosphere simulation model. The nested grids technique is a kind of adaptive mesh refinement (AMR) technique,[2] and it uses multiple simulation boxes. With multiple simulation boxes, multiple timesteps can be treated in the same simulation. Thus, a short timestep is used for the simulation of the region around the Earth, while in other cases, a normal timestep is used. Thus, it is possible to set the inner boundary at $1R_E$ to reduce the time required for calculations. In addition, there is no need to run the same simulation model in each box. For example, we can run the MHD code in one box, while simultaneously running the Hall MHD code in another box. This means that we can choose the simulation models that we wish to run in each box depending on the phenomena that we want to study. It should be noted that the nested grids form just the grid model and not the numerical calculation model. Hence, the application of nested grids does not require any changes to be made to the numerical model, and the grids can be easily introduced in various simulation models.

Figure 1 shows the calculation mechanism of the nested grids. The black grid denotes the outer box ($f_{\mathrm{out}}(x, y)$) with a rough mesh, and the red grid denotes the inner box ($f_{\mathrm{in}}(x, y)$) with a small mesh. First, the blue data shown in the figure are transferred from the outer box to the inner box. Then, the green data are interpolated with the blue and yellow data.

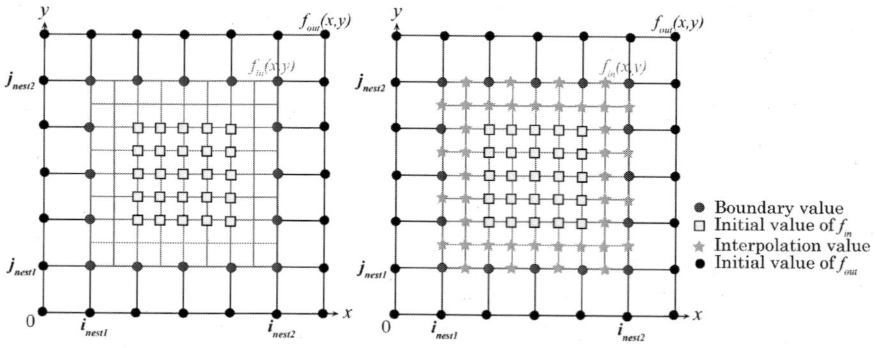

Fig. 1. Mechanism of the nested grids model shown in 2 dimensions.

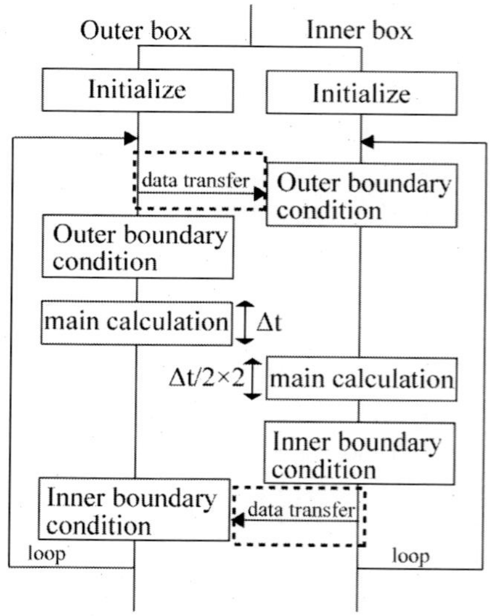

Fig. 2. Flowchart of the calculation with nested grids.

Following this, calculations start in each box. After the calculations are completed, the blue data in the inner box is transferred to the outer box, and the sequence is repeated. Figure 2 summarizes the calculation sequences using a flowchart of the actual simulation. The additional calculations due to the use of the nested grids are indicated by the dashed lines in Fig. 2.

The relation between the outer box and inner box is defined below. If $\Delta x_{\rm out}$ is the grid spacing of the outer box, $\Delta x_{\rm in}$ is that of the inner box, and $1 \leq x \leq x_{\rm in}$, $1 \leq y \leq y_{\rm in}$ is the size of inner box, then,

$$\begin{cases} x_{\rm in} = 1 + \dfrac{\Delta x_{\rm out}}{\Delta x_{\rm in}}(i_{\rm nest2} - i_{\rm nest1}) \\ y_{\rm in} = 1 + \dfrac{\Delta x_{\rm out}}{\Delta x_{\rm in}}(j_{\rm nest2} - j_{\rm nest1}) \end{cases} \tag{2}$$

where $i_{\rm nest1}, i_{\rm nest2}, j_{\rm nest1}$, and $j_{\rm nest2}$ are the positions in $f_{\rm out}(x, y)$ indicated in Fig. 1.

2.2. *Magnetic field interpolation*

First we used the linear interpolation of the magnetic field, and then we observed that an abnormal increase occurred in the plasma velocity after a few calculations. The magnetic field is required to satisfy the condition $\nabla \cdot \mathbf{B} = 0$. However, in general, a linearly interpolated magnetic field does not satisfy the solenoidal condition. Not satisfying the solenoidal condition may generate the numerical sources in MHD simulation code.[3] To overcome this, several techniques to preserve $\nabla \cdot \mathbf{B} = 0$ in MHD code are developed and evaluated.[4] In addition for the AMR and particle trajectory trace, some interpolation models which satisfy $\nabla \cdot \mathbf{B} = 0$ are developed.[5–8] Our motivation of this study is to set the inner boundary at $1\mathrm{R_E}$ without increase of calculation time in the global magnetospheric simulation. Thus, in our model, the interpolation of conservation for $\nabla \cdot \mathbf{B} = 0$ with short calculation time[5] is used. This interpolation is based on spline interpolation, and its details are described in the study by *Shimizu and Ugai.*[5] Their interpolation model is in three dimensions with a second order, in their paper, however, the interpolation model is applied to the two-dimensional MHD data. In this paper we briefly describe the functions derived in their study. The interpolation area is set as $(i, j, k) \sim (i+1, j+1, k+1)$, and x', y' and z' are defined as

$$\begin{aligned} x' &= \frac{2}{\Delta x}\left\{x - \left(i + \frac{1}{2}\right)\Delta x\right\} \\ y' &= \frac{2}{\Delta y}\left\{y - \left(j + \frac{1}{2}\right)\Delta y\right\} \\ z' &= \frac{2}{\Delta z}\left\{z - \left(k + \frac{1}{2}\right)\Delta z\right\} \end{aligned} \tag{3}$$

x, y, and z indicate the interpolated position and Δx, Δy, and Δz are the grid spacing in each direction. Then, the interpolation function for

x-component of magnetic field B'_x is given as

$$B'_x(x, y, z) = \sum_{l=1}^{3} \sum_{m=1}^{2} \sum_{n=1}^{2} a_{lmn} e_l(x') g_m(y') g_n(z') \tag{4}$$

$$\begin{aligned} a_{111} &= \frac{(B_x)_{i+1,j,k} - (B_x)_{i-1,j,k}}{2\Delta x} \\ a_{211} &= \frac{(B_x)_{i,j,k} + (B_x)_{i+1,j,k}}{2} \end{aligned} \tag{5}$$

$$\begin{aligned} e_1(\xi') &= \frac{-(\xi' - 1)^2 \Delta\xi}{8} \\ e_2(\xi') &= 1 \\ e_3(\xi') &= \frac{(\xi' + 1)^2 \Delta\xi}{8} \\ g_1(\xi') &= \frac{(1 - \xi')}{2} \\ g_2(\xi') &= \frac{(1 + \xi')}{2} \end{aligned} \tag{6}$$

Here, $a_{121}, a_{112}, a_{122}, a_{311}, a_{321}, a_{312}$, and a_{322} obey a_{111}, while a_{212}, a_{221}, and a_{222} obey a_{211}, however when $l = 3$, i is replaced by $i + 1$, while when $m = 2$, j is replaced with $j + 1$. Further, when $n = 2$, k is substituted with $k + 1$. B_x is the simulation data. $\Delta\xi$ is the grid interval in each x-, y- and z-direction. Other interpolation function for y- and z-component of magnetic field B_y and B_z are given by changing the subscript in the above expressions circularly. In addition, the other physical values (plasma density, velocity, and pressure) are interpolated linearly.

2.3. *Filter*

However, there are some problems associated with the use of nested grids. As described in Sec. 2.1, the simulation boxes are connected to each other to enable data transfer. However, these connected regions cause numerical oscillations due to the difference in the mesh sizes of the boxes.[9] In order to remove or inhibit numerical oscillations, it is necessary to introduce a filter in the simulation model. *Sugimura et al.*[9] proposed the use of a smoothing function as given below.

$$f(r) = a(r)f_{\rm in}(r) + (1 - a(r))f_{\rm out}(r) \tag{7}$$

Here, r is the distance from the Earth, and $f_{\rm in}(r)$ and $f_{\rm out}(r)$ are the inner grid and outer grid, respectively. The smoothing function $a(r)$ is given by

$$a(r) \equiv \frac{a_0 h^2}{a_0 h^2 + 1} \tag{8}$$

where $a_0 = 5$ and

$$h \equiv \frac{r^2}{r_a^2} - 1 \quad (r \geq r_a) \tag{9}$$

$$h \equiv 0 \quad (r < r_a) \tag{10}$$

with $r_a = 4\,\mathrm{R_E}$.

Using the above formula, the connected regions are merged and the numerical oscillations are reduced.

3. Application of Nested Grids to Magnetosphere Simulation

3.1. *Simulation parameters*

In order to simulate the magnetosphere using nested grids, two simulation boxes were used. The first was the outer box, whose size was $(x_{\rm out}, y_{\rm out}, z_{\rm out}) = (300, 200, 200)$ with a $0.2\,\mathrm{R_E}$ grid interval, and the other was the inner box, whose size was $(x_{\rm in}, y_{\rm in}, z_{\rm in}) = (99, 99, 99)$ with a $0.1\,\mathrm{R_E}$ grid interval. The gird number of the inner box was one-twelfth that of the outer box. The nested region in the outer box was given by $76 < x, y, z < 127$. The timesteps of the outer and inner boxes were set in the outer-to-inner box ratio of $2 : 1$. The inner boundary was set at $1\,\mathrm{R_E}$. In this study, we used the MHD simulation model described by *Ogino et al.*[10] However, other simulation models can also be used as the nested grids do not depend on any particular simulation model.

The simulation parameters are discussed in detail in the study by *Ogino et al.*[10] Here, we only provide a brief description of the different parameters of the model. A magnetized solar wind with a velocity of $v_{sw} = 400\,\mathrm{kms^{-1}}$, a density of $n_{sw} = 5\,\mathrm{cm^{-3}}$, and a temperature of $T_{sw} = 2 \times 10^5\,\mathrm{K}$ was launched from the upstream boundary of the simulation box. The IMF (interplanetary magnetic field) was a pure northward IMF of $B_Z = 5\,\mathrm{nT}$.

In addition, the plasmas around the Earth were assumed to corotate with the Earth's rotation.[11] The speeds of the plasmas were much less than that of the solar wind, and thus they were useful to check any abnormal increase in the velocity.

3.2. *Test results*

The results of the magnetosphere simulation with the nested grids are showed in Figs. 3 and 4. The left panel is the inner box and the right is the outer box in Figs. 3 and 4. These panels show the magnitude of plasma velocity on the equatorial plane. The grid numbers are written along the X and Y axes. Non- colored regions show data over the maximum value of color bar. The solar wind comes from the left boundary of the outer box. The Earth is located at the center of the inner box. Figure 3 is the state after 2.7 hours simulation starts. The solar wind does not reach the inner box and in this case it is possible to check the simulation results on the steady state. From this result the data transfer between two boxes and interpolation in the inner box (in particular, of the magnetic field) for the Nested grids work well because the non-physical oscillations and the abnormal enhancement of velocity which appeared in the initial results do not appear.

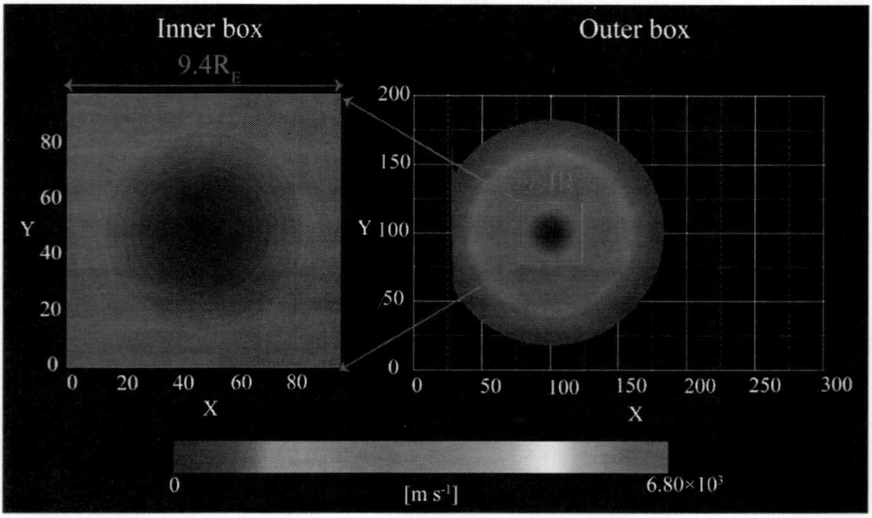

Fig. 3. Results of the nested grids model after 2.7 h of simulation.

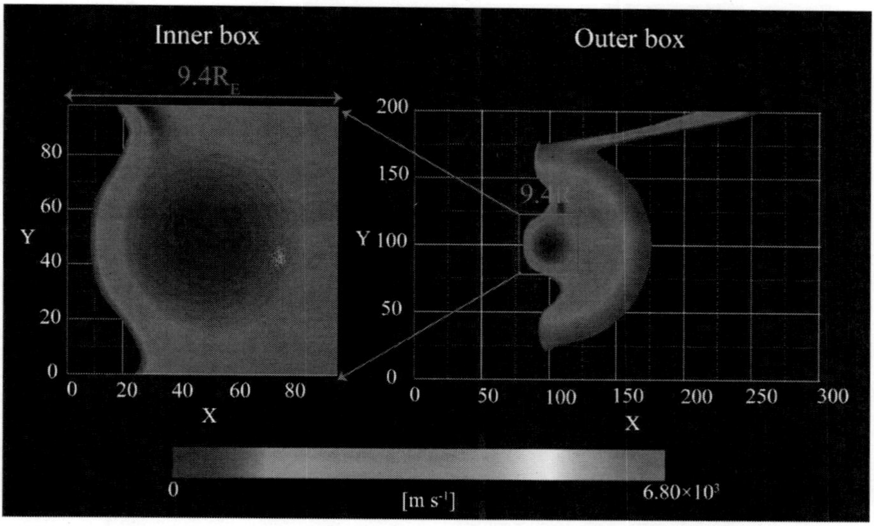

Fig. 4. Results of the nested-grids model after 27 h of simulation.

Figure 4 shows the result after 27 hours. The solar wind reaches the inner box. In this case we can see the wave propagation test results from the outer box to the inner box. It can be seen that the solar wind propagates into the inner box through the outer box in this figure and there are no apparent numerical fluctuations.

3.3. *Difference of nested grids model from no nested grids*

The comparison of the results of the simulation using nested girds and those of the simulation with no nested grids is shown in Fig. 5. The format in this figure is the same as that in the Figs. 3 and 4. The upper panel shows the results of the simulation without nested grids, while the lower panel shows the results of the simulation with nested grids. The inner boundary of the model with no nested grids was set at $3\,\mathrm{R_E}$ and the inner boundary with nested grid is set at $1\,\mathrm{R_E}$. The figures show the magnitude of plasma velocity on the equatorial plane. As the simulations are identical except for the use of the nested grids, the overall results for the two simulations are quite similar. However, it can be observed that some noise is caused along the inner boundary ($3\,\mathrm{R_E}$) around the Earth ($(x, y) = (100, 100)$) in the case of the simulation without nested grids. At the inner boundary we connect the simulation data with the initial data. Thus the noise appeared

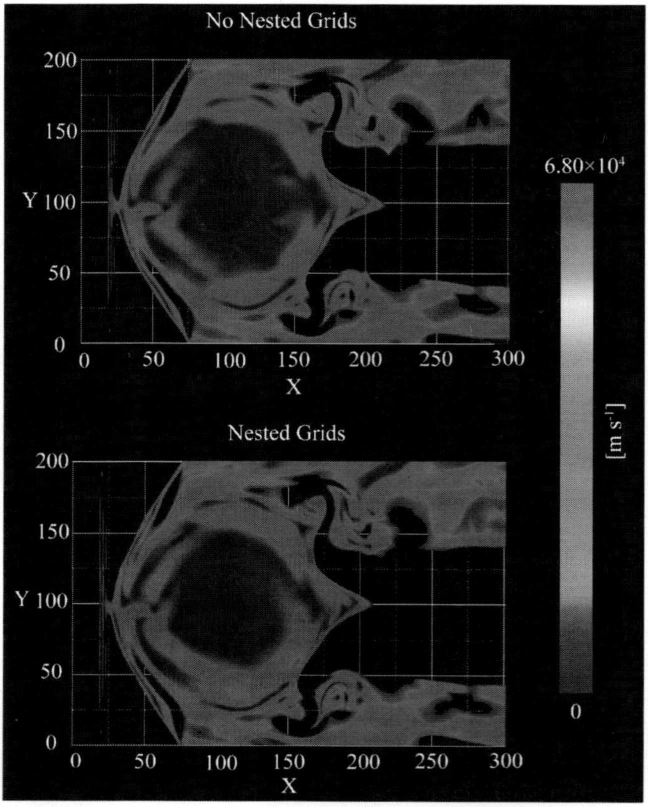

Fig. 5. Comparison of the results of the simulation with no nested grids with those of the simulation with nested grids.

around the inner boundary. On the other hand, when the nested grids were used, the distribution of velocity around the Earth was smooth, as can be observed in the lower panel of Fig. 5.

The calculation time of the simulation with nested grids was 1.2 times that of the simulation performed without nested grids. If the model without nested grids is assumed to have the same grid interval and timestep as the inner box of the nested grids model then its calculation time increases by 16 times. Therefore, it can be said that the nested grids model is, in a sense, an "eco" grids model that reduces calculation time.

The key stages in the simulation with nested grids are data transfer and interpolation. Thus, it appears that the nested grids model works as a connector between several simulation models. This implies that the nested grids model has the potential to play an important role in coupling various regions such as the magnetosphere and ionosphere.

4. Summary

In this study, we have developed a new model for real-time magnetosphere simulation using nested grids. All the calculations in this model are performed in parallel using the interpolation of conservation for $\nabla \cdot \mathbf{B} = 0$. Using the nested grids, the grid sizes are defined (manually) before the start of the simulation thus the load balance of calculation is easier to design in the case of nested grids than in AMR. The grids are connected to each other to ensure smoothing function. By applying the nested grids technique to the MHD magnetospheric model, the inner boundary can be set at $1\,\mathrm{R_E}$ with only a slight increase in the calculation time. Thus, this model is extremely suitable for real-time magnetosphere simulations.

Acknowledgments

This work was supported by a Grant-in-Aid from CREST, the Japan Science and Technology Agency (JST). Computational and visualization support was provided by the National Institute of Information and Communications Technology (NICT) and the Information Technology Center of Nagoya University through the joint research program of the Solar-Terrestrial Environment Laboratory, Nagoya University.

References

1. M. Den, *et al.*, *Space Weather* **4** (2006) S06004.
2. K. G. Powell, *et al.*, *J. Comp. Physics* **154** (1999) 284.
3. J. Brackbill and D. Barnes, *J. Comput. Phys.* **35** (1980) 426.
4. G. Tóth, The $\nabla \cdot \mathbf{B} = 0$ constraint in shock-capturing magnetohydrodynamics codes, *J. Comput. Phys.* **161** (2000) 605.
5. T. Shimizu and M. Ugai, *Trans. Jpn. Soc. Ind. Appl. Math.* **5**(4) (1995) 423.
6. G. Tóth and P. Roe, Divergence- and curl-preserving prolongation and restriction formulas, *J. Comput. Phys.* **180** (2002) 736.
7. D. Balsara, Second-order-accurate schemes for magnetohydrodynamics with divergence-free reconstruction, *Astrophys. J. Suppl. Ser.* **151** (2004) 149.
8. F. Mackay, R. Marchand and K. Kabin, Divergence-free magnetic field interpolation and charged particle trajectory integration, *J. Geophys. Res.* **111** (2006) A06208.

9. T. Sugimura, K. Takahashi and H. Sakagami, *Nagare* **25** 247 (2006).
10. T. Ogino, R. J. Walker and M. Ashour-Abdalla, *IEEE Trans. Plas. Sci.* **20**(6) (1992) 817.
11. K. Fukazawa, T. Ogino and R. J. Walker, *J. Geophys. Res.* **111** (2006) A10207.

Advances in Geosciences
Vol. 21: Solar & Terrestrial Science (2008)
Ed. Marc Duldig

THE FATE OF LANGMUIR WAVES: WIND AND STEREO OBSERVATIONS

P. J. KELLOGG
University of Minnesota, Minneapolis, MN, USA
kellogg@waves.space.umn.edu

The earth's foreshock has a rich collection of Langmuir waves useful for the study of their generating instabilities as well as processes important in electromagnetic wave generation as in type II and type III solar radio bursts. These processes include parametric decay to another Langmuir wave and an ion acoustic wave, possible other parametric processes, condensation to a long wavelength state when such decay is no longer possible, and trapping in pre-existing density depressions. Some processes suggested in the past, such as collapse, for example, have been shown by Wind observations not to take place in the Earth's foreshock. In his paper studies of Langmuir waves after they have been generated are presented.

1. Introduction

Langmuir waves occur prominently in space in two places, in the Earth's foreshock and in the electron streams that create Type III solar bursts. In both cases, the characteristic energy of the beams seems to be in the range 2 to 20 keV.[1,2] The Langmuir waves seem to be generated by a time-of-flight mechanism which depends on having regions of space which can only be reached by higher energy particles. The principle of this is shown for foreshock waves in Fig. 1, reproduced by permission from Filbert and Kellogg.[3] Note that there is no unique beam energy in this picture.

What happens to these waves once they have been generated? Sometimes they seem to persist for a long time, many hours. Examples are shown in Fig. 2, data from the Waves experiment on Wind.[4] In this figure, a type III burst is already in progress at the beginning of the day. It began at 2,200 on the previous day and is still visible at 1,200, so it appears that the generating Langmuir waves, whose frequency is slightly above the local plasma frequency as shown by the line of intermittent quasi-thermal noise, lasted for 14 hours. Another type III burst starts at about 1,545, and

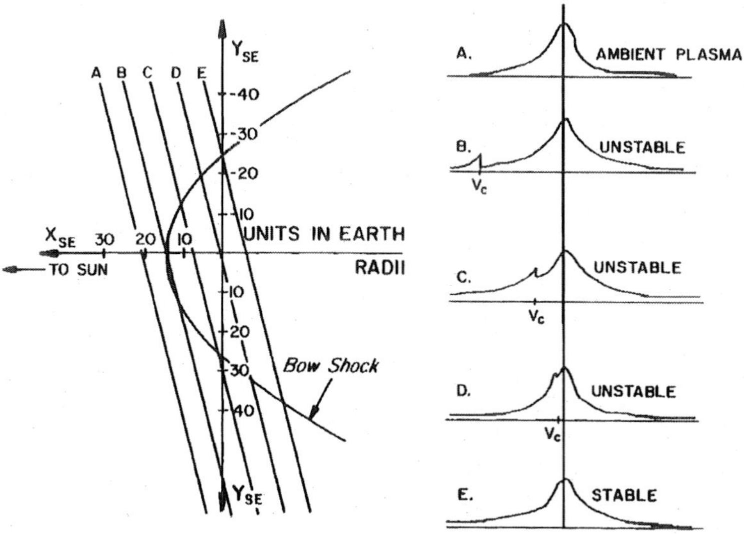

Fig. 1. Principle of generation of a bump-on-tail instability by time-of-flight. Reproduced by permission from Filbert and Kellogg.[3]

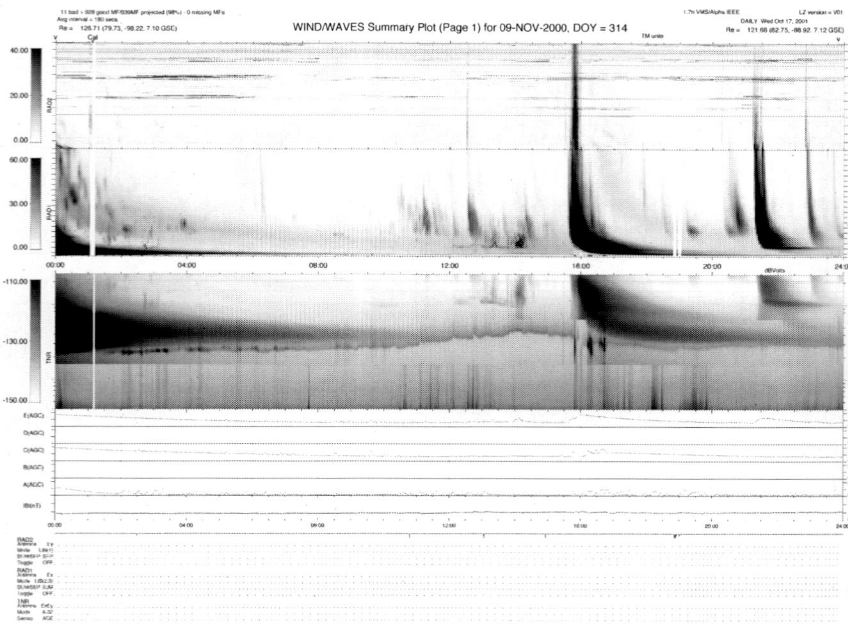

Fig. 2. Several type III solar radio bursts, illustrating long duration at certain frequencies.

leaves waves lasting about 4 hours at about 70 kHz. At about 2,115 a type III burst begins which terminates at about 100 kHz, probably by interaction with a shock as suggested by MacDowall.[5] These long duration emissions are somewhat rare. It may be that they are formed in an unusually turbulent plasma and the electromagnetic emission takes a long time to diffuse out of the generation region.[1]

2. Properties in a Uniform Plasma

If the solar wind plasma were uniform, a calculation of the wave vector of Langmuir waves generated by an instability can be made and some other properties after the Langmuir wave leaves the unstable region and finds itself in a plasma where the wave is stable can be estimated. The wave vector, k is taken to be resonant with the beam, velocity v_b:

$$kv_b = \omega \tag{1}$$

where ω is given by the dispersion relation

$$\omega^2 = \omega_p^2 + 3k^2 v_e^2 \tag{2}$$

with $v_e^2 = kT_e/m$.

The Landau damping rate is given by the well known:

$$\gamma = \omega_r \frac{\pi \omega_p^2}{2k^2} \frac{df_0(u)}{du}_{|u=\frac{\omega_r}{k}} \tag{3}$$

The damping of foreshock waves and Type III burst waves might be different because the foreshock waves are travelling toward the sun and will resonate with the back, the core of the ambient electrons, while the Type III waves will resonate with the more energetic halo and strahl. For the resonant k above, and a Maxwellian f_0, the Maxwellian exponent gives a factor $\exp(-(v_b/v_e)^2)$ which for a beam energy v_b of 2 keV and an electron temperature (halo) of 80 eV, gives $\exp(-25)$ and for higher beam energy of 20 keV and core temperature 10 eV gives $\exp(-2000)$. So this damping on ambient electrons is not very important for Type III beams and quite negligible for foreshock waves.

The phase velocity v_ϕ of the Langmuir waves is the beam speed, and the group velocity v_g satisfies:

$$v_\phi v_g = v_e^2 \tag{4}$$

so that the group speed for a 2 keV beam and 10 eV plasma temperature would be 130 km/s. This is relative to the plasma, so such foreshock Langmuir waves would be swept back to the earth, even if they were not also carried back by being trapped in density depressions. A more important difference is then that the foreshock waves are carried into a region where they will be damped by the higher energy parts of the beam distribution, but Type III waves are carried out of their region. Damping by this process seems to be one of the most important processes determining the fate of Langmuir waves.

3. Langmuir Waves in a Nonuniform Plasma

The solar wind is, however, seriously nonuniform in the sense that some things cannot be understood without considering the density fluctuations. Figure 3, taken from Kellogg[1] shows the relation between typical density fluctuations, expressed as plasma frequency fluctuations, and the difference between the plasma frequency and the resonant frequency for typical beam energies. Clearly Langmuir waves may be trapped, reflected if they go into

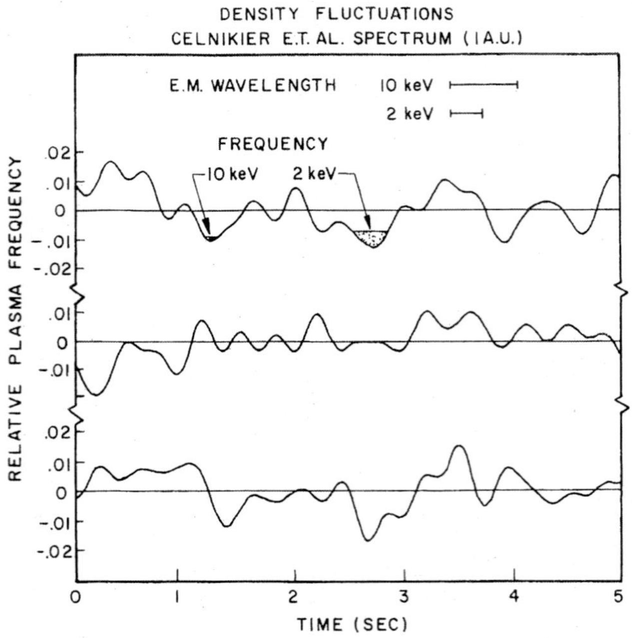

Fig. 3. Relation of observed plasma frequency variations to the resonant frequency for different beam energies. Reproduced by permission from Kellogg.[1]

regions of higher density, or may go into regions of lower density. The damping rate, which increases rapidly as the ratio of frequency to plasma frequency increases, will become larger for those waves going into lower density, and conversely, smaller for those approaching reflection.

Robinson, Cairns, and their group (Robinson and Cairns[6] and references therein) have used the randomness of the fluctuations and their effect on the growth rate of the generating instability to develop a theory, "Statistical Growth Theory" which has been successful in explaining many of the properties of these Langmuir waves. Here, however, we try to understand the various things that could happen to an individual wave. For starters, though, it seems likely that Langmuir waves tend to be generated at maxima or minima of the density, where the resonance at a fixed k will last longer. At minima the waves will be trapped. Trapping is observed. Figure 4 shows an intense soliton-like Langmuir wave, taken from

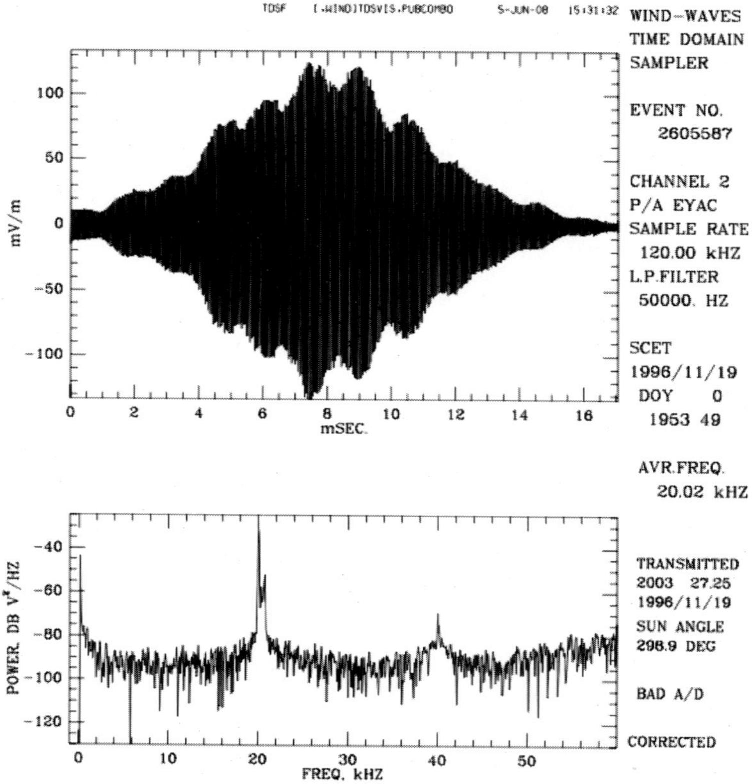

Fig. 4. Short Langmuir wave bursts, assumed to be trapped in density depressions.

Wind-Waves data. At one time it was thought that such solitons would have sufficient ponderomotive pressure to deepen the cavity, thus confining the Langmuir wave to a smaller space and increasing its amplitude, eventually leading to collapse and emission of energetic electrons. Kellogg *et al.*[7] have shown that the amplitudes required for such collapse do not occur, either in the Earth's foreshock or in Type III burst streams. Ergun *et al.*[8] have shown that the Langmuir waves occupy eigenmodes in density cavities which are presumed to be preformed.

As these eigenmodes are standing waves, representing oppositely directed waves, they will radiate electromagnetic radiation at the harmonic, 2 f_p by a nonlinear process. The time for this to occur, evaluated from Eq. 20 in Smith[9] is of the order of 10^4 sec for an amplitude of 100 mV/m, and proportional to $1/E^2$. This means that the amplitude will decrease, but the soliton will not completely disappear in a reasonable time. At 30 mV/m for example, the lifetime would be 10^5 sec, comparable to times in Fig. 2.

4. Decay

An important process which limits the amplitudes is decay of Langmuir wave into an oppositely directed wave and an ion acoustic wave. This happens, as shown in Fig. 5, data from the SWaves experiment on STEREO.[10] Events of this type are interpreted to be the beating of a Langmuir wave with another wave due to decay of the parent wave to a daughter Langmuir wave and an ion acoustic wave. In SWaves, the frequency response of the Time Domain Sampler was correctable and useable down to 10–20 Hz so that the daughter ion acoustic wave is also recorded. The ion acoustic wave is difficult to observe nevertheless, since it is quickly damped if it leaves the unstable region. The top panel shows the full signal on one antenna. By filtering out the Langmuir frequencies, the ion acoustic wave is displayed in the second panel, in its generation region. The lower two panels show the full frequency spectrum and the ion acoustic part of the spectrum.

The antennas respond to both density changes and to electric fields. For low frequency ion acoustic waves, the response to density is much larger than the response to the electric field of the ion acoustic wave. But there is also seen the electric field response to the ponderomotive force of the Langmuir waves, which causes the apparent broad minimum in the traces.

Since the Langmuir waves and the ion acoustic wave have comparable wave vector magnitudes, unlike the case for decay into an ion acoustic wave

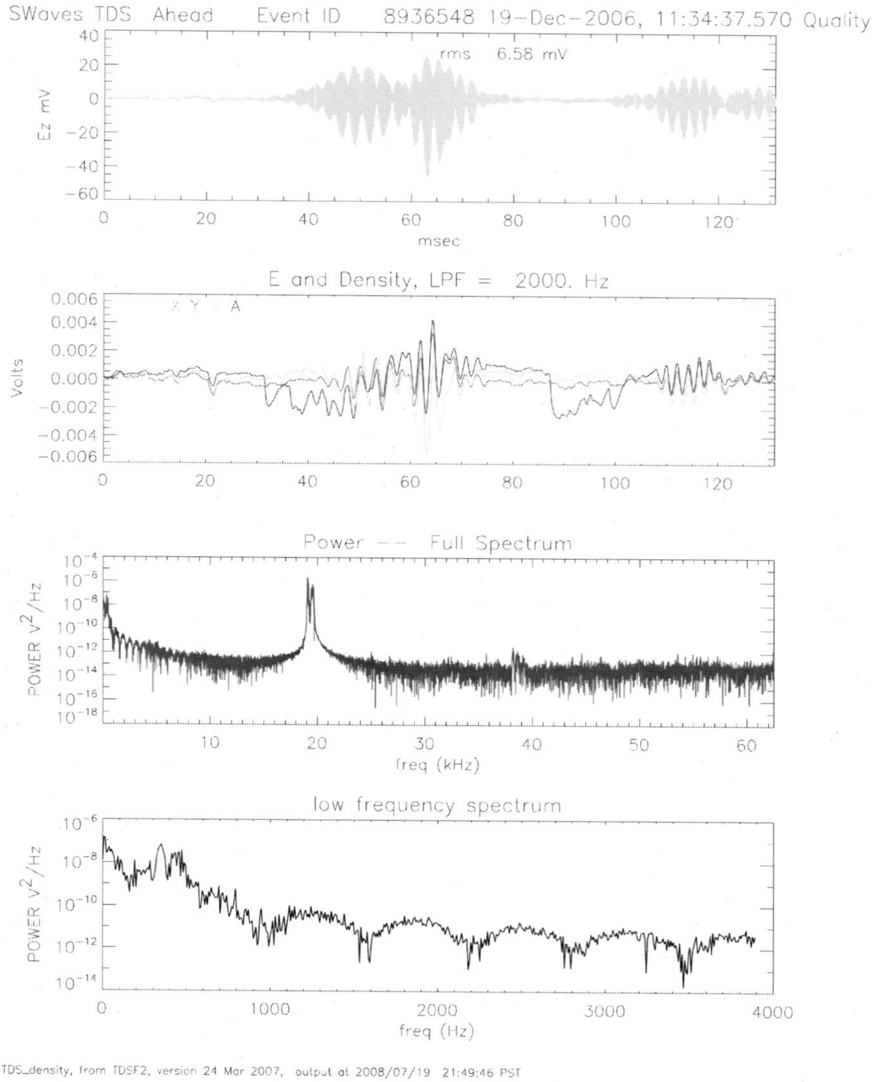

Fig. 5. Strongly modulated Langmuir wave, interpreted as decay.

and an electromagnetic wave, the daughter wave may have a wide range of directions. Hence the daughter Langmuir wave may not resonate with the remainder of the beam and then may persist for a long time. Figure 6 shows a hodogram of a Langmuir wave at a large angle to the magnetic

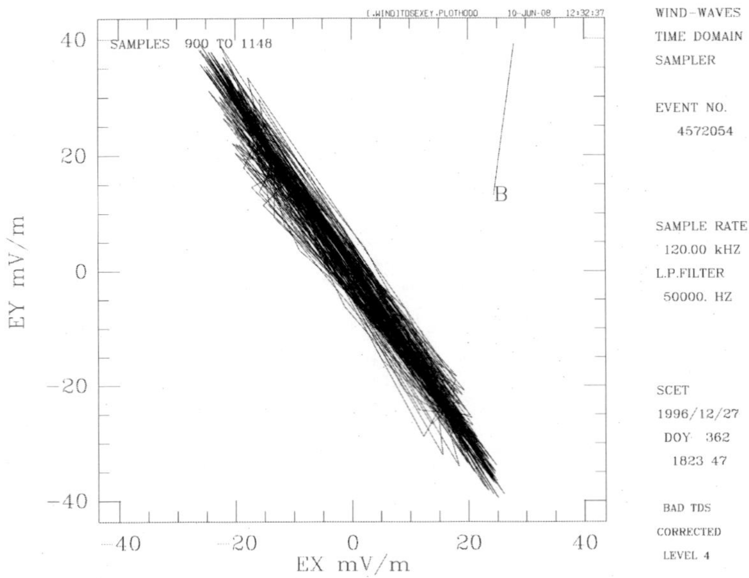

Fig. 6. Hodogram showing non-field-aligned Langmuir wave.

field. The time series looks just like any other foreshock Langmuir wave, but the hodogram shows that E makes a large angle to B.

5. Angular Distributions

A significant fraction of observed Langmuir waves make large angles with the magnetic field. Figure 7 shows histograms of the angles of a number (554) of Langmuir waves, from STEREO SWaves data. There is a central peak, as expected for waves generated parallel to B, which has a width (variance) of about 7 deg. There are 186 Langmuir waves at angles greater than 18 deg., ∼3 standard deviations, which we interpret either as daughter waves from the decay process or as Langmuir waves obliquely reflected from a density increase, but in any case, they should be out of the damping resonance. The width of the central peak is larger than one might expect from the dispersion relation with a beam. I believe this is due to Langmuir waves which have propagated to regions where the direction of B is different from that of their birth.

It is somewhat surprising that there are so few large angle waves, only a third of the total, given that their lifetimes should be longer, as they

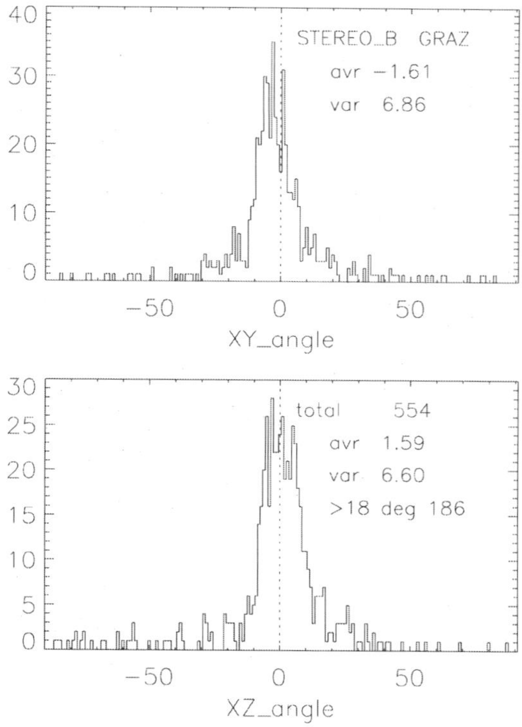

Fig. 7. Angular Distribution of foreshock Langmuir waves.

are taken out of resonance with the decreasing distribution function of the beam.

6. Conclusions

It is shown that some foreshock waves are out of resonance. Such foreshock and type III Langmuir waves may be long lived and may be found at some distance from their place of generation. The formula for Landau damping implies that foreshock waves might be rapidly damped by interaction with higher energy parts of the beam which created them, but the data indicate that this is not important. There are many waves which appear to be resonant also.

About 1/3 of the waves observed have electric fields at large angles to the magnetic field, indicating that they have become out of resonance, probably either because they have been reflected by a density increase or because they are the daughter waves of a parametric decay.

It is expected that many waves are trapped in density depressions. These will radiate by emission of electromagnetic waves at 2 f_p, but this will only reduce their amplitude to a few tens of mV/m, and will not eliminate them entirely. At such amplitudes, they can persist for a long time.

Acknowledgments

This work was partially supported by the National Aeronautics and Space Administration under grants NNX07AM97G and NNX07AF23G.

References

1. P. J. Kellogg, Observations concerning the generation and propagation of Type III solar bursts, *Ast. and Astrophys.* **169** (1986) 329.
2. S. D. Bale, D. E. Larson, R. P. Lin, P. J. Kellogg, K. Goetz and S. J. Monson, On the beam speed and wavenumber of intense electron plasma waves near the foreshock edge, *J. Geophys. Res.* **105**(A12) (2000) 27353.
3. P. C. Filbert and P. J. Kellogg, Electrostatic Noise at the Plasma Frequency Beyond the Earth's Bow Shock, *J. Geophys. Res.* **84** (1979) 1369.
4. J.-L. Bougeret, M. L. Kaiser, P. J. Kellogg, R. Manning, K. Goetz, S. J. Monson, N. Monge, L. Friel, C. A. Meetre, C. Perche, L. Sitruk and S. Hoang, WAVES: The radio and plasma wave investigation on the Wind spacecraft, *Space. Sci. Rev.* **71** (1995) 231.
5. R. J. MacDowall, Effects of interplanetary shocks on kilometric type III radio bursts, *Geophys. Res. Lett.* **16** (1989) 923.
6. P. A. Robinson and I. H. Cairns, Stochastic growth of localized plasma waves, *Physics of Plasmas* **8** (2001) 2394.
7. P. J. Kellogg, K. Goetz, S. J. Monson and S. D. Bale, A search for Langmuir solitons in the Earth's foreshock, *J. Geophys. Res.* **104**(A4) (1999) 6751.
8. R. E. Ergun *et al.*, Eigenmode Structure in Solar Wind Langmuir Waves, *Phys. Rev. Lett.* **101** (2008) 051101.
9. D. F. Smith, Second Harmonic Radiation and Related Nonlinear Phenomena in *Type III Solar Radio Bursts* **216** (1977) L53.
10. J. L. Bougeret *et al.*, S/WAVES: The radio and plasma wave investigation on the STEREO mission, *Space Sci. Rev.* **136** (2008) 487–528.

Advances in Geosciences
Vol. 21: Solar & Terrestrial Science (2008)
Ed. Marc Duldig

FORESHOCK TURBULENCE

YASUHITO NARITA
Institute of Geophysics and Extraterrestrial Physics,
Technical University Braunschweig,
Mendelssohnstr. 3, D-38106 Braunschweig, Germany
y.narita@tu-bs.de

Dispersion relations, wave number spectra, and a bispectrum are presented using measurements of four Cluster spacecraft in the Earth's foreshock region. These analyses suggest that the foreshock is in various turbulent states.

1. Introduction

Foreshock is a region upstream of the collisionless shock wave connected by the magnetic field. This region is typically characterized by large amplitude fluctuations of magnetic field and plasma and serves as a natural laboratory of plasma turbulence. The ultimate origin of the fluctuations in the foreshock is the ions that are specularly reflected at the shock, which provides a shock dissipation mechanism in collisionless plasma. The reflected ions form a beam streaming along the magnetic field toward upstream against the incoming flow. Therefore in the foreshock one finds a counter-streaming ion beam configuration, the incoming and the reflected ions,[1] which is unstable to various kinds of instabilities, exciting electromagnetic waves and scattering particles. One of the most effective instabilities in such an ion beam system is the right-hand ion beam resonance that excites right-handed waves (the same rotation sense as the electron gyration around the magnetic field) propagating parallel or anti-parallel to the magnetic field.[2] This mode has a transition to the fast magnetosonic mode in the low frequency, magnetohydrodynamic limit and the whistler mode in the high frequency limit.

Recent developments of multi-point measurements in space provide a great deal of information on physics of waves and turbulence in the foreshock. With multiple measurement points one can determine not only spatial properties of wave structures, wavelengths, but also dispersion relations, wave number spectra, and bispectra directly in the wave number domain. They are useful analysis tools to study what kind of turbulence states the foreshock reaches. Here we present measurements of the four Cluster spacecraft[3] in the Earth's foreshock region.

2. Dispersion Relations

One of the applications of the multi-point measurements is to determine dispersion relations directly from observations. Examples of the dispersion analysis can be found in Refs. [4–5]. We determine the wave power as a function of frequency and wave vector $E(\omega, \vec{k})$, where ω and $\vec{k}$ denote the spacecraft frame frequency and the wave vector, respectively. The wave telescope method[7–9] enables to compute the power $E(\omega, \vec{k})$. By searching peaks of wave energy distribution in the frequency and wave vector domain, and identifying the frequencies and the wave vectors associated with the peaks, one can experimentally determine the dispersion relation. Furthermore, the frequencies can be transformed into the plasma rest frame frequency ω_{re} (co-moving with the plasma bulk flow) using the Doppler relation $\omega_{re} = \omega - \vec{k} \cdot \vec{V}$, where $\vec{V}$ is the flow velocity vector.

We apply the dispersion analysis to the measurements of magnetic field fluctuation in the foreshock provided by the Cluster spacecraft.[10] Fig. 1 displays three distinct examples of the dispersion relations for the wave numbers in the direction to the magnetic field, $\omega_{re}(k_z)$. The frequencies and the wave numbers are normalized to the proton cyclotron frequency, and the inverse of the ion inertial length, respectively.

In the case A (Feb. 16, 2002, 0700-0730 UT) almost all of the waves are identified in the direction away from the shock. Only few waves propagate in the opposite direction at very low frequencies ($\omega_{re} \simeq 0$, here frequency is normalized). The dispersion branch starts at $(\omega_{re}, k_z) = (0, 0)$ and extends solely in the anti-parallel direction, keeping the phase speed almost at the Alfvén speed V_A at low frequencies, $\omega_{re}/k_z \simeq V_A$. At $\omega_{re} \sim 0.5$ the dispersion branch starts to be bent and deviates from the linear branch toward higher frequencies. This is characteristic to the low frequency part of the magnetosonic/whistler waves.

The case B (Apr. 27, 2002, 0200-0230 UT) exhibits waves that are counter-propagating at low frequencies. Most of the identified waves propagate in the $+z$ direction at various frequencies ($\omega_{re} \leq 3$), but some waves propagate in the opposite direction at low frequencies ($\omega_{re} \leq 0.4$).

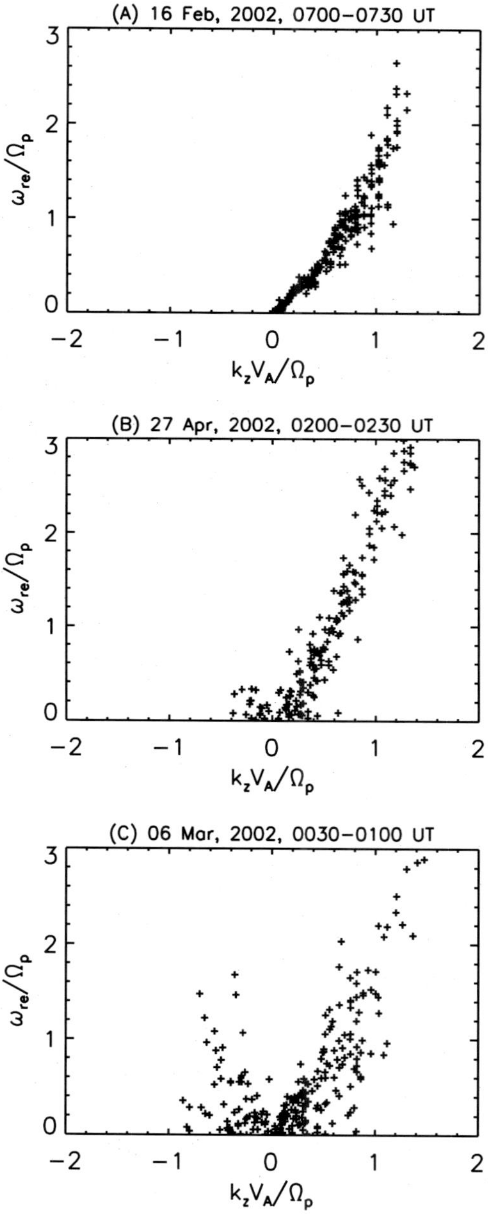

Fig. 1. Dispersion relations in the plasma rest frame. Wave numbers are aligned with the mean magnetic field. The $+z$ direction is away from the shock toward upstream.[6]

The dispersion branch in the $+z$ direction, can be approximated by a straight line (linear dispersion relation), but the phase speed does not agree with the Alfvén speed ($\omega_{re} \geq k_z V_A$). In the $-z$ direction it is not clear if the waves follow any dispersion relation, since only few waves are identified.

The case C (Mar. 6, 2002, 0030-0100 UT) exhibits an example of the enhanced counter-propagating waves. Compared to the case B, the identified waves look rather scattered in the dispersion diagram. One branch starts at $(\omega_{re}, k_z) \simeq (0, 0)$ and extends in the $+z$ direction to $(\omega_{re}, k_z) = (3, -1.5)$, while the branch in the $-z$ direction stops at $\omega_{re} \simeq 1.5$. The scatter in the dispersion diagram makes it difficult to identify the dispersion relation, but roughly speaking, the waves follow the magnetosonic/whistler branch.

The dispersion relation in the case A shows that fluctuations represent a set of waves on the magnetosonic/whistler branch, whereas those in the cases B and C exhibit deviations from this branch and scattering of wave components in the dispersion diagram, suggesting that waves of non-normal mode exist. The fluctuations in the foreshock are therefore case dependent — in some cases they represent a set of normal wave modes but in other cases they are a mixture of normal and non-normal modes.

3. Wave Number Spectra

The second application of the multi-point measurements is the wave number spectra (energy spectra in the wave number domain). Examples are found in Refs. [6, 11, 12]. We determine two kinds of wave number spectra: $E^+(k_z)$ and $E^-(k_z)$. They represent the energy for the magnetic field fluctuations perpendicular to the background field, propagating along the magnetic field away from the shock (E^+), and toward the shock (E^-), respectively.

Figure 2 displays the wave number spectra for the three cases A, B, and C in Fig. 1. The wave numbers are again scaled to the ion inertial length and the energy spectra are scaled to $E_0 = B_0^2 V_A / \Omega_p$. Qualitatively the three wave number spectra exhibit common features, while they are quantitatively different from one another. There are two characteristics in common. One is that E^+ (solid line) is always greater than E^- (dotted line), that is the foreshock fluctuations contain more energy in the waves propagating away from the shock than that for the waves toward the shock. Another point is that the spectra exhibit humps, which makes identification of an inertial range (power-law spectrum over decades of wave numbers) difficult. One interpretation is that the humps are an indication of the

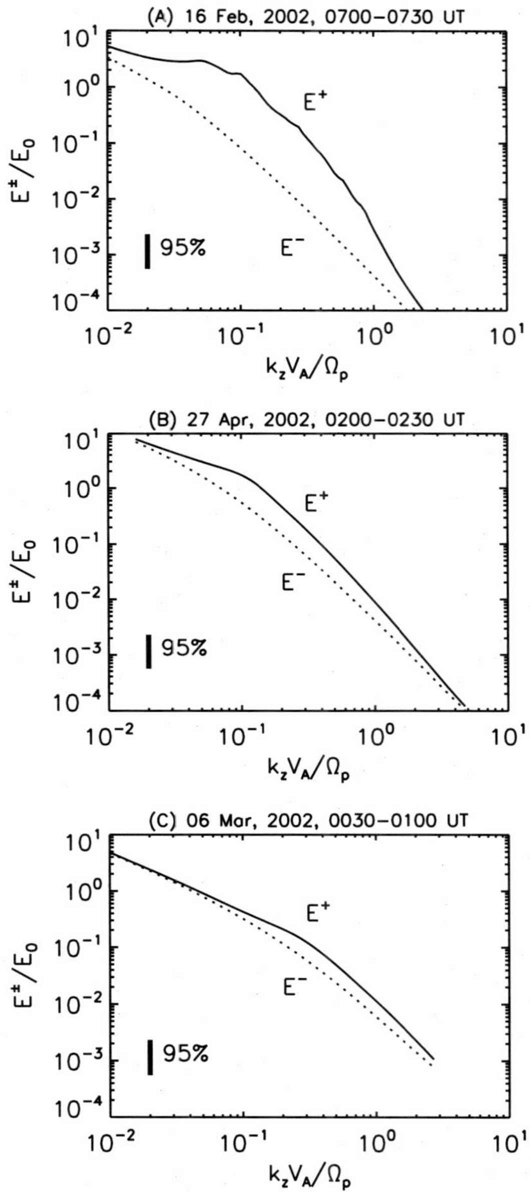

Fig. 2. Energy spectra of magnetic field fluctuations perpendicular to the mean field in the wave number domain. The wave numbers are in the direction away from the shock (solid curve) and toward the shock (dotted curve).[6]

energy injection on intermediate scales. In the foreshock region shock-reflected ions excite waves by the beam instabilities. The humps can be seen only in E^+, while E^- shows only monotonously decaying curves toward the larger wave numbers. The injection scale is, however, slightly different from case to case. It becomes shifted gradually toward the larger wave numbers from the case A to the case C. A closer inspection yields the values of the normalized wave numbers for the injection scales: $k_z \approx 0.05 - 0.1$ (case A); $0.1 - 0.2$ (case B); and 0.3 (case C). The difference between E^+ and E^- becomes also smaller from the case A to the case C.

Various interpretations are possible in the wave number spectra. One is that the fluctuations are superposition of fluctuations in the solar wind and the foreshock, and the energy cascade process, if any, is altered due to the foreshock wave excitation. On the other hand, two-dimensional turbulence may also explain different energy cascade processes at once.

4. Bispectrum

As the dispersion relations in Fig. 1 display, the foreshock fluctuations are not monochromatic (i.e. single wave) but a mixture of different waves. One idea to explain the existence of multiple waves is wave-wave couplings. The concept of bispectrum is useful to investigate the wave coupling processes in the observations. The bispectrum is given as a statistical average of product of three fields, $\langle A_1 A_2 A_3^* \rangle$, where $A_1, \ldots, A_3$ denote the wave fields and the asterisk the complex conjugate. The bispectrum becomes non-zero if the three waves are in resonance in the frequency and the wave number domain, e.g. $\omega_1 + \omega_2 = \omega_3$ and $k_1 + k_2 = k_3$.

Bispectrum is one of the methods in higher order statistics and of great importance in studying nonlinear processes of turbulence. They reflect the shape of probability distribution functions of fluctuations. Theoretical treatments of turbulent fields often assume the Gaussian or almost-Gaussian distribution (called the quasi-normal approximation) to solve the closure problem of turbulence. In the frame of weak turbulence the breakdown of the Gaussian distribution can be interpreted as a sign of phase coherence, e.g. coherent three-wave or four-wave couplings. Estimating the bispectrum provides the very means to measure the magnitude of three-wave couplings (See Ref. 13). Gaussian features in the probability distribution function of fluctuation assume the existence of random phase, i.e. multi-wave couplings may occur but the phases of the generated waves soon become random or incoherent. In laboratory and space plasma physics the method of the bispectrum has been widely applied: in ionospheric radio emission;[14] in laboratory plasmas;[15–17] and upstream waves of the Earth's quasi-parallel bow shock.[18] The bispectrum investigations have been made solely in the frequency domain, as one uses single point measurements and therefore time series data. Also, Taylor's hypothesis is often applied in a flowing medium and the bispectra in the frequency domain reflect that in the wave number domain. However, the wave-wave couplings should be examined in the wave number domain, too, which motivates us to estimate the bispectrum directly in the wave number domain using only a limited number of measurement points. For example, with Taylor's hypothesis one cannot measure the wave number perpendicular to the flow.

The bispectrum estimator in the wave number domain is applied to the fluctuation in the Earth foreshock region in order to examine if wave–wave coupling processes are present.[19] We use the magnetic field measurement[10] and the electron density measurement[20] of Cluster on Feb. 18, 2002, 0805–0835 UT. We use the perpendicular component of the magnetic field and the electron density fluctuation associated with wave vectors parallel to the magnetic field. The result is displayed in Fig. 3. A peak is identified at the pair of wave numbers about $k_b = 3.5 \times 10^{-4}\,\mathrm{km}^{-1}$ and

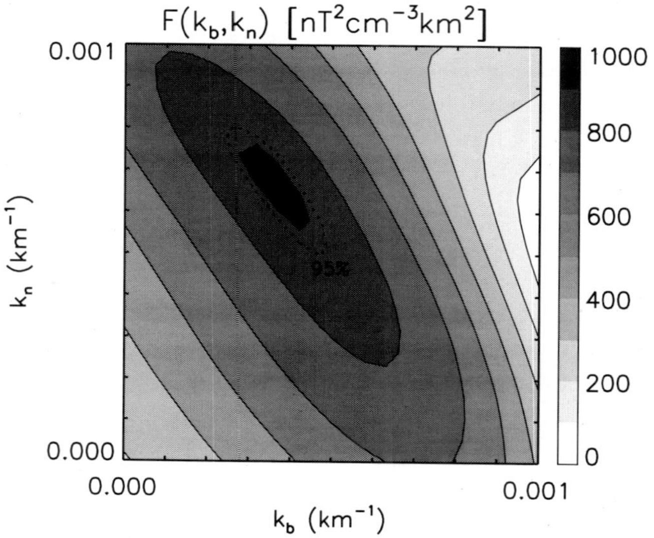

Fig. 3. Bispectrum in the wave number domain. The x- and y-axis are the wave number of magnetic field and electron density fluctuation, respectively. The dotted line is 95% confidence level of the peak.[19]

$k_n = 6.5 \times 10^{-4}\,\mathrm{km}^{-1}$ for the magnetic field and the density fluctuations, respectively. The distribution of the bispectrum is elongated and tends to be aligned to a resonance line (straight line with the slope -1).

The bispectral analysis provides useful information about wave–wave interactions. However, one should be careful in that the bispectrum is a necessary but not a sufficient condition for having three-wave interactions. A nonlinearity in the detector, for example, may give rise to a non-zero bispectrum,[21] or one may have four-wave interactions. Also, the higher order moments may be remnants of wave–wave interactions that took place somewhere upstream and were convected by the flow. However, the bispectrum method can successfully identify a peak within the limit of the confidence level, which is an indication that the three-wave coupling may really be present in the foreshock. The bispectrum study with the Cluster data set is an on-going study and the physical interpretation of Fig. 3 needs more close studies (detailed comparison with the dispersion relation, bispectrum in the frequency).

5. Summary

Multi-point measurements provide powerful tools to study wave and turbulence phenomena. Although the number of measurement points is too small to perform the Fourier transform from the spatial coordinate into wave numbers, it is possible to derive dispersion relations, energy spectra, and bispectra. The Cluster observations suggest that the foreshock is in various turbulent states. Some represent a set of linear, normal modes with a clear dispersion curve but some include non-normal modes and the dispersion relation becomes distorted. Most of wave components propagate in the direction away from the shock. The bispectrum in the wave number domain gives a hint that wave-wave interactions are indeed occurring. These tools will be used to compare the observations with models of wave-wave interactions (decay and modulational instabilities).[22–24]

Acknowledgments

This work was financially supported by Bundesministerium für Wirtschaft und Technologie and Deutsches Zentrum für Luft- und Raumfahrt, Germany, under contract 50OC0103.

References

1. G. Paschmann, N. Sckopke, I. Papamastorakis *et al.*, *J. Geophys. Res.* **86**, (1981) 4355.
2. S. P. Gary, *Theory of Space Plasma Microinstabilities* (Cambridge, 1993).
3. C. P. Escoubet, M. Fehringer and M. Goldstein, *Ann. Geophys.* **19** (2001) 1197.
4. Y. Narita, K.-H. Glassmeier, S. Schäfer *et al.*, *Geophys. Res. Lett.* **30** (2003) SSC 43.
5. Y. Narita and K.-H. Glassmeier, *J. Geophys. Res.* **110** (2005) A12215.
6. Y. Narita, K.-H. Glassmeier, M. Fränz *et al.*, *Nonlin. Processes Geophys.* **14** (2007) 361.
7. J. L. Pinçon and F. Lefeuvre, *J. Geophys. Res.* **96** (1991) 1789.
8. U. Motschmann, T. I. Woodward, K.-H. Glassmeier *et al.*, *J. Geophys. Res.* **101** (1996) 4961.
9. K.-H. Glassmeier, U. Motschmann, M. Dunlop *et al.*, *Ann. Geophys.* **19** (2001) 1439.
10. A. Balogh, C. M. Carr, M. H. Acuña *et al.*, *Ann. Geophys.* **19** (2001) 1207.
11. Y. Narita, K.-H. Glassmeier and R. A. Treumann, *Phys. Rev. Lett.* **97** (2006) 191101.
12. Y. Narita, K.-H. Glassmeier, S. P. Gary *et al.*, *J. Geophys. Res.* submitted (2008).
13. T. Dudok de Wit, *Space Plasma Simulation*, (eds.) J. Büchner, C.T. Dum, and M. Scholer (Springer, 2003).
14. D. Lagoutte, F. Lefeuvre and J. Hanasz, *J. Geophys. Res.* **94** (1989) 435.
15. Y. C. Kim and E. J. Powers, *IEEE Trans. Plasma Sci.* **PS-7** (1979) 120.
16. Ch. P. Ritz, E. J. Powers and R. D. Bengtson, *Phys. Fluids B* **1** (1989) 153.
17. F. Brochard, T. Windisch, O. Grulke and T. Klinger, *Phys. Plasmas* **13** (2006) 122305.
18. T. Dudok de Wit and V. V. Krasnosel'skikh, *Phys. Plasmas.* **2** (1995) 4307.
19. Y. Narita, K.-H. Glassmeier, P. M. E. Décréau *et al.*, *Ann. Geophys. Comm.* **26** (2008) 3389.
20. P. M. E. Décréau, P. Fergeau, V. Krasnoselskikh *et al.*, *Ann. Geophys.* **19** (2001) 1241.
21. S. N. Walker, M. A. Balikhin, I. Bates and R. Huff, *Adv. Space Res.* **30** (2002) 2815.
22. M. Longtin and B. Sonnerup, *J. Geophys. Res.* **91** (1986) 798.
23. T. Terasawa, M. Hoshino, J.-I. Sakai and T. Hada, *J. Geophys. Res.* **91** (1986) 4171.
24. H. K. Wong and M. L. Goldstein, *J. Geophys. Res.* **91** (1986) 5617.

Advances in Geosciences
Vol. 21: Solar & Terrestrial Science (2008)
Ed. Marc Duldig

THE PREDOMINANT FREQUENCIES OF THE ATMOSPHERIC PERTURBATIONS AND GEOMAGNETIC OSCILLATIONS

M. MATSUMURA*, T. IYEMORI, M. NOSE,
M. UTSUGI and Y. ODAGI
Graduate School of Science,
Kyoto University, Kyoto, Japan
**mitsuru@kugi.kyoto-u.ac.jp*

Y. TANAKA
Faculty of Engineering,
Setsunan University, Neyagawa, Japan

D. HAN
Polar Research Institute of China,
Shanghai, China

N. OSHIMAN
Disaster Prevention Research Institute,
Kyoto University, Uji, Japan

H. SHINAGAWA
National Institute of Information and Communications Technology,
Koganei, Japan

Y. TABATA
Research Institute for Sustainable Humanosphere,
Kyoto University, Uji, Japan

It has been expected that the atmospheric perturbation causes the acoustic resonance between the lower atmosphere and the thermosphere. The main resonance frequencies theoretically reported are around 3.7, 4.4-4.6 and 5.1-5.3 mHz. We calculated mean power spectral densities and the occurrence rates of spectral peaks for the pressure perturbations and geomagnetic variations and found that the local maxima appeared in both data sets around the resonance frequencies. These maxima appear most clearly in August to October for pressure and in every season for geomagnetic fields. These results confirm the existence of the acoustic resonance coupling between the lower atmosphere and the ionosphere and suggest that the geomagnetic oscillations are frequently caused not only through magnetospheric plasma processes but also by lower atmospheric disturbances.

1. Introduction

Acoustic gravity waves generated in the lower atmosphere can propagate to the thermosphere and cause the resonance between the lower atmosphere and the thermosphere. The main resonance frequencies theoretically reported are around 3.7, 4.4–4.6 and 5.1–5.3 mHz.[1–4] Since the amplitude of acoustic gravity waves is inversely proportional to the square root of the mass density of the atmosphere, even a small atmospheric perturbation near the surface results in a considerably large displacement in the thermosphere. Therefore, the acoustic gravity waves could cause appreciable electromagnetic variations through ionospheric processes.

A low frequency Pc 5 pulsation was observed at Phimai in Thailand, shortly after the origin time of the Sumatra earthquake on December 26, 2004.[5] The localized nature and the frequency of oscillations suggest that the low frequency magnetic pulsation was generated by dynamo action in the ionosphere, which was caused by the acoustic resonance set up by the earthquake.

However, the resonance itself has not been confirmed directly with atmospheric pressure observations.

2. Observations

Triggered by the magnetic pulsation event at the 2004 Sumatra earthquake, we have started atmospheric pressure observation since the summer of 2006 at Shigaraki (34.8°N, 136.1°E), Japan. The specification of the barometric sensor resolution is about 1.6Pa and the output signal is A/D converted with 490Hz sampling and 0.4Pa resolution. One second averaged data are recorded, and the one second sampled data from a fluxgate magnetometer at Aso (32.9°N, 131°E) are used for comparison.

In this paper data in 2007 are analyzed, though the data of magnetic components are missing from September 27 to December 11.

For removal of low frequency variations a high-pass filter of Gaussian type is applied with a cutoff frequency around 1.5 mHz. Figure 1 shows an example of filtered data of atmospheric pressure and magnetic H and D components on August 21, 2007 for 05:00UT-09:00UT. The top panel

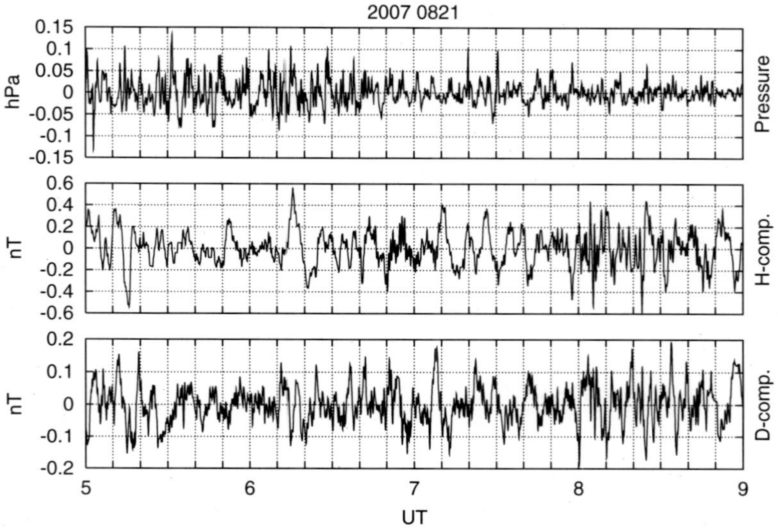

Fig. 1. Filtered data of atmospheric pressure and magnetic H and D components on August 21, 2007, 05:00UT-09:00UT. The top panel shows the variation of the atmospheric pressure perturbation, and the middle and the bottom show those of magnetic H and D components, respectively.

shows the variation of the atmospheric pressure perturbation, and the middle and the bottom show those of magnetic H and D components, respectively.

Figure 2 shows power spectral density (PSD) calculated with MEM (Maximum Entropy Method) from the data on the same day for 06:30UT-07:30UT. The maximum lag in MEM calculation is set 1200 (i.e. 20 minutes). The same method and parameters are used for spectral calculation throughout this paper. The top panel shows the PSD of the atmospheric pressure perturbation, and the middle and bottom show those of magnetic H and D components, respectively. Although there is no clear correlation between the wave forms shown in Fig. 1, common spectral peaks appear at about 3.7, 4.7, 5.7 and 6.6 mHz, and the first two frequencies correspond to the theoretically reported resonance frequencies.

In order to check whether such phenomena appear statistically or not, the PSDs are averaged in logarithmic scale, as shown in Fig. 3.

The data for one hour are used to calculate each spectrum. The left panel shows that of the pressure perturbations, and the middle and the right show those of magnetic H and D components, respectively. In each panel,

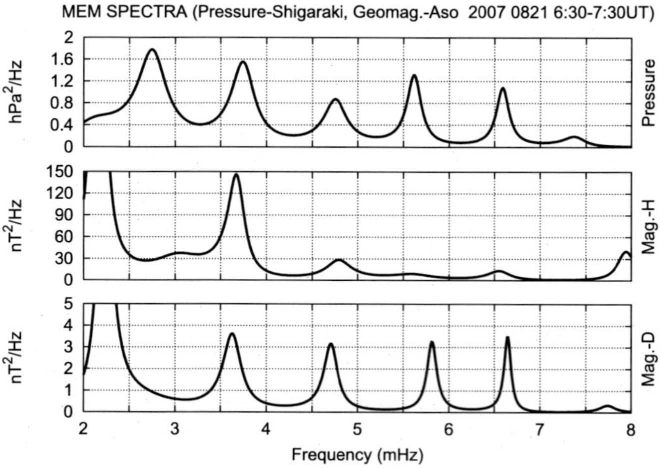

Fig. 2. The power spectra calculated with MEM (Maximum Entropy Method) for the data on August 21, 2007, 06:30UT-07:30UT. The top panel shows the spectrum of the atmospheric pressure perturbation, and the middle and the bottom show those of magnetic H and D components, respectively. Common peaks appear around theoretically expected frequencies of fundamental mode (3.7 mHz) and the first overtone (4.4–4.6 mHz). See, e.g., Nishida (2000).

PSDs in twelve months are shown. To avoid overlapping of the lines, PSD for each month is shifted by 10^1 from the next month. Error bars show standard error, i.e. standard deviations divided by the square root of the number of observations. The shadings show the theoretically reported resonance frequencies. Peaks appear around 3.7 mHz, one of the theoretically expected frequencies mainly in August, September and October for pressure, though their power is within the standard error. Clear peaks are not seen for magnetic components.

Next, local time variations of PSDs are investigated. Figure 4 to 15 show local time variations of PSDs for the pressure perturbations in January to December in 2007. The figure in each month shows PSDs in every hour. during day (9:00LT-15:00LT), evening (15:00LT-21:00LT), night (21:00LT-3:00LT) and morning (3:00LT-9:00LT). Clear peaks around 3.7 mHz in August to October, shown in Fig. 3, appear all day long and the frequency of peaks does not change.

These figures show that PSDs at all frequencies are enhanced in daytime and PSDs are lower in June to October than those in other months.

To show the existence of the resonance peaks clearly for magnetic components too, we developed a software to search spectral peaks and

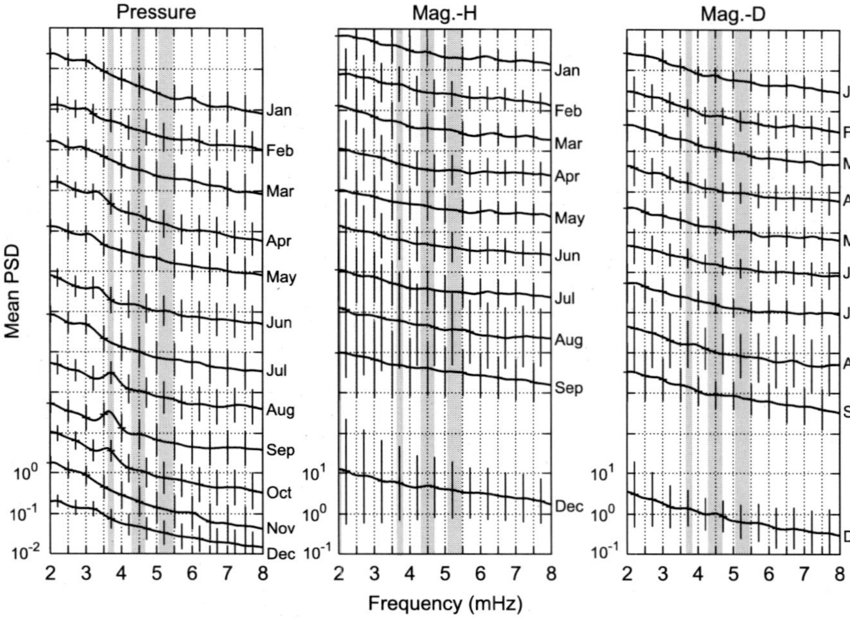

Fig. 3. The mean power spectral density of the pressure perturbations in logarithmic scale. The left panel shows that of the pressure perturbations, and the middle and the right show those of magnetic H and D components, respectively. For magnetic fields, the spectra in the case of geomagnetic ap index ≤ 9 are used to decrease the effect of geomagnetic pulsation caused by magnetospheric plasma processes. In each panel, spectra in twelve months are shown. To avoid overlapping of the lines, PSDs for each month is shifted by 10^1. Error bars show standard error, i.e. standard deviations divided by the square root of the number of observation. The shadings show the theoretically reported resonance frequencies. Peaks appear around the 3.7 mHz, one of the theoretically expected frequencies, mainly in August, September and October for pressure.

calculate the occurrence rate of spectral peaks of the pressure perturbations and geomagnetic oscillations. The procedure of searching the spectral peaks is as follows: The interval between a point and the next is 0.01 mHz, i.e. the PSDs are calculated at frequencies of every 0.01 mHz. Among continuous five points, if only the center point is a local maximum and its power is 10% greater than the average of the two side end points, i.e. the average of the first and the fifth points, it is defined as a peak. The occurrence rate is defined as the total number of peaks at the frequency divided by the total number of calculated spectra.

The occurrence rates were smoothed by applying low-pass filter of Gaussian type with a cutoff frequency around 0.7 mHz.

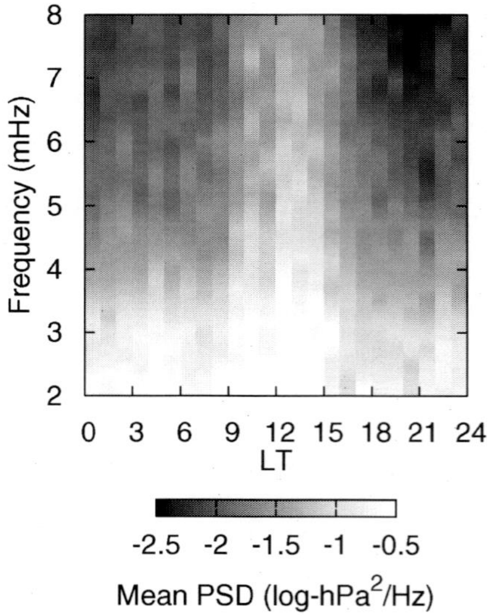

Fig. 4. January, 2007.

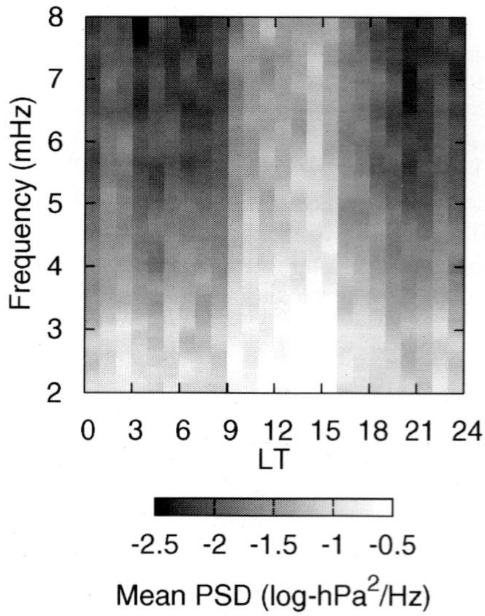

Fig. 5. February, 2007.

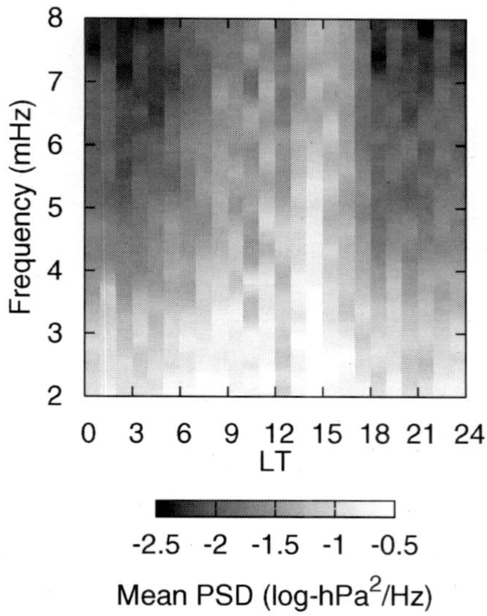

Fig. 6. March, 2007.

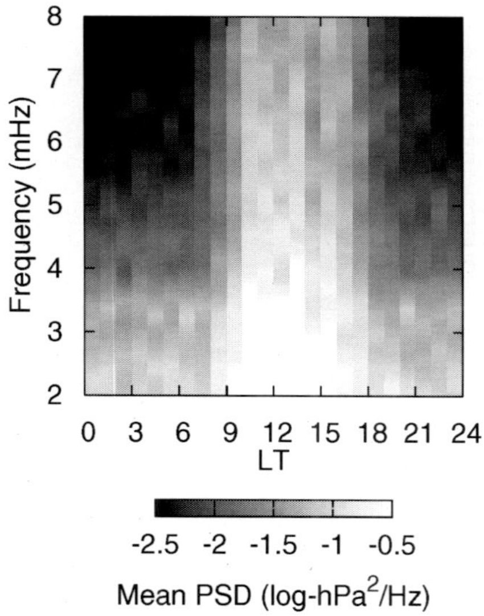

Fig. 7. April, 2007.

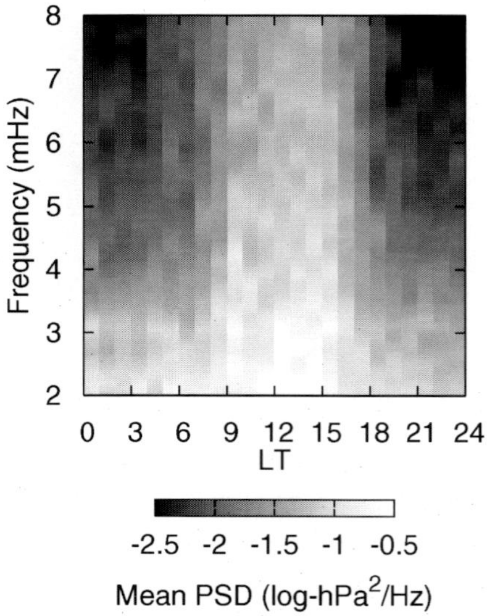

Fig. 8. May, 2007.

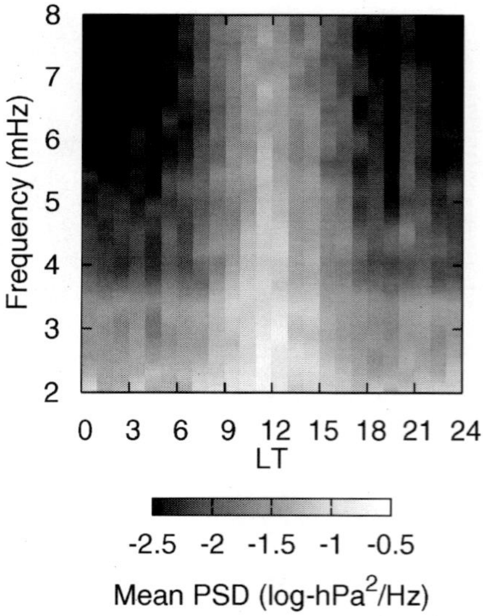

Fig. 9. June, 2007.

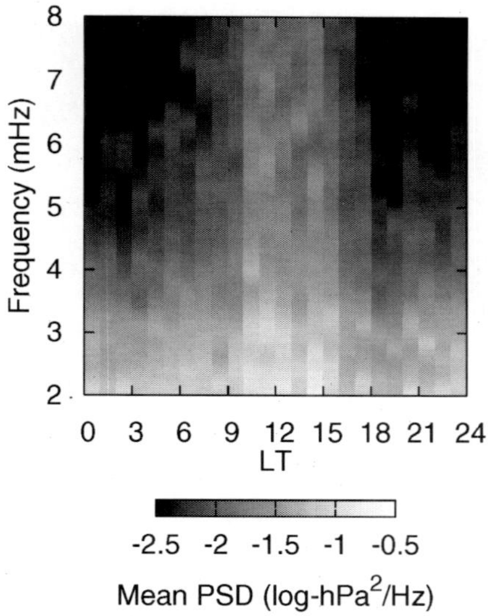

Fig. 10. July, 2007.

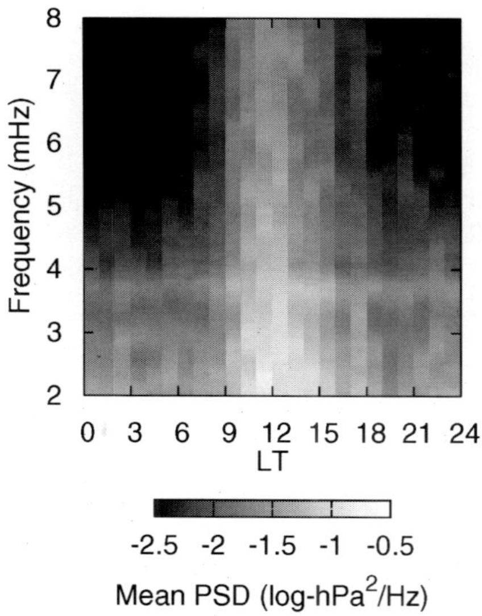

Fig. 11. August, 2007.

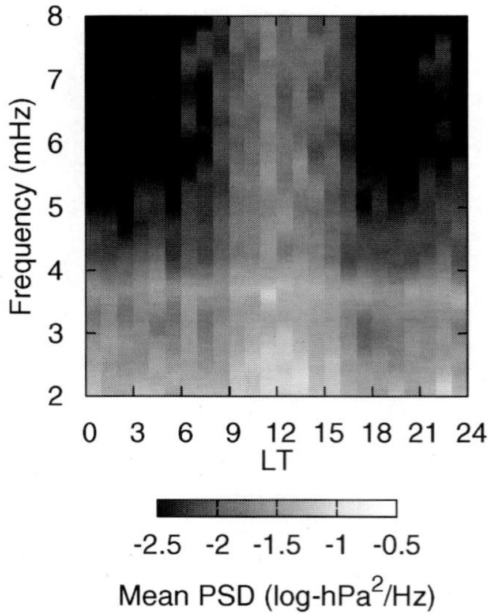

Fig. 12. September, 2007.

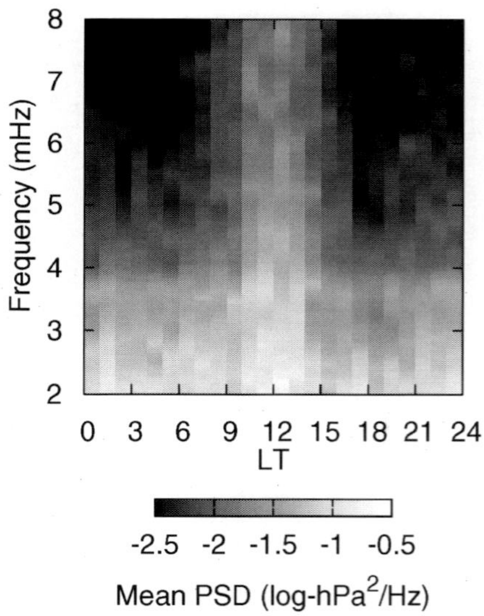

Fig. 13. October, 2007.

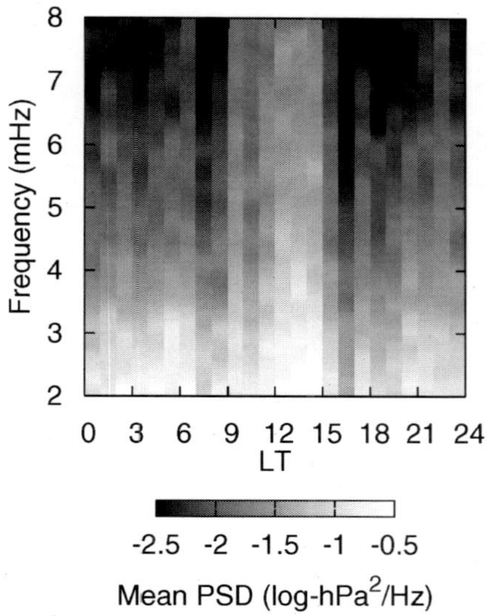

Fig. 14. November, 2007.

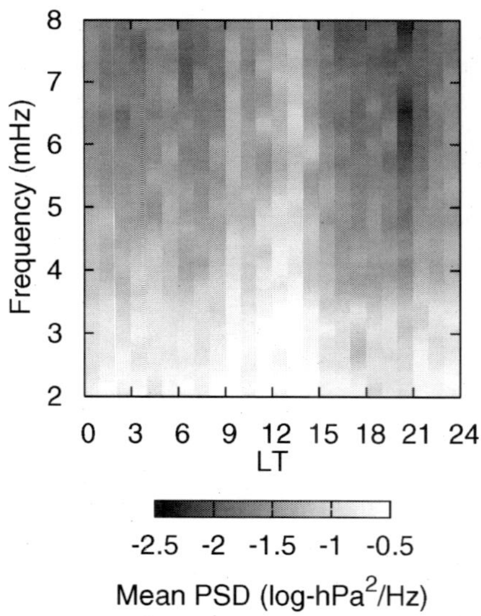

Fig. 15. December, 2007.

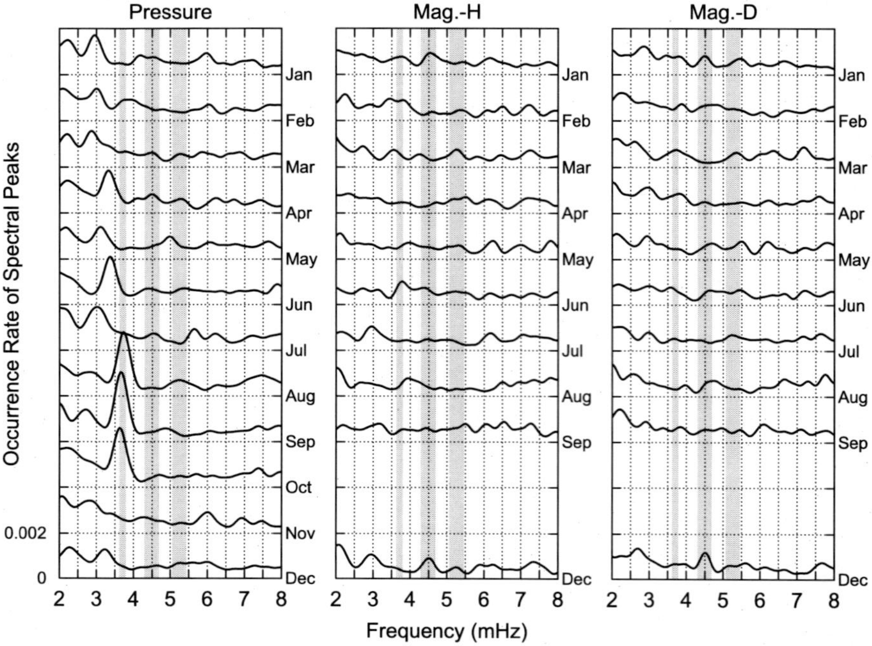

Fig. 16. The smoothed occurrence rate of power spectral peaks in twelve months. The left panel shows that of the pressure perturbations, and the middle and the right show those of magnetic H and D components, respectively. For magnetic fields, we used the spectra in the case of geomagnetic ap index ≤ 9 to decrease the effect of geomagnetic pulsation caused by magnetospheric plasma processes. The base line of each plot is shifted by 0.002 from the next month. The shadings show the theoretically expected resonance frequencies. Local maxima appear around the expected resonance frequencies, mainly in August to October for pressure, and in some months for magnetic components.

Figure 16 shows the average of them for each month in 2007. The left panel shows that of the pressure perturbations, and the middle and the right panels show those of magnetic H and D components, respectively. The base line of each plot is shifted by 0.002 from the next month. In this figure, the local time variation is not considered because Fig. 3 shows it is not so important, as is the case with magnetic H and D components though they are not shown here. For magnetic fields, we used the spectra in the case of geomagnetic ap index ≤ 9 to decrease the effect of geomagnetic pulsations caused by magnetospheric plasma processes. Local maxima appear around the expected resonance frequencies, mainly in August to October for pressure, and in some months for magnetic components.

The theoretically expected peaks are more clearly seen in Fig. 16 than those in Fig. 3. For pressure, clear local maxima appear at lower frequencies than 3.7 mHz in months other than August to October.

3. Discussion

The correspondence of the peak frequencies of pressure and magnetic components with theoretically expected resonance frequencies strongly suggests the existence of acoustic resonance and generation of magnetic pulsations by the resonance.

A possible explanation why local maxima of the pressure perturbations are clearer in August to October than those in other seasons is as follows.

First, the sources of acoustic gravity waves are strong in this term because of typhoon. Next, the local turbulence is weak and the noise level is low at Shigaraki station in this term. As shown in Fig. 4 to 15, PSDs in this term are lower than those in other months. The strong signal due to typhoon and the low noise level lead to higher S/N ratio, and clear peaks appears.

As for the geomagnetic oscillations, local maxima around the expected resonance frequencies appear in every season. So, if the geomagnetic oscillations around the resonance frequencies are caused by the acoustic resonance of atmosphere, the resonance is caused through the year. Clear local maxima at lower frequencies than 3.7 mHz in months other than August to October in Fig. 16 also follow this. Since the temperature profile of the atmosphere is different amongst seasons, resonance frequencies also varies with seasons.

One unexpected finding is that we can see local maxima not only at the theoretically expected frequencies but also at higher frequencies such as 6 mHz, which is true for both of the pressure perturbations and geomagnetic oscillations in all seasons. So, effective resonance frequencies might exist in addition to the theoretically reported three major frequencies.

There exists no clear correlation in wave forms between pressure and magnetic field as shown in Fig. 1 although common spectral peaks often appear. This might be explained by the localized nature of resonance.

To understand the characteristics of the resonance and its effects, investigation of the ionospheric perturbation caused by acoustic waves from lower atmosphere would be helpful.

Acknowledgments

The observations at Shigaraki are supported by the Shigaraki MU radar observatory. We thank the staff there for their help. This study was partially supported by Grant-in-Aid 19654073 and the fund for the 21st Century COE Program (Kyoto University, G3). We also thank Dr. K. Nishida and Prof. R. Yamamoto for providing us useful theoretical and observational information.

References

1. R. M. Jones and T. M. Georges, Infrasound from convective storms. III. Propagation to the ionosphere, *J. Atmos. Sci.* **59**(4) (1976) 765–779.
2. K. Nashida, Earth's background free oscillations, Ph.D. thesis, University of Tokyo (2000).
3. H. Shinagawa, T. Iyemori, S. Saito and T. Murayama, A numerical simulation of ionospheric and atmospheric variations associated with the Sumatra earthquake on December 26, 2004, *Earth Planets Space* **59** (2007) 1015–1026.
4. M. Tahira, Acoustic resonance of the of the atmosphere at 3.7 mHz, *J. Atmos. Sci.* **52** (1995) 2670–2674.
5. T. Iyemori *et al.*, Geomagnetic pulsations caused by the Sumatra earthquake on December 26, 2004, *Geophys. Res. Lett.* **32** (2005) L20807, doi:10.1029/2005GL024083.

Advances in Geosciences
Vol. 21: Solar & Terrestrial Science (2008)
Ed. Marc Duldig

NEAR-REAL TIME MONITORING OF TEC OVER JAPAN AT NICT (RWC TOKYO OF ISES)

W. MIYAKE* and H. JIN
National Institute of Information and Communications Technology, 4-2-1, Nukui-kitamachi, Koganei, Tokyo 184-8795, Japan
**Department of Aeronautics and Astronautics, Tokai University, 1117 Kitakaname, Hiratsuka, Kanagawa 259-1292, Japan*
wmiyake@keyaki.cc.u-tokai.ac.jp

The world wide use of global navigation satellite systems such as GPS offers unique opportunities for a permanent monitoring of the total electron content (TEC) of the ionosphere. We have developed a system of the rapid derivation of TEC from GEONET (a dense GPS receiver network in Japan). In addition to a previous plot of TEC temporal variation over Japan, we have recently developed a near-real-time two-dimensional TEC map and have used it for the daily operation of Space Weather Forecast Center at NICT (Regional Warning Center Tokyo of International Space Environment Service). The TEC map can be used to continuously monitor the ionospheric disturbances over Japan, including spatial and temporal development of ionospheric storms, large-amplitude traveling ionospheric disturbances, and plasma bubbles intruding over Japan, with high time resolution. The development of the real-time monitoring system of TEC enables us to monitor large ionospheric disturbances, ranging from global- to small-scale disturbances, expected in the next solar maximum. The plot and maps are open to the public and are available on http://wdc.nict.go.jp/IONO/index_E.html.

1. Introduction

Ionospheric disturbances have a variety of effects on radio communication and satellite positioning and navigation systems. HF communications are controlled by the ionosphere behavior and respond to space weather phenomena such as solar flares and geomagnetic storms. Large perturbation of the ionization density at the F-region height during geomagnetic storms can cause either positive or negative ionospheric storms. The maximum usable frequency (MUF) for oblique propagation of HF radio waves is largely determined by the peak electron density in the F-region. Correlated with the peak electron density is the total electron content (TEC), which significantly affects the propagation delay of higher frequency satellite-to-ground signals. If the TEC along a signal path from a satellite to a ground receiver varies significantly in either time or space, due to the presence of ionospheric irregularities, the phase of the radio waves varies rapidly, which causes difficulties in precise positioning of the receiver.

Although ionospheric storms and irregularities have been studied extensively, it is still difficult to predict whether or not certain space weather conditions, such as the conditions during the onset of a geomagnetic disturbance may lead to any kind of ionospheric disturbances. Current efforts are largely focused on providing early warning of the disturbances developing in the ionosphere. Therefore, real-time monitoring of the ionosphere is essential to the current operation system of the space weather prediction and warnings.

The National Institute of Communications and Informations Technology (NICT) runs the Regional Warning Center (RWC) Tokyo of the International Space Environment Service (ISES)[1] and issues HF radio propagation warnings and forecasts based on real-time ionospheric observation in Japan. NICT has four stations installed for ionosonde observations in Japan (see Fig. 1 of Miyake[2]). Radio echoes from the F-region provide considerable information about the state of the ionosphere; however, these ionosonde observations are subject to some limitations. The intensity of the reflected radio waves tends to be weak during the summer because of large absorption, and the weak signals are sometimes swamped in the interference of HF communications. Moreover, sporadic E-layers that mask the F-region from the ground often develop during the summer. High-frequency radio signals are lost when solar X-ray flares increase the electron density in the D-region. A sparse distribution of the ionosonde observation is also not suitable for monitoring small-scale or local phenomena such as ionospheric irregularities.

The world-wide use of global navigation satellite systems such as GPS offers unique opportunities for a permanent monitoring of the TEC of the ionosphere. The GPS Earth Observation Network (GEONET) is a dense network of GPS receivers that was set up in Japan by the Geographical Survey Institute of Japan.[3] A quick distribution of the GEONET data (within a few hours of its acquisition) has started on the WEB (http://terras.gsi.go.jp/ja/index.html). A newly developed system for rapidly calculating TEC from the GEONET data[2] is used to

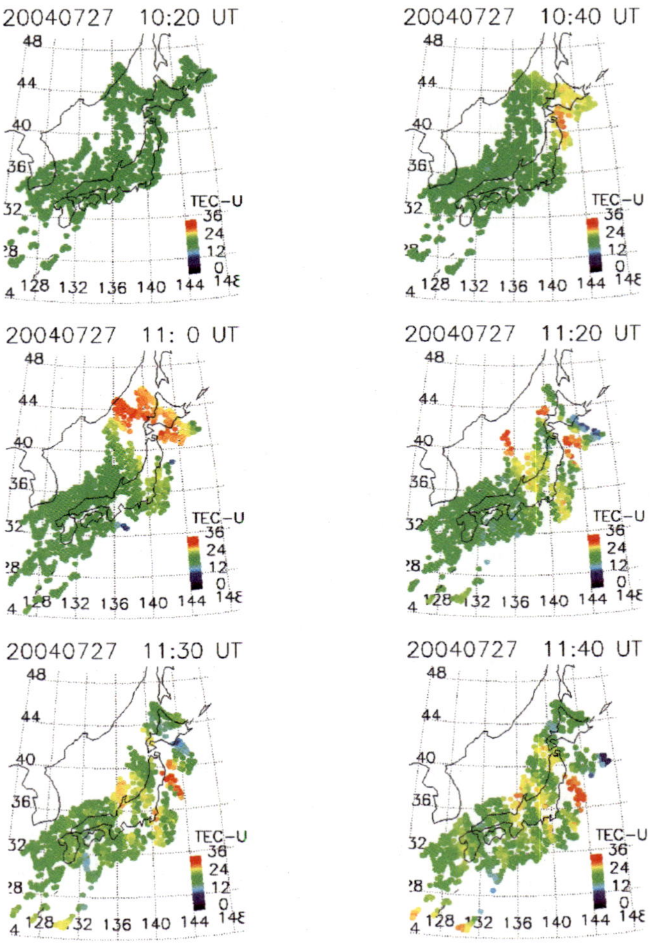

Fig. 1. Example of large-amplitude TID events propagating towards the south-west direction. The coordinate reference system uses geodetic longitude and latitude. The color indicates the vertical TEC values expressed in TEC unit($\times 10^{17}$ electrons/m^2).

the daily operation of the RWC Tokyo of ISES. The system provides a real-time plot of TEC and is very useful for continuously monitoring the progress of ionospheric storms under any ionospheric conditions. We have recently expanded the system toward creating seamless two-dimensional maps of TEC over Japan and monitoring small-scale phenomena, like traveling ionospheric disturbances and plasma bubbles, which have a large spatial gradient in TEC and significantly affect satellite-to-ground radio signals (e.g. Warnant *et al.*[4]). A similar product of real-time TEC maps over US has been developed.[5] In this report, we present the first real-time TEC maps over Japan in operation and demonstrate the capability of the expanded system to monitor local irregularities in the ionosphere.

2. Real-time TEC Maps

2.1. *Method*

Assuming that the TEC is uniform over a small area, Ma and Maruyama[6] divided the ionosphere over Japan into a grid comprising 32 areas; each grid is a $2° \times 2°$ latitude/longitude grid. The absolute values of TEC and both the satellite and receiver biases were derived, simultaneously. They arranged unknown TECs and biases in 1-day unit into a set of 35,000 equations and determined the unknowns by using the least-squares fitting methods. It took about 8 hours to carry out the whole process on a personal computer. Although the method estimated TECs accurately, it is obviously unable to provide timely information because it needs to accumulate a entire day's worth of data before beginning calculations that take hours.

By examining the day-to-day variations in the biases, Miyake[2] found that the variation in the instrumental biases over a few days is generally small (less than 2 TEC units ($\times 10^{17}$ electrons/m^2)) and introduced a rapid method to derive TEC by using the instrumental biases obtained a few days before obtaining the latest GEONET data. After determining the bias for all the satellites and receivers, we can directly obtain the vertical TEC (TECV) from the observed slant TEC (TECS), as follows:

$$\mathrm{TECV} = (\mathrm{TECS} - \mathrm{Br} - \mathrm{Bs}) \cos \chi, \qquad (1)$$

where Br and Bs denote the receiver and satellite biases, respectively, and χ denotes the zenith angle. Details of the method, including the accuracy of the TEC derivation, are described elsewhere.[2]

After determining all the biases, we are free from the assumption that the TEC is uniform in the $2° \times 2°$ grid. This is the key point to derive a high-resolution TEC map. A TECV is set for each receiver-satellite pair and can be projected to the ionospheric height along the signal path. The receiver distribution of GEONET is so dense that collection of the projected TECV produces a seamless high-resolution map of TEC over Japan.

A real-time TEC map is created every 5 min, and a set of maps for the latest 24 h have been open to the public since January 2008 on http://wdc.nict.go.jp/IONO/index_E.html. Since the interval is at the solar minimum, major disturbances have rarely occurred since the new expanded system has been in operation. Therefore, we apply the system to the GEONET data obtained during the recent solar maximum, and we investigate the capability of the system to monitor small-scale phenomena during major ionospheric disturbance events. A few GPS satellites are usually observed at the same time; hence, the TEC map covers an area wider than Japan. The maps only indicate latitudinal variations at quiet times and are not very impressive. However, we find them very useful under active conditions. The maps indicate large spatial gradients of TEC with small-scale disturbances over Japan. We consider three examples of disturbances, for which the TEC maps are quite powerful for monitoring their development, movement, and decay.

2.2. *Monitoring travelling ionospheric disturbances by using TEC maps*

Traveling Ionospheric Disturbances (TIDs) usually show slight variations in TEC (for example, refer Saito *et al.*[7]), but it is not always the case. Some TIDs under active geomagnetic conditions have very large amplitudes with steep spatial gradient; this causes severe problems in navigation systems. We present such an example from the 2003/07/27 event. The TEC maps provide a clear view of what is happening over Japan.

Heating at high latitudes in the auroral electrojet causes expansion of the neutral atmosphere; this expansion is often rapid during large geomagnetic disturbances. When the heating events are impulsive, the expansion often takes the form of equatorward surges or traveling atmospheric disturbances (e.g. Hocke and Schlegel[8]). These large-scale gravity waves can penetrate into low latitudes, and they are observed as traveling ionospheric disturbances, which have been observed as a wavy pattern of substantial height increase of the ionosphere, occasional enhancement of airglow emission intensity, and enhancement of TEC.

Figure 1 shows an example of a large-amplitude TID event. The coordinate reference system uses geodetic longitude and latitude. The color indicates the vertical TEC values expressed in TEC unit ($\times 10^{17}$ electrons/m^2), as shown by the color bar. It must be noted that this is the first near-real time TEC map of Japan, though a TEC map in a similar format was provided previously.[7] The event was observed during a prolonged period of geomagnetic disturbances. A large TEC portion first appeared in northern Japan at 10:40 UT. The ionospheric disturbance then expanded rapidly toward the south-west direction. After 11:20, it was clear that the disturbance had a propagating wavy structure with the wave-front directed along the north-west to south-east directions. The disturbances shown in the figure are medium-scale TIDs with wave-lengths of the order of hundreds of kilometers. The TEC variation is then times larger than that of the usual MSTID.

2.3. *Monitoring progress of ionospheric storm by using TEC maps*

One of the other examples is a probable Storm Enhanced Density (SED) (e.g. Foster[9]) over Japan during an intense geomagnetic storm on 2004/11/08, which also had a steep spatial gradient. The magnetic storm began on 2004/11/07. The Dst index started to decrease at 21:00 UT and reached its maximum depression of −373 nT at 06:00 UT on 2004/11/08. The significant enhancement in TEC of the positive ionospheric storm was observed all over Japan during the daytime.

The enhancement in TEC was once deceased all over Japan, but another enhancement was intruded over northern Japan from east after sunset. Figure 2 shows the progress of the TEC enhancement event. The time of sunset was approximately 09:00 UT. The first indication of the TEC enhancement was observed at around 10:30 UT. The abnormal enhancement became clear at 11:00 UT and decayed gradually later. In contrast with the daytime enhancement in TEC with a wide spatial extent, the enhancement after sunset was highly localized, and a steep spatial gradient of TEC was clearly observed on the TEC maps.

Maruyama[10] reported this event and discussed a probable connection with the storm enhanced density. Storm enhanced density is associated with a positive storm at lower latitudes, and it is separated by a density trough from the low latitude density enhancement. It is supposed that lower latitude dense plasma is a source of SED plasma and that it is redistributed through some transport processes in the inner magnetosphere and ionosphere during the disturbance period.[9,11,12] The TEC enhancement appears to move from east to west, as shown in the TEC maps in Fig. 2. This movement is consistent with E×B drift in the disturbance dynamo electric field.[13] The westward plasma drift was actually observed by the DMSP H15 satellite passing over Japan during the event.[10]

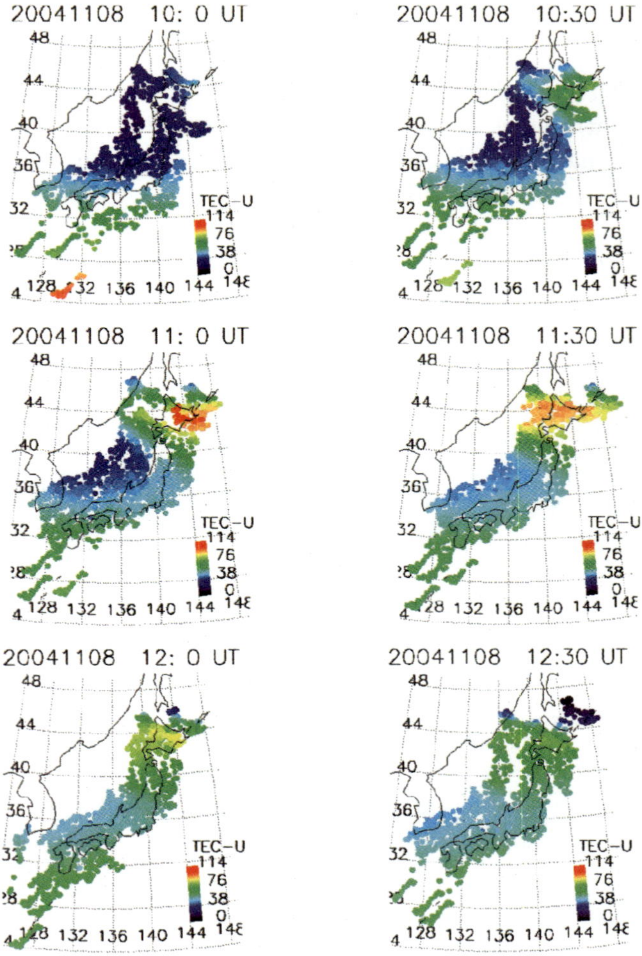

Fig. 2. Probable SED (Storm Enhanced Density) over Japan during an intense geomagnetic storm on 2004/11/07-08. The TEC enhancement moved over northern Japan from the east.

2.4. *Detecting a plasma bubble intruding over Japan by using TEC maps*

Plasma bubbles extending from the equatorial region also have a steep spatial gradient of TEC. Plasma bubbles originate near the equatorial ionosphere in association with the non-linear Rayleigh-Taylor process at the bottomside of the F-region after sunset. The bubbles develop as a field-aligned structure and rise to the topside ionosphere.[14]

Under certain conditions, the bubbles can reach high altitudes (greater than 2,000 km at the equator) ; therefore, their foot point along the field line can reach latitudes as high as those of Japan.[15,16] The steep density gradient along their flank can be the source region for cascades of density irregularities and therefore results in severe scintillation of satellite transmission signals, which is one of the major hazards of adverse space weather conditions.[17,18] Here, we provide an example of a probable bubble extension over Japan. One of the events was observed on 2003/07/16 under enhanced TEC conditions during a positive storm.

Figure 3 shows an example of a plasma bubble extension over Japan. TEC enhancement started all over Japan in the afternoon (after 06:00 UT) on 2003/0716. The bubble was found during the prolonged period of the TEC enhancement after the sunset. At around 12:30 UT–12:45 UT, a small area, where no TEC values were obtained, and surrounded with sparse and low TEC regions, appeared at 34° in latitude and 137° in longitude. Rapidly fluctuating radio signals hindered the calculation of TEC values. The area moved toward the north to north-west direction and appeared as a TEC depletion region evidently after 13:15 UT. It was observed that the region was a low-density plasma bubble penetrating deeply into the midst of Japan. The background TEC decreased rapidly after 13:30 UT, and the TEC depletion region was submerged in the decreased background.

NICT has ionosonde stations at Wakkanai (45.4°N, 141.7°E), Kokubunji (35.7°N, 139.5°E), Yamagawa (31.2°N, 130.6°E), and Okinawa (26.7°N, 128.2°E). None of these ionosonde observatories detected the intrusion of the bubble. The spatial scale of a bubble is too small to be always detected by the sparse ionosonde stations. The TEC depletion region fell into the space between the Kokubunji and Yamagawa ionosonde stations. A seamless two-dimensional TEC map created from a network of dense GPS receivers is the only method to detect the intrusion of a plasma bubble at any region in Japan.

3. Summary and Discussion

We have developed real-time maps of TEC over Japan, which is open to the public on http://wdc.nict.go.jp/IONO/index_E.html. We have tested the capability of the map for some small-scale disturbances during the latest solar maximum. We found the maps quite powerful for monitoring the spatial and temporal development of ionospheric storms, large-amplitude TID, and plasma bubbles intruding over Japan. All these phenomena

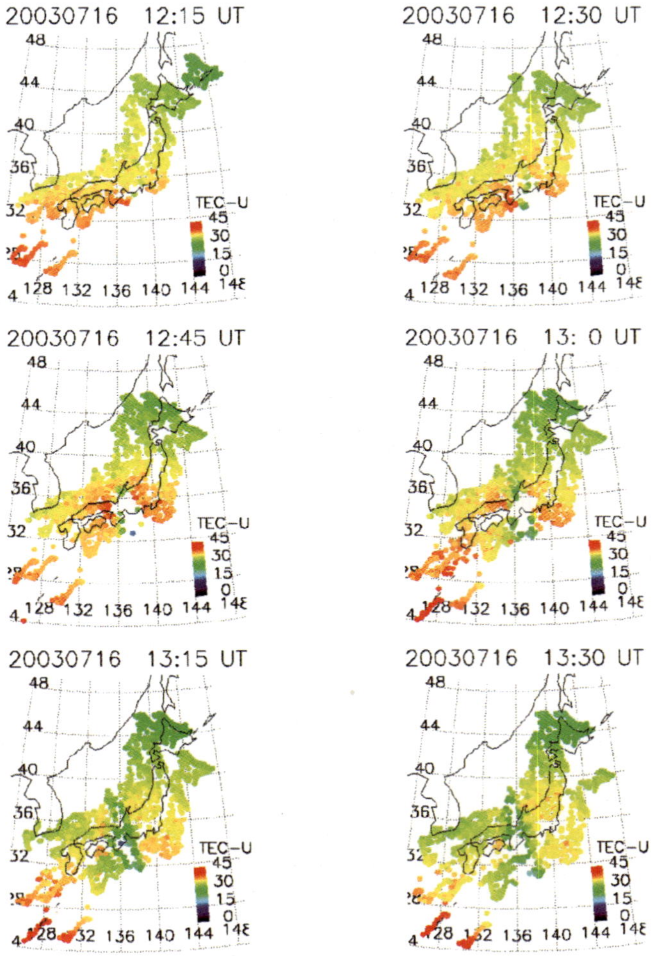

Fig. 3. Example of a plasma bubble extending over Japan. The bubble was observed during the prolonged period of TEC enhancement after the sunset. At around 12:30 UT–12:45 UT, a small area, where no TEC values were obtained, and which was surrounded with sparse and low TEC regions, appeared at 34° in latitude and 137° in longitude. The region moved toward north later and appeared as a TEC depletion region.

are small-scale or very local, but they have steep density gradient along the boundary. They cannot be monitored adequately by means of either sparse ionosonde observations or previous TEC observations in $2° \times 2°$ grids. The high-resolution TEC maps presented here provided us with the full view of the occurrence, movement, and decay of these local phenomena.

The current services of ionospheric information provided at NICT are mainly for domestic users of HF communications and VHF broadcasters, so that ionosonde observations have been a primary tool for monitoring F-region ionospheric storms and sporadic E-layers. The growing field of concern is, however, the influence of ionospheric disturbances on satellite precise positioning and navigation systems. In this regard, local phenomena with steep TEC gradients are becoming more important. The real-time TEC maps with high spatial resolution are quite essential for future services at space weather forecast centers.

The GEONET data are released for the public every 3 h, which means that a few hours of delay will occasionally remain between the latest TEC map and the real TEC distribution. The local phenomena generally have a short duration, and in the worst case, the disturbance may almost cease, when the TEC maps are created. It is still beneficial to observe within a short delay the events that have occurred, and analyze whether or not the disturbance is further in progress or it has almost ceased. Near-real time TEC maps are a powerful tool for monitoring local ionospheric phenomena; this technology enables us to monitor major ionospheric disturbances expected during the next solar maximum.

Acknowledgments

This study was conducted as one of the authors (W. M.) was a visiting researcher of NICT. We thank all the members of the Space Environment Group of NICT for their extensive support.

References

1. J. Kunches, *Space Weather* **4** (2006) S02005.
2. W. Miyake, *J. Atmos. Sol. Terr. Phys.* **69** (2007) 803.
3. S. Miyazaki *et al.*, *Bullet. Geographical Survey Institute* **43** (1997) 23.
4. R. Warnant, S. Lejeune and M. Bavier, *Astrophysics and Space Science library* **344** (Springer), *Space Weather* **129** (2007).
5. T. Fuller-Rowell *et al.*, *Radio Sci.* **41** (2006) RS6003.
6. G. Ma and T. Maruyama, *Annales Geophysicae* **21** (2003) 2083.
7. A. Saito *et al.*, *Geophys. Res. Lett.* **28** (2001) 689.
8. K. Hocke and K. Schlegel, *Ann. Geophysicae* **14** (1996) 917.
9. J. C. Foster, *J. Geophys. Res.* **98** (1993) 1675.
10. T. Maruyama, *Geophys. Res. Lett.* **33** (2006) L20111.
11. M. J. Buonsanto, *Space Sci. Rev.* **88** (1999) 563.
12. J. C. Foster and W. Rideout, *Geophys. Res. Lett.* **32** (2005) L12S04.

13. M. Blanc and A. D. Richmond, *J. Geophys. Res.* **85**(A5) (1980) 1669.
14. R. F. Woodman and C. La Hoz, *J. Geophys. Res.* **81** (1976) 5447.
15. Y. Otsuka, K. Shiokawa, T. Ogawa and P. Wilkinson, *Geophys. Res. Lett.* **29**(15) (2002) 1753.
16. T. Ogawa *et al.*, *Earth Planets Space* **57** (2005) 385.
17. G. Ma and T. Maruyama, *Geophys. Res. Lett.* **33** (2006) L21103.
18. M. Kelly, J. Makela and O. de la Beaujardiere, *Space Weather* **4** (2006) S02C04.

Advances in Geosciences
Vol. 21: Solar & Terrestrial Science (2008)
Ed. Marc Duldig

IMPROVE THE TOPSIDE IONOSPHERIC MODEL USING COSMIC ELECTRON DENSITY PROFILE DATA

X. SHI*, P. GUO and J. S. PING
Shanghai Astronomical Observatory, Chinese Academy of Sciences,
Graduate University of Chinese Academy of Sciences,
Beijing 100049, China
**shixian@shao.ac.cn*

The topside ionosphere has been studied for decades, but due to the insufficiency of spatial and temporal coverage of the previous data, analysis and modeling of this part of the ionosphere remain difficult. The development of radio occultation (RO) technique between Global Positioning System (GPS) satellites and Low Earth Orbit (LEO) satellites is able to provide ionospheric data with denser and more uniform coverage in both space and time. COSMIC is one such project that investigates the terrestrial neutral atmosphere and ionosphere using RO technique. In this study, the electron density profile (EDP) data from the COSMIC project is analyzed, and some main characteristics concerning the F2 layer peak density and height are investigated including diurnal variations, annual, semiannual, and winter anomalies, as well as equatorial anomaly. Descriptions of these characteristics in previous study are further confirmed through the analysis. Based on the advantages of the COSMIC database, a statistical model is built for the topside ionosphere using a Chapman-type function and spherical harmonic analysis, taking into account diurnal and seasonal variations. By comparing the topside total electron content (TEC) derived from this new model and from the International Reference Ionosphere (IRI) model, the feasibility of this new model is basically confirmed.

1. Introduction

The terrestrial ionosphere is the ionized portion of the upper atmosphere.[1] It has long been studied for its close relationship with human beings as well as its complicated variations in both space and time. Generally, the ionosphere is divided into four layers D, E, F_1, and F_2 (see Fig. 1). The topside ionosphere is defined to be the region above the F region peak.[1] The transition of the dominant ion from O^+ to H^+ determines the density structure of the topside ionosphere.

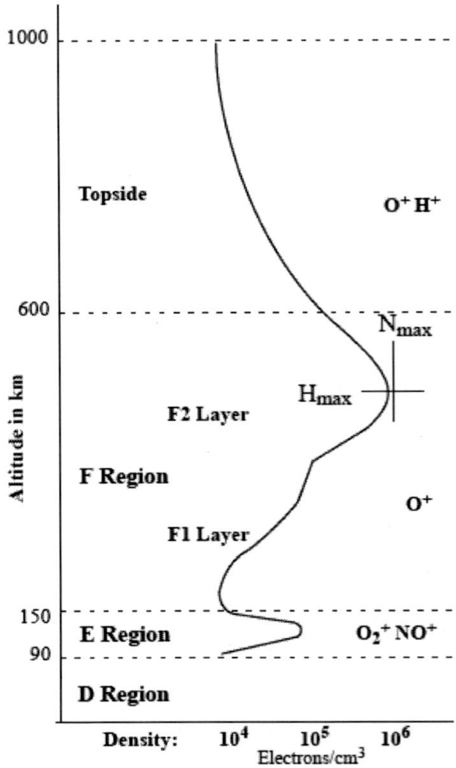

Fig. 1. The four-layer electron density profile of the terrestrial ionosphere (www.noaa.gov).

Developing an adequate model for the topside ionosphere has always been one of the central topics in ionospheric research. Though many techniques have been developed to reveal the characteristics of the ionosphere and its variations, only some of them are able to provide information of the topside ionosphere, which has made modeling of this region a tough task. Besides the limitation in quantity, the poor spatial and temporal coverage of previous data is also an obstacle in developing topside models.

In the middle of 1960s, the radio occultation (RO) technique began to develop in the Jet Propulsion Laboratory (JPL). This new method was firstly used in the studies of planetary atmosphere and later in the research of terrestrial neutral atmosphere and ionosphere. The

Constellation Observing System for Meteorology, Ionosphere and Climate (COSMIC) mission, launched on April 15th 2006, is one of the most recent projects using RO technique to provide measurements of terrestrial neutral atmosphere and ionosphere.[2,3] With its six low earth orbit (LEO) satellites, COSMIC is able to provide ionospheric RO measurements with a spatial and temporal coverage far better than any previous project before.

Taking advantage of the good quantity and quality of the ionospheric data from COSMIC satellites, the objective of our work has two respects. Firstly, an analysis of certain global characteristics of the ionosphere was carried out to confirm the ionospheric theory as well as verify the data. Secondly, an attempt to improve the topside ionospheric model with statistical methods was done. A comparison between this statistical model and the International Reference Ionosphere (IRI) 2007 model shows the feasibility of the new model.

2. Database

2.1. *Brief introduction of COSMIC*

Each LEO satellite in the COSMIC constellation carries three science payloads: GPS occultation receiver, Tiny Ionospheric Photometer, and Triband Beacon transmitters. Table 1 shows some of the COSMIC science requirements.[4]

The data from COSMIC mission is expected to address some of the most interesting yet difficult problems in the studies of Earth sciences.

Table 1. Some of the COSMIC science requirements.[4]

Data parameter	Science requirement	Comment
Vertical GPS occ. resolution	0.3–1.5 km	0.3 km near surface, 1.5 km at 45 km
Horizontal GPS occ. resolution	300–600 km	Smaller for fronts with large slope
Bending angle profile	1.5×10^{-6} rad	Limited mainly by residual ionospheric error and data noise
Electron density profile	$< 1 - 20\%$	Profile quality near the F_2 peak can be worse elsewhere; Limited by horizontal gradients
GPS *TEC*	0.001 TECU	Relative *TEC* (absolute ~3–5 TECU)
LEO position	~ 10 cm	3-D rms position error (no critical)
LEO velocity	< 0.1 mm/s	Relative velocity error for paired satellites

The great amount of high quality ionospheric measurements will play an important role in understanding and modeling the ionosphere as well as improving space weather forecasting.[4]

2.2. *The RO measurements of the ionosphere in COSMIC mission*

A lot of techniques have been developed to investigate the terrestrial ionosphere, among which several can provide information of the topside region, for example, *in-situ* measurements, ISR probing, and topside sounding. However, all these techniques have deficiencies in temporal or spatial coverage. With the constellation of Global Positioning System (GPS), the RO technique which was first used in investigating planetary atmosphere is now able to provide information of the terrestrial ionosphere using following algorithms (see Fig. 2).

The total electron content (*TEC*) between the LEO satellite and the GPS satellite is achieved in the first place by[5]:

$$TEC = \frac{1}{d}\frac{f_1^2 f_2^2}{(f_1^2 - f_2^2)}(L_1 - L_2) \quad (1)$$

Here, f_1 and f_2 are the frequencies of the two broadcasting signals of the GPS satellites with L_1 and L_2 as their excessive phases in the unit of meter. For the resulted *TEC* in the unit of TECU (TECU $= 10^{16}$ electron/m^2), $d = 40.3 \times 10^{16}$.

The derived *TEC* is then calibrated by subtracting the *TEC* between the GPS satellite altitude and the LEO satellite altitude:

$$TEC' = TEC - TEC_0 \quad (2)$$

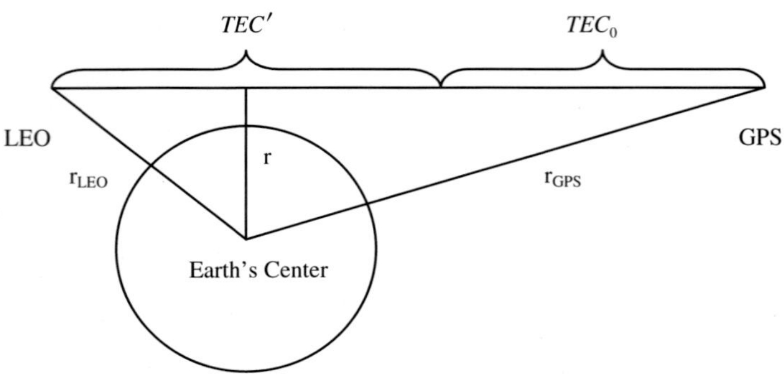

Fig. 2. Illustration of the geometry of an RO event between a GPS and a LEO satellite.

On the assumption of spherically symmetric atmosphere, using Abel integral transform, the electron densities $N(r)$ can be obtained from the calibrated *TEC'* by[6]:

$$N(r) = -\frac{1}{\pi}\int_{r}^{r_{\mathrm{LEO}}} \frac{\mathrm{d}TEC'(r')/\mathrm{d}r'}{\sqrt{r'^2 - r^2}}\mathrm{d}r' \tag{3}$$

where r is the radius of the tangent point, r_{LEO} is the radius of the LEO satellite. The EDP data derived from every RO event forms one of the second-level databases provided by COSMIC Data Analysis and Archival Center (CDAAC). The comparisons between the COSMIC retrieved electron density profile (EDP) data and the measurements from incoherent scatter radars (ISR) and ionosondes show agreement.[7]

2.3. *Ionospheric parameters retrieved from the EDP data*

In this paper, two characteristics were mainly investigated concerning the topside ionosphere, the F_2 layer peak height $h_m F_2$, and the F_2 layer peak density $N_m F_2$. For every EDP, a parabolic fitting is done using data points in the range of ± 20 km around the measured maximum electron density point. The height and electron density correspond to the vertex of the parabola are defined as the $h_m F_2$ and $N_m F_2$ of this EDP.

2.4. *Data distribution*

In order to take into account seasonal variations, data used in this paper are the EDP data from CDAAC between the 36th day of year 2007 and the 35th day of year 2008. The database is divided into four seasons, centering on the corresponding solstice/equinox day (see Table 2).[8]

All the EDP data in the above time range passes the fitting process to form sets of statistical results for four seasons. A subsequent quality control procedure eliminates problematic data which may be caused by failures in observation or inversion process. A total of 669,433 EDPs are fit, out of

Table 2. Definition of four seasons in the northern hemisphere (DOY: day of year).

Spring	Summer	Autumn	Winter
DOY: 036–127 (2007)	DOY: 128–218 (2007)	DOY: 219–127 (2007)	DOY: 311–365 (2007) DOY: 001–035 (2008)

Table 3. The total and used amount of EDP data in every season (referring to northern hemisphere).

	Spring	Summer	Autumn	Winter
Total	214,359	182,994	139,181	132,899
Used	179,691	140,201	111,800	103,560

which 535,252 profiles (~80%) are left after the quality control. Detailed data distribution is listed in Table 3.

3. Statistical Modeling of the Topside Ionosphere

3.1. *Method*

Along with the development of new techniques and the new missions, the model for topside ionosphere has been improved a lot. The traditional data sources for any empirical model mainly come in three types: incoherent scatter radars which provide topside EDP data with good spatial coverage but limited locations; satellite topside sounder and *in-situ* instruments which offer global coverage but only cover a period of time.[9] As mentioned above, the ionospheric RO data has the advantage in both spatial and temporal coverage, which make it one of the best data source for modeling the topside ionosphere. After entering its final configuration, the constellation has been providing more than 2,500 EDP every day. Considering the features of the COSMIC EDP data, a statistical method is adopted to improve the topside ionospheric model in this paper.

First, every profile is fit into a Chapman type model:

$$N(h) = N_m\mathrm{F}_2 \exp[c(1 - Z - e^{-Z})]$$
$$Z = \frac{h - h_m\mathrm{F}_2}{h_0} \tag{4}$$

where $N(h)$ is the electron density varying with mean sea level height h; c is a parameter which equals to 0.5 or 1 for a Chapman α- or β-function; h_0 is the F_2 layer scale height .

Second, spherical harmonic analysis up to 20th degree and order is selected to obtain the numerical mapping of parameter c and scale height h_0 with reference to local time (LT) and geographic latitude θ:

$$c(\theta, LT) = \sum_{l=0}^{20} \sum_{m=-l}^{l} c_l^m Y_l^m(\theta, LT)$$
$$h_0(\theta, LT) = \sum_{l=0}^{20} \sum_{m=-l}^{l} h_{0l}^m Y_l^m(\theta, LT) \tag{5}$$

The spherical harmonic functions are defined as:

$$Y_l^m(\theta, LT) = \sqrt{\frac{(2l+1)}{4\pi}\frac{(l-m)!}{(l+m)!}}P_l^m(\cos\theta)e^{imLT} \quad (6)$$

where $P_l^m(\cos\theta)$ are the associated Legendre functions.

Finally, a comparison between our new model and the IRI2007 model is carried out. The topside TEC derived from both models are compared with the measurements from COSMIC mission.

3.2. *Modeling results*

Figure 3 shows the distribution of two parameters before and after modeling. The definition of seasons is referring to northern hemisphere. Daily and seasonal anomalies could be seen in the distribution of the scale height. For day time, the scale height has a depression in the winter hemisphere.

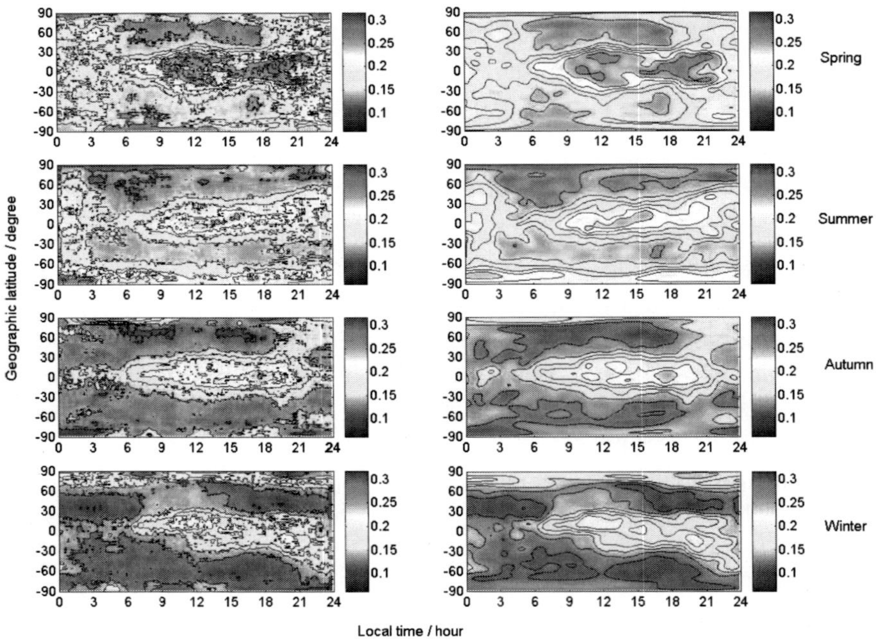

(a)

Fig. 3. The diurnal-latitudinal variation of fitting parameters c (a) and h_0 (b) before and after the statistical modeling.

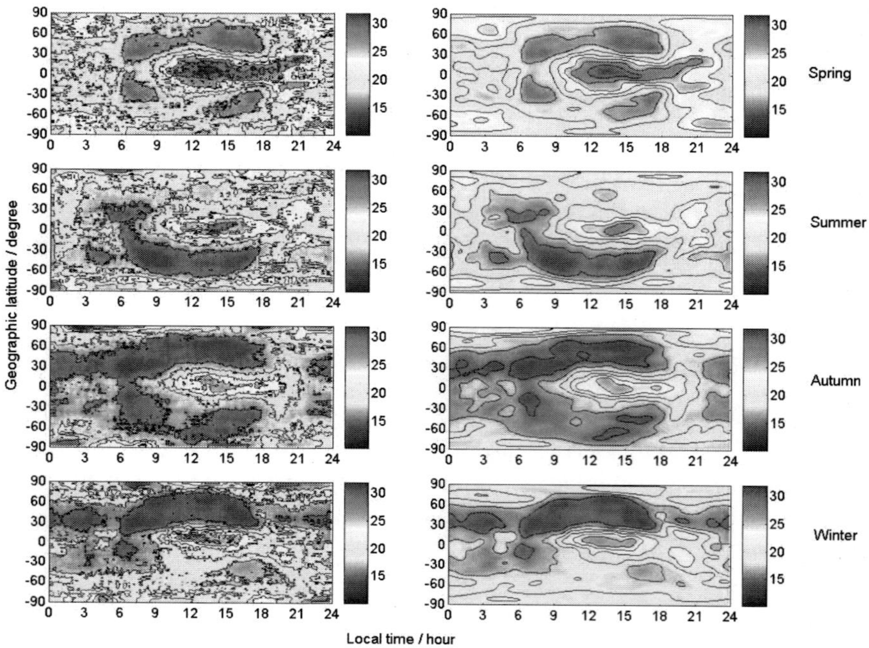

(b)

Fig. 3. (*Continued*)

3.3. *Comparison with IRI2007 model*

A comparison between this statistical model and the topside part of the IRI2007 model[10] is done taking the COSMIC EDP data as standard. $N_m F_2$ and $h_m F_2$ are considered as input parameters for both models. Figure 4 shows four examples of the comparison of EDP between models and observation. It can be seen that IRI2007 model fits well at high altitudes, while the new statistical model shows better performance from F_2 peak height to approximately 400 km.

Taking topside *TEC* as an indicator, a comparison of the distribution of differences between both models and the observed data is shown in Fig. 5, using the following equations:

$$\Delta TEC_{\mathrm{IRI}} = (TEC_{\mathrm{IRI}} - TEC_{\mathrm{COSMIC}})/TEC_{\mathrm{COSMIC}} \times 100\%$$

$$\Delta TEC_{\mathrm{NEW\ MODEL}} = (TEC_{\mathrm{NEW\ MODEL}} - TEC_{\mathrm{COSMIC}})/TEC_{\mathrm{COSMIC}} \times 100\% \quad (7)$$

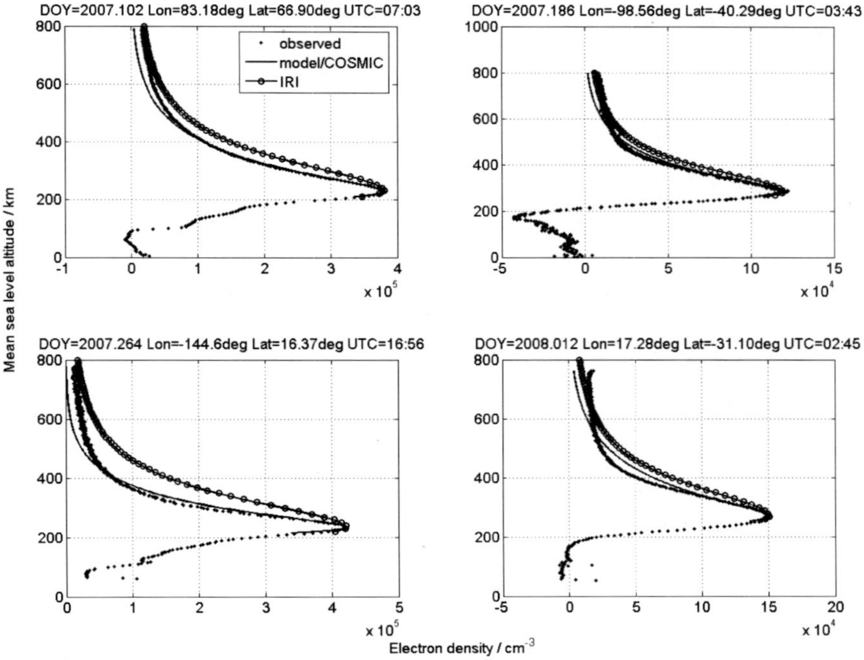

Fig. 4. Four EDPs from COSMIC observed data, IRI2007 model, and the new statistical model.

It can be seen that the topside *TEC* of IRI2007 model is 10–40% higher than the COSMIC measurements on average, while the topside *TEC* of the new statistical model is 5–15% lower than the COSMIC measurements. The IRI2007 model and the COSMIC measurements have their greatest divergence at around *LT* 7:00, and show couple of characteristics of equatorial anomaly with a minimum on the geomagnetic equator and maximum in the subtropics. Large differences between the new statistic model and the COSMIC measurements appear in the polar regions due to lack of data of latitude above ±85°.

To test the forecasting ability of this new model, the model parameters derived from the Spring of year 2007 are used to forecast the topside *TEC* data in the Spring of year 2008.

As shown in Fig. 6, the differences are rather similar to the ones in the Spring of 2007. The topside *TEC* derived from the new model is approximately three times closer to the COSMIC measurements than the IRI2007 model.

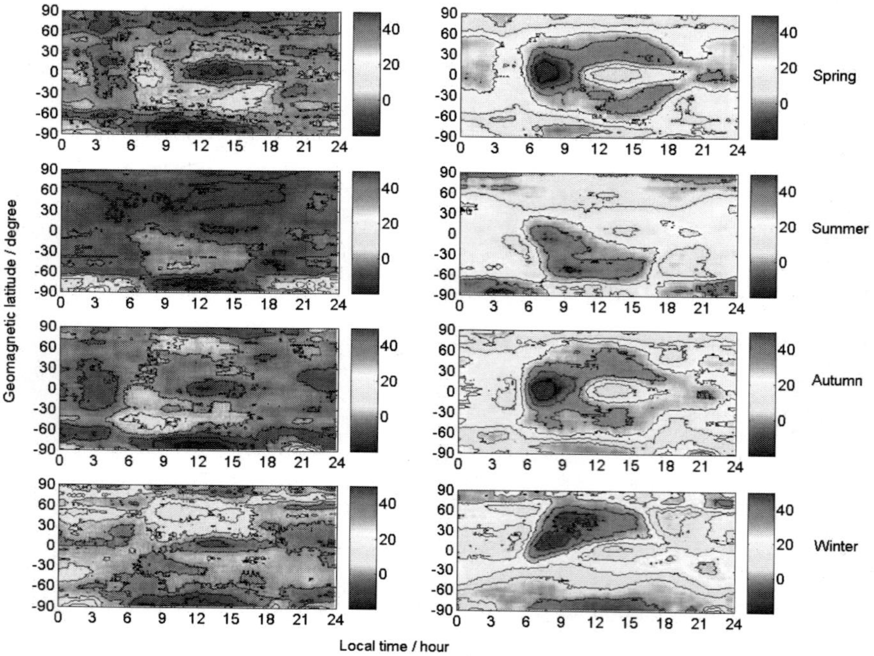

Fig. 5. Difference (in percentage) of topside *TEC* (in TECU) between new statistical model and COSMIC measurements (left) as well as between IRI2007 model and COSMIC measurements (right) in the four seasons of year 2007.

4. Conclusion and Discussion

Presented in this paper were analysis and modeling of the topside ionosphere based on the RO data from COSMIC. Analysis of the $N_m\mathrm{F}_2$ and $h_m\mathrm{F}_2$ retrieved from the COSMIC EDP data reveals some characteristics of the ionosphere on a global scale.

Diurnal variations of both $N_m\mathrm{F}_2$ and $h_m\mathrm{F}_2$ at mid-latitudes can be clearly seen from the study. Annual and semiannual anomaly of the ionosphere is further confirmed in this study as well as the winter anomaly though rather vague due to low solar activity. Also, the distribution of $N_m\mathrm{F}_2$ and $h_m\mathrm{F}_2$ with respect to geomagnetic latitude and *LT* displays the pattern of equatorial anomaly.

Considering that the spatial and temporal coverage of COSMIC data is one of the best ever achieved, a process of statistical modeling was carried out to get a better description of the topside ionosphere. The

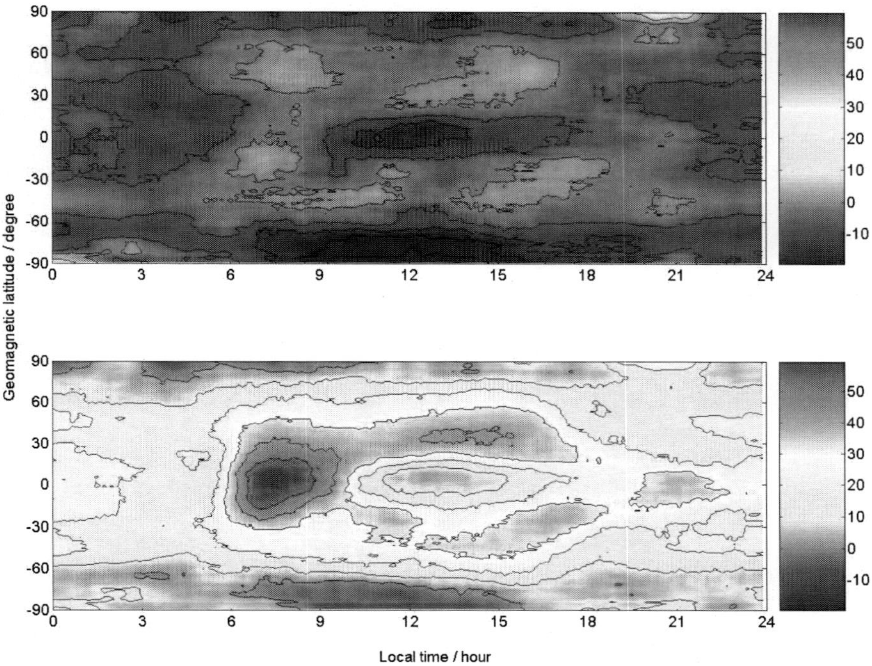

Fig. 6. Difference (in percentage) of topside *TEC* (in TECU) between new statistical model and COSMIC measurements (left) as well as between IRI2007 model and COSMIC measurements (right) in the Spring of year 2008.

Chapman-function-based profile has two fitting parameters c and h_0 which are expanded as a series of spherical harmonic functions. Diurnal and seasonal variations are included in the modeling. By comparing the topside *TEC* result in year 2007 retrieved from this new model and the IRI2007 model, it was demonstrated that the new model has a better approximation to the real measurement, and a more smooth distribution of error. The maximum of error in polar areas of the new model is mainly due to the insufficiency of RO data acquired in that region. The forecasting result of the new model in the Spring of year 2008 further confirmed the feasibility of the new model.

With the growth of the COSMIC RO database, the new model is still being upgraded. The influence of solar activity will be included when the temporal coverage of the database is sufficient. Till now, only tranquil ionosphere is considered in this global model, and it is planned to introduce storm-time changes into the model in the future.

Acknowledgments

The authors acknowledge the COSMIC Data Analysis and Archive Center (CDAAC) at the University Corporation for Atmospheric Research (UCAR) for providing the COSMIC ionospheric profiles. This study was supported by the National HighTechnology Research and Development Program("863"Program) of China (2009AA12Z319).

References

1. R. W. Schunk and A. F. Nagy, *Ionospheres Physics, Plasma Physics, and Chemistry*, Cambridge, Cambridge University Press, (2000).
2. C. Z. Cheng, Y. H. Kuo, R. A. Anthes and L. Wu, Satellite Constellation Monitors Global and Space Weather, *EOS* **87** (2006) 166.
3. W. Schreiner, C. Rocken, S. Sokolovskiy, S. Syndergaard and D. Hunt, Estimates of the precision of GPS radio occultations from the COSMIC/FORMOSAT-3 mission, *Geophys. Res. Lett.* **34** (2007) L04808.
4. C. R. Rocken, Y. H. Kuo and W. Schreiner *et al.*, COSMIC System Description. Terrestrial, *Atmospheric and Oceanic Sciences* **11**(1) (2000) 21–52.
5. G. A. Hajj, L. C. Lee, X. Q. Pi *et al.*, COSMIC GPS Ionospheric Sensing and Space Weather. Terrestrial, *Atmospheric and Oceanic Sciences* **11**(1) (2000) 235–272.
6. W. S. Schreiner, S. V. Sokolovskiy and C. Rocken, Analysis and validation GPS/MET radio occultation data in the ionosphere, *Radio Sci.* **34**(4) (1999) 949.
7. J. H. Lei, S. Syndergaard and A. G. Burns, *et al.*, Comparison of COSMIC ionospheric measurements with ground-based observations and model predictions: Preliminary results, *J. Geophys. Res.* **112**(A7) (2007), doi:10.1029/2006JA012240.
8. S. M. Stankov and N. Jakowski, Topside ionospheric scale height analysis and modeling based on radio occultation measurements, *J. Atm. Solar Terr. Phys.* **68** (2006) 134.
9. D. Bilitza, B. W. Reinisch, S. M. Radicella *et al.*, Improvements of the International Reference Ionosphere model for the topside electron density profile, *Radio Sci.* **41** RS5S15 (2006), doi:10.1029/2005RS003370.
10. D. Bilitza and B. Reinisch, International Reference Ionosphere 2007: Improvements and new parameters, *J. Adv. Space Res.* **42** (2008) 599.

Advances in Geosciences
Vol. 21: Solar & Terrestrial Science (2008)
Ed. Marc Duldig

DERIVATION OF THERMOSPHERE DENSITY VARIATION RESPONSES TO SPACE WEATHER EVENTS DURING APR. 2002 TO JUL. 2007 USING FY1D SATELLITE TRACK DATA

CONG HUANG, DAN-DAN LIU*, JING-SONG WANG
and TAO YU
National Center For Space Weather,
Beijing, 100081, China
** liudd@cma.gov.cn*
www.spaceweather.gov.cn

We analysed Chinese meteorological satellite-FY1D (at the altitude of 870 km) track data to derive thermosphere density variations and investigate the space weather events in the past five years (2002–2007). Our results confirm previous studies that thermosphere density variations have good responses to short-term space weather events such as geomagnetic storms and thermosphere density has multiple changes in amplitude during these events. We also found that thermosphere density will be influenced by solar radio flux (F10.7 Index) evidently and periodically. In this work, we compared the NRLMSISE-00 model density results (at the altitude of 870 km) with the density variations calculated by the FY1D satellite track data and marked the differences in amplitude of the responses.

1. Introduction

It is well known that thermosphere density variations cause great influences on the atmospheric drag force to LEO (Low Earth Orbit) satellites and the changing of the air drag can strongly affect the LEO satellites' orbit status, especially the altitude and cycle time of the satellites' orbit.[1]

There are two main energy sources of thermosphere: one is solar EUV (extreme ultra-violet) radiation, the other is the energy brought by solar wind particles. Solar EUV radiation is the largest contributor to thermosphere heating. 80% of thermosphere's energy input is solar EUV radiation.[2] The remainder of the energy input in thermosphere is controlled by Joule power heating associated with the geomagnetic activities caused by high energy solar wind particles and Joule power is determined by the level of geomagnetic activities.[3] Solar activities influence the energy input in thermosphere and drive perturbations of thermosphere density. Solar EUV radiation gives a long-term disturbance in thermosphere,[4,5] while Joule heating effect makes significant contributions to thermosphere density variations during strong geomagnetic storms. Thermosphere density can increase as much as 134% with the increasing of the Kp index from 1 to 6.[6]

LEO satellites' orbit parameters are always used to investigate the correlation between space weather events and thermosphere density variations. Kim *et al.*[5] analysed atmospheric drag effects using KOMPSAT-1 satellite track data and drew a conclusion that Joule heating driven by geomagnetic storms makes more effects to thermosphere density increase than solar EUV radiation. Park *et al.*[7] compared the KOMPSAT-1 drag derived density with MSISE model results to show that MSISE model's results underestimate thermosphere density during the strongly disturbed geomagnetic intervals. In this paper, we adopt FY-1D satellite's track data to examine the relationship between space weather events and thermosphere density variations. We also compare the satellite drag derived density variations with NRLMSIS-00 model's results to see how the model's results go with solar activities.

2. Observation Data and Inversion Method

FY1D satellite is Chinese meteorological satellite which was launched on May 15, 2002. Its orbit is a sun-synchronous circular orbit with an inclination of 98° and orbit altitude is about 870 km. FY1D track data are first used to study thermosphere density variations. FY1D track data we derived cover three period: May 15, 2002 to September 22, 2003; November 20, 2003 to March 25, 2006; June 12, 2006 to June 29, 2007. The F10.7 index[a] (the 10.7-cm solar radio index) is applied as a proxy for solar EUV radiation, this index is usually used as a conventional proxy to represent solar EUV radiation.[8] We adopt Ap index[b] to indicate the level of global geomagnetic activities. In order to avoid the influence of 12-day's regression time in track data, we do 12-day's moving average to FY1D's cycle time, corresponding F107 index and Ap index (see Fig. 1).

[a]http://www.sec.noaa.gov
[b]http://swdcwww.kugi.kyoto-u.ac.jp

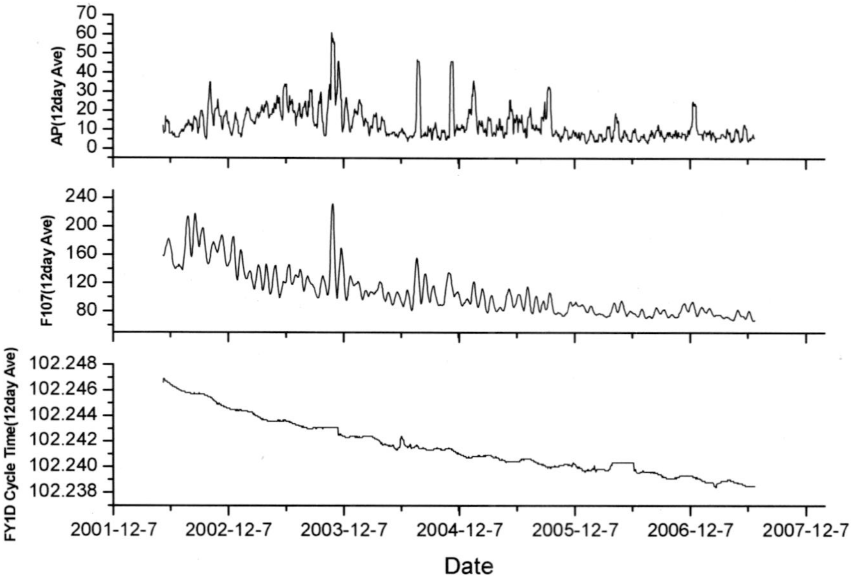

Fig. 1. 12 day's average of ap index, F10.7 index and FY1D cycle-time.

According to the paper of Montenbruck and Gill,[9] the air drag force on a satellite is given by:

$$F_d = -\frac{1}{2}\rho v^2 C_d S \tag{1}$$

where ρ is the atmospheric density, v is the satellite's orbital velocity relative to the atmosphere, C_d is the dimensionless drag coefficient and S is the satellite's cross-sectional area perpendicular to the direction of the motion. Applying the Law of Conservation of Energy and Keplers Laws, Eq. 1 will be turned into the form like this:

$$\frac{dT}{dt} = \frac{3\pi C_d S r}{m}\rho \tag{2}$$

where r is the radius of the satellite's orbit and m is the satellite's mass. To a certain satellite, $\frac{3\pi C_d S}{m}$ is a constant. So the atmosphere density is only determined by the radius of the satellite's orbit and variation of satellite's cycle time. We use FY1D's track data to derive the atmosphere density results, these density results are only relative variation results for we did not get the exact value of the constant, $\frac{3\pi C_d S}{m}$. We plot the density variation

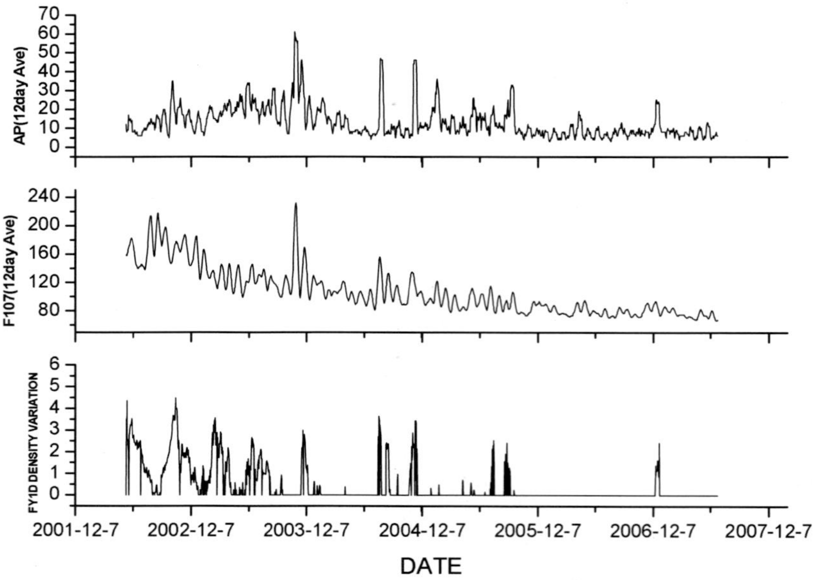

Fig. 2. Thermosphere density variations derived from FY1D track data, Ap index and F107 index.

results in Fig. 2, it does not mean zero but "absence of significant results" where the density variation values are null.

In next section, we divide the density variation results into three period parts: high solar activities year, geomagnetic storm days, low solar activities year. these three period will give us a scene to show how thermosphere density variations make responses to the space weather events.

3. Results and Discussions

We plot three periods of density variation results in Fig. 3. The top panel is the density variations in high solar activities year: from May 15, 2002 to September 22, 2003. In top panel, the peak values of density variations are decreasing with F107 index decreasing. It indicates that the solar EUV radiation contributes to thermosphere density enhancement and the solar EUV radiation makes accumulative effects to density variations. We found an interesting result that large peaks of density variations in top panel split to several small peaks and each small peak's period is about 27 days. Maybe it implies that thermosphere density variations are influenced by solar rotation period. In the middle panel of Fig. 3, we plot

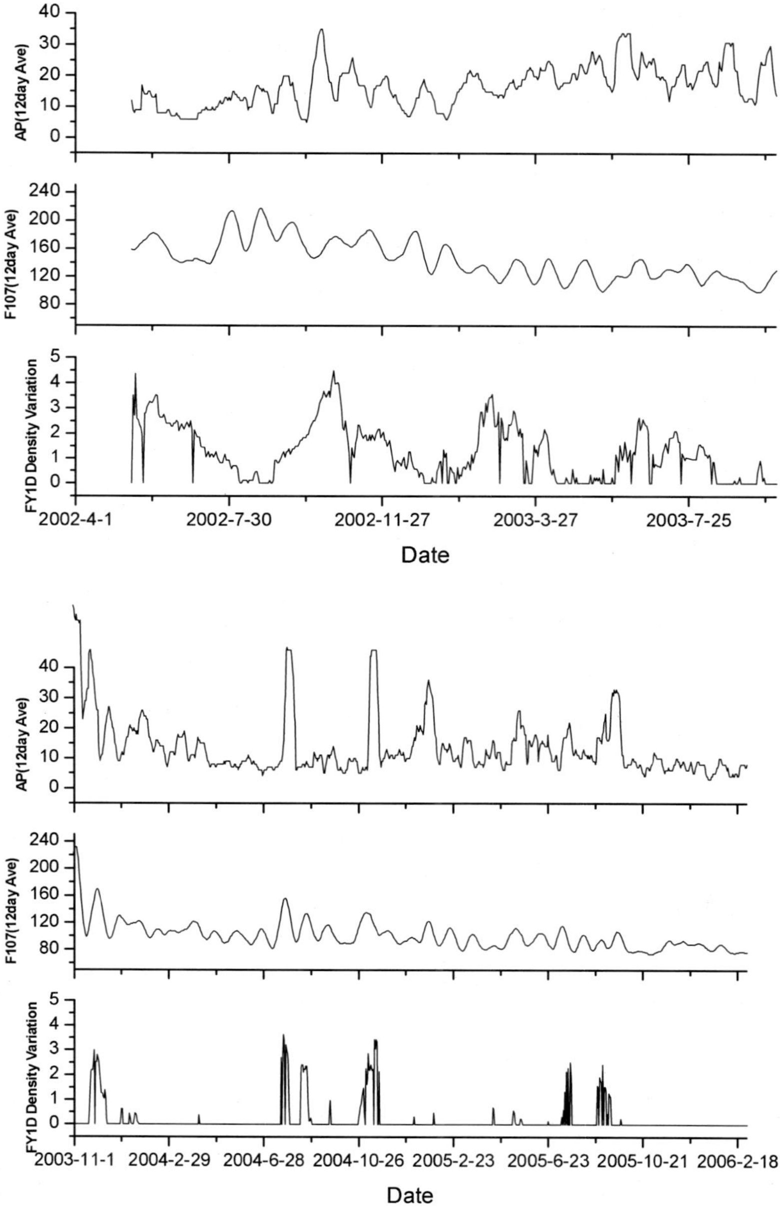

Fig. 3. Top panel: Density variations in high solar activities year; Middle panel: Density variations in geomagnetic storm days; Bottom panel: Density variation response to the space weather events in December 2006.

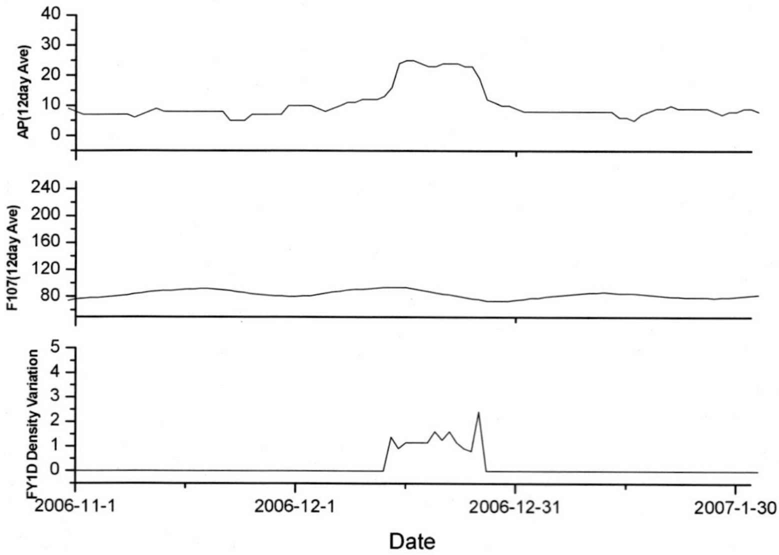

Fig. 3. (*Continued*)

thermosphere density variations versus intense geomagnetic storms. We found that each intense geomagnetic storm almost can cause an obvious density variation and the amplitudes of density variations are similar as the peak values of those density variation in high solar activities year. That means the contributions of intense geomagnetic storms to thermosphere density variations are as much as solar EUV radiation of high solar activities year. We plot the space weather events of December, 2006 and the corresponding density variations in the bottom panel of Fig. 3 to investigate the density variations' responses to space weather events in low solar activities year. In this panel, we found a good correlation between Ap index and density variations. It can be concluded that thermosphere density variations are sensitive to strong geomagnetic disturbances in low solar activities year.

In order to compare thermosphere density variations derived from air drag with model's results, we adopt NRLMSIS-00 model to calculate the air density of 870 km height and plot the results with Ap index and F107 index in Fig. 4. The MSIS (Mass-Spectrometer-Incoherent-Scatter) models such as MSIS-90 and NRLMSIS-00 model describe the neutral temperature and densities in the upper atmosphere (above 100 km), the data sources of NRLMSIS-00 model include total mass density from satellite

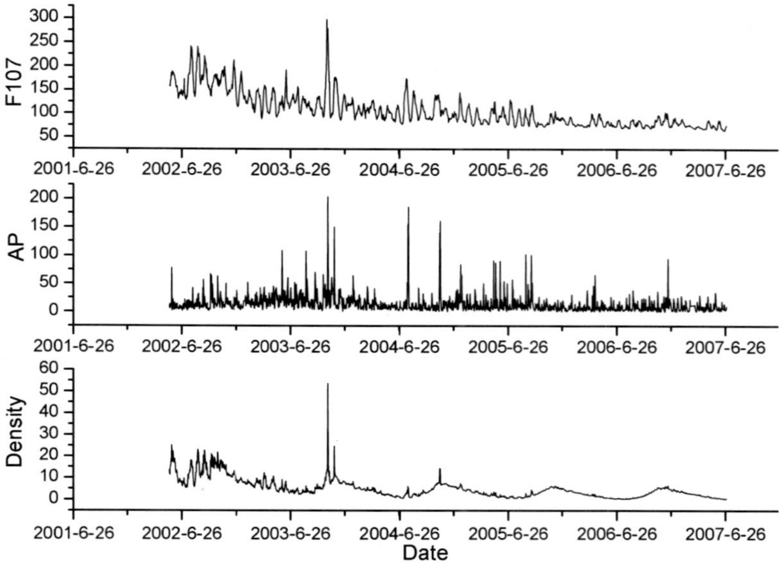

Fig. 4. NRLMSIS-00 model density results, Ap index, F107 index.

accelerometers as well as orbit determination and it makes NRLMSIS-00 model has good performances of forecasting atmosphere density in elevated geomagnetic activities.[10] However, the MSIS models are statistical models and underestimate atmosphere density in strong geomagnetic storm.[4,7,11–13] In Fig. 4, the model's density results reveal good correlations with solar EUV radiation and non-sensitive responses to intense geomagnetic storms. We plot NRLMSIS-00 model's density results versus Ap and F107 index in Fig. 5. We also plot our density variation results versus Ap and F107 index to do comparison with model's results in Fig. 5. In two top panels of Fig. 5, model's results reveal good relationship with F107 index and no obvious tendency with Ap index. Due to the interaction of solar EUV radiation and geomagnetic storms, it shows large dispersions in the bottom panels. Despite of these dispersions in bottom-right panel of Fig. 5, there exist more large density variations than small ones in the region of high Ap index ($Ap > 35$) otherwise than the model's results. The situation in Fig. 5 implies that the density variations derived from air drag are more sensitive to intense geomagnetic disturbances than the MSIS model and the model's density results give good reflections of atmosphere density with long-term solar activities.

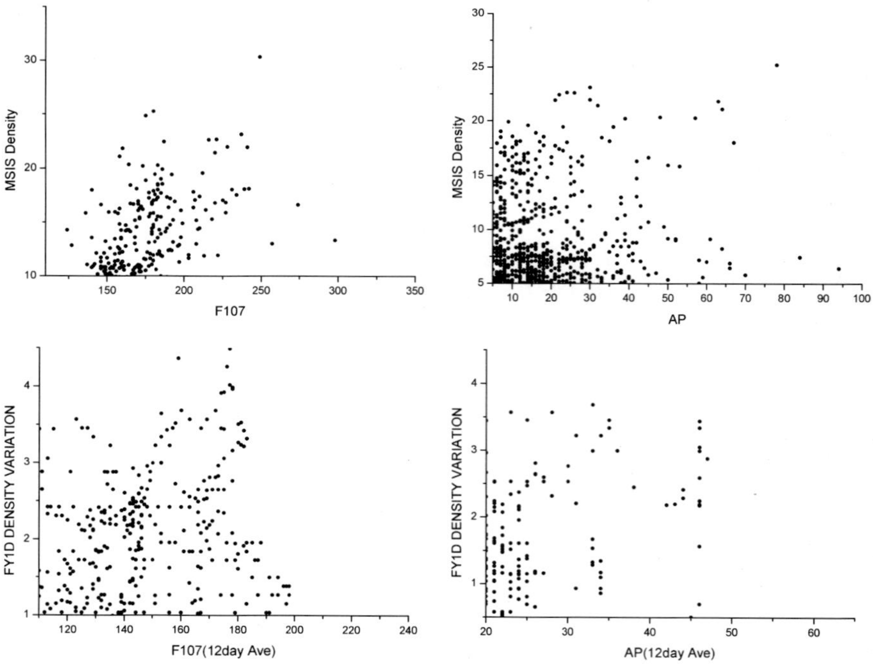

Fig. 5. Top-left Panel: NRLMSIS-00 model density results versus F107 index; Top-right Panel: NRlMSIS-00 model density results versus Ap index; Bottom-left Panel: FY1D density variations versus F107 index; Bottom-right Panel: FY1D density variations versus Ap index

4. Summary and Conclusion

In this study, we calculate thermosphere density variations derived from FY1D satellite's drag and analyse how thermosphere density variations respond to the space weather events in the past five years (2002–2007). Our results using new data set support previously published finding that thermosphere density variations have good responses to intense space weather events.[6,14] Solar EUV radiations show long-term effects to density variations and these variations are possibly influenced by sun's 27 days rotation period. In strongly disturbed geomagnetic interval, thermosphere density has intense perturbations. We also compared density variations with NRLMSIS-00 model's results. The comparison indicates that the density variations derived from air drag are more sensitive to intense geomagnetic storms than the MSIS model and the model's density results underestimate atmosphere density variations during the short-term of strong geomagnetic disturbance. Our results confirm that the current atmospheric models are statistical and underestimate thermosphere density in short-period intense geomagnetic storms.[7,11]

However, we do 12-day moving average to FY1D orbit data to exclude the effects of 12-day regression time. It makes us lose details of thermosphere density variations and provides only qualitative analysis of density variation responses to space weather events. Fortunately, the launch of FY3A satellite (the altitude of 830 km, sun-synchronous circular orbit) will provide us opportunity to derive new satellite's orbit data with the time resolution of 1 second. To apply FY3A orbit data, we desire new studies on quantitative analysis of thermosphere density variation responses to space weather events.

References

1. W. K. Tobiska, E10.7 Use for Global Atmospheric Density Forecasting in 2001, AIAA/AAS *Astrodynamics Specialist Conference and Exhibit*, Monterey, California, Aug. 5-8, AIAA-2002-4892 (2002).
2. D. J. Knipp, W. K. Tobiska and B. A. Emery, Direct and indirect thermosphere heating sources for solar cycle 21–23, *Sol. Phys.* **224** (2004) 495.
3. F. K. Chun, D. J. Knipp, M. G. McHang, *et al.*, Polar cap index as a proxy for hemispheric Joule Heating, *Geophys. Res. Lett.* **26** (1999) 1101.
4. H. Liu and H. Lühr, Strong disturbance of the upper thermospheric density due to magnetic storms: CHAMP observations, *J. Geophys. Res.* **110** A09S29, doi:10.1029/2004JA010908 (2005).
5. K. -H. Kim, Y. -J. Moon, K. -S. Cho *et al.*, Atmospheric dra effects on the KOMPSAT-1 satellite during geomagnetic superstorms, *Earth Planets Space* **58** (2006) e25–e28.
6. E. A. Rhoden, J. M. Forbes and F. A. Marcos, The influence of geomagnetic and solar variabilities on lower thermosphere density, *J. Atmos. Terr. Phys.* **62** (2000) 999.
7. J. -Y. Park, Y. -J. Moon, K. -H. Kim, *et al.*, Comparison between the KOMPSAT-1 drag derived density and the MSISE model density during strong solar and/or geomagnetic activies, *Earth Planets Space* **60** (2008) 1–6.
8. R. F. Donnelly, D. F. Heath, J. L. Lean and G. J. Rottman, Differences in the temporal variations of solar UV flux, 10.7-cm solar radio flux, sunspot number, and Ca-K plage data caused by solar rotation and active region evolution, *J. Geophys. Res.* **88** (1983) 9883–9888.
9. O. Montenbruck and E. Gill, *satellite orbits: Models, methods and applications* (Springer Press, Berlin 2001) 83.
10. J. M. Picone, A. E. Hedin, D. P. Drob and A. C. Aikin, NRLMSISE-00 empirical model of the atmosphere: Statistical comparisons and scientific issues, *J. Geophys. Res.* **107** CiteID 1468, doi:10.1029/2002JA009430 (2002).

11. H. Koskinen, E. Tanskanen, R. Pirjola, *et al.*, *Space Weather Effects Catalogue, ESA Space Weather Study in Pressed* (2001).
12. H. Liu, H. Lühr, V. Henize and W. Köhler, Global distribution of the thermospheric total mass density derived from CHAMP, *J. Geophys. Res.* **110** A04301, doi:10.1029/2004JA010741 (2005).
13. W. J. Burke, C. Y. Huang, F. A. Marcos and J. O. Wise, Interplanetary contral of thermospheric densities during large magnetic storms, *J. Atmos. Solar-Terr. Phys.* **69** (2007) 279–287.
14. S. H. Knowles, J. M. Picone, S. E. Thonnard and A. C. Nicholas, The effect of atmospheric drag on satellite orbits during the bastille day event, *Solar Physics* **204** (2001) 387–397.

Advances in Geosciences
Vol. 21: Solar & Terrestrial Science (2008)
Ed. Marc Duldig

OBSERVATIONS OF OH AIRGLOW AND LIGHTNING EXCITATIONS FROM THE FORMOSAT-2 SATELLITE

JAN B. NEE* and TING-HUNG PENG†
Department of Physics, National Central University,
No. 300, Jhongda Road, Jhongli City,
Taoyuan County, ROC 32054, Taiwan
**jbnee@cc.ncu.edu.tw*
†*93229008@cc.ncu.edu.tw*

OH airglow was measured by the Imager of Sprites and Upper Atmospheric Lightnings (ISUAL) instruments on board the FORMOSAT-2 satellite in 2004 and 2007. OH airglow is found to have an equatorial maximum and the peak latitude varied between ±15°. The peak height and emission rates are derived for these observations. OH airglow is also measured by using a broad band filter covering the wavelengths 630–750 nm associated with lightning events. An enhancement in OH airglow is observed and possible mechanisms are discussed.

1. Introduction

OH airglow in the mesosphere was first reported by Meinel[1] and studied by Bates and Nicolet.[2] The later proposed a reaction of hydrogen and ozone as a mechanism to produce the airglow emission. Because this chemical mechanism involves ozone molecule in the upper mesosphere, OH airglow reveals useful information about the chemistry and dynamics in the mesosphere. OH airglow has been extensively studied by ground based instruments,[3,4] rockets[5,6] and satellites.[7–11] Satellite measurement is useful for understanding the global variation and height distribution. Previous studies based on HRDI or WINDII on board UARS showed OH airglow has a peak height at about 88 km with an equatorial maximum caused by tidal activities as verified by GCM model.[9–11]

With a highly sensitive ICCD camera, ISUAL (Imager of Sprites and Upper Atmospheric Lightnings) instruments on board the FORMOSAT-2 satellite can measure several types of airglow emissions including O (green line and red line), OH Meinel bands, and O_2 A band. OH Meinel band

Table 1. Filters for CCD imager and their central wavelengths, bandwidth (Full width at half maximum or FWHM) and molecular emissions detectable by these filters.

Filter #	Wavelength λ (nm)	FWHM (nm)	Molecular emission type and height	
1	630–750	125	OH Meinel bands and N_2 (B → A)	87 km
2	762	7	O_2 (b → X) A band	90 km
3	630	7	O ($^1D \rightarrow {}^3P$)	200–300 km
			OH (9,3) band	87 km
4	557.7	6	O ($^1S \rightarrow {}^3D$)	90 km
5	427.8	6	O_2 (A → X) Herzberg band or N_2^+ (B → X)	200 km/95 km
6	Open	450*		Non-specific

*Filter 6 is a blank window with transmission in all visible wavelengths. The bandwidth is mainly determined by the response functions of the detector and atmospheric transmission which is roughly estimated to be 450 nm.

airglow has been observed with a filter (#3) peaked at the (9,3) band at 630 nm or with a broadband filter (#1) covering wavelengths 630–750 nm (There are six filters used to measure various optical events as shown in Table 1). Because of large field of view and high spatial resolution, ISUAL is able to provide detailed latitudinal distributions. Many other satellites have also measured airglow images showing spatial distributions, but often with less spatial resolution because of the need to bin data to improve signal to noise ratio.

Filter #1 has high transmission in the red wavelength region as shown in Table 1 and has been extensively used in observing lightning activities. Images measured by this filter show a background OH airglow layer. It is interesting that enhanced emissions in the OH airglow layer are observed following some lightning events. These observations are similar to the previous reports about lightning induced brightening in the D-region based on images measured by television cameras on board the Space Shuttle.[13,14] The past Shuttle missions have found single event or limited cases of enhanced airglow during the flight. In contrast, ISUAL has routinely observed many cases of enhanced airglow with an occurrence rate about a few percent. For example, on Oct. 21, 2004, there were seven cases of enhanced airglow images out of 114 total lightning events. ISUAL has also observed enhancement in O_2 airglow by using filter #2 centered at 762 nm.

2. ISUAL Instruments and FORMOSAT-2 Satellite

FORMOSAT-2 was launched on 21 May 2004 to a height of 891 km with a sun-synchronous orbit at an inclination angle of 98.9°. It has 14 orbits everyday such that the satellite sees the same region twice daily. More information of the scientific objectives, the payload instrument, and operation modes of the ISUAL project were given by Chern *et al.*[15]

The imaging instrument is nominally pointing 27.5° down from the satellite horizontal plane with a field of view of 20° horizontal and 5° vertical as shown in Fig. 1. The ISUAL instrument measures a limb view of the Earth over a range of 3370 km with a window about 80 km × 1200 km corresponding to 128 × 524 pixels in the CCD image with a spatial resolution of about 2.14 km for each pixel based on geometric factors. Although, a lower resolution may be more realistic. Pre-flight radiometric calibration of the CCD in the laboratory allows conversion of the CCD signal count to Rayleigh units. The imager can be set to observe airglow emissions or lightning events with two different modes — the SPRITE (SPR) mode and the AURORA (AUR) mode. In the SPR mode, it records six consecutive images following the lightning trigger. In figures shown in this paper, the first image (img 0) is the pre-lightning record. But this operation depends on the operational command. The exposure time for each image is 29 milliseconds with one millisecond in between any two neighboring images. In the AUR mode, it takes only one picture using 999 milliseconds of exposure time. AUR mode was used for airglow measurements in 2004 and SPR mode was used to capture transient lightning of lightning related processes. However, the airglow measurements in 2007 were carried out by SPR mode also, but with only one image for each record (instead of 6 as in lightning events).

Fig. 1. The first image out of 129 images measured by using filter 3 (630 nm) on October 7, 2004. The intensity is brighter in the left hand side than the right hand side because there are two CCD chips joined together in the camera. At the central part of the image the airglow layer has vertical pixel positions at about #90–100 and the earth limb is at about pixel #70. The image consists of 128 vertical pixels and 524 horizontal pixels.

Table 2. Summary of OH observations for 2004 and 2007 data. In the second and third column, the averages and errors of the peak height and emission rate are determined from retrieved profiles for the measurement period. The fourth column of the intensity variation indicates the percentage errors shown in column 3. The fifth column is the satellite orbit where measurements were taken. The last column shows the variation of the peak latitude for the measurement period.

Observation period	Average peak height and error (km)	Average emission rate and error (photon/cm^3-s)	Intensity variation	orbit	Peak latitude range
Oct. 7–17, 2004	86.3 ± 2.8	574.4 ± 94.4	16.4%	10	$-5° \sim +5°$
Apr. 9–22, 2007	89.2 ± 1.0	692.2 ± 47.1	6.8%	10	$-15° \sim +10°$
Aug. 7–16, 2007	89.8 ± 0.9	504.2 ± 137.1	27.2%	11	$-10° \sim +10°$
Oct. 6–14, 2007	89.0 ± 2.2	587.0 ± 95.5	16.3%	11	$-10° \sim +15°$
Nov. 10–17 2007	90.1 ± 2.4	553.5 ± 113.3	20.5%	4	$-10° \sim +15°$

The airglow images are processed to recover the height profile. We have employed an onion-peeling retrieval process. Results show OH airglow has a peak emission height at about 86–90 km which is consistent with previous satellite observations. Table 2 shows a summary of OH emission rate and peak height observed by FORMOSAT-2 during several flights in a specific orbit. The details are discussed in the following section.

3. OH Airglow Observations

Figure 1 shows the first record out of total 129 records of airglow images measured at UT16:22 October 7, 2004 using filter 3 with the AUR mode. The image was taken when the satellite passed from south to north along an orbit at longitude of about 100°E (#10) with the instrument looking to the side or east direction (perpendicular to the satellite's path from south to north). The observation covers the upper atmosphere over Indonesia, Taiwan and nearby area. The total observation time was about 3 minutes generating 129 images with each image about 1 s of time. The whole observation covered a latitude range from about $-8°$ to $+8°$ at longitudes of about 92° to 115°. The CCD imager was shut down at higher latitudes (~40°N). The airglow images had extensive overlapped latitudes between neighboring images (this was a learning period for instrument operation in 2004). This operation changed in 2007 with reduced overlapped latitudes between neighboring images so that a total of 40 images were measured to cover the same equatorial latitudes which required 129 images in 2004). In Fig. 1, we can see an asymmetric intensity between the left hand side and the right hand side of the picture. This is because the CCD camera jointed two CCD chips in the image plane, with each chip having different sensitivity. However, the difference in the response can be corrected to produce a uniform intensity through software and pre-flight calibration data. There is a concern about the uniform response of the 524 horizontal pixels especially for pixels near the edges. Calibration data show most of the central region of the images is uniform, but reduced responses in the left and right edges. We have therefore processed the images by using the central 10 pixels (covering latitudes of about 0.16° at an average latitude). By binning these 10 pixels and then combine all 129 images, we can produce a contour diagram of airglow intensity as functions of latitude and height as shown in Fig. 2. The choice of central 10 pixels has the additional advantage of reducing overlapped latitudes.

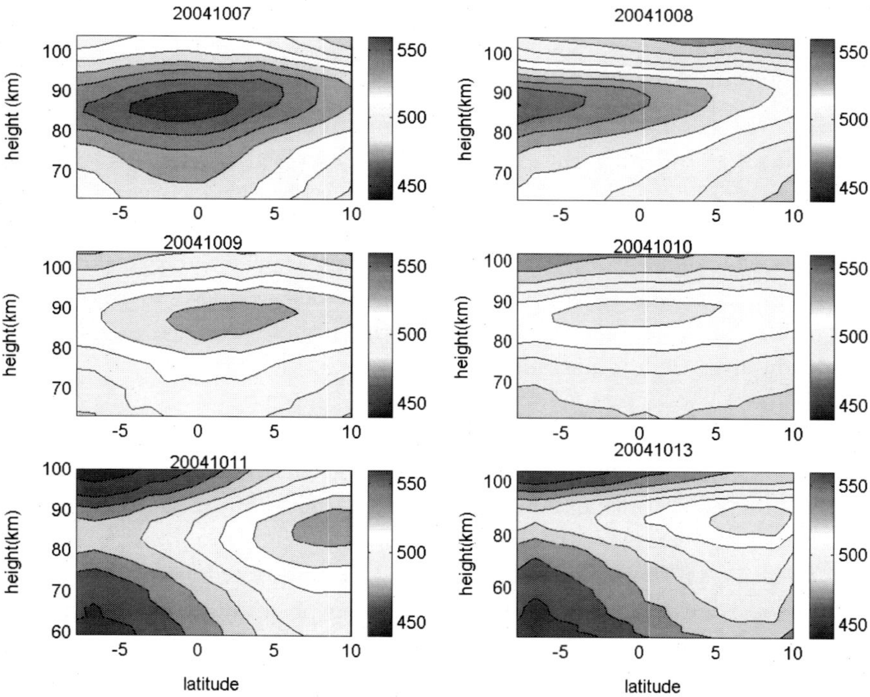

Fig. 2. The contour diagram of latitude and height for OH emissions observed on Oct. 7–13, 2004 with the exception of Oct. 12 where no observation was made. These 6 contour diagrams show intensity distribution of height versus latitude, with the intensity in photon counts as shown in the color bars. These contours are derived directly from 129 images similar to Fig. 1. Software was used to convert the pixel position (central pixels in this case) to latitude. In order to show performance of the instrument (for example, signal to noise ratio which can be read from the plots), the background signals were not removed. The peak airglow heights were set at 88 km.

Figure 2 shows contour diagrams of OH airglow intensity distribution in terms of latitude and height for 6 days in October, 2004. Each contour diagram consists of 129 images generated by method describe above. In the processing of height information, we have set the maximum intensity of the airglow at 88 km and calculated the height above and below the airglow layer by 2.14 km/pixel. Figure 2 shows the emission peak consistently occurred within a band of 15° to −15° latitude but varied from day to day. This variation of peak intensity has an oscillation with an apparent period of two to three days which will be discussed in a future publication.

OH airglow was again measured in 2007 with periods of 10–14 days in April, July, October, and November. Similar to Fig. 2, 2007 data show equatorial maximum with the emission peak latitude varied from day to day.

4. Transient Airglow Enhancement Produced by Lightning

ISUAL was designed to capture transient luminous events produced by the lightnings by using the SPR mode as described before. An example of transient airglow enhancement is shown in Fig. 3 which consists of 6 consecutive images taken during a lightning event that occurred on Oct 21, 2004. As we can see in Fig. 3, all 6 images show a bright OH airglow layer at the top part. An enhanced airglow layer was produced in the second image. In the third to sixth images, the OH airglow returned to its stable condition after the diminishing of the lightning.

Figure 4 shows two more events in 2004 and 2005. In Fig. 4 only the second image of the six is shown here. Figure 5 shows the height profiles for Fig. 3. From these figures, we can see a background OH airglow measured by this filter. The transmission curve of filter 1 (630–750 nm) actually covers several Mienel bands of OH emissions corresponding to vibrational levels v" = 6 − 9. When the lightning strikes, the OH airglow enhanced by a factor of 10 over a horizontal range of about 500 km produced in the 2nd image only which has a time less than 30 ms. All records show enhancement occurs only in the 2nd images whose rise and decay times are consistent with the few millisecond radiative lifetime of OH Meinel bands. The height in Fig. 5 was derived by referencing to airglow height at 88 km (pixel ∼95) and the earth limb (pixel ∼65) then calculated the heights for pixels in between. This calibration produces an effective vertical resolution 2.8 km/pixel which is larger than the 2.14 km/pixel given by the spec. The derived height positions are not very accurate in the lower region near the earth surface.

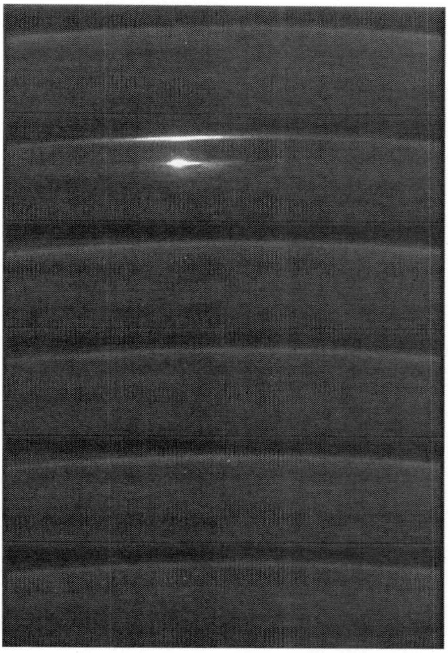

Fig. 3. Transient airglow enhancement observed on Oct. 21, 2004 record 28 in the second image captured by the CCD camera by using filter 1 (630–750 nm). Note the first image was the pre-lightning record.

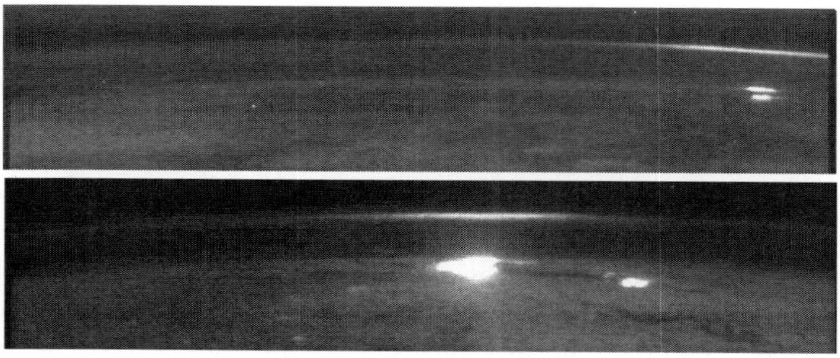

Fig. 4. More images of the OH airglow enhancement with the upper panel measured on March 22, 2005 and lower one on September 1, 2004.

For example, the lightning was at 20 km in the figure instead of 10 km in reality.

ISUAL has observed many lightning events but only a few events showing enhanced OH airglow. For example, the event rates for Oct. 21–25,

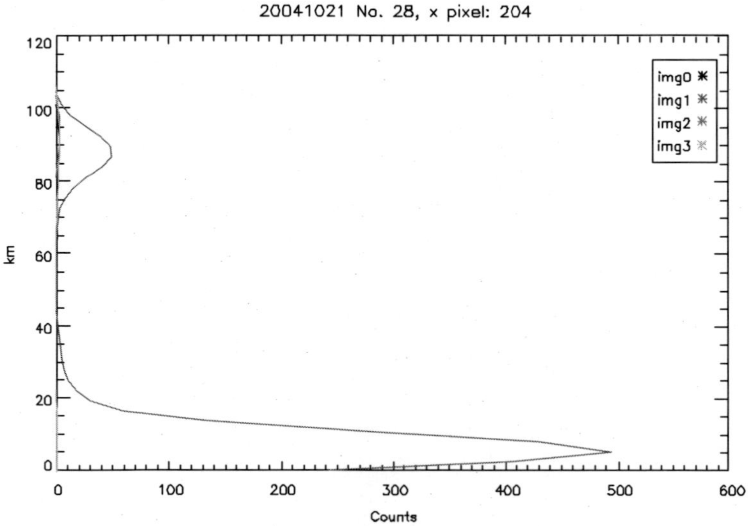

Fig. 5. Intensity profiles of transient airglow images showing in Fig. 3. The height information is determined by reference to the airglow height set at 88 km and earth limb at 0 km. Airglow enhancement can be seen in the second image (img 1:red line). Image 0 is pre-lightning which shows a background OH airglow with small intensity, and in the images 2 and 3, the airglow resume to its background condition.

2004 are 6.1%, 2.6%, 6.6%, 3.1%, and 1.2%, indicating only some particular types of EM pulses can produce such phenomenon.

5. Discussion

5.1. *Mesospheric OH airglow*

The equatorial maximum of OH airglow has been consistently reported by many satellite observations and is attributed to the diurnal tides which can vary the density or temperature conditions in the atmosphere.[8–11] The peak height of 88 km is consistent with many other space observations.[16–18] Mende *et al.* observed 630 nm OH airglow from space shuttle with limb view.[17] Their results are very similar to this one. Our studies of latitudinal variation show peak OH latitudes which are located within a range of $\pm 10°$ about the equator. This variation of peak latitude may be related with the ozone and temperature conditions in the mesosphere which determine the OH airglow production rate but are themselves affected by the dynamics. The (9,3) band OH Meinel band detected by filter 3 is the highest vibrational level ($v' = 9$) which can only be produced by the reaction

$H + O_3$. Other OH observations utilizing bands of lower vibrational levels of OH may involve cascading transitions from upper vibrational levels from which collisions between OH and the ambient atmosphere is important.

Two most important reactions related with OH airglow are given below.

$$\mathrm{H} + \mathrm{O_3} \rightarrow \mathrm{OH}^*(v' \leq 9) + \mathrm{O_2} \quad k_1 = 1.4 \times 10^{-10} \exp(-470/\mathrm{T}) \mathrm{cm}^3/\mathrm{s} \tag{1}$$

$$\mathrm{O} + \mathrm{O_2} + \mathrm{M} \rightarrow \mathrm{O_3} + \mathrm{M} \quad k_2 = 6.0 \times 10^{-34} (300/\mathrm{T})^{2.3} \mathrm{cm}^3/\mathrm{s} \tag{2}$$

These rate constants are based on De More *et al.*[12] Both reactions are temperature dependent and the first reaction shows higher rate at warmer temperature while the second reaction has a negative temperature dependence meaning higher rate at lower temperature. Simulation shows that stronger OH airglow intensity is produced at heights where the temperature is warmer. Recently, Marsh *et al.*[19] used the TIMED-SABER data to study the variation of OH airglow with temperature in the mesosphere, and they reported a positive correlation between both. ISUAL results found more detailed latitudinal information than previous studies. This variation could be used to understand the dynamics in the equatorial region.

5.2. *Lightning enhanced OH airglow*

Boeck *et al.*[13] reported enhanced airglow luminosity based on space shuttle observations. The event suggested a coupling between the lower and upper atmospheres over a distance of about 80 km. A subsequent report by Israilevich *et al.*[14] also discussed this luminous event in terms of ELVES (Emissions of Light and Very low frequency) produced by the electromagnetic pulse (EMP).[20] The EMPs are generated by positive or negative cloud to ground lightning events which can cause the strongest absorption of the EMP at the altitude of ~85 km.[20–22] Emissions of excited states of nitrogen and oxygen molecules can be produced by intense EMPs lasting 50 μs which have been studied by Taranenko *et al.*[22] and Inan *et al.*[23] These studies found enhanced nitrogen and oxygen emissions at the height about 85 km by simulation involving excited states of oxygen and nitrogen but without OH related processes. In agreement with these simulations, the oxygen emission in O_2 A band at 762 nm was indeed observed by ISUAL using filter 2.

The OH chemistry was not included in the study of Tarenenko *et al.*[22] To simulate the OH airglow, one needs to include chemistry of hydrogen and oxygen species. Lightning can produce transient excited states including O^* and $O_2{}^*$ to initiate chemical reaction. Detailed results involving excited states of oxygen by processes such as $O(^1D) + H_2$ will be reported later. Alternatively, a study involving temperature increase to enhance the OH airglow production rate has been discussed in Huang.[24]

6. Summary

The ISUAL instrument has measured mesospheric OH airglow in the (9,3) band with a filter centered at 630 nm. The airglow layer has a peak height at about 85–90 km and shows an equatorial maximum near midnight, which is consistent with previous studies. Filter #1 in the visible wavelengths (630–750 nm) has observed cases of transient airglow produced by lightning. The enhanced airglow shows an intensity increase by a factor about 10 in a time of less than 30 ms. We have shown observations and analysis of events of production of transient airglow enhancement. The mechanism for producing the OH airglow is briefly discussed.

Acknowledgments

The authors are thankful to the National Space Program Office, Taiwan for research grant with No.94-NSPO(B)-ISUAL-FA09-01. The authors wish to thank National Cheng Kung University (NCKU), Taiwan for providing necessary facilities to carry out research work. We also acknowledge useful discussions with Prof. G. Shepherd.

References

1. A. B. Meinel, *Ap. J.* **111** (1950) 555.
2. D. R. Bates and M. Nicolet, *J. Geophys. Res.* **55** (1950) 301.
3. J. R. Hecht, R. L. Walterscheid and M. N. Ross, *J. Geophys. Res.* **99**(A6) 11,449, doi: 10.1029/94JA00584 (1994).
4. H. Takahashi, Y. Sahai, B. R. Clemesha, D. M. Simonich, N. R. Teixeira, R. M. Lobo and A. Eras, *Planet. Space Sci.* **37**(649) doi: 10.1016/0032-0633(89)90035-4 (1989).
5. D. J. Baker and A. T. Jr. Stair, *Phys. Scr.* **37** (1988) 611.
6. B. R. Clemesha and H. Takahashi, *Advances in Space Res.* **17** (1996) 81.
7. V. J. Abreu, and J. H. Yee, *J. Geophys. Res.* **94** (1989) 11949.

8. S. M. L. Melo, R. P. Lowe and H. Takahashi, *J. Geophys. Res.* **104** 24, 657, doi: 10.1029/1999JA900291 (1999).
9. J. H. Yee, G. Crowley, R. G. Roble, W. R. Skinner, M. D. Burrage and P. B. Hays, *J. Geophys. Res.* **102** 19, 949, doi: 10.1029/96JA01833 (1997).
10. S. P. Zhang and G. G. Shepherd, *Geophys. Res. Lett.* **26** 529, doi: 10.1029/1999GL900033 (1999).
11. S. P. Zhang, R. G. Roble and G. G. Shepherd, *J. Geophys. Res.* **106** 21, 381, doi: 10.1029/2000JA000363 (2001).
12. W. B. DeMore, C. J. Howard, S. P. Sander, A. R. Ravishankara, D. M. Golden, C. E. Kolb, R. F. Hampson, M. J. Molina and M. J. Kurylo, Chemical Kinetics and Photochemical Data for Use in Stratospheric Modeling Evaluation #12 January 15, 1997, NASA, Jet Propulsion Laboratory.
13. W. L. Boeck, O. H. Jr. Vaughan, R. Blakeslee, B. Vonnegut and M. Brook, *Geophys. Res. Lett.* **19** (1992) 99.
14. P. L. Israilevich, Y. Yair, A. D. Devir, J. H. Joseph, Z. Levin, I. Mayo, M. Moalem, C. Price, B. Ziv and A. Sternlieb, *Geophys. Res. Lett.* **31** L06124, doi: 10.1029/2003GL019110 (2004).
15. J. L. Chern, R. R. Hsu, H. T. Su, S. B. Mende, H. Fukunishi, Y. Takahashi and L. C. Lee, *J. Atmos. Sol. Terr. Phys.* **65** (2003) 647.
16. G. Liu and G. G. Shepherd, *Geophys. Res. Lett.* **33** L09805, doi: 10.1029/2005GL025297 (2006).
17. S. B. Mende, G. R. Swenson, S. P. Geller, R. A. Viereck, E. Murad and C. P. Pike, *J. Geophys. Res.* **98**(19) (1993) 117.
18. G. G. Shepherd, G. Liu and R. G. Roble, *Advances in Space Res.* **35** 1945. doi: 10.1016/j.asr.2004.12.036 (2005).
19. D. R. Marsh, A. K. Smith, M. G. Mlynczak and J. M. Russell, *J. Geophys. Res.* **111** A10S05, doi: 10.1029/2005JA011451 (2006).
20. H. Fukunishi, Y. Takahashi, M. Kubota, K. Sakanoi, U. S. Inan and W. A. Lyons, *Geophys. Res. Lett.* **23** 2157, doi: 10.1029/96GL01979 (1996).
21. U. S. Inan, *Geophys. Res. Lett.* **17** (1990) 729.
22. Y. N. Taranenko, U. S. Inan and T. F. Bell, *Geophys. Res. Lett.* **20** (1993) 2675.
23. U. S. Inan, T. F. Bell and J. V. Rodriguez, *Geophys. Res. Lett.* **18** (1991) 705.
24. T. Y. Huang, *J. Geophys. Res.* **111** A11311, doi: 10.1029/2006JA011791 (2006).

Advances in Geosciences
Vol. 21: Solar & Terrestrial Science (2008)
Ed. Marc Duldig

SOLAR CYCLE LENGTH AND SOLAR ACTIVITY

SHINICHI WATARI
National Institute of Information and Communications Technology, Nukuikita 4-2-1, Koganei, Tokyo 184-8795, Japan

Length of solar cycle is a measure of solar activity. Average of cycle length is approximately 11 years. However, it varies from 9 to 13.6 years even during 23 cycles since 1755. We studied the relation between cycle length and maximum sunspot number. Maximum sunspot number shows a good negative correlation with length of the previous cycle. We forecasted maximum sunspot number of Cycle 24 of 91 ± 20 using this relation. Our forecast suggests that amplitude of Cycle 24 is smaller than that of Cycle 23.

1. Introduction

Sunspot numbers (SSNs) are a useful index to study solar activity. Because there is a long reliable record of SSNs since 1749 by Wolf.[1] Schove[2] found that solar activity varies with approximate 11-year period in variation of SSNs. A period between adjacent solar minima are called a solar cycle and Cycle 1 started in 1755. Usually maximum SSNs are used as an indicator of amplitude of each solar cycle. Solar radio flux of wavelength 10.7 cm is also used as an index of solar activity. However, the 10.7 cm flux measurement started in 1947 and covers only 6 solar cycles.

A measurement of carbon-14 covers long time periods. Miyahara *et al.*[3,4] made a measurement of carbon-14 content in tree rings between 1631 and 1739 using a 382-year-old Japanese cedar tree from Yakushima. They detected the periods around 13–15 and 24–29 years in the carbon-14 content. They pointed out that the length of the solar cycle was approximately 14 years during the Maunder Minimum (1645–1715)[5] although the Sun's magnetic field continued to reverse its polarity through the analysis of carbon-14 variation. Their result suggests that the length of solar cycle is an important factor to determine solar activity of each solar cycle. Eddy *et al.*[6] noted that the low latitude photospheric layers were rotating rapidly than at present during the Maunder Minimum. Those results suggest that solar dynamo was in a special mode in Grand minima, such as the Maunder Minimum.

As another aspect, forecasting amplitude of solar cycle is important for space weather.[7] Because severe space storms frequently occur in the solar cycles with large amplitude. Those storms have a potential to affect on manmade systems such as satellites, Global Navigation Seattleite System (GNSS), power grids, and so on. There are many methods to forecast amplitude of a solar cycle.[8–10] There are large variations in the forecasts of Cycle 24 using those methods.[10]

In this paper, we statistically studied parameters characterizing solar cycles using a past record of 23 cycles. We forecasted the maximum SSN of Cycle 24 using the result of our statistical study.

2. Data

We used the table of parameters characterizing solar cycles produced by the NOAA/National Geophysical Data Center (see Table 1). We made a statistical study on these parameters. Table 1 shows years of minimum, minimum SSNs, years of maximum, maximum SSNs, rise times, fall times, and cycle lengths.

3. Analysis

We calculated the Pearson's correlation coefficients between parameters characterizing solar cycles using Table 1.

Table 2 shows the correlation coefficients between the parameters of the same cycle. We tested significance of those correlation coefficients using t-values computed by Eq. (1).

$$t = \frac{|r|\sqrt{N-2}}{\sqrt{1-r^2}} \tag{1}$$

where r is the Pearson's correlation coefficient and $N-2$ is the degree of freedom for the Student's t-distribution.

According to the t-test of probability level of 0.05 for Table 2, the correlation coefficients are not equal to 0 for the cycle length and rise time, the cycle length and fall time, the cycle length and min. SSN, the rise time and fall time, the rise time and max. SSN, and the max. SSN and min. SSN.

From Table 2, we find that (1) the rise time and max. SSN show the best negative correlation of -0.821, (2) the fall time and cycle length and

Table 1. Parameters charactering solar cycles.

No.	Year of min.	Min. SSN	Year of max.	Max. SSN	Rise time (year)	Fall time (year)	Cycle length (year)
1	1755.2	8.4	1761.5	86.5	6.3	5.0	11.3
2	1766.5	11.2	1769.7	115.8	3.2	5.8	9.0
3	1775.5	7.2	1778.4	158.5	2.9	6.3	9.2
4	1784.7	9.5	1788.1	141.2	3.4	10.2	13.6
5	1798.3	3.2	1805.2	49.2	6.9	5.4	12.3
6	1810.6	0.0	1816.4	48.7	5.8	6.9	12.7
7	1823.3	0.1	1829.9	71.7	6.6	4.0	10.6
8	1833.9	7.3	1837.2	146.9	3.3	6.3	9.6
9	1843.5	10.5	1848.1	131.6	4.6	7.9	12.5
10	1856.0	3.2	1860.1	97.9	4.1	7.1	11.2
11	1867.2	5.2	1870.6	140.5	3.4	8.3	11.7
12	1878.9	2.2	1883.9	74.6	5.0	5.7	10.7
13	1889.6	5.0	1894.1	87.9	4.5	7.6	12.1
14	1901.7	2.6	1907.0	64.2	5.3	6.6	11.9
15	1913.6	1.5	1917.6	105.4	4.0	6.0	10.0
16	1923.6	5.6	1928.4	78.1	4.8	5.4	10.2
17	1933.8	3.4	1937.4	119.2	3.6	6.8	10.4
18	1944.2	7.7	1947.5	151.8	3.3	6.8	10.1
19	1954.3	3.4	1957.9	201.3	3.6	7.0	10.6
20	1964.9	9.6	1968.9	110.6	4.0	7.6	11.6
21	1976.5	12.2	1979.9	164.5	3.4	6.9	10.3
22	1986.5	12.3	1989.6	158.5	2.8	6.8	9.6
23	1996.4	8.0	2000.3	120.8	3.9	—	—
Mean values:		6.1		113.2	4.3	6.7	11.0

Table 2. Correlation coefficients between parameters of the same solar cycle.

	Cycle length	Rise time	Fall time	Max. SSN	Min. SSN
Cycle length	1.000	0.451	0.526	−0.377	−0.547
Rise time	0.451	1.000	−0.521	−0.821	−0.341
Fall time	0.526	−0.521	1.000	0.421	−0.194
Max. SSN	−0.377	−0.821	0.421	1.000	0.510
Min. SSN	−0.547	−0.341	−0.194	0.510	1.000

the max. SSN and min. SSN show a good positive correlation, (3) the cycle length and min. SSN and the rise time and fall time show a good negative correlation.

Table 3 shows the correlation coefficients between the parameters of adjacent cycles. We tested significance of the correlation coefficients in Table 3 using Eq. (1).

Table 3. Correlation coefficients between parameters of adjacent solar cycles.

Next cycle	Cycle length	Rise time	Fall time	Max. SSN	Min. SSN
Cycle length	0.139	−0.182	0.296	0.180	−0.096
Rise time	0.680	0.202	0.458	−0.258	−0.307
Fall time	−0.500	−0.362	−0.132	0.412	0.190
Max. SSN	−0.647	−0.359	−0.277	0.349	0.559
Min. SSN	−0.293	0.001	−0.270	0.191	0.433

According to the t-test of possibility level of 0.05 for Table 3, the correlation coefficients are not equal to 0 for the cycle length and rise time of the next cycle, the cycle length and fall time of next cycle, the cycle length and max. SSN of the next cycle, and the fall time and rise time of the next cycle.

From Table 3, we find that (1) the cycle length and rise time of the next cycle show the best positive correlation of 0.680, (2) the min. SSN and max. SSN show a positive correlation, (3) the cycle length and fall time of the next cycle and the cycle length and max. SSN of the next cycle show a good negative correlation.

We show several plots of a pair of the parameters in Tables 2 and 3. They are useful for forecasting the next cycles.

Figure 1 is plots of (a) cycle length and max. SSN of the same cycle and (b) cycle length and max. SSN of the next cycle. The solid lines show the result of the least square fittings. The dashed curves are the 95% confidence intervals for expected values. This figure shows that cycle length and max. SSN of the next cycle have a good negative correlation of −0.647 while the correlation between length and max. SSN of the same cycle is low. This suggests that longer cycle length is a signature of smaller max. SSN of the next cycle.

The least square fitting to the data presented in Fig. 1(b) yields the equation of the form:

$$\text{Max. SSN of the next cycle} = -21.6 \times \text{Cycle length} + 352.3 \qquad (2)$$

Figure 2 is a plot of rise time and max. SSN for the same cycle. According to this figure, there is a good negative correlation between them (−0.821). Faster rising cycle tends to have lager max. SSN. This is known as "Waldmeier Effect".[11,12] Eq. (3) is the result of the fitting for Fig. 2.

$$\text{Max. SSN} = -27.7 \times \text{Rise time} + 233.1 \qquad (3)$$

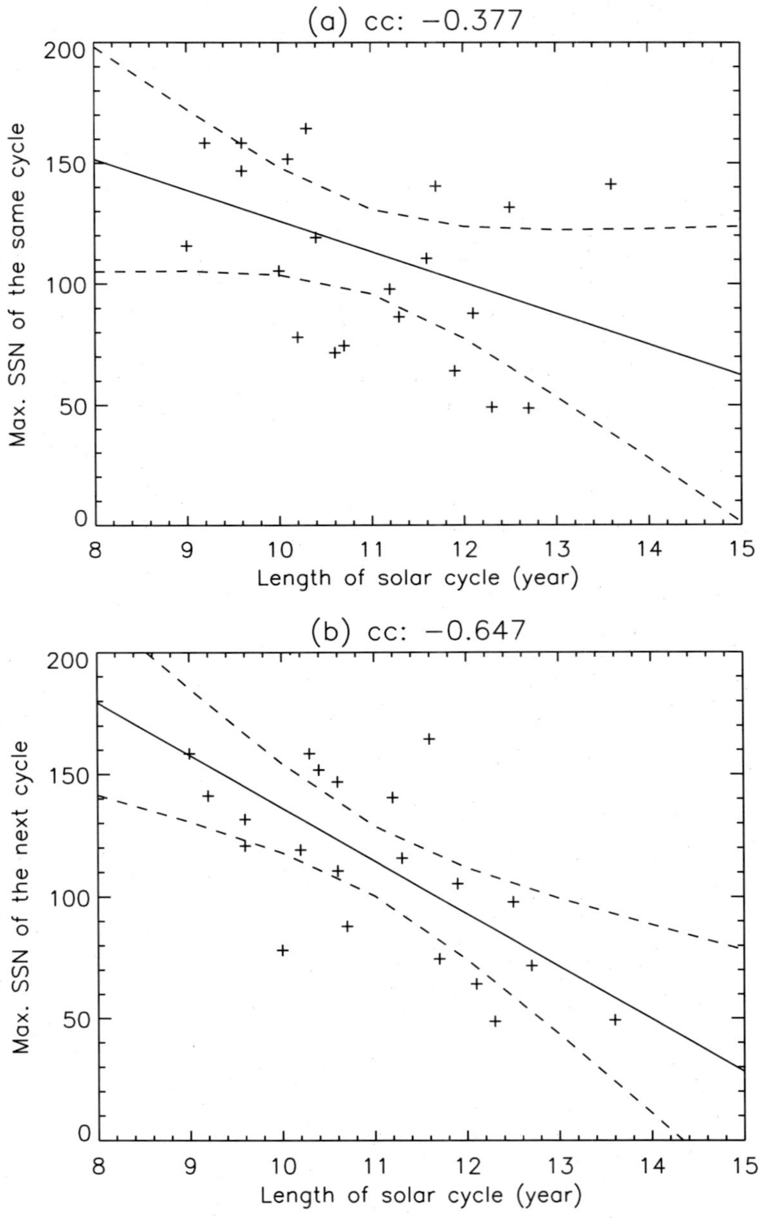

Fig. 1. Plots of (a) length of solar cycle and max. SSN of the same cycle and (b) length of solar cycle and max. SSN of the next cycle. Solid lines show the least square fittings. Dash lines show the 95% confidence intervals for expected values.

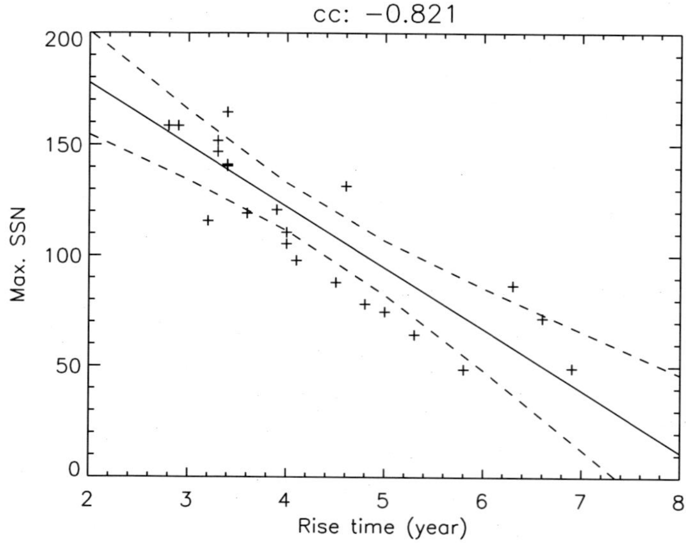

Fig. 2. Plot of rise time and max. SSN.

Figure 3 is a plot of cycle length and rise time of the next cycle. There is a good positive correlation of 0.680 between them as expected from Figs. 1(b) and 2. Rise time of the next cycle becomes longer after the cycle with longer length. Equation (4) is the result of the fitting for Fig. 3.

$$\text{Rise time of the next cycle} = 0.6 \times \text{Cycle length} - 2.7 \tag{4}$$

On the other hand, there is a low correlation between the cycle length and rise time of the same cycle (cf. Table 2).

Figure 4 is a plot of rise time and fall time for the same cycle. There is a good negative correlation between them (−0.521). Cycles with shorter rise time tends to have longer fall time. Equation (5) is the result of the fitting for Fig. 4.

$$\text{Fall time} = -0.5 \times \text{Rise time} + 9.0 \tag{5}$$

4. Results and Discussion

According to Fig. 1, a longer period between adjacent minima (cycle length) is a signature of a smaller max. SSN of the next cycle. This result supports the longer cycle length during the Maunder Minimum.[3,4] The cycle length might be an appearance of condition of the solar dynamo. Miyahara *et al.*[3,4]

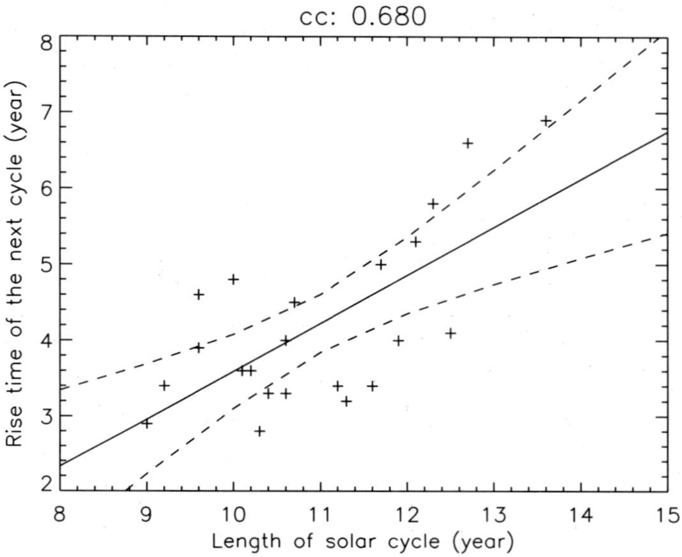

Fig. 3. Plot of length of solar cycle and rise time of the next cycle.

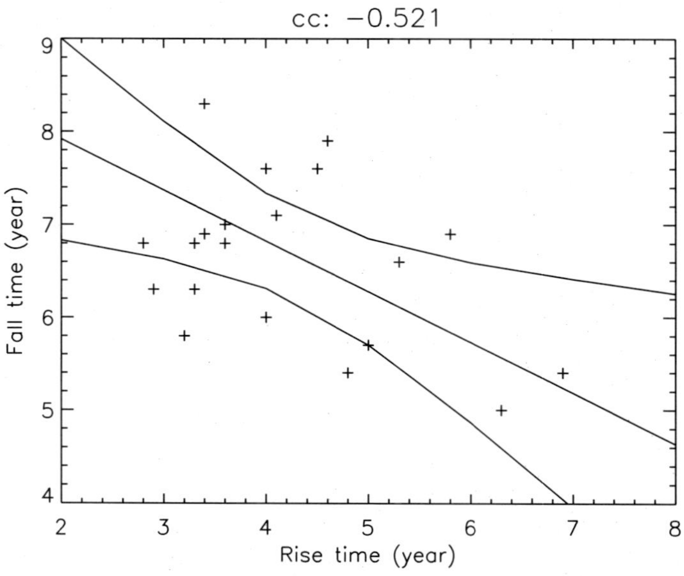

Fig. 4. Plot of rise time and fall time.

noted that the magnetic Hale cycle kept working even during the Maunder Minimum.

The result of our statistical study can be used for forecasting the time of minimum of Cycle 23 and the time and SSN at maximum of Cycle 24. We got the fall time of Cycle 23 of 7.1 years when we inputted the rise time of cycle 23 of 3.9 years to Eq. (5). The forecasted time of the minimum is in May, 2007. However, the SSN is still decreasing in the upper half of 2008.

We can forecast the max. SSN and the time of maximum of Cycle 24 using Eqs. (2) and (4) and the length of Cycle 23. If we assume the time of minimum of Cycle 23 is in Jun, 2008, the length of Cycle 23 is 12.1 years. Inputting this value to Eqs. (2) and (4), we got the max. SSN of Cycle 24 of 91 ± 20 and the rise time of 4.6 years. The expected time of the maximum of Cycle 24 is in Feb, 2013. Our result suggests that the amplitude of Cycle 24 is smaller than that of Cycle 23 of 120.8. According to the condition of 2008, the cycle length seems to become longer than our assumption. This suggests that the max. SSN becomes less than 91 ± 20.

We forecasted the max. SSN of Cycle 24 of 91 ± 20 in Feb, 2013 using the result of our statistical study. According to our forecast, activity of Cycle 24 will be smaller than that of Cycle 23. There is the solar Cycle 24 prediction panel organized by the NASA and NOAA. This panel made two predictions for Cycle 24 in Apr, 2008. One is 140 ± 20 in Oct, 2011 and another is 90 ± 10 in Aug, 2012. Our forecast is close to the lower prediction by the panel.

Figure 5 is a plot of number of no-sunspot days around minimum and max. SSN of the next cycle. The number of no-sunspot days is counted in two-year interval centered the solar minimum using daily sunspot number reported from the SIDC, the Royal Observatory of Belgium. This figure shows that there is a good negative correlation of -0.663 between the number of no-sunspot days and the SSN at maximum of the next cycle. Equation (6) is the result of the fitting for Fig. 5.

$$\begin{aligned} &\text{Max. SSN of the next cycle} \\ &\quad = -0.2 \times \text{Number of no-sunspot days} + 187.2 \end{aligned} \tag{6}$$

From Eq. (6), we got the max. SSN of Cycle 24 of 101 ± 15 for number of no-sunspot days of 429 days between Jan, 2007 and Dec, 2008. This result is slightly larger than the prediction computed from Eq. (1).

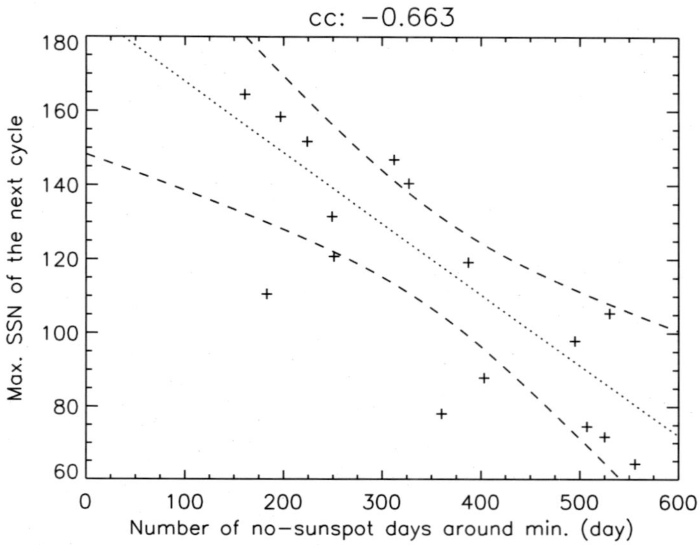

Fig. 5. Plot of number of no-sunspot days around minimum and max. SSN of the next cycle.

5. Summary

We showed that cycle length is an important parameter to describe characteristics of solar cycles through our study. There is a good negative correlation between cycle length and max. SSN of the next cycle. This supports the longer cycle length during the Maunder Minimum obtained by the analysis of the carbon 14 variation.[3,4] Babcock[13] proposed the solar dynamo process as a driver of solar cycle variation. The weakening of the solar dynamo process might produce a longer and weaker cycle.

According to our prediction, the amplitude of Cycle 24 is smaller than that of Cycle 23. There are many predictions in the literature that have predicted ycle 24 to be weaker than Cycle 23.[14] Schatten[15] expected the max. SSN of 120 ± 40 peaking in Apr. 2011 ± 3 months based on a solar dynamo amplitude index. Svalgaard *et al.*[16] predicted 75 ± 8 in 2011 using average polar field value during approximately three years before sunspot minimum. Hiremath[17] predicted 110 ± 11 based on the model of a forced and damped harmonic oscillator. Choudhuri *et al.*[18] suggested that Cycle 24 will be approximately 35% weaker than Cycle 23 using their solar dynamo model. However, prediction remains a controversial issue as some authors have suggested Cycle 24 to be stronger than Cycle 23.[19]

Acknowledgments

We acknowledge the NOAA/National Geophysical Data Center for the table of parameters characterizing solar cycles. We would like to thank the SIDC-team, World Data Center for the Sunspot Index, Royal Observatory of Belgium for daily sunspot numbers.

References

1. R. Wolf, *Astron. Mitt., Zurich* **24** (1868) 111.
2. D. J. Schove, *J. Geophys. Res.* **60**(2) (1955) 127.
3. H. Miyahara, K. Masuda, Y. Muraki, H. Furuzawa, H. Menjo and T. Nakamura, *Sol. Phys.* **224**(1–2) (2004) 317.
4. H. Miyahara, D. Sokoloff and I. G. Usoskin, *Advances in Geosciences* **2**(ST) (2006) 1.
5. J. A. Eddy, *Science* **192**(4245) (1976) 1189.
6. A. Eddy, P. A. Gilman and D. E. Trotter, *Sol. Phys.* **46** (1976) 3.
7. L. J. Lanzerotti, P. Song, H. Singer and G. Siscoe (eds.), *Space Weather, Geophys. Monogr. Ser.* **125** (2001) 11.
8. D. H. Hathaway, R. M. Wilson and E. J. Reichmann, *J. Geophys. Res.* **104**(A10) (1999) 22375.
9. D. H. Hathaway and R. M. Wilson, *Sol. Phys. J. Geophys. Res.* **224**(1–2) (2004) 5.
10. R. P. Kane, *Solar Phys.* **243**(2) (2007) 205.
11. M. Waldmeier, *Astron. Mitt. Zurich* **14**(133) (1935) 105.
12. M. Waldmeier, *Astron. Mitt. Zurich* **14**(138) (1939) 439–470.
13. H. W. Babcock, *Astrophys. J.* **133**(2) (1961) 572.
14. K. Schatten, *J. Geophys. Res.* **107**(A11) (2002) SSH 15–1H.
15. L. Svalgaard, E. W. Cliver and Y. Kamide, *Geophys. Res. Lett.* **32**(1) (2005) L01104.
16. A. R. Choudhuri, P. Chatterjee and J. Jiang, *Phys. Rev. Lett.* **98**(13) (2007) 131103.
17. K. M. Hiremath, *Astrophys. and Space Science* 314(1–3) (2008) 45.
18. V. N. Obridko and B. D. Shelting, *Sol. Phys.* **248**(1) (2008) 191.
19. M. Dikpati and P. A. Gilman, *Astrophys. J.* **649**(1) (2006) 498.

Advances in Geosciences
Vol. 21: Solar & Terrestrial Science (2008)
Ed. Marc Duldig

DRIFT EFFECTS AND THE AVERAGE FEATURES OF COSMIC RAY DENSITY GRADIENT IN CIRS DURING SUCCESSIVE TWO SOLAR MINIMUM PERIODS

A. FUSHISHITA, T. NARUMI, C. KATO,
S. YASUE and K. MUNAKATA*
Physics Department, Shinshu University,
Matsumoto, Nagano 390-8621, Japan
**kmunaoo@shinshu-u.ac.jp*

Y. OKAZAKI
Department of Geophysics, Tohoku University,
Sendai, Miyagi 980-0861, Japan

T. KUWABARA, J. W. BIEBER and P. EVENSON
Bartol Research Institute and
Department of Physics and Astronomy,
University of Delaware, Newark, DE 19716, USA

M. R. DA SILVA and A. DAL LAGO
National Institute for Space Research (INPE),
12227-010 Sao Jose dos Campos, SP, Brazil

N. J. SCHUCH
Southern Regional Space Research Center (CRS/INPE),
P.O. Box 5021, 97110-970, Santa Maria, RS, Brazil

M. TOKUMARU
Solar Terrestrial Environment Laboratory,
Nagoya University, Nagoya, Aichi 464-8601, Japan

M. L. DULDIG
Australian Antarctic Division,
Kingston, Tasmania 7050, Australia

J. E. HUMBLE
School of Mathematics and Physics,
University of Tasmania, Hobart,
Tasmania 7001, Australia

I. SABBAH
Astronomy Department, Faculty of Science,
King Abdulaziz University, Jeddah 21589, Saudi Arabia
on leave from Physics Department,
Faculty of Science,
University of Alexandria, Egypt

J. KÓTA
Lunar and Planetary Laboratory,
University of Arizona, Tucson, AZ 87721, USA

We deduce on hourly basis the spatial gradient of the cosmic ray density in three dimensions from the directional anisotropy of high-energy (~50 GeV) galactic cosmic ray (GCR) intensity observed with a global network of muon detectors on the Earth's surface. By analyzing the average features of the gradient in the corotational interaction regions (CIRs) recorded in successive two solar activity minimum periods, we find that the observed latitudinal gradient (G_z) changes its sign from negative to positive on the Earth's heliospheric current sheet (HCS) crossing from the northern to the southern hemisphere in A $<$ 0 epoch, while it changes from positive to negative in A $>$ 0 epoch. This is in accordance with the drift prediction. We also find a negative enhancement in G_x after the HCS crossing in both A $<$ 0 and A $>$ 0 epochs, but not in G_y. This asymmetrical feature of G_x and G_y indicates significant contributions from the parallel and perpendicular diffusions to the the gradient in CIRs in addition to the contribution from the drift effect.

1. Introduction

The drift model of cosmic ray transport in the heliosphere predicts a bidirectional latitude gradient of the GCR density, pointing in opposite directions on opposite sides of the HCS.[1,2] The predicted spatial distribution of the GCR density has a minimum along the HCS in the "positive" polarity period of the solar polar magnetic field (also referred as the A $>$ 0 epoch), when the interplanetary magnetic field (IMF) directs away from (toward) the Sun in the northern (southern) hemisphere, while the distribution has the local maximum on the HCS in the "negative" period (A $<$ 0) with the opposite field orientation in each hemisphere. The field orientation reverses every 11 years around the maximum period of solar activity. By analyzing the GCR density measured with the anticoincidence guard on board *ACE* and *Helios* satellites and by ground-based neutron monitors (NMs), Richardson *et al.*[3] and Richardson[4] investigated GCR modulation by CIRs, which are formed at the leading edges of corotating high-speed solar wind streams originating in coronal holes on the solar surface. From statistical analyses of variations in many CIRs with and without the HCS, the above authors concluded that the IMF sector boundary (i.e. HCS) does not organize the GCR density, contrary to expectations from the simple drift model. On the other hand, they also found observational evidence that the amplitude of the CIR-related modulations of the GCR density around the solar minimum is ~50% larger in A > 0 epochs than in A < 0 epochs. The GCR modulation in CIRs is still an open question.

These analyses were based on a one dimensional spatial distribution of the GCR density measured as the temporal variation of the GCR count rate recorded by each detector as it sweeps across the CIR. A single detector measurement of the density cannot, however, observe the spatial distribution separately from the temporal variation of the density at a particular location in space. The spatial distribution sampled by a single detector also does not reflect the global distribution surrounding the detector, above and below the detector's viewing path. It is also affected by local irregularities in the solar wind. This can be serious, particularly when analyzing satellite measurements of low-energy (sub-GeV) GCRs.

The spatial distribution of GCRs also can be inferred from the GCR spatial gradient in three dimensions, which provides us with additional information on the distribution around the detectors viewing paths. The diffusive flux equation describes the cosmic ray anisotropy in terms of the solar wind convection and the GCR density gradient. By inversely solving the diffusive flux equation, including the drift, Okazaki *et al.*[5] (hereafter referred to as paper I) deduced the gradient from the anisotropy that is derived from the observation made by the Global Muon Detector Network (GMDN). The GMDN has a major response to 50GeV GCRs whose Larmor radii are as large as 0.2AU in 5nT IMF. They analyzed the three geocentric solar eclipic coordinate (GSE) components of the gradient observed in association with CIRs in a single solar rotation period (Carrington Rotation CR 2043) in 2006, around the solar activity minimum of A < 0 epoch. It was found that the temporal evolution of the gradient before and after the HCS crossing is consistent with the prediction of the drift model, that the GCR distribution has a local maximum on the HCS.[1,2] As far as the gradient in this particular rotation period concerned, on the other hand, there is no clear signature of the significant modulation in association with enhancements in the solar wind velocity and the IMF magnitude in the CIR. Examining the drift effect solely from the gradient in a single rotation in A $<$ 0 epoch, however, is not easy, as the local maximum of the GCR density on the HCS is expected not only from the drift effect, but also from the minimum solar wind velocity on the HCS in a CIR. In the present paper, therefore, we examine the drift effect by analyzing the average gradient of multiple CIRs observed during the minimum periods of the solar activity in A $<$ 0 and A $>$ 0 epochs.

2. Analysis and Result

2.1. *Superposition analysis of CIRs at the HCS crossing*

We obtain the average feature of the GCR density gradient by superposing multiple CIRs at the timing of the HCS crossing in both A $<$ 0 and A $>$ 0 epochs. Throughout this paper, we use the hourly mean solar wind velocity and IMF data from the Omni2 data for A $>$ 0 epoch and from *ACE*-Level2 data for A $<$ 0 epoch, respectively.[a,b] We use *ACE*-Level2 data lagged by one hour, as a rough correction for the solar wind transit time between *ACE* and the Earth. By using IMF components (B_x,B_y,B_z) in GSE coordinates, the IMF sector is designated *toward* (T) if $B_x > B_y$ and *away* (A) if $B_y > B_x$. We define the HCS crossing at the timing when the sector polarity (*toward* or *away*) designated by 23-hours central moving average of B_x and B_y alternates. We then choose HCS crossings associated by the enhancement of the solar wind velocity with the maximum velocity exceeding 500 km/s and the IMF magnitude with the maximum exceeding 10 nT. According to this selection criteria, we chose 42 CIRs observed in 2006–2008 for A $<$ 0 period and 38 CIRs in 1995–1998 for A $>$ 0 epoch, respectively. The number of CIRs superposed for A $>$ 0 period is limited mainly due to many data gaps in the Omni2 data. Figure 1 displays the solar wind velocity and the source surface synoptic charts (Radial 250 model) by the Wilcox Solar Observatory. The HCS crossings identified in this paper are also indicated by the vertical solid line in this figure. The average HCS tilt angle in A $<$ 0 epoch analyzed in this paper is 15.2°, while it is 9.0° in A $>$ 0 epoch. The superposed IMF and solar wind velocity are displayed in Figs. 2a–e as functions of time for ±5 days from the HCS crossing. Clear signatures of CIR around the timing of the HCS crossing, such as enhancements of the IMF magnitude and the solar wind velocity (Figs. 2a and 2d) and

[a]Omni2 data are available at http://omniweb.gsfc.nasa.gov/html/ow_data.html
[b]*ACE*-Level2 data are available at http://www.srl.caltech.edu/ACE/ASC/

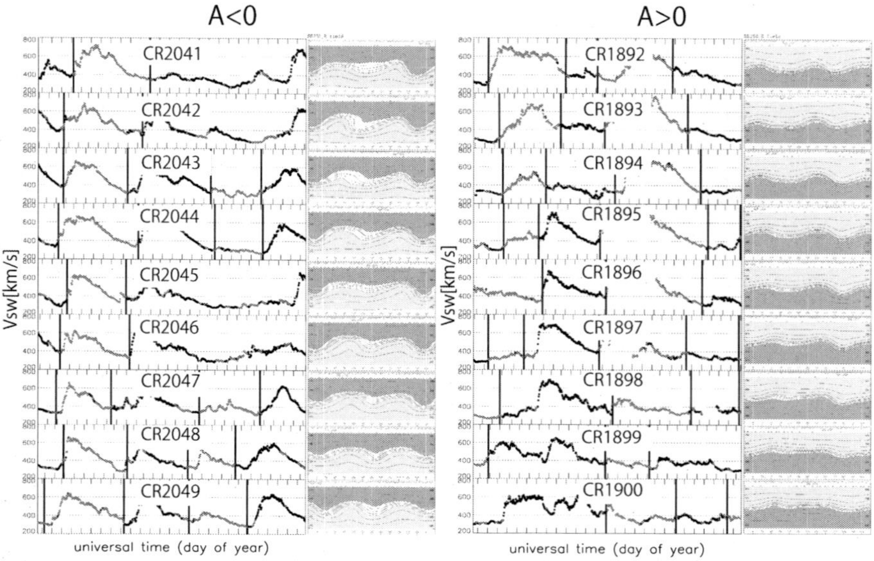

Fig. 1. Sample HCSs superposed in this paper for A < 0 (left) and A > 0 (right) epochs. Each panel displays the solar wind velocity in a solar rotation period as a function of the universal time (day) on the horizontal axis. The time of the Earth's HCS crossing is indicated by the vertical solid line, while the away and toward IMF sectors are indicated by black and gray data points, respectively. Attached on the right side in each figure is the source surface synoptic charts (Radial 250) in the corresponding rotation by the Wilcox Solar Observatory.[c]

reversals of the azimuthal solar wind flow angle (Fig. 2e), are clearly seen in this figure. We also note that the superposed IMF longitude in *away* (*toward*) sector in Fig. 2b is close to 135° (315°) and the average latitude in Fig. 2c is close to 0° regardless the IMF sector. This implies that the nominal Parker spiral field with the spiral angle of 45° is a good approximation of the observed IMF as long as the average field is concerned.

2.2. *Analysis of GCR density gradient in three dimensions*

We deduce the GCR density gradient in three dimensions on hourly basis from the anisotropy observed with the network of muon detectors. Readers can refer to paper I for each muon detector in the network and also for the

[c] The source surface synoptic charts and the HCS tilt angle are available at http://wso.stanford.edu/

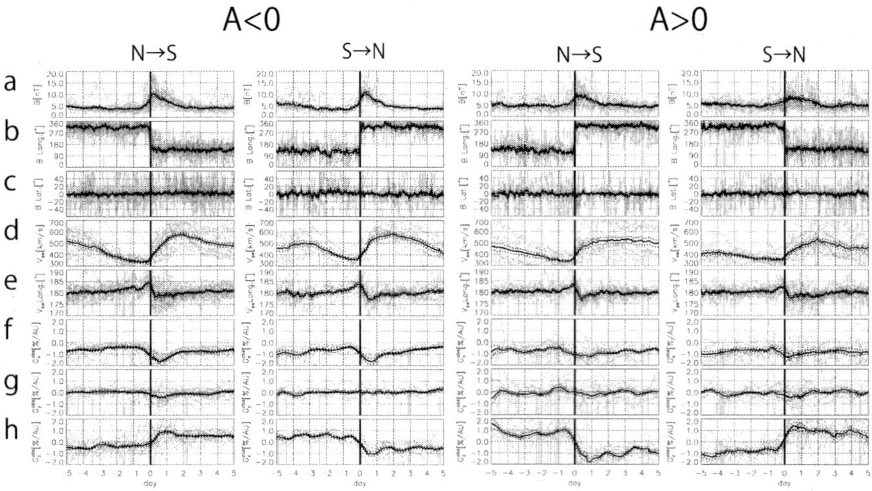

Fig. 2. From top to bottom, each panel shows the IMF magnitude (a), the GSE-longitude and latitude of the IMF (b-c), the solar wind velocity (d), the GSE-longitude of the solar wind (e) and three GSE components of the observed gradient (f-h). CIRs are superposed at the HCS crossing in A $<$ 0 (left) and A $>$ 0 (right) epochs. Thin gray curves in each panel display individual CIRs superposed, while the black curve shows the average as a function of time from the HCS crossing indicated by a vertical solid line. The dotted curves above and below the black curve bound the error of the average evaluated from the standard deviation of superposed CIRs. Only $\pm$5 days period from the HCS crossing is plotted. The superposition at the HCS crossing from the northern (southern) hemisphere to the southern (northern) hemisphere is indicated by "N→S" ("S→N") at the top of each figure.

detail analysis method to derive the gradient from the observed anisotropy. In this paper, we analyze the average gradient in multiple CIRs observed in 2006–2008 with the GMDN during the minimum period of the solar activity in A $<$ 0 epoch. We also derive the average gradient in 1995–1998 in A $>$ 0 epoch by using the observation with a couple of muon detectors at Nagoya and Hobart and examine the dependence of the gradient on the polarity of the solar magnetic field.

We first correct the observed anisotropy vector in GSE coordinates for the solar wind convection anisotropy and the Compton-Getting anisotropy arising from the Earth's 30 km/s orbital motion. We then divide the corrected anisotropy vector $\boldsymbol{\xi}(t)$ into components parallel and perpendicular to the IMF as

$$\boldsymbol{\xi}(t) = \boldsymbol{\xi}_{\parallel}(t) + \boldsymbol{\xi}_{\perp}(t). \tag{1}$$

These components are given as

$$\boldsymbol{\xi}_{\|}(t) = R_L(t)\alpha_{\|}\mathbf{G}_{\|}^{Diff}(t), \tag{2}$$

$$\boldsymbol{\xi}_{\perp}(t) = R_L(t)[\alpha_{\perp}\mathbf{G}_{\perp}(t) - \mathbf{b}(t) \times \mathbf{G}_{\perp}(t)], \tag{3}$$

where $\mathbf{G}_{\|}^{Diff}(t)$ and $\mathbf{G}_{\perp}(t)$ are respectively the density gradient components parallel and perpendicular to the IMF, $R_L(t)$ is the particle's effective Larmor radius of 50 GeV protons in the observed IMF which is 0.2 AU in 5nT IMF, $\mathbf{b}(t)$ is the unit vector pointing in the direction of the IMF, and $\alpha_{\|}$ and $\alpha_{\perp}$ are the dimensionless mean free paths ($\lambda_{\|}(t)$ and $\lambda_{\perp}(t)$) of GCR scattering by magnetic irregularities, defined as

$$\alpha_{\|} = \lambda_{\|}/R_L(t), \tag{4}$$

$$\alpha_{\perp} = \lambda_{\perp}/R_L(t). \tag{5}$$

In this paper, we adopt an ad hoc choice of $\alpha_{\|} = 7.2$ and $\alpha_{\perp} = 0.36$ which were also assumed in paper I. Qualitative features would remain similar for a wide range of reasonable parameters. Inversely solving Eqs. (2) and (3) for $\mathbf{G}_{\|}^{Diff}(t)$ and $\mathbf{G}_{\perp}(t)$, we obtain

$$\mathbf{G}_{\|}^{Diff}(t) = \frac{1}{\alpha_{\|}R_L(t)}\boldsymbol{\xi}_{\|}(t), \tag{6}$$

$$\mathbf{G}_{\perp}(t) = \mathbf{G}_{\perp}^{Diff}(t) + \mathbf{G}_{\perp}^{Drift}(t) \tag{7}$$

where,

$$\mathbf{G}_{\perp}^{Diff}(t) = \frac{\alpha_{\perp}}{(1+\alpha_{\perp}^2)R_L(t)}\boldsymbol{\xi}_{\perp}(t) \tag{8}$$

and,

$$\mathbf{G}_{\perp}^{Drift}(t) = \frac{1}{(1+\alpha_{\perp}^2)R_L(t)}\mathbf{b} \times \boldsymbol{\xi}_{\perp}(t). \tag{9}$$

Finally $\mathbf{G}(t)$ can be expressed in terms of two distinct contributions respectively from the diffusion and drift effects as,

$$\mathbf{G}(t) = \{\mathbf{G}_{\|}^{Diff}(t) + \mathbf{G}_{\perp}^{Diff}(t)\} + \mathbf{G}_{\perp}^{Drift}(t), \tag{10}$$

where the first two terms in the bracket on the right hand side express the contribution from the parallel and perpendicular diffusions respectively, while the second term denotes the contribution from the drift effect.

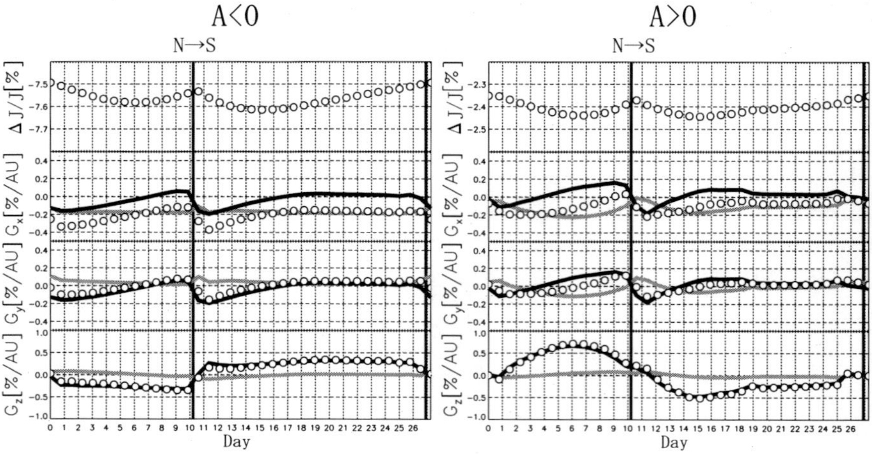

Fig. 3. The numerical solution of a simple drift model for 50 GeV GCRs. From top to bottom, each panel displays the GCR density and three GSE components of the gradient as a function of day in a solar rotation period in A $<$ 0 (left) and A $>$ 0 (right) epochs. The numerical values are shown by open circles. Contribution from the diffusion effect is shown by gray curve, while the contribution from the drift effect is displayed by black curve. The HCS crossing is indicated by a vertical black line. See paper I for the detail of this numerical model.

Three GSE-components of the superposed $\mathbf{G}(t)$ are displayed in Figs. 2f–h, while the numerical prediction of a simple drift model are shown in Fig. 3. This model assumes that the solar wind velocity and IMF magnitude are constant without any enhancement seen as CIR signatures. This model is identical to that which paper I compared with the observation and readers can refer to paper I for more detail of this model. The observed gradient $\mathbf{G}(t)$ in Eq. (10) shows a large fluctuation due to the large fluctuation of the IMF orientation (b_x, b_y, b_z). To show the systematic variation of each component, we filter this fluctuation with a central 23 hour moving average of hourly data. In Fig. 2, the gradient in each CIR event is displayed by a gray curve, while the average gradient of all CIR events is plotted as the black curve. It is clear that the average GSE-z component of the gradient (G_z) in Fig. 2h is negative (positive) above (below) the HCS in A $<$ 0 epoch, while it is positive (negative) above (below) the HCS in A $>$ 0 epoch. This is consistent with the drift model prediction in Fig. 3 at least qualitatively. We confirm this for the first time with the average gradient in multiple CIRs observed. It is also seen that the GSE-x component of the average gradient (G_x) remains negative in Figs. 2f implying that the ambient radial density gradient is positive and the density increases away from the Sun in response to the outward convection by the solar wind.

In Fig. 2f, we also note the negative enhancement in G_x observed following the HCS crossing. Such enhancements were first observed in both G_x and G_y in a sample rotation period in paper I and attributed to the drift effect causing the bidirectional density gradient toward the HCS in the ecliptic plane, together with the contribution from the ambient radial gradient away from the Sun. Similar enhancements are also seen in both G_x and G_y in the drift model prediction in Fig. 3. The observed gradient in Fig. 2, however, shows the enhancement only in G_x, but not in G_y. This asymmetry is interpreted in terms of the contribution from the diffusion effect as follows. Figure 4 shows three components of the average $\mathbf{G}(t)$ by dotted curves, together with contributions from the diffusion effect $(\mathbf{G}_{\|}^{Diff}(t) + \mathbf{G}_{\perp}^{Diff}(t))$ by gray curves and the drift effect $(\mathbf{G}_{\perp}^{Drift}(t))$ by black curves in Eq. (10). In A < 0 epoch, $\mathbf{G}_{\|}^{Diff}(t)$ due to the inward diffusion along the IMF is dominant as well as $\mathbf{G}_{\perp}^{Drift}(t)$. $\mathbf{G}_{\perp}^{Drift}(t)$ shows the negative enhancement in both G_x and G_y, while $\mathbf{G}_{\|}^{Diff}(t)$ shows the negative enhancement in G_x, but the positive enhancement in G_y. This is because of the geometry of the nominal Parker field in the GSE coordinate system. This asymmetric contribution from $\mathbf{G}_{\|}^{Diff}(t)$ to G_x and G_y results in the

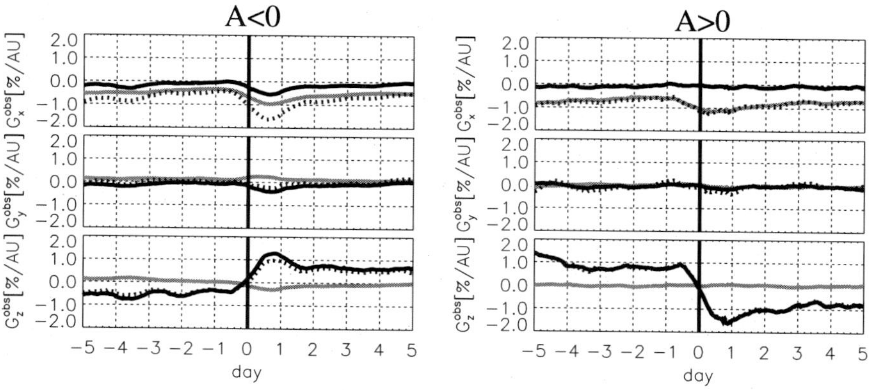

Fig. 4. Contributions from the diffusion and drift effects to three GSE components of the average gradient $\mathbf{G}(t)$. From top to bottom, dotted curve in each panel displays the average G_x, G_y and G_z in Fig. 2, while gray and black curves show respectively the contributions to $\mathbf{G}(t)$ from the diffusion $(\mathbf{G}_{\|}^{Diff}(t) + \mathbf{G}_{\perp}^{Diff}(t))$ and drift $(\mathbf{G}_{\perp}^{Drift}(t))$ effects, in A < 0 epoch (left) and A > 0 epoch (right). G_x and G_y in this figure are obtained by averaging those in "N→S" and "S→N" in Figs. 2f and 2g. We also calculate the average for G_z in Fig. 2h after reversing sign of G_z in "S→N" in Fig. 2h.

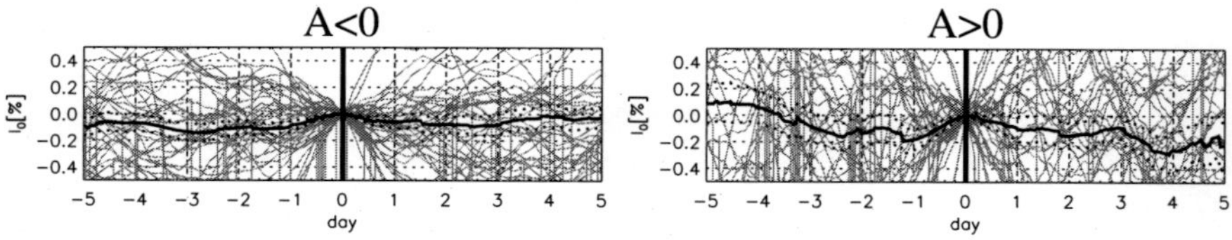

Fig. 5. GCR density superposed at the HCS crossing in A < 0 (left) and A > 0 (right) epochs. This figure is plotted in the same format as Fig. 2, except that the derived density is normalized to 0% at the HCS crossing to clarify the variation associated with the HCS crossing.

asymmetry seen in Fig. 2. This asymmetry, on the other hand, is less evident in A > 0 epoch, because of another contribution from $\mathbf{G}_{\perp}^{Diff}(t)$ due to the inward perpendicular diffusion which results in negative enhancements in both G_x and G_y. These contributions to the observed $\mathbf{G}(t)$ from the parallel and perpendicular diffusions are not clarified by a sample rotation analysis in paper I and first revealed in this paper by analyzing the average gradient in multiple CIR events. We are now preparing for a quantitative analyses of each contribution from $\mathbf{G}_{\parallel}^{Diff}(t)$, $\mathbf{G}_{\perp}^{Diff}(t)$ and $\mathbf{G}_{\perp}^{Drift}(t)$.

In Fig. 5, we finally display the GCR density which is derived as the omnidirectional intensity, together with the anisotropy vector, from the best-fitting to the observed muon count rates in various directional channels (see paper I). In this figure, we normalize the density to 0% at the HCS crossing to clarify the variation associated with the HCS crossing. It is seen that the amplitude of the average variation is small and the fluctuation is large even for 50 GeV GCRs recorded with muon detectors, easily masking the average systematic variation seen in Fig. 3 by the numerical model. This difficulty due to the fluctuation can be serious, particularly when analyzing low-energy GCRs, which have smaller Larmor radii and are more sensitive to small-scale structures in the solar wind. The variation in this figure is much less significant than the average gradient variation in Fig. 2. This implies how difficult is to deduce the GCR spatial distribution solely from single detector measurement of the GCR density.[3,4]

3. Discussions

In Table 1, we compare the average GCR gradient and density with the model prediction. We first note that the magnitude of G_z is 50% larger in A > 0 epoch in accordance with the model prediction. The magnitude of negative G_x, on the other hand, is larger in A > 0 epoch, while the model

Table 1. GCR density gradient (G_x,G_y,G_z), density (I_0), solar wind velocity (V_{SW}) and IMF magnitude (B) averaged over a period in Fig. 2 in A $<$ 0 and A $>$ 0 epochs. The upper lines for the gradient and density represent the observed average, while the lower lines represent the model prediction. The average G_z is calculated after reversing the sign of G_z in *toward* sectors. The period between $\pm$1 day from the HCS crossing is excluded from the average calculation. Errors for the observed values are deduced from the standard deviation of CIRs used for the calculation, while errors for the model prediction are derived from the standard deviation of data points in Fig. 3.

Heliomagnetic polarity	A $<$ 0	A $>$ 0
$G_x[\%/AU]$	$\mathbf{-0.706 \pm 0.006}$	$\mathbf{-0.781 \pm 0.009}$
	-0.215 ± 0.016	-0.105 ± 0.017
$G_y[\%/AU]$	$\mathbf{-0.017 \pm 0.005}$	$\mathbf{-0.060 \pm 0.009}$
	-0.005 ± 0.015	-0.003 ± 0.016
$G_z[\%/AU]$	$\mathbf{0.585 \pm 0.005}$	$\mathbf{0.890 \pm 0.009}$
	0.239 ± 0.017	0.358 ± 0.047
$I_0[\%]$	$\mathbf{-0.077 \pm 0.004}$	$\mathbf{-0.112 \pm 0.008}$
	-0.042 ± 0.005	-0.039 ± 0.006
$V_{SW}[km/s]$	$\mathbf{482.74 \pm 1.18}$	$\mathbf{449.01 \pm 1.32}$
B[nT]	$\mathbf{4.373 \pm 0.021}$	$\mathbf{5.112 \pm 0.029}$

predicts G_x almost twice in A $<$ 0 epoch in an oppsite sense. We also note the average density (I_0) being 50% lower in A $>$ 0 epoch. As we normalized I_0 to 0% at the HCS crossing, this implies the amplitude of the density variation before and after the HCS crossing is 50% larger in A $>$ 0 epoch, in accordance with Richardson *et al.*[4] reporting that the amplitude of the recurrent GCR intensity variations in solar rotation periods are 50% larger in A $>$ 0 epoch than in A $<$ 0 epoch.

In concluding anything from these observed features about the drift model prediction, however, we should be aware of that the gradient in A $>$ 0 epoch presented in this paper might be biased, because it is derived from an incomplete network before the GMDN started operation in March, 2006. To check such a possible bias, we analyzed A $<$ 0 epoch by using only two detectors (at Nagoya and Hobart) used for A $>$ 0 in the GMDN and confirmed that the derived gradient can be 10 $\sim$ 30% larger than that derived from the entire GMDN depending on values of $\alpha_{\|}$ and $\alpha_{\perp}$ assumed. In order to precisely clarify the difference of the gradient in A $<$ 0 and A $>$ 0 epochs, therefore, we need to improve the analysis method further and correct the observed gradient in A $>$ 0 for this bias effect.

Acknowledgments

This work is supported in part by U.S. NSF grant ATM-0527878 and NASA grant NNX07-AH73G, and in part by Grants-in-Aid for Scientific Research from the Ministry of Education, Culture, Sports, Science and Technology in Japan and by the joint research program of the Solar-Terrestrial Environment Laboratory, Nagoya University. The observations with the Kuwait University muon detector are supported by the Kuwait University grant SP03/03. We thank N. F. Ness for providing *ACE* magnetic field data via the *ACE* Science Center.

References

1. J. R. Jokipii and J. Kóta, *Geophys. Res. Lett.* **9** (1982) 656.
2. J. Kóta and J. R. Jokipii, *Astrophys. J.* **265** (1983) 573.
3. I. G. Richardson *et al.*, *J. Geophys. Res.* **101** (1996) 13483.
4. I. G. Richardson, *Space Sci. Rev.* **111** (2004) 267.
5. Y. Okazaki *et al.*, *Astrophys. J.* **681**(paper I) (2008) 693.

Advances in Geosciences
Vol. 21: Solar & Terrestrial Science (2008)
Ed. Marc Duldig

SKY-MAPS OF THE SIDEREAL ANISOTROPY OF GALACTIC COSMIC RAY INTENSITY AND ITS ENERGY DEPENDENCE

K. MUNAKATA*, N. MATSUMOTO, S. YASUE†,
C. KATO‡ and S. MORI
Physics Department, Shinshu University,
Matsumoto, Nagano 390-8621, Japan
**kmuna00@.shinshu-u.ac.jp*
†shyasue@.shinshu-u.ac.jp
‡ckato@shinshu-u.ac.jp

M. TAKITA
Institute for Cosmic Ray Research,
University of Tokyo, Kashiwa, Japan
takita@icrr.u-tokyo.ac.jp

M. L. DULDIG
Australian Antarctic Division, Kingston,
Tasmania 7050, Australia
marc.duldig@aad.gov.au

J. E. HUMBLE
School of Mathematics and Physics,
University of Tasmania, Hobart,
Tasmania 7001, Australia
john.humble@utas.edu.au

J. KÓTA
Lunar and Planetary Laboratory,
University of Arizona, Tucson, Arizona, USA
kota@lpl.arizona.edu

We analyze the sidereal daily variations observed between 1985 and 2006 at Matsushiro, Japan (MAT) and between 1993 and 2005 at Liapootah, Tasmania (LPT). These stations comprise the two hemisphere network (THN) of underground muon detectors in Japan and Australia. Yearly mean harmonic vectors at MAT and LPT are more or less stable without any significant change in phase and amplitude in correlation with either the solar activity- or magnetic-cycles. In this paper, therefore, we analyze the average anisotropy over the entire observation periods, i.e. 1985–2006 for MAT and 1993–2005 for LPT. We apply to the THN data a best-fitting analysis based on a model anisotropy in space identical to that adopted by Amenomori *et al.*[1] for Tibet III data. The median energies of primary cosmic rays recorded are ~0.5 TeV for THN and ~5 TeV for the Tibet III experiment. It is shown that the intensity distribution of the best-fit anisotropy is quite similar to that derived from Tibet III data, regardless of the order of magnitude difference in energies of primary particles. This, together with the THN observations, confirms that the analysis by Amenomori *et al.*[1] based on the Tibet III experiment in the northern hemisphere is not seriously biased. The best-fit amplitudes of the anisotropy, on the other hand, are only one third or less of those reported by the Tibet III experiment, indicating attenuation due to solar modulation. The rigidity dependence of the anisotropy amplitude in the sub-TeV region is consistent with the spectrum reported by Hall *et al.*,[2] smoothly extending to the Tibet III result in the multi-TeV region. The amplitude at higher energies appears almost constant or gradually decreasing with increasing rigidity. The rigidity spectrum indicating the solar modulation also supports the conclusion first implied by the Super Kamiokande deep underground experiment that the large scale anisotropy observed by Tibet III is due to the charged component of primary cosmic rays, and not due to high energy gamma rays to which underground muon detectors have negligible response.

1. Introduction

Long term observations of the count rates of cosmic rays recorded by underground muon telescopes have consistently reported the existence of an average diurnal variation in sidereal time.[3–7] Historically, this variation has been studied in an attempt to elucidate the origin of galactic cosmic rays (GCRs) with energies between a few hundred and a few thousand GeV. The observations from underground muon telescopes also provide interesting comparisons with those made from air-shower experiments, which respond to primary cosmic rays with energies more than an order of magnitude higher.[8,9] Most investigations have concluded that the variation is small (~0.1%) and the apparent right ascension of intensity maximum is somewhere in the early hours of the local sidereal day. Presently, no common consensus exists about the production mechanism of this variation in the local region of the galaxy. Information on the declination of the anisotropy can only be obtained by making multiple observations of various regions of the sky. This was not possible until recently. With multi-directional telescopes, the amplitude of this variation was shown to be distributed asymmetrically about the celestial equator.[5] It has now been conclusively demonstrated that the sidereal daily variation will be recorded with the largest amplitude in its first (diurnal) harmonic when viewed from around mid-latitudes in the southern hemisphere.[10,11] This is called the North-South (NS) asymmetry of the sidereal diurnal variation.

Aiming to reveal detail features of this NS asymmetry for understanding the physical mechanism responsible for such an asymmetry, the two-hemisphere network (THN) of underground muon telescopes at Matsushiro (at a vertical depth of 220 meters of water equivalent: m.w.e.) in central Japan and at Liapootah (154 m.w.e.) in Tasmania, Australia started operation in December, 1991 and continuously monitored the high energy GCR intensity until observations at Liapootah ended in March 2006. By using the initial four years of data from the THN, together with other underground muon data available, Hall *et al.*[2,12] used non-linear iterative fitting procedures to model the observed daily variations by a superposition of two Gaussian functions. Based on the best-fit parameters, they plotted the observed intensity distributions in a contour map on the celestial sphere and found some notable features, particularly in the intensity excess region. First, this region had its maximum shifted significantly southward over the equator and, second, its distribution is not symmetric around the maximum direction, appearing "squashed" along a plane which is not aligned with either the solar or galactic equatorial planes. The physical implications of these features are still unknown.

By analyzing GCRs recorded by the Tibet III air shower experiment, Amenomori *et al.*[13] first presented a precise sky-map of GCR intensity in the multi-TeV region. In the sky-map representing the relative intensity as a function of the declination and right ascension of the incident direction, the large-scale structure of the anisotropy is evident with a statistical significance exceeding 10 sigma. Amenomori *et al.*[1] (hereafter referred to as paper I) proposed a model interpreting this large-scale anisotropy in terms of the local interstellar structure around the heliosphere. The heliosphere is located inside the local interstellar cloud (LIC) very close to the inner edge of the LIC.[14] If the GCR population is lower inside the LIC than outside, uni-directional and bi-directional flows (UDF and BDF) are both expected from the parallel diffusion of GCRs into the LIC along the local interstellar magnetic field (LISMF) connecting the heliosphere with the region outside the LIC, where the GCR population is higher. In addition, the UDF perpendicular to the LISMF is also expected from the $\mathbf{B} \times \nabla n$ drift anisotropy driven by a spatial gradient of GCR density (n) in the LISMF ($\mathbf{B}$). If the weak scattering regime applies, the contribution from the cross-field diffusion to the perpendicular UDF should be much smaller than the drift anisotropy. It is found that the LISMF orientation deduced from the best-fit direction of the BDF axis is almost parallel to the galactic plane and is more consistent with the suggestion of Frisch[15] than with that of Lallement *et al.*[16]

The Tibet III experiment is currently the world's highest precision measurement of GCR intensity in this energy region, utilizing both high count rates and good angular resolution of the incident direction. The GCR anisotropy in this energy region is free from solar modulation, while it is still sensitive to the local magnetic field structure with a spatial scale comparable to the Larmor radius of GCR particles, about 0.002 pc or 300 AU for 5 TeV protons in the 3 μG LISMF. The analysis of Tibet III data by paper I, on the other hand, can in principle be biased, because the experiment cannot observe the mid-latitude sky in the southern hemisphere, where a large anisotropy is reported as a consequence of the NS asymmetry of the sidereal diurnal variation.[2,12] In order to remove such a bias, it is necessary to analyze the existing data in the southern hemisphere together with the Tibet AS data.

In this paper, we analyze the THN data collected during its total observation period from 1993 to 2005. We apply to the THN data a best-fitting analysis identical to that adopted by paper I for the Tibet III data and derive the energy dependence of the anisotropy in space by comparing the best-fit parameters from the THN observations in the sub-TeV region and the Tibet III experiment in the multi-TeV region.

2. Analysis

2.1. *Long term variation of the sidereal anisotropy*

We analyze the sidereal daily variations observed with a pair of muon detectors at Matsushiro (MAT) between 1985 and 2006 in the northern hemisphere and at Liapootah (LPT) between 1993 and 2005 in the southern hemisphere. These two detectors are both multidirectional and capable for measuring muon intensities in a total of 34 directional channels simultaneously. The median primary energy for muons recorded by these detectors covers the sub-TeV region ranging from 0.565 to 0.861 TeV in MAT and from 0.454 to 0.984 TeV in LPT, respectively, while the effective viewing latitude corrected for geomagnetic bending ranges from 61.3°N to 15.0°S in MAT and from 13.6°N to 57.6°S in LPT. For detailed information of these detectors, readers can refer to Hall *et al.*[2,12] We first obtain the first and second harmonics, $A_{1\,i,j}$, $B_{1\,i,j}$, $A_{2\,i,j}$ and $B_{2\,i,j}$, by expanding the yearly mean sidereal daily variation $D_{i,j}(t)$ in the j-th directional channel ($j = 1$ to 34) of the i-th detector ($i = 1$ for MAT and $i = 2$ for LPT), as

$$D_{i,j}(t) = \Sigma_{m=1}^{2}(A_{m\,i,j} \cos m\omega t_j + B_{m\,i,j} \sin m\omega t_i),$$

where t_i is the local sidereal time for the i-th detector and $\omega = 2\pi/24$. Figure 1 displays $A_{m\,i,j}$ and $B_{m\,i,j}$ in five directional channels viewing respectively the vertical (V), north (N), south (S), east (E) and west (W) directions as harmonic vectors. The amplitude of each vector is represented by its length from its origin, while the phase is represented by the angle measured clockwise from the vertical (y) axis, i.e. the local times of $+y$, $+x$, $-y$, and $-x$ directions in Figs. 1a-1b are, respectively, 00:00, 06:00, 12:00, and 18:00. To demonstrate the long-term variation, we plot in this figure "vector summation dials", in which yearly mean vectors are summed one by one from the first year, i.e. 1985 for MAT and 1993 for LPT. To show the difference in the average amplitudes by MAT and LPT free from the difference in observation periods, we scaled each annual vector by dividing it by the number of years of observation (22 for MAT and 13 for LPT). The final data point in each summation dial, therefore, represents the harmonic vector averaged over the total observation period. The errors attached to the vertical vector (V) are deduced from the dispersion of the yearly values of $A_{m\,i,j}$ and $B_{m\,i,j}$. It is seen first in this figure that the average harmonic vectors by MAT and LPT are both statistically significant and persistent resulting in the summation dial extending toward almost the same local time. It is also seen, in common for both detectors that the phase of the vector in the east viewing channel (E) is systematically earlier than that in the west viewing channel (W) with the phase in the vertical channel (V) being intermediate. This is consistent with the daily variation due to the anisotropy recorded in directional channels on the spinning Earth. On the other hand, there are also notable features appearing differently in MAT in the northern hemisphere and LPT in the southern hemisphere. The average vector in the south viewing channel (S) is much larger than that in the north viewing channel (N) in MAT. There is actually a geometrical effect so that one does expect larger anisotropy for MAT-S than for MAT-N, but the difference is larger and cannot be explained by the effect. Furthermore, all five vectors (V, N, S, E, W) have almost the same amplitude in LPT, different to those vectors in MAT. This is the signature of the north-south

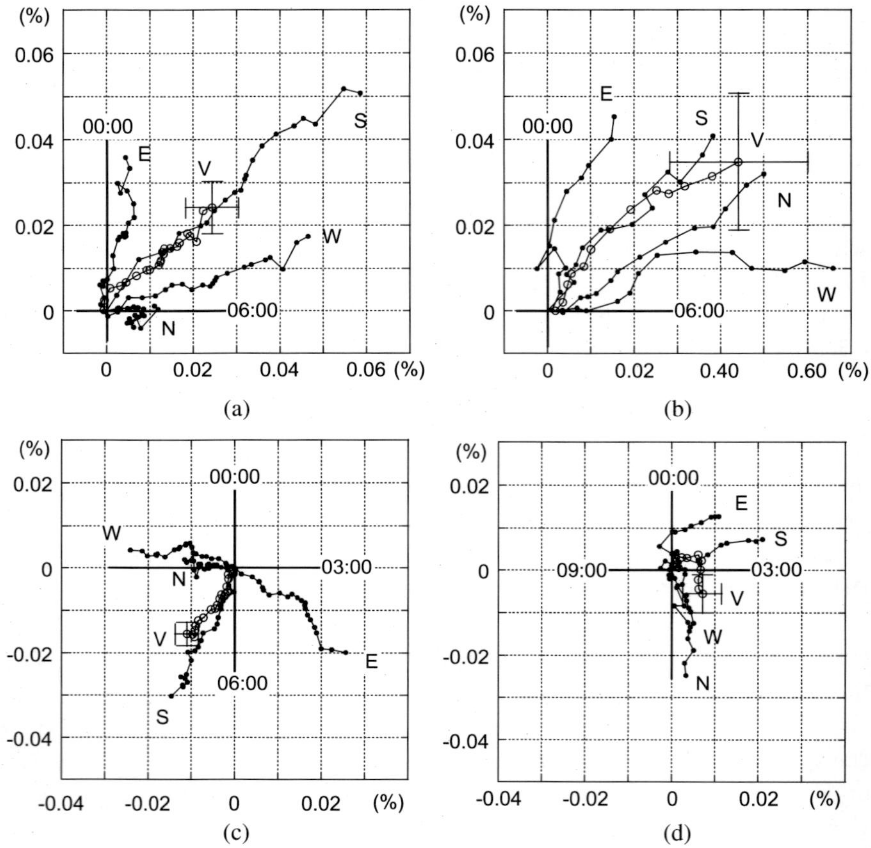

Fig. 1. Summation dials of the yearly mean harmonic vectors observed by five directional channels of MAT and LPT. Directional channels are indicated by alphabetical characters, "V", "N", "S", "E", "W". The upper panels (a), (b) display the diurnal vectors ($m = 1$), while the lower panels (c), (d) show the semi-diurnal vectors ($m = 2$). The left panels (a), (c) show observations by MAT during the 22 years between 1985 and 2006, while the right panels (b), (d) show observations by LPT during the 13 years between 1993 and 2005. To allow direct comparison of the average vectors for MAT and LPT, free from the difference in observation periods, each annual vector is scaled by the number of years of observation for that station. The error attached to each vector labeled "V" represents the standard deviation of the long-term mean value, calculated from the dispersion of the annual vectors.

asymmetry of the sidereal anisotropy as observed with multidirectional detectors.[10,11]

As shown in Fig. 1, for all the directional channels at both MAT and LPT there are few, if any, significant changes in amplitude and phase from year to year. In this paper, therefore, we analyze the average anisotropy over the whole observation period, i.e. 1985–2006 for MAT and 1993–2005 for LPT. We will analyze the temporal variation of the anisotropy possibly due to solar modulation correlated with solar activity- and magnetic-cycles in a separate paper.

2.2. *Best-fitting with the model anisotropy*

Following paper I, we assume the GCR intensity in free space consisting of the bi-directional and uni-directional flows (BDF and UDF) along a reference axis in the direction of the right ascension ϕ_R and co-latitude θ_R in celestial coordinates. We also assume another UDF along the direction $(\theta_R^{\perp}, \phi_R^{\perp})$ perpendicular to the axis. In this model, the anisotropic GCR intensity $F(\theta_J, \phi_J; \theta_R, \phi_R, \theta_R^{\perp}, \phi_R^{\perp})$ expected in a directional channel viewing right ascension ϕ_J and co-latitude θ_J is given as

$$\begin{aligned} &F(\theta_J, \phi_J; \theta_R, \phi_R, \theta_R^{\perp}, \phi_R^{\perp}) \\ &\quad = \eta_1^{\perp} P_1^0(\cos\chi^{\perp}) + \eta_1^{\parallel} P_1^0(\cos\chi) + \eta_2^{\parallel} P_2^0(\cos\chi), \end{aligned} \tag{1}$$

where $\eta_2^{\parallel}$ is the amplitude of the BDF, $\eta_1^{\parallel}$ and $\eta_1^{\perp}$ are respectively amplitudes of UDFs parallel and perpendicular to the BDF axis and χ and $\chi^{\perp}$ are angles of the viewing direction measured from the BDF axis and the perpendicular UDF axis, respectively. P_1^0 and P_2^0 in Eq. (1) are the Schmidt semi-normalized associated Legendre functions, defined as

$$P_n^m(\cos\theta) = \begin{cases} P_{n,m}(\cos\theta) & \text{for } m = 0 \\ \dfrac{\sqrt{2(n-m)!}}{(n+m)!} p_{n,m}(\cos\theta) & \text{for } m \neq 0 \end{cases} \tag{2}$$

and

$$P_{n,m}(\cos\theta) = \sin^m\theta \frac{d^m P_n(\cos\theta)}{d(\cos\theta)^m}. \tag{3}$$

Angles χ and $\chi^{\perp}$ in Eq. (1) are expressed in terms of the right ascensions (ϕ_R and $\phi_R^{\perp}$) and and co-latitudes (θ_R and $\theta_R^{\perp}$) of the BDF and the perpendicular UDF axes, as

$$\cos\chi = \cos\theta_R \cos\theta_J + \sin\theta_R \sin\theta_J \cos(\phi_J - \phi_R), \tag{4}$$

$$\cos\chi^{\perp} = \cos\theta_R^{\perp} \cos\theta_J + \sin\theta_R^{\perp} \sin\theta_J \cos(\phi_J - \phi_R^{\perp}). \tag{5}$$

By introducing Eqs. (4) and (5) into (1) and eliminating terms which are independent of ϕ_J, we get the anisotropy responsible for the daily variation, as

$$f(\theta_J, \phi_J; \theta_R, \phi_R, \theta_R^{\perp}, \phi_R^{\perp}) = (x_1^1 \cos\phi_J + y_1^1 \sin\phi_J) P_1^1(\cos\theta_J) + (x_2^1 \cos\phi_J + y_2^1 \sin\phi_J) P_2^1(\cos\theta_J) + (x_2^2 \cos 2\phi_J + y_2^2 \sin 2\phi_J) P_2^2(\cos\theta_J), \qquad (6)$$

where x_1^1, y_1^1, x_2^1, y_2^1, x_2^2 and y_2^2 are called space harmonic components and given, as

$$x_1^1 = \eta_1^{\perp} P_1^1(\cos\theta_R^{\perp}) \cos\phi_R^{\perp} + \eta_1^{\|} P_1^1(\cos\theta_R) \cos\phi_R, \qquad (7a)$$

$$y_1^1 = \eta_1^{\perp} P_1^1(\cos\theta_R^{\perp}) \sin\phi_R^{\perp} + \eta_1^{\|} P_1^1(\cos\theta_R) \sin\phi_R, \qquad (7b)$$

$$x_2^1 = \eta_2^{\|} P_2^1(\cos\theta_R) \cos\phi_R, \qquad (7c)$$

$$y_2^1 = \eta_2^{\|} P_2^1(\cos\theta_R) \sin\phi_R, \qquad (7d)$$

$$x_2^2 = \eta_2^{\|} P_2^2(\cos\theta_R) \cos 2\phi_R, \qquad (7e)$$

$$y_2^2 = \eta_2^{\|} P_2^2(\cos\theta_R) \sin 2\phi_R, \qquad (7f)$$

The daily variation $d_{i,j}(t)$ expected for the j-th directional channel of the i-th detector from $f(\theta_J, \phi_J; \theta_R, \phi_R, \theta_R^{\perp}, \phi_R^{\perp})$ in Eq. (6) is given, as

$$d_{i,j}(t) = \Sigma_{m=1}^{2} (a_{m\,i,j} \cos m\omega t_i + b_{m\,i,j} \sin m\omega t_i), \qquad (8)$$

where $a_{m\,i,j}$ and $b_{m\,i,j}$ are the expected harmonic components, related to the space harmonic components by the coupling coefficients, $c_{ni,j}^m$ and $s_{ni,j}^m$, as

$$a_{1i,j} = c_{1i,j}^1 x_1^1 + s_{1i,j}^1 y_1^1 + c_{2i,j}^1 x_2^1 + s_{2i,j}^1 y_2^1, \qquad (9a)$$

$$b_{1i,j} = -s_{1i,j}^1 x_1^1 + c_{1i,j}^1 y_1^1 - s_{2i,j}^1 x_2^1 + c_{2i,j}^1 y_2^1, \qquad (9b)$$

$$a_{2i,j} = c_{2i,j}^2 x_2^2 + s_{2i,j}^2 y_2^2, \qquad (9c)$$

$$b_{2i,j} = -s_{2i,j}^2 x_2^2 + c_{2i,j}^2 y_2^2 \qquad (9d)$$

We calculate the coupling coefficients by utilizing the response function of the atmospheric muons to the primary particles by Murakami *et al.*[17] and assuming $P^{0.7}$ dependence of the anisotropy on the rigidity P of the primary GCR reported from the analysis of the initial THN data.[2,12]

By comparing $a_{m\,i,j}$ and $b_{m\,i,j}$ with the observed harmonic components $A_{m\,i,j}$ and $B_{m\,i,j}$, we obtain six best-fit parameters $(\eta_1^{\perp}, \eta_1^{\|}, \eta_2^{\|}, \theta_R^{\perp}, \theta_R, \phi_R)$

minimizing the residual S defined, as

$$S = \Sigma_{i,j} \left\{ \frac{(A_{1i,j} - a_{1i,j})^2}{\sigma_{A^2_{1i,j}}} + \frac{(B_{1i,j} - b_{1i,j})^2}{\sigma_{B^2_{1i,j}}} \right.$$

$$\left. + \frac{(A_{2i,j} - a_{2i,j})^2}{\sigma_{A^2_{2i,j}}} + \frac{(B_{2i,j} - b_{2i,j})^2}{\sigma_{B^2_{2i,j}}} \right\}, \quad (10)$$

where $\sigma_{A_{mi,j}}$ and $\sigma_{B_{mi,j}}$ are respectively errors of $A_{mi,j}$ and $B_{mi,j}$ deduced from standard deviations of yearly mean values. Note that, if $\theta_R^\perp$, θ_R, ϕ_R, are selected, $\phi_R^\perp$ is uniquely given by the orthogonal condition. For a selected set of $\theta_R^\perp$, $\phi_R^\perp$, θ_R, ϕ_R, $\eta_1^\perp$, three amplitude parameters $\eta_1^\parallel$, $\eta_2^\parallel$ can be determined by a simple linear least-square method. It is noted that this analysis gives us the best-fit anisotropy in space, being corrected for the geomagnetic and atmospheric effects by utilizing the coupling coefficients in Eqs. (9a)–(9d). This is an important difference from the previous analysis by Hall *et al.*,[2,12] who presented the model anisotropy best-fit to the observed daily variation itself tentatively ignoring terrestrial effects.

2.3. *Best-fit result*

The best-fit parameters to the THN data are listed and compared with those by paper I in Table 1. It is seen in this table that the best-fit orientation of the reference axis of the anisotropy is quite similar for the two independent observations at 0.5 TeV by the THN and at 5 TeV by the Tibet III experiment, with the angular difference equal to or less than 10°. This implies, indirectly, that the conclusion derived by paper I from the best-fit analysis to Tibet III data covering only the northern hemisphere is not seriously biased.

The amplitudes of the best-fit anisotropy seen by the THN, on the other hand, are only one third or less of those observed by Tibet III, indicating an attenuation probably due to solar modulation effects. It is

Table 1. Best-fit parameters for THN data at the median primary rigidity (P_m) of 0.5 TV (upper line). The parameters derived by paper I for Tibet III data at $P_m = 5.0$ TV are also listed for comparison (lower line). Errors of amplitudes are only stastical.

P_m (TV)	$\eta_1^\perp$ ($\times10^{-2}$)	$\delta_R^\perp$ (°)	$\phi_R^\perp$ (°)	$\eta_1^\parallel$ ($\times10^{-2}$)	$\eta_2^\parallel$ ($\times10^{-2}$)	δ_R (°)	ϕ_R (°)	χ/ d.o.f.
0.5	0.020 ± 0.001	-12.5	2.5	0.016 ± 0.001	0.020 ± 0.001	-17.1	96.4	1.72
5.0	0.108 ± 0.001	-22.5	357.5	0.096 ± 0.002	0.095 ± 0.001	-22.5	97.4	2.44

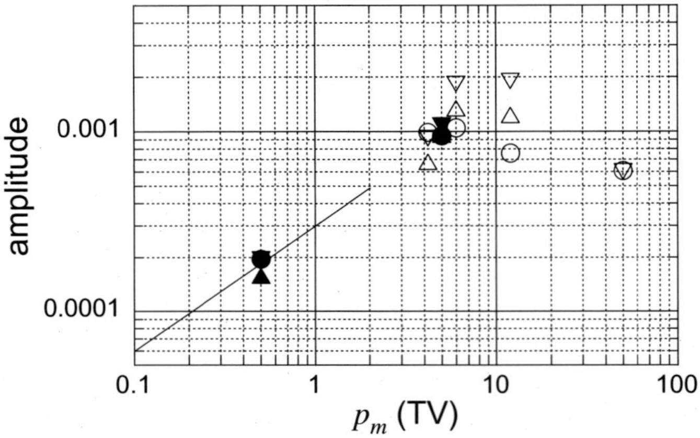

Fig. 2. Rigidity spectra of the best-fit anisotropy. Best-fit amplitudes $\eta_1^{\parallel}$, $\eta_1^{\perp}$ and $\eta_2^{\parallel}$ are respectively displayed by full inverted triangles, full triangles and full circles, each as a function of the median primary rigidity (P_m) which is 5.0 TV for Tibet III experiment and normalized to 0.5 TV for THN. Each open symbol displays corresponding amplitude derived from best-fitting to each of five subset of Tibet III data with different median rigidities of 4.2, 6.0, 12.0 and 50.0 TV. The thin solid line indicates the spectrum ($P_m^{0.7}$) reported by Hall *et al.*[2]

seen in Fig. 2, showing the amplitude as a function of the median primary rigidity (P_m), that the rigidity spectrum ($P_m^{0.7}$) reported by Hall *et al.*[2] and used in this paper is fairly consistent with the Tibet III experiment at 5 TeV. The amplitude at higher rigidities appears almost constant or gradually decreasing above 10 TeV, as shown by open symbols in the figure indicating the best-fit amplitudes to five subset data with different energies.

In Fig. 3, we show the best-fit performance of the model anisotropy for sidereal daily variations observed with the THN. Each panel of this figure displays the sidereal daily variation recorded in one of eight directional channels in the THN. It is seen first that the observed daily variation in each directional channel is well reproduced by the Fourier series in Section 2, i.e. a superposition of the diurnal ($m = 1$) and semi-diurnal ($m = 2$) variations displayed by the dotted curve. The solid curve, which indicates the reproduction by the best-fit model described in the preceding section, also shows reasonable agreement with the data, particularly in directional channels viewing the equatorial region with smaller declination. For directional channels with larger declination, on the other hand, there is notable deviation of the best-fit model from the data. For directional

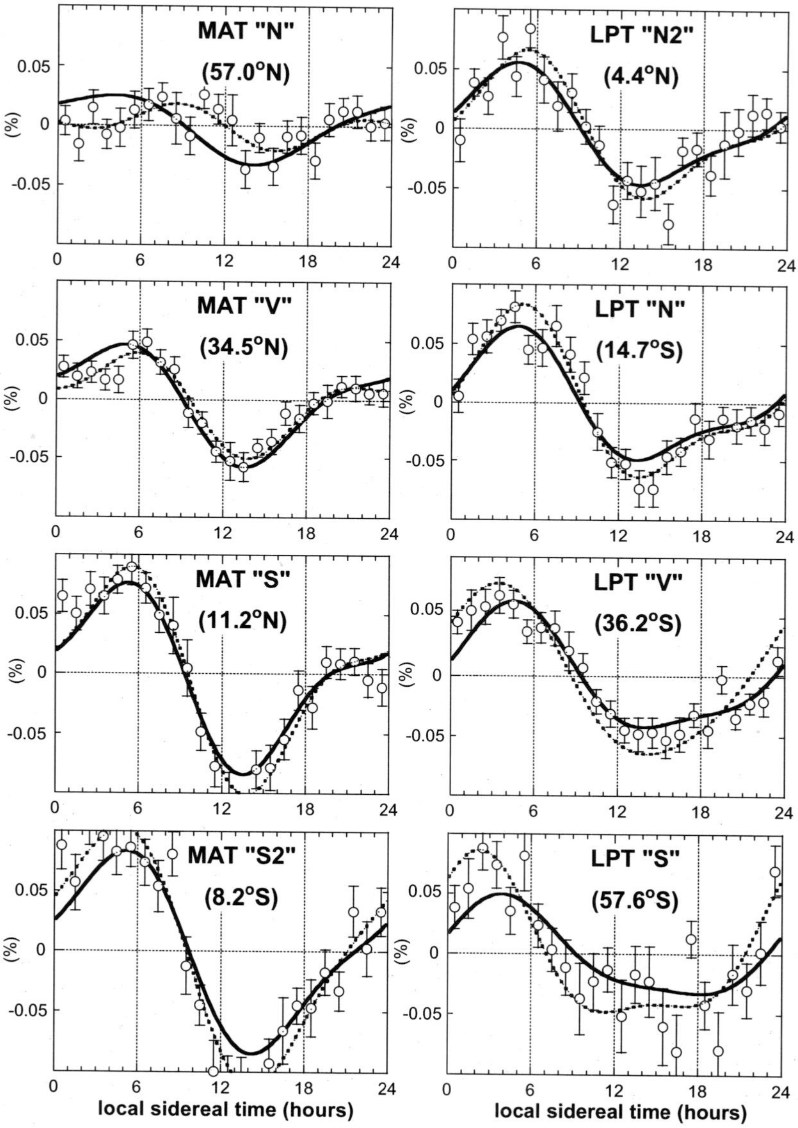

Fig. 3. Best-fit performance of the model anisotropy for sidereal daily variations observed with the THN. Each panel of this figure displays the sidereal daily variation in % recorded in each of eight directional channels of MAT (left) and LPT (right). The name of directional channel and its average declination of viewing are indicated in each panel. Error bars are deduced from the standard deviation of yearly mean count rate at each local sidereal time. The dotted curve displays the superposition of the diurnal ($m = 1$) and semi-diurnal ($m = 2$) variations, while the solid curve displays the reproduction by the best-fit model described in the text.

channels viewing high northern (southern) latitudes, the phase of the solid curve deviates to earlier (later) hours from the data and the dotted curve. This is due to the observed intensity maximum around ~06:00 local sidereal time, which has the maximum phase shifting to earlier hours as the viewing latitude moves southward over the equator, as first reported by Hall *et al.*[2,12] This feature is also reported with a great significance by the Tibet III experiment[18] and cannot be reproduced properly by the simple model anisotropy assumed in this paper (see also paper I).

3. Summary and Discussion

It is now well established by experiments in the multi-TeV region that the amplitude of the anisotropy is about 0.1% or 10^{-3}. Based on the diffusive propagation of GCRs throughout the large-scale galactic magnetic field, it has often been discussed that the amplitude ($\eta_{\parallel}$) is given as

$$\eta_{\parallel} \sim 3\kappa_{\parallel}/(Lc), \tag{11}$$

where $\kappa_{\parallel}$ is the diffusion coefficient, L is the scale length of the region where the diffusive propagation takes place and c is the speed of light. Using 10^{-3} for $\eta_{\parallel}$ and $10^{29}\,\mathrm{cm}^2/\mathrm{s}$ for $\kappa_{\parallel}$ for 1 TeV GCRs,[19] we obtain $L \sim 10^{22}\,\mathrm{cm} = 3.2\,\mathrm{kpc}$ which is comparable to the scale length of the entire Galaxy. A difficulty in interpreting Tibet III's anisotropy in terms of this large-scale diffusion picture is that the observed angular separation between the excess and the deficit is only ~120°, which is much smaller than the 180° expected from a uni-directional flow (UDF) due to simple diffusion. The separation angle, on the other hand, is significantly larger than the 90° expected from a bi-directional flow (BDF) which is expected in the direction along the magnetic field from the adiabatic focusing or loss-cone effect. Only a combination of UDF and BDF can achieve a reasonable fit to the observed anisotropy, as reported by paper I.

As shown in Table 1, the best-fit analysis indicates that there is also a UDF with amplitude ($\eta_{\perp}$) of $\sim 10^{-3}$ (at 5 TeV) perpendicular to the BDF, i.e. perpendicular to the magnetic field, if the orientation of BDF is parallel to the field. The UDF perpendicular to the magnetic field is generally produced either from cross-field diffusion or the drift effect. The mean free path ($\lambda_{\parallel}$) of the pitch-angle scattering responsible for the diffusion is given as a function of $\kappa_{\parallel}$, as

$$\lambda_{\parallel} = 3\kappa_{\parallel}/c \sim 3\,\mathrm{pc} \gg R_L = 2 \times 10^{-3}\,\mathrm{pc}, \tag{12}$$

where R_L is the Larmor radius of 5 TeV protons in a 3 μG magnetic field. This implies a very large Bohm factor, which is defined as the ratio of mean free path to Larmor radius, suggesting that the cross-field diffusion is negligibly smaller than the drift effect. In this case, the perpendicular UDF is produced solely from the drift flux, as

$$\eta_\perp \sim R_L/L', \tag{13}$$

giving an alternative scale length $L' \sim 2$ pc with $\eta_\perp \sim 10^{-3}$. To interpret the anisotropy seen by Tibet III, therefore, we need a structure with scale of $L' \sim 2$ kpc, which is much smaller than that of $L \sim 3.2$ kpc for entire Galaxy. This is why paper I proposed a LISMF model based on a local structure associated with the local interstellar cloud (LIC) with a radius of ~3 pc.[14]

In this paper, we analyzed the sidereal anisotropy of GCR intensity in the sub-TeV region observed with the THN of underground muon detectors. The THN can observe not only the northern hemisphere, but also the entire sky in the southern hemisphere which is not observed by the Tibet III experiment monitoring the multi-TeV GCR intensity. By applying a best-fit analysis based on the same model anisotropy in space applied for Tibet III data by paper I, we find that the best-fit orientation of reference axis of the anisotropy is quite similar for each experiment, regardless of the order of magnitude difference in GCR energy. This is apparent in Fig. 4, which displays celestial sky-maps derived from best fitting the same model anisotropy to the THN data (upper panel) and Tibet III data (lower panel). This similarity implies, at least indirectly, that the conclusion derived by paper I based on the Tibet III data only in the northern hemisphere is not seriously biased. The best-fit amplitudes of the anisotropy, on the other hand, are only one third or less of those from Tibet III, indicating an attenuation possibly due to solar modulation.[20] The rigidity dependence of the anisotropy amplitude in the sub-TeV region is consistent with the spectrum ($P_m^{0.7}$) reported by Hall *et al.*,[2] smoothly extending to the Tibet III result in the multi-TeV region. The amplitude at higher energies appears almost constant or gradually decreasing with the increasing rigidity.

The air shower experiment has a response not only to primary nuclei component in GCRs but also to high energy gamma rays, to which underground muon detectors have only negligible response. It is possible, therefore, that the anisotropy by Tibet III contains a significant contribution from primary gamma rays. The Super Kamiokande experiment, a deep underground detector capable for observing multi-TeV

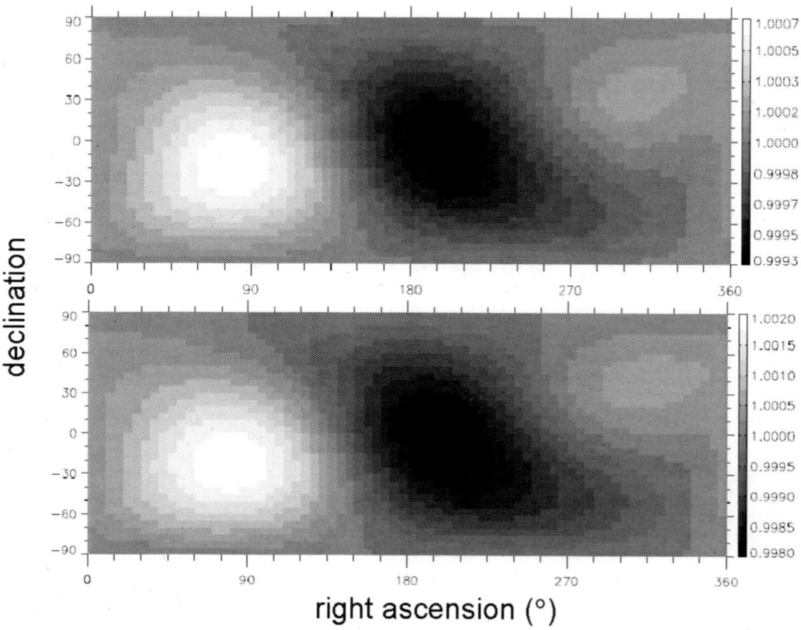

Fig. 4. Sky maps of the model anisotropy derived from best-fitting to sidereal daily variations observed with the THN (top) and the Tibet III experiment (bottom). Each panel displays the GCR intensity in $5^\circ \times 5^\circ$ pixels as a function of the right ascension on the horizontal axis and the declination on the vertical axis in a gray-scale format. The saturations in the top and bottom panels occur at ± 0.0007 and ± 0.002 respectively, as indicated by color code bars on the right side. Note that the average in each 5° declination band is subtracted from the intensity in each pixel.

GCRs by measuring muons, however, has recently reported a large-scale anisotropy in agreement with the Tibet III result.[21] This implies that the large-scale anisotropy observed by Tibet III is not due to high energy gamma rays, but due to the anisotropy of the charged component of primary cosmic rays. This is also consistent with the attenuation of amplitude in the sub-TeV region deduced in this paper, as the attenuation due to solar modulation cannot be expected for an anisotropy due to primary gamma rays.

The model anisotropy with the best-fit parameters derived in this paper, on the other hand, does not fully reproduce the NS asymmetric feature of the observed anisotropy. The observed phase of intensity maximum shifts to earlier hours as the viewing latitude moves southward over the equator, as reported earlier from the THN and Tibet III experiments.[2,12,18] A localized region of excess 10 TeV cosmic rays is also seen in Tibet III data at around the right ascension of ~70° and the declination of ~15° in agreement with a recent report by Milagro experiment.[22] These observed features need to be modeled further.

Acknowledgments

The THN observations were supported by research grants from the Japanese Ministry of Education, Science, Sports and Culture and the Australian Research Council. The authors gratefully acknowledge the help received from the Tasmanian Hydro-Electric Commission for their invaluable assistance during periods of maintenance work at Liapootah. They also appreciate financial support from the Mitsutoyo Co. The observations at Matsushiro have been supported by Shinshu University, while the observations at Liapootah were maintained by the University of Tasmania and the Australian Antarctic Division.

References

1. M. Amenomori *et al.*, *AIP Conf. Proc.* **932** (2007) 283 (paper I).
2. D. L. Hall *et al.*, *J. Geophys. Res.* **104** (1999) 6737.
3. A. G. Fenton and K. B. Fenton, *Proc. 14th Int. Cosmic Ray Conf. (Munich)* **4** (1975) 1482.
4. D. J. Cutler *et al.*, *Astrophys. J.* **248** (1981) 1166.
5. H. Ueno *et al.*, *Proc. Int. Symp. Cosmic Ray Modulation in the Heliosphere*, Iwate Univ., Morioka, 349 (1984).
6. D. J. Cutler, D. M. Groom, *Astrophys. J.* **376** (1991) 322.
7. K. B. Fenton *et al.*, *Proc. 24th Int. Cosmic Ray Conf. (Rome)* **4** (1995) 635.
8. K. Nagashima *et al.*, *IL. Nuovo Cimento*, **12C**(6) (1989) 695.
9. M. B. Aglietta *et al.*, *Astrophys. J.* **470** (1996) 501.
10. S. Mori *et al.*, *J. Geomag. Geoelectr.* **47** (1995) 1097.
11. K. Munakata *et al.*, *J. Geomag. Geoelectr.* **47** (1995) 1103.
12. D. L. Hall *et al.*, *J. Geophys. Res.* **103** (1998) 367.
13. M. Amenomori *et al.*, *Science* **314** (2006) 439.
14. S. Redfield and J. L. Linsky, *Astrophys. J.* **534** (2000) 825.
15. P. C. Frisch, *Space Sci. Rev.* **78** (1996) 213.
16. R. Lallement *et al.*, *Science,* **307** (2005) 1447.
17. K. Murakami *et al.*, *IL. Nuovo Cimento* **2C**(5) (1979) 635.
18. M. Amenomori *et al.*, *Astrophys. J. Lett.* **626** (2005) L29.
19. I. V. Moskalenko *et al.*, *Astrophys. J.* **565** (2002) 280.
20. K. Nagashima *et al.*, *Planet. Space Sci.* **30** (1982) 879.
21. G. Guillian *et al.*, *Phys. Rev. D* **75** (2007) 062003-1.
22. A. A. Abdo *et al.*, *Phys. Rev. Lett.* **101** (2008) 221101-1.

Advances in Geosciences
Vol. 21: Solar & Terrestrial Science (2008)
Ed. Marc Duldig

TOWARD A NEW ERA OF CANADIAN GROUND-BASED RESEARCH OF GEOSPACE SCIENCE

WILLIAM LIU*, JOHN MANUEL and RICHARD GIROUX
Space Science Branch, Canadian Space Agency,
6767 route de l'Aéroport Saint-Hubert,
Québec J3Y 8Y9 Canada
**william.liu@space.gc.ca*

Ground-based investigation of geospace has a long history of contributing to our understanding of space plasma physics affecting the near-Earth space environment. In Canada, as well as elsewhere, a coordinated effort is afoot to perfect the network approach to ground-based geospace observations. This paper gives an overview of the Canadian GeoSpace Monitoring program and its value to several outstanding problems of current interest.

1. Introduction

This paper arose from an invited talk at the Future Space Missions and Instrumentation for Space and Planetary Science Session, 5th Asia-Oceania Geoscience Society meeting in Busan, Korea. The original talk described the instrumentation of the Canadian GeoSpace Monitoring (CGSM) array. In this expanded version, the science behind the instrumentation is also discussed. Due to the scope of CGSM and limitation of space, the present discussion is not exhaustive but serves the purpose of giving the reader a glimpse of the world's largest ground-based geospace observing network. More detailed information can be obtained by consulting the references or contacting the scientists engaged in the CGSM program.

There is a historical recognition of the importance of ground-based observations. Physics in geospace occurs in a spatiotemporal continuum. Multiscale structures are believed to be the products of nonlinear plasma processes. In space plasma physics, coupling has become a dominant theme of research. In geospace, regions of the overall system couple (i.e. between the plasma sheet and ionosphere), the particles and fields couple, and structures on different scales apparently also couple. Satellites and constellations of satellites do not provide the requisite scale coverage to tackle the cross-scale coupling problem. Coordinated ground-based observations have been long exploited to fill this gap.

A geospace process typically has a number of observable features. For example, a field-aligned current generates geomagnetic perturbations, ionospheric convection, and aurora. Within each family, there is the possibility of different sub-distributions. Take nightside aurora as an example. Aurora is a result of particle precipitation from the magnetosphere.[1] The dozens of auroral lines respond to different particles at different energies and have different emission heights. Therefore, the 630 nm electron aurora has a very different distribution from the 486 nm proton aurora, while the 558 nm aurora is further different from both. Aurora is also notorious for its multiscale structuring. An All-Sky Imager (ASI) optimized for wide field of view is often suboptimal for detecting small-scale structures. Fast auroral telescope such as Portable Auroral Imager (PAI), on the other hand, sacrifice context to capture details of small-scale dynamics.[2] Furthermore, some particles, such as electrons in the 30 to 100 keV range, do not have unique auroral fingerprint, but nonetheless play significant roles in magnetospheric physics. These particles are tracked by riometers. Using the multitude of instruments is analogous to hyperspectral imaging. Each instrument array produces an image. If the inner working of the magnetosphere is compared to a black box, an image is a projection of the internal structures of the black box on a plane. Projections on different planes at different angles give more information to figure out how the black box works.

The primary observables in the ionosphere, and secondary quantities that can be readily deduced from the primary observables are geomagnetic perturbations, ionospheric currents, ionospheric convection, auroral emissions in different wavelengths, and reflection, refraction and absorption of radio-frequency waves. Ground-based observations consist, first, in forming high-quality maps of these quantities in the whole or parts of the ionosphere, and as a function of time, and second, in linking these images logically to form or test theories. Compared to spatially constrained satellite observations, this is a remarkable luxury and advantage. Coupled with models specially designed for ground-based data, we are now at the verge of being able to represent multiscale magnetospheric dynamics by virtue of real-time observations. Through a basic introduction to CGSM and some of its recent scientific advances, we give a description of the observational capabilities at our disposal in this endeavor.

2. From CANOPUS to CGSM

The landmass of Canada covers seven hours of magnetic local time and has the world's largest stretch of unbroken terrain to observe aurora and geomagnetic perturbations. In this paper, we introduce the ground-based program in Canada. This program is anchored by the Canadian GeoSpace Monitoring arrays, with complementary special-purpose arrays such as THEMIS GBOs and major facilities such as AMISR.

CGSM is a successor to Canadian Auroral Network for OPEN Program Unified Study (CANOPUS). Planning for CANOPUS began in the late 70's as Canada's contribution to what later became ISTP. Primary sites were identified in 1982, and the so-called the CANOPUS Yellow Book was published in 1986 (see http://cgsm.ca/doc/CANOPUS_yellow_book.pdf), marking a new era of ground-based multi-instrument array. CANOPUS was a significant construction project for that time, with 13 fluxgate magnetometers, 13 riometers, 4 meridional scanning photometers, 2 bistatic radars, and 1 all-sky camera. The instruments were most heavily populated in Central Canada, along the renowned Churchill Line. The network was supported by a real-time data collection system via ANIK communications satellites, an impressive technological feat for that time and also a demonstration that ground-based observations can be used for space weather nowcasting and forecasting. See Rostoker *et al.*[3] for a detailed description of the CANOPUS instrument array.

By today's standard of rhetoric, the scientific objectives of CANOPUS were somewhat understated:

1. Evaluation of ionospheric Joule and particle heating in the auroral oval and cleft in relation to events observed by magnetospheric spacecraft;
2. Studies of dynamic morphology of proton and electron auroral substorms with detailed CANOPUS data supplemented by synoptic satellite imager data;
3. Studies of the relation between pulsations and magnetospheric boundary oscillations directly observed by spacecraft;
4. Studies of Pc 3, 4, and 5 pulsations processes and their latitude variation on the ground together with in-situ observations on the corresponding field lines in the magnetosphere;
5. Rapid production of catalogues of substorm onset times and magnetic indices to facilitate magnetospheric studies.

Notwithstanding the unassuming list, CANOPUS has been credited with a number of important discoveries. Of particular note was the combined use of magnetometer and optical data to study the role of field-line resonances (FLRs) in magnetospheric dynamics.[4–7] CANOPUS data were also used in establishing the empirical correlation between global ULF waves and increase of relativistic electrons in the outer radiation belt.[8]

The arrival of the new Millennium prompted the Canadian scientific community to look for ways to take CANOPUS to the 21st century. Year 2000 marked the launch of two landmark spacecraft, Cluster and IMAGE. Cluster was the first successful implementation of satellite constellation in space and significantly expanded our ability to probe micro-scale plasma physics. IMAGE, on the other hand, was the first composite imaging satellite using radio, FUV, EUV, and ENA techniques to render the magnetosphere in 3D. As the two missions pushed the envelope on both ends, issues of cross-scale coupling became paramount. The Canadian community quickly came to grip with the idea that ground-based observations would provide the needed information to characterize and quantify cross-scale coupling that moves energy not only from point to point, but also across the scale hierarchy. The CANOPUS 2000 report (http://cgsm.ca/doc/canopus_2000.pdf) by Samson, Donovan and Voronkov (2000) defined the overall theme of Canada's ground-based program as "Energy and mass transport and scaling in the Earth's ionosphere and magnetosphere". In 2003, this overall theme was elaborated into a list of specific scientific objectives, as articulated by Donovan *et al.* in 2003 (see http://aurora.phys.ucalgary.ca/donovan/publications_d.html): 'The primary CGSM scientific objective is to elucidate the fundamental processes that cause and control

1. Convection within and energy injection into the global magnetosphere.
2. Magnetotail instabilities and flows.
3. Auroral particle acceleration.
4. Energization, transport and loss of energetic magnetospheric particles.
5. Injection, transport, and loss of low energy magnetospheric particles'.

Compared to its CANOPUS progenitor, the new CGSM mantra demonstrated a keener sensitivity to physical problems behind the phenomena. However, between declaring goals and actually reaching them is a vast terrain of "walking the walk". How can a continental-sized, multi-instrument array help advance our understanding of the above problems? The answer was sought through a Canadian Space Agency AO in 2007.

Eleven projects were selected through a competitive process, forming the modern backbone of Canada's solar-terrestrial research program.

3. CGSM Components

The observational elements of CGSM comprise a network of networks, consisting of the following:

- Canadian Super/Polar Dual Auroral Radar Network
- Canadian High Arctic Ionospheric Network (CHAIN)
- Canadian Array for Real-Time Investigations of Magnetic Activity (CARISMA)
- CGSM Riometer Network
- Multi-Spectral Auroral Imaging Network.

Complementing the observational networks are several areas of activity that use the CGSM data for research and application purposes. These are:

- Facility for Data Analysis and Modeling
- Canadian Space Weather Forecast Centre
- Solar drivers of space weather: Contributions to Forecasting Technology
- Space weather now- and forecasting with ULF wave radial diffusion models.

In this paper, we give synoptic descriptions of these components of CGSM. Due to the variety and scope of the program, the descriptions are more illustrative than detailed. In our presentation, we will give a basic introduction to the instruments and the primary data they collect, and then highlight the scientific value or potential through a few examples of recent advances.

3.1. *Canadian superDARN*

SuperDARN is an international collaboration among 11 countries with 20 HF radars deployed in both hemispheres (http://superdarn.jhuapl.edu/). Four of the SuperDARN radars are Canadian.

The instrument concept of SuperDARN was first proposed by Greenwald *et al.* [9]. High-frequency radio beams in the ~10 Mhz range are transmitted to the ionospheric F region. Backscatter from ambient electron irregularities is received. The Doppler shift of the received signal gives a measure of the line-of-sight speed of the reflecting irregularities. Intersecting

Table 1. CGSM Instruments [M = MSI, P = MSP, F = Fluxgate, I = Induction Coil, R = Riometer, C = CADI, T = F10.7 solar radio telescope, S = HF-radar, G = GPS, √ = deployed, ○ = planned]. (*) Non-operational sites.

Name	Abbrev.	Lat.	Long.	M	P	F	I	R	C	S	T	G
Ann Arbor	ANNA	42.49	276.22			√						
Alert	ALE	82.50	297.65	○	○	√		√	○			○
Athabasca	ATHA	54.72	246.72	√								
Back	BACK	57.72	265.83			○						
Baker Lake	BLC	64.32	263.99			√						
Blue River	BLUE	52.11	240.69			○						
Brandon	BRN	49.87	260.03			√		√				
Calgary	CALG	51.00	246.00			○						
Cambridge Bay	CBB	69.12	254.97	○		√		√	√			√
Contwoyto Lake*	CONT	65.75	248.75	○		√		√				
Dawson	DAWS	64.05	220.89			√	√	√				
Eskimo Point (Arviat)*	ESKI	61.11	265.95			√		√				
Eureka	EUR	80.00	274.10		○	√		√	√			√
Fort Chipewyan	FCHIP	58.70	248.87			√						
Fort Churchill	FCHU	58.76	265.92			√	√	√				
Fort McMurray	FMCM	56.66	248.79			√		√				
Fort Simpson	FSIM	61.76	238.77	√		√		√				
Fort Smith	FSMI	60.02	248.05	√	√	√	√	√				
Gillam	GILL	56.38	265.36	√	√	√	√	√				
Gull Lake	GULL	50.06	251.74			√						
Hall Beach		68.46	278.85					√	○			√
Inuvik	INUV	68.30	226.70	○		√		○	√	√		
Iqaluit	IQA	63.75	291.48	○	○	√		√	○			√
Island Lake	ISLL	53.85	265.33			√	√	√				
Little Grande Rapids	LGRR	51.97	264.96			○						
London Ontario	LON-C	42.98	278.77						√			
Meanook	MEA	54.62	246.65			√		√				
Ministik Lake	MSTK	53.35	247.03			√	√					√
Norman Wells	NORM	65.28	233.20			√						
Olney	OLNE	48.28	245.06			○						
Ottawa	OTT	45.40	284.45			√		√				
Osakis	OSAK	45.87	264.92			√						
Oxford House	OXHO	55.13	265.26			○						
Pentincton	PEN							√			√	
Pinawa	PINA	50.25	263.96	√	√	√	√	√				
Pond Inlet		72.41	282.42					√	○			√
Prince George	PRGE	53.55	237.51							√		
Qikiqtarjuaq		67.33	295.99					√				√
Rabbit Lake	RABB	58.22	256.32	√		√		√				
Rankin Inlet	RANK	62.82	267.89	√	○	√		√	√	√		

(*Continued*)

Table 1. (*Continued*)

Name	Abbrev.	Lat.	Long.	M	P	F	I	R	C	S	T	G
Resolute Bay	RES	74.69	265.11	○		√		√	√			√
Sachs Harbor*	SACHS	71.98	234.77	○		○						
Sanikiluak	SNK	56.53	280.80			√		√				√
Saskatoon	SASK	52.16	253.47	○				√	√	√		
St Johns	STJ	47.60	307.32			√						
Taloyoak	TALO	69.54	266.45	√		√		√				√
Thief River Falls	THRF	48.03	263.64			√	√					
Victoria	VIC	48.52	236.58			√						
Weyburn	WEYB	49.60	256.00			√						
Yellowknife	YKC	62.48	245.52			√		√				

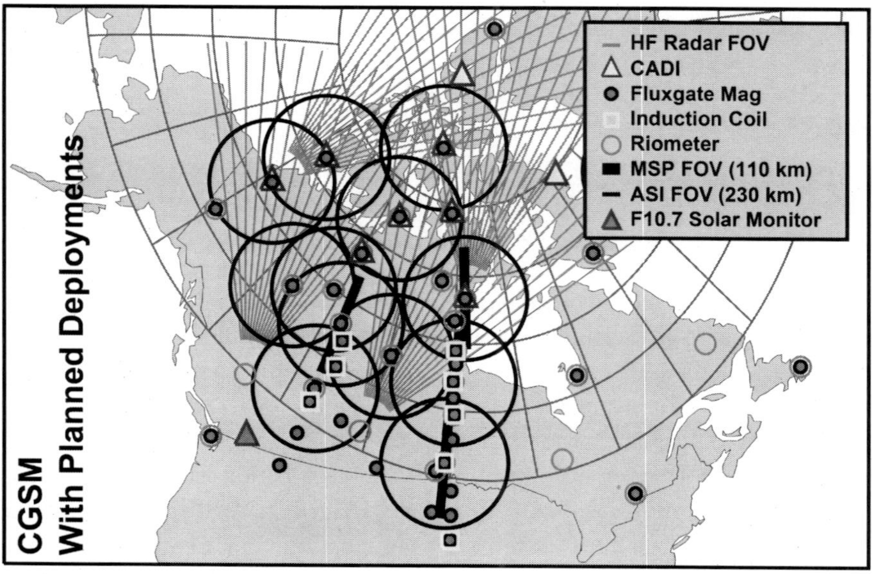

Fig. 1. CGSM arrays as of 31 December 2007. Instrument legends are indicated in the figure.

beams from two radars give the velocity vector in the reflecting plane, which is largely perpendicular to the magnetic field. The global convection pattern can be reconstructed through spherical harmonic analysis[10].

The Canadian SuperDARN radar consists of a transmitter and receiver. Most of the original SuperDARN use analogue phasing matrix, but newer ones such as PolarDARN use digital phasing matrix, allowing their beams to be steered in any direction with a 1.5-degree resolution. The new Canadian

SuperDARN at Rankin Inlet PolarDARN was the first to use the new twin terminated dipole antenna, a design which drastically reduced the construction costs without performance penalty. The primary SuperDARN data are the autocorrelation and cross-correlation functions between the transmitted and received signals. From these are derived the power, line-of-sight speed, and spectral width. Line of sight speeds from two offset beams give an estimate of convection velocity at the intersecting point.

The original SuperDARN was deployed with a focus on the auroral oval. The polar cap was poorly covered. Evidence has shown that the polar cap has a dynamic quite distinct from the oval, especially when the IMF is northward or dominated by the y or x component. Early work on the cellular structures of convection in the polar cap (e.g. Ref. [11]) did not anticipate fully the importance of polar cap's own dynamics. Change in polar cap convection will not only affect the delay of solar wind perturbations, but can also create enclaves of dynamics isolated from the oval. Watanabe and Sofko[12] proposed that IMF clock angle $<30°$, an interchange reconnection cycle is possible, whereby reconnection takes place between different flux populations in the two hemispheres, and the resulting interchange creates a distinct polar cap convection pattern shown in Fig. 2.

An interesting implication of the interchange reconnection cycle is that convection cells can form on closed field lines in one hemisphere as reciprocal effect of IMF-lobe reconnection in another. This would have important effect on north-south asymmetry of the magnetosphere and retention and transport of plasma in the dayside region. PolarDARN is uniquely positioned to investigate the existence of interchange reconnection, as well as a number of other reconnection-driven convection signatures under different IMF conditions. Coupled with CHAIN and AMISR observations (discussed later), the much improved coverage of the polar cap will allow us to tackle unsolved problems such as transpolar arcs and ionospheric patches and their dependence on solar wind conditions.

3.2. *Chain*

The Canadian High-Arctic Ionospheric Network (CHAIN) constitutes another thrust in the expansion to the polar cap. While PolarDARN will address global aspects such as the cellular structure of polar-cap convection and transit delay of solar wind perturbations, CHAIN will focus on the micro- and mesoscale structure of the ionosphere.

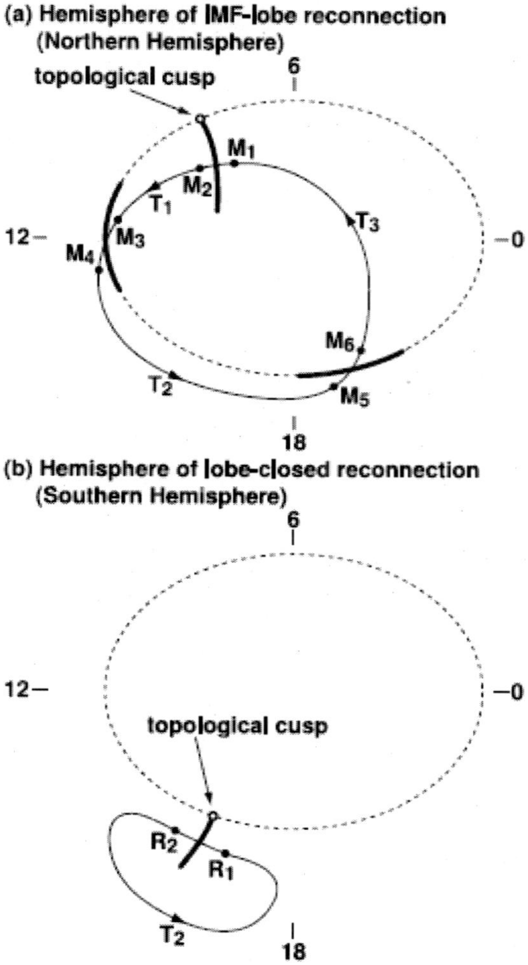

Fig. 2. The convection cell equatorward of the open-closed flux boundary in the southern hemisphere is a product of the interchange reconnection cycle. PolarDARN provides the best chance to verify this structure, hence the interchange reconnection hypothesis. Credit *Geophys. Res. Lett.*

The radio science component of CHAIN consists of 6 Canadian Advanced Digital Ionosondes (CADIs) and 10 GPS Ionospheric Scintillation and TEC monitors. CADI is a digitized ionosonde and operates at 95 frequency steps from 1 to 20 MHz. Interference between the four receiving antennas gives an estimate of electron drift at 30-s cadence. A 95-frequency ionogram is produced every minute. The GPS receiver uses the L1 band to measure amplitude and phase scintillation, and L1 and L2 bands to measure total electron contents.

CHAIN has unique and standalone value for polar-cap research. Electron density profile and total electron content (TEC) have been poorly sampled in the polar cap. Consequently, most reference ionospheric models show the greatest deficiency there. As convection topology changes, hot central plasma sheet particles can intrude the polar cap as evidenced by transpolar arcs.[13] Energetic solar protons have relative unfettered access to the polar cap, making it particularly vulnerable to solar flares. How the polar-cap ionosphere behaves under the influence of hot plasmas and energetic particles has no ready answer. There is statistical evidence that about half of substorms are triggered by northward turning of IMF.[14] Whatever the reason, this trigger has to propagate through the polar cap, accentuating the need to understand its behavior under all IMF conditions.

The statistical study of Drayton *et al.*[15] indicates that convection speed estimated with SuperDARN is about 30% less than the ion driftmeter estimate from DMSP satellites. One of the possible explanations of this discrepancy is that backscattering may become de-localized at large range. Six of the CHAIN sites have collocated CADI and GPS. This combination would provide a useful check on how actual ionospheric physics affect the accuracy of convection measurement using HF radars. Observations of TEC and electron density profile in altitude are important steps toward understanding how the polar cap can be changed by intrusions of CPS plasmas and solar energetic protons.

3.3. *CARISMA and CANMOS*

Spanning the Canadian sector, CARISMA and CANMOS (Canadian Magnetic Observatory System) provide the most extended measurement of geomagnetic activity in the auroral oval in the world. The 42 fluxgate magnetometers (FGM) of CARISMA and CANMOS are based on the standard ring-core design. The dynamic range is +/− 70,000 nT and resolution is 0.025 nT. The primary data from CARISMA is the 8-Hz measurements. Higher-level data products include auroral indices CE, CU, and CL (Canadian versions of AE, AU, and AL), dynamic spectrograms, ULF wave power, and magnetospheric density. In connection with algorithms such as AMIE, CARISMA and CANMOS data are used to produce global ionospheric current distribution.

The expansion of CARISMA was motivated in part by the idea to optimize magnetometer spacing in the Churchill line so that the cross-phase technique can be applied accurately to diagnose density structures in the magnetosphere (e.g. Ref. [16]). As well, a second Alberta line is formed to better determine longitudinal propagation of substorm disturbances. Milling *et al.*[17] and Murphy *et al.*[18] elaborated a method which can track the propagation of substorm disturbances in the ionosphere with ∼30 s resolution. Figure 3 is an event adapted from Murphy *et al.*[18] and underscored the importance of having a wide-area network with a sufficient number of magnetometers to time substorm propagation. The addition of 8 pairs of induction coil magnetometers (ICM) to CARISMA was a response to the need to understand the role of EMIC waves in the loss of radiation-belt electrons. The equatorward extension of CARISMA links with magnetometers in the US to form diagnostic chains to study inner magnetospheric physics of the ring current, radiation belts, and plasmasphere.

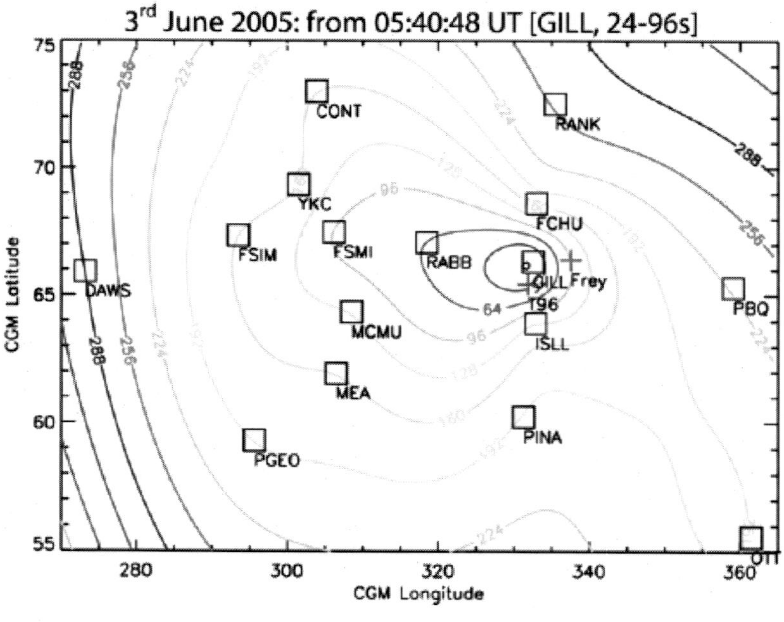

Fig. 3. Propagation of ULF wave power in the June 3, 2005 substorm (contours are 32 s apart). The onset starts at the Gillam station and expanded to multiple CGSM and CANMOS sites (marked in squares) in 4 min. Substorm onset determined by IMAGE is marked by the cross near Gillam at epoch time 0. Credit *J. Geophys. Res.*

The construction of the expanded CARISMA array makes it an attractive contextual tool for satellite missions. The ORBITALS (*Outer Radiation Belt Injection, Transport, Acceleration, Loss, Satellite*, www.orbitals.org) mission concept under study at the Canadian Space Agency phases its orbit in such a way that the satellite would spend most of its time dwelling over the CGSM array.

3.4. *CGSM MSI and MPA arrays*

Auroral emissions tell an interesting story of geospace. The 630 nm, 558 nm, 486 nm and other lines give ionospheric projections of different plasma populations in the magnetosphere. Multispectral imaging, and projection of auroral distributions in a sufficiently large area is the driving force behind the MSI array, with 16 imagers deployed in Canada.

A CGSM MSI uses a fish-eye lens to project light from a $\sim$160° field-of-view onto a CCD detector, which has linear response over a wide dynamic range and close to 100% quantum efficiency. The instrument is operated at 12-s cadence. Imaging is performed at wavelengths 427.8 nm. 557.7 nm, and 630.0 nm. Filtering is performed through filter wheels in 9 MSIs, and microfilters directly deposited on the CCD in 7 others. The related meridional-scanning photometer array (MPA) has 3 instruments deployed. The photometer uses a rotating fold mirror to scan the sky in 0.225° increment. A scan is completed once every 30 seconds, forming a latitudinal distribution of aurora brightness in wavelength selected by a filter wheel. Three new photometers were purchased under the CHAIN project discussed earlier. Photometry, due to its spectral cleanliness, allows more quantitative studies of aurora. Foremost among them is the proton aurora at 486 nm, which cannot be reliably measured with alternative techniques. Data products from the MSI and MPA are images of electron aurora in key visible wavelengths, and corroborating latitudinal profiles, including that of proton aurora. These images are generally different, and their overlay and comparison give a powerful means to establish where a particular process occurs in relation to regions of the magnetosphere.

The system-level view of the magnetosphere proposes that, in a large and complex system, most degrees of freedom cancel each other out on the macroscopic level. Only a few "emerged" degrees of freedom are observable. Some have argued that the magnetosphere is in a self-organized critical state (SOC) (e.g. Ref. [19]). In this state, details of interactions between parts of the system become unimportant — the system feeds back on itself so that, from a dynamical point of view, it does not change. Empirical evidence does not seem to support the radical notion of a SOC magnetosphere, but it is legitimate to ask how close the magnetosphere is to such a state. Put it differently, how efficiently do different scales couple in the magnetosphere? Aurora is a quintessential multiscale process: Large-scale auroral features break up into small-scale ones, and small-scale features coalesce to form large-scale features we see during substorms (see Ref. [20] for a cellular automaton simulation of multiscale structuring seen in auroras). Aurora is also a diagnostic tool of magnetospheric processes. Wanliss *et al.*[21] and Liu *et al.*[22] illustrated how proton auroral observations obtained with MPA can be used to characterize the magnetotail current sheet.

Another important question that has been with us since the beginning is the connection between aurora and neutral line(s) in the magnetosphere. Figure 4 shows the sudden onset of 558 nm aurora (greenline) due to 1–10 keV electrons in the mist of proton aurora (grey background) captured on March 3, 1997, by the Gillam MSP. Toward the end of the event, a very sharp electron flash with half-width 0.1° lasted for about a min amid a much broader proton aurora marking the central plasma sheet. Whether or not the flash is a signature of magnetic reconnection, this example illustrates the value of cross-spectral comparison.

3.5. *CGSM riometer array*

Riometer is a simple passive radio receiving device. The operational principle is based on the idea that cosmic radio noises are absorbed by the ionosphere. The ratio of absorption is modified by precipitating electrons in the energy range 30 to 100 keV. Therefore tracking the fluctuations of the absorption against a carefully calibrated baseline represents a relative measure of electron flux in this energy range. During periods of substorm or state of strongly enhanced magnetospheric convection, electrons are accelerated to >10 keV energy and injected from the central plasma sheet to the ring current. Knowledge of this population is important for 1) monitoring energy input into the inner magnetosphere and the development of magnetic storms, and 2) understanding the seed population of pre-heated electrons which are ultimately accelerated to the relativistic energy of >1 MeV. The CGSM riometer array consists of the 13 original CANOPUS units plus 14 acquired by Geomagnetic Laboratory, NRCan, giving a rough but valuable way to monitor plasma injection into the ring current.

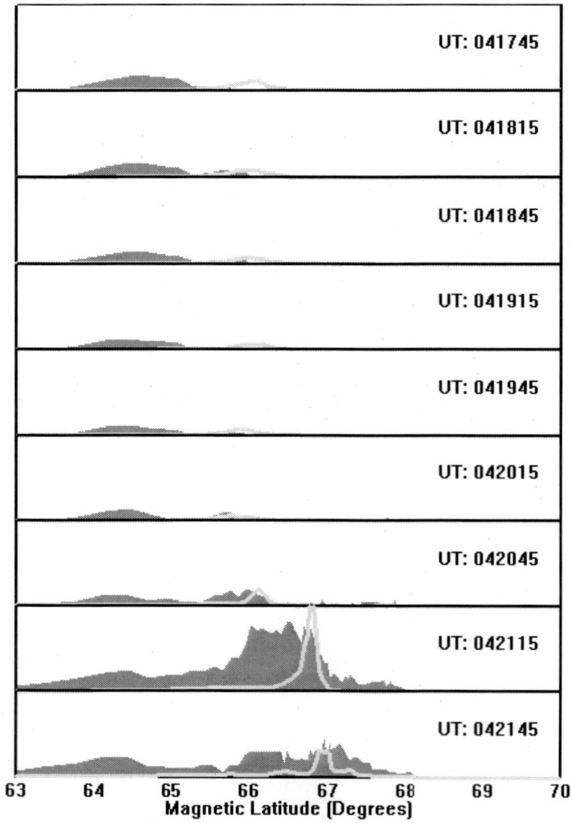

Fig. 4. Sudden brigtening of aurora on March 7, 1997. Toward the last minute of the event, the central plasma sheet represented by the grey proton aurora emission expanded rapidly, while the 558 nm aurora (green line) produced by 1–10 keV exhibits a brief flash lasting less than 1 min. This picture is consistent with the inferred ionospheric effects of magnetotail reconnection.

Early studies of riometer absorption have shown that it is closely related to particle injection in the ring current.[23] Spanswick *et al.*[24] showed disperisonless injection events (which occur in proximity to the source where the electrons are accelerated) observed by CANOPUS riometers and CRRES satellite correlated well (Fig. 5). They further showed that a network of sparsely based riometers give a good global picture of the evolution of injection in space. Liu *et al.*[25] developed a theory that correlates flux of injection (which is largely perpendicular to the magnetic field) with that of precipitation (which is directly measured by riometers), as well as

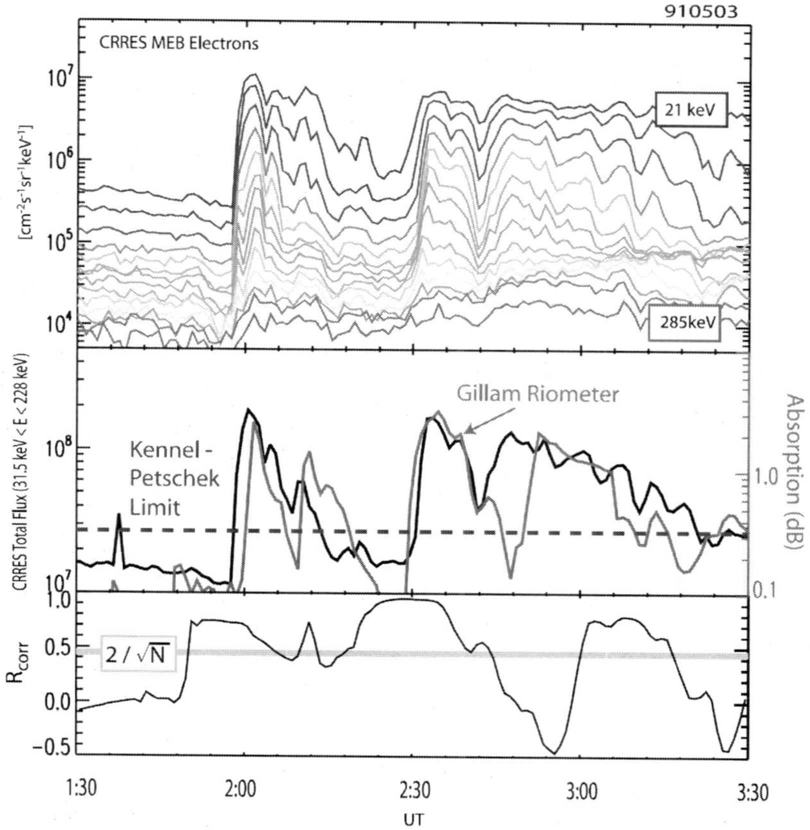

Fig. 5. MEP integrated electron flux above 21 keV (black curve, middle panel) matches the riometer absorption (red curve, middle panel). Credit Geophysical Research Letters.

with the magnitude of electric field enhancement in the magnetosphere. Using an extended version of the theory, Liang *et al.*[26] explained the difference between the dispersionless injection and so-called riometer spikes.[27]

3.6. *CGSM theory and modeling*

CGSM generates an overwhelming amount of data that cannot be easily sorted, much analyzed, by the human mind. Within the CGSM program envelop, a Facility for Data Analysis and Modeling (FDAM) provides mission-oriented support to turn multiperspective imaging observation into improved knowledge. Conversely, different model-based algorithms are developed to help interpret and manipulate imaging data. Much of the FDAM activity centers on the origin and dynamics of aurora. Alfvén waves play a dominant role in magnetosphere-ionosphere coupling. A focus of FDAM has been what roles Alfvén waves play in auroral dynamics, particularly in the inertial regime. Papers by Lu *et al.*[28] and Watt and Rankin,[29] for example, have elucidated how multiscale structures might develop as Alfvén waves cascade in the inertial regime, and how electrons can be accelerated to energies that produce aurora. Figure 6 is a case adapted from the latter work. In partnership with the University of Michigan, FDAM and its high-performance computers aim to perform near real-time simulation of magnetospheric dynamics. Characterizing Alfvénic modes in the simulated magnetosphere[30] is another example of value-added model products.

3.7. *Space weather*

Three activities are supported under the space weather module of CGSM. Through *Space weather now- and forecasting with ULF wave radial diffusion models*, the important problem of predicting relativistic electron flux in the outer radiation belt is tackled. CANOPUS played an important role

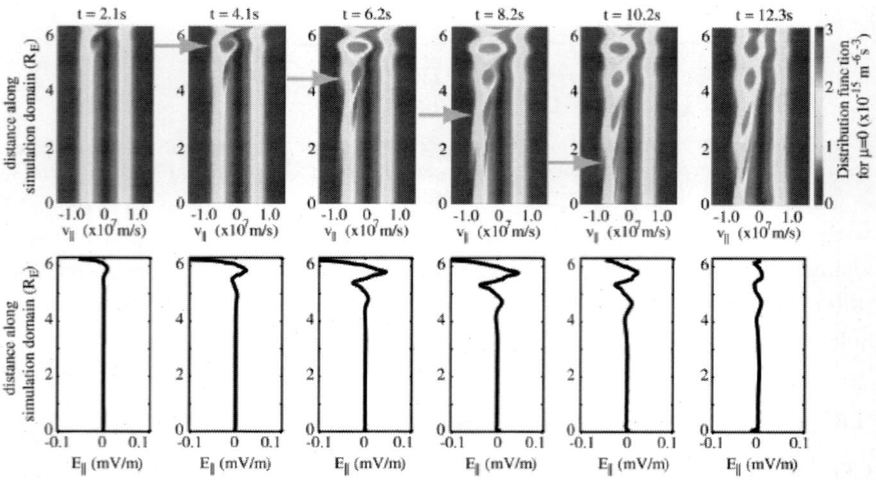

Fig. 6. 1D simulation of an Alfvén wave propagating through a diple field reveals that electrons are initially energized by the wave to keV energy at higher altitudes than previously thought. The trapped electrons are freed when the Alfvén speed becomes less than the electron thermal speed, giving the appearance that the electrons are accelerated at the lower altitudes. Credit Phys. Rev. Lett.

in establishing the correlation of relativistic electrons and ULF power.[8] Several theoretical and simulation studies have made the connection from a physical point of view.[31–34] This activity combines first-principle models and parameterized diffusion coefficients to arrive at a practical forecast algorithm. A key input to the models is the distribution of ULF power in the radiation belt latitude.

Through Solar drivers of space weather: Contributions to Forecasting Technology, computational techniques and codes will be developed on one of Canada's supercomputing clusters, SHARNET. The current focus is the propagation of CMEs in interplanetary medium through an implementation of 3D MHD code on an unstructured mesh. Real-time input from the STEREO satellites will be used as input to drive and verify the forecast model.

The Canadian Space Weather Forecast Center uses CGSM, as well as other data, to provide operational space weather information to the user community (see http://spaceweather.ca/index_e.php). The forecast is performed mainly through empirical and statistical methods, with targeted research to improve the performance of forecasting algorithms. Targeted research on space weather effects and effect mitigation is performed in collaboration with airline, pipeline, powergrid, communication satellites operators, and other affected industrial sectors. The Canadian F10.7 solar radio flux monitoring program, operated by the National Research Council of Canada, is also integrated under the service.

4. Related Ground-based Programs

In addition to the CGSM arrays, there are several special-purpose arrays and observatories in Canada that give the Canadian ground-based research program added capability and versatility.

4.1. *THEMIS GBO array*

Much of the controversy surrounding the substorm problem stems from imprecise knowledge of timing. Instabilities at play in substorm initiation erupt in 1–10 seconds. The resulting disturbances propagate through the system in less than 2 minutes.[35] In comparison, the 30–120 second time resolution of spaceborne imagers onboard Polar and IMAGE proves to be inadequate to separate temporally cause from effect. Another shortcoming in substorm study is the lack of conjunction between space and ground-based observational platforms. Accidental conjunctions are few, and ground-based and in-situ observations made on non-conjugate field lines often do not lend themselves to direct comparison. The GBO component was added to the THEMIS mission to optimize the probability that substorms are observed on the same field lines in the magnetosphere and ionosphere.[36,37] A network of 20 white-light ASIs operating at 3-s cadence was deployed across Canada and Alaska to precisely time the onset of auroral substorms. A complementary magnetometer array, including 5 magnetometers from CARISMA and 2 magnetometers from CANMOS, is also deployed.

Figure 3 is adapted from Liang *et al.*,[38] who found that most of the substorm onsets observed by THEMIS ASIs originate from a wave-like structure with the azimuthal m number between 100–300, corresponding to azimuthal wavelength 1000–3000 km in the magnetosphere. The structure grows in intensity and expands longitudinally without moving much initially in latitude. After ∼1 min, the growth reaches the breakup stage, and the auroral bulge starts to evolve with rapid poleward expansion. Figure 3 gives the $\boldsymbol{k}$-distribution of the dominant unstable mode, as well as pegging the growth time at ∼10 s. In a parallel study, Liu *et al.*[39] performed a wavelet transform of perturbations near a THEMIS satellite around the time of a local current disruption. They found that there is a dominant ω mode with period at ∼30 s. The two papers put the substorm onset wave mode in a narrow constraint and have prompted the proposal and theoretical calculation that the onset wave is a cross-scale mode in the two-fluid regime.[40]

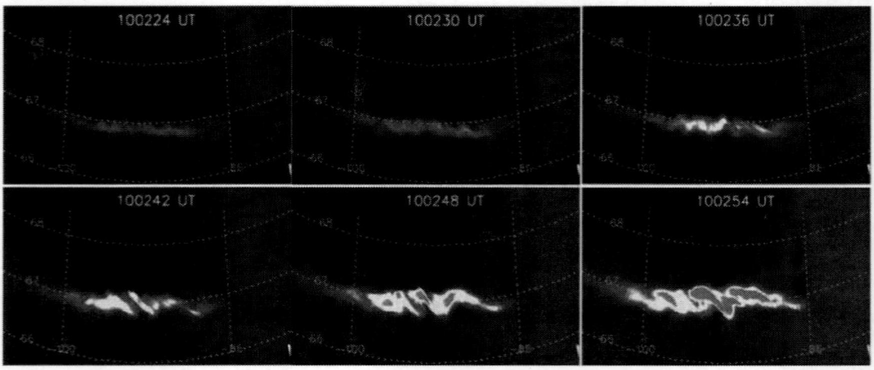

Fig. 7. An event study of THEMIS ASI showing wave-like activation of auroral breakup. Credit Geophys. Res. Lett.

THEMIS ASI and CGSM MSA and MPA arrays have complementary measurement capabilities. The THEMIS ASIs have higher cadence and a wider longitudinal coverage (10 h MLT) than CGSM, but lack spectral resolution to distinguish precipitating particles. CGSM MSI has a cadence of 12 s, which put them somewhat outside the margin of resolving onset wave growth reported earlier. On the other hand, with its ability to discern between warm and energetic electrons, as well as between electrons and protons with MPA, CGSM auroral imaging can specify important magnetospheric boundaries such as open-close flux boundary and peak of ion plasma sheet pressure.

4.2. *Advanced modular incoherent scatter radar (AMISR)*

AMISR is a US National Science Foundation-funded major facility for studies of high-latitude ionosphere and magnetosphere-ionosphere coupling. The principle of incoherent radar has been known and used since the 60's. What makes AMISR unique and valuable is the use of phased-array technology to achieve steerable beams instead of a monolithic antenna. AMISR has three faces. The first, Poker Flat Incoherent Radar (PFISR) was constructed in 2006 and has generated exciting new results. The remaining two faces are to be built in Resolute Bay, Canada. At the time of writing, the Canadian space physics community is bidding for $20 M in funding from Canada Foundation for Innovation (CFI) for approximately half of the Resolute AMISR project costs.

Each AMISR face is 32 square-meter in area and comprises of 128 panels. Phasing between the panels creates highly steerable beams (in milliseconds) to track fast moving ionospheric and atmospheric features due to internal dynamics and/or coupling with the magnetosphere. One outstanding feature of AMISR lacking in ground-based auroral imaging is its ability to go *in-situ* into the region where aurora is actually produced. This gives a three-dimensional view on how precipitating particles deposit their energy. The ability to remote sense electron and ion temperatures and probe electron density profile up to 1000's km in altitude allows AMISR to investigate wave-particle interactions that, among other things, accelerate ionospheric particles upward to the magnetosphere. On the other end, AMISR is able to probe the mesosphere and lower thermosphere and address the outstanding question how the ionosphere, especially under storm and substorm conditions, couple to the neutral atmosphere. The Poker Flat facility is located in the nominal auroral oval, whereas the

Resolute Bay facility will be located at the nominal open-closed flux boundary. The latter location is particularly valuable to the question of polar cap-auroral oval coupling, as discussed in previous sections.

5. Conclusions

Today, although satellite missions are commonplace and constellations of a dozen spacecraft are being seriously planned, one can hardly think of any significant magnetospheric problem, which does not require ground-based observations for closure. This is the most apt testimony to the staying power and considerable value of this long-standing endeavor. Just as satellite missions are becoming more focused on logically defined problems and less dependent on chance discovery, ground-based observations have experienced a shift in direction and strategy. Before, ground-based observations were best known for producing indices such as AE and giving evidence that something has happened, without much to say about why. While providing contextual and indicial information will continue to be important, ground-based research has become more sophisticated, both in terms of synergistic use of complementary instruments, and coordinated studies in conjunction with satellites. Use of multi-instrumental network over a large area such as Canada promises to provide crucial data to understand the multiscale behavior of geospace.

References

1. A. Vallance Jones, *Aurora*, Geophys. and Astrophys. Monogr. D. Reidel, Boston, 1974.
2. T. S. Trondsen and L. L. Cogger, High-resolution television observations of black aurora, *J. Geophys. Res.* **102**(A1) (1997) 363–378.
3. G. Rostoker, J. C. Samson, F. Creutzberg, T. J. Hughes, D. R. McDiarmid, A. G. McNamara, A. Vallance Jones, D. D. Wallis and L. L. Cogger, CANOPUS, A ground-based instrument array for remote sensing the high-latitude ionosphere during ISTP/GGS program, *Space Sci. Rev.* **71** (1995) 743–760.
4. J. C. Samson, B. G. Harrold, J. M. Ruohoniemi, R. A. Greenwald and A. D. M. Walker, Field-line resonances associated with MHD waveguides in the magnetosphere, *Geophys. Res. Lett.* **19** (1992) 441.
5. J. C. Samson, L. R. Lyons, P. T. Newell, R. F. Creutzberg and B. Xu, Proton aurora and substorm intensifications, *Geophys. Res. Lett.* **21** (1992) 2167–2170.
6. B. L. Xu, J. C. Samson, W. W. Liu, F. Creutzburg and T. J. Hughes, Observations of modulated auroral arcs by resonant Alfven waves, *J. Geophys. Res.* **98** (1993) 11531.

7. W. W. Liu, B. L. Xu, J. C. Samson and G. Rostoker, Theory and observation of auroral substorms: A magnetohydrodynamic approach, *J. Geophys. Res.* **100** (1995) 79.
8. G. Rostoker, S. Skone and D. N. Baker, On the origin of relativistic electrons in the magentosphere, *Geophys. Res. Lett.* preprint, 1998.
9. R. A. Greenwald, W. Weiss, E. Nielson and N. R. Thomson, STARE: A new radar auroral backscatter experiment in northern Scandinavia, *Radio Sci.* **13** (1978) 1109.
10. J. M. Ruohoniemi and K. B. Baker, Large-scale imaging of high-latitude convection with Super Dual Auroral Radar Network HF radar observations, *J. Geophys. Res.* **103**(A9) (1998) 20797–20811.
11. P. H. Reiff and J. L. Burch, IMF By-dependent plasma flow and Birkeland currents in the dayside magnetosphere — 2. A global model for northward and southward IMF, *J. Geophys. Res.* **90** (1985) 1595–1609.
12. M. Watanabe and G. J. Sofko, The interchange cycle: A fundamental mode of magnetic flux circulation for northward interplanetary magnetic field, *Geophys. Res. Lett.* **36** (2009) L03107, doi:10.1029/2008GL036682.
13. S.-W. Chang, J. D. Scudder, J. B. Sigwarth, L. A. Frank *et al.*, A comparison of a model for the theta aurroa with observations from Polar, Wind, and SuperDARN, *J. Geophys. Res.* **103** (1998) 17367–17390.
14. T. S. Hsu and R. L. McPherron, Average characteristics of triggered and nontriggered substorms, *J. Geophys. Res.* **109** (2004) A07208, doi:10.1029/2003JA009933.
15. R. A. Drayton, A. V. Koustov, M. R. Hairston and J.-P. Villain, Comparison of DMSP cross-track ion drifts and SuperDARN line-of-sight velocities, *Annales Geophys.* **23** (2005) 2479–2486.
16. Z. C. Dent, I. R. Mann, J. Goldstein, F. W. Menk and L. G. Ozeke, Plasmaspheric depletion, refilling, and plasmapause dynamics: A coordinated ground-based and IMAGE satellite study, *J. Geophys. Res.* **111** (2006) A03205, doi:10.1029/2005JA011046.
17. D. K. Milling, I. J. Rae, I. R. Mann, K. R. Murphy, A. Kale, C. T. Russell, V. Angelopoulos and S. Mende, Ionospheric localisation and expansion of long-period Pi1 pulsations at substorm onset, *Geophys. Res. Lett.* **35** (2008) L17S20, doi:10.1029/2008GL033672.
18. K. R. Murphy *et al.*, Wavelet-based ULF wave diagnosis of substorm expansion phase onset, *J. Geophys. Res.* **114** (2009) A00C16, doi10.1029/2008JA013549.
19. V. M. Uritsky, A. J. Klimas, D. Vassiliadis, D. Chua and G. Parks, Scale-free statistics of spatiotemporal auroral emissions as depicted by POLAR UVI images: Dynamic magnetosphere as an avalanching system, *J. Geophys. Res.* **107** (2002) 1426, doi:10.1029/2001JA000281.
20. W. W. Liu, P. Charbonneau, K. Thibault and L. Morales, Energy avalanches in the central plasma sheet, *Geophys Res. Lett.* **33** (2006) L19106, doi:10.1029/2006GL027282.
21. J. A. Wanliss, J. C. Samson and E. Friedrich, On the use of photometer data to map dynamics of the magnetotail current sheet during substorm growth phase, *J. Geophys. Res.* **105**(A12) (2000) 27673–27684.

22. W. W. Liu, E. F. Donovan, J. Liang, I. Voronkov, E. Spanswick, P. T. Jayachandran and M. Meurant, On the equatorward motion and fading of proton aurora during substorm growth phase, *J. Geophys. Res.* **112** (2007) A10217, doi:10.1029/2007JA012495.
23. D. N. Baker, P. Stauning, E. W. Hones Jr., P. R. Higbie and R. D. Belian, Near equatorial, 202 high-resolution measurements of electron precipitation and $L'' = 6.6$, *J. Geophys. Res.* **86** (1981) 2295.
24. E. Spanswick, E. Donovan, R. Friedel and A. Korth, Ground based identification of dispersionless electron injections, *Geophys. Res. Lett.* **34** (2007) L03101, doi:10.1029/2006GL028329.
25. W. W. Liu, J. Liang, E. Spanswick and E. F. Donovan, Remote-sensing magnetospheric dynamics with riometers: Observation and theory, *J. Geophys. Res.* **112** (2007) A05214, doi:10.1029/2006JA012115.
26. J. Liang, W. W. Liu, E. Spanswick and E. F. Donovan, Azimuthal structures of substorm electron injection and their signatures in riometer observations, *J. Geophys. Res.* **112** (2007) A09209, doi:10.1029/2007JA012354.
27. E. Spanswick, E. Donovan, W. Liu, D. Wallis, A. Aasnes, T. Hiebert, B. Jackel and M. Henderson, Substorm associated spikes in high-energy particle precipitation, *Physics and Modeling of the Inner Magnetosphere, AGU Book Series*, edited by T. I. N. A. Pulkkinen Tsyganenko and R. H. Friedel, p. 227, 2005.
28. J. Y. Lu, R. Rankin, R. Marchand, V. T. Tihonchuk and J. Wanliss, Finite elements modeling of nonlinear dispersive field line resonances: trapped shear Alfvén waves inside field aligned density structures, *J. Geophys. Res.* **108** (2003) 1394, doi:10.1029/2003JA010035.
29. C. E. J. Watt and R. Rankin, Electron trapping in shear Alfvén waves that power the aurora, *Phys. Rev. Lett.* PRL 102, (2009) 045002, doi:10.1103/PhysRevLett102045502.
30. R. Rankin, K. Kabin and R. Marchand, Alfvénic field line resonances in arbitrary magnetic field geometry, *Adv. Space Res.* **38** (2006) 1720–1729, doi:10.1016/j.asr.2005.09.034.
31. W. W. Liu, G. Rostoker and D. N. Baker, Magnetic pumping acceleration of relativistic electrons by large-amplitude ULF pulsations, *J. Geophys. Res.* **104** (1999) 17391.
32. S. R. Elkington, M. K. Hudson and A. A. Chan, Acceleration of Relativistic Electrons Via Drift-Resonant Interaction with Toroidal-Mode Pc-5 ULF Oscillations, *Geophys. Res. Lett.* **26**(21) (1999) 3273–3276.
33. L. G. Ozeke and I. R. Mann, Energization of radiation belt electrons by ring current ion-driven ULF waves, *J. Geophys. Res.* **113** (2008) A02201, doi:10.1029/2007JA12468.
34. A. W. Degeling, L. G. Ozeke, R. Rankin, I. R. Mann and K. Kabin, Drift resonant generation of peaked relativistic electron distributions by Pc 5 ULF waves, *J. Geophys. Res.* **113** (2008) A02208, doi:10.1029/2007JA012411.
35. E. Friedrich, J. C. Samson, I. Voronkov and G. Rostoker, Dynamics of the substorm expansive phase, *J. Geophys. Res.* **106** (2001) 13145.

36. E. F. Donovan, S. Mende, B. Jackel, H. Frey, M. Syrjasuo, I. Voronkov, T. Trondsen, L. Peticolas, V. Angelopoulos, S. Harris, M. Greffen and M. Connors, The THEMIS all-sky-imaging array — system design and initial results from the prototype imagers, *J. Atm. Solar Terr. Phys.* **68** (2006) 1472.
37. S. B. Mende, S. E. Harris, H. U. Frey, V. Angelopoulos, E. Donovan, B. Jackel, M. Greffen, C. T. Russell and L. M. Peticolas, A large array of ground based observatories for the study of auroral substorms, *Space Sci. Rev.*, in press, 2008.
38. J. Liang, E. F. Donovan, W. W. Liu, B. Jackel *et al.* Intensification of pre-existing auroral arc at substorm expansion onet: Wave-like disruption during the first tens of seconds, *Geophys. Res. Lett.* (2008) L17S19, doi :10.1029/2008GL033666.
39. W. W. Liu, J. Liang and E. F. Donovan, Interaction between kinetc ballooning mode and thin current sheet: Quasi-electrostatic field, local onset, and global characteristics, *Geophys. Res. Lett.*, in press (2008).
40. W. W. Liu and J. Liang, Disruption of magnetospheric current sheet byquasi-electrostatic fieldAnnales Geophys., **27** (2009) 1941–1950.

Advances in Geosciences
Vol. 21: Solar & Terrestrial Science (2008)
Ed. Marc Duldig

MICRO/MINI SATELLITES FOR EARTHQUAKE STUDIES — TOWARD INTERNATIONAL COLLABORATION

K.-I. OYAMA*, Y. KAKINAMI, J. Y. LIU and C. Y. CHEN
National Central University, No. 300 Jhongda Rd.
Jhongli City, Taoyuan 32001, Taiwan
**oyama@jupiter.ss.ncu.edu.tw*

T. KODAMA
Japan Aerospace Exploration Agency, Tukuba,
Ibaraki, Japan, 305-8505, Japan

Past studies show that the effects of earthquakes appear in the ionosphere prior to large earthquakes, and recently the number of the papers which report the precursor effect has rapidly increased and the content of these papers is becoming more compelling. However the results are still not fully convincing, because the number of the events, which are studied from ground as well as satellite based measurements, are not sufficient to deduce the morphology of the effects; the data presented is still not fully persuasive. In the first part of this paper, we describe the precursor behavior obtained by the Japanese satellite "HINOTORI" as well as the US satellite "Dynamic Explorer–2". We then discuss the problems, which we encountered in deducing the morphology associated with earthquake precursors. To establish he morphology is the first step in solving the mechanism for earthquakes to influence the ionosphere. Since the data which have been reported suggest a high probability for large earthquake to be forecast, we propose collaborating to launch micro-/mini-satellites, especially among the countries who suffer from earthquake disasters, and to share the mission costs. Key issues, which should be taken into account for the mission, are also discussed.

1. Introduction

The possible effects of earthquakes on the ionosphere have been reported by many scientists[1] Recently the number of the report has rapidly increased in scientific journals. Reduction of total electron content (TEC) produced prior to earthquake occurrence has been reported by Liu *et al.*[2] Devi *et al.*[3] reported reduction of TEC as well as increase of TEC at the anomaly crest region. They also reported the bite out phenomena of NmF2, which they claim to appear prior to earthquakes. In addition to the earthquake effects on low latitude ionosphere, there are several reports, which study the middle latitude ionospheric f_oF_2 and f_bE_s.[4,5] Zakharenkova *et al.*[6] reported precursory phenomena observed in the TEC measurements before the Hokkaido earthquake of September 25, 2003 ($M = 8.3$).

Recently, Pulinets *et al.*[7] conducted an intensive analysis of the Irpinia earthquake which occurred on November 23, 1980 using seismic data, radon emanation, hydrological anomalies, ground based ionosonde network data, thermal infrared irradiance and Intercosmos-19 satellite topside sounding. They concluded that air ionization by radon, which was emitted during the earthquake build up could explain all atmospheric and ionosphere parameters. However, most ionosphere researchers are still not convinced that earthquake precursor effects exist.[8] This attitude is quite understandable, because the number of the events which we can study, and the quality and quantity of the data we can use, is limited, depending on the countries, and therefore the arguments in most of the manuscripts are far from convincing. Even when we have sufficient information, it needs a lot of time and effort for us to draw conclusions on the effect of earthquakes on the ionosphere because this is a new field and established methods do not exist. Most scientists do not want to take risks by spending time on this un-established field.

In this paper we describe our experience detecting peculiar features of the ionosphere and show two phenomena which we believe are a precursor effect of earthquakes on the ionosphere. After we present the ionosphere phenomena associated with earthquakes, we discuss the ionosphere parameters which should be measured in order to understand the physical mechanism of the effect of earthquakes on the ionosphere. We propose launching 6 micro-satellites for earthquake study as the first step. We discuss the size of the satellites to be launched and basic items to be included during the planning. We also stress that the mission should be an international collaboration, especially among Asian countries, to share the cost and effort.

2. Procedure to Detect Earthquake Related Features

Finding the effects of the earthquakes on the ionosphere requires four characteristics. These processes are actually the history until one of the authors (Oyama) was convinced the existence of the effect(s) of earthquake on ionosphere. First of all, the instruments should be reliable and accurate. For example electron temperature (Te) measurement still has problems even for recently launched satellites.

It should be stressed that researchers should be experienced in the measurement of Te, including the effects of electrode contamination.[9] Secondly after the data are retrieved from the satellite, we need to obtain general features of Te/Ne, depending on local time, season, solar flux, latitude and longitude,[10–12] as well as the magnetic disturbance, and to understand the physics of these various features, such as Te in plasma bubbles,[13] the effect of the electric field on the morning overshoot,[14] Te behavior of equatorial ionization anomaly,[15,16] annual behavior of Ne/Te,[17] and the effect of neutral wind upon Te/Ne related to the tilted magnetic meridian.[18,19] Thirdly a model of Te[20] as well as of Ne,[21] which describes the normal and average behaviors of the ionosphere, should be constructed and verified with the various observations. We tried to check the model by studying nighttime's increase of Te during geomagnetic disturbance.[22]

Based on this modeling, we derive the earthquake effect on the ionosphere by calculating the deviation of Te from the model value (Δ Te). Variations of the ionospheric Te during one large earthquake are shown as an example. In order to find as much data as possible which support the existence of the effects of the earthquake on the ionosphere, we extended our study to include data from the Dynamic Explorer–2 (DE-2) satellite, which was in a polar orbit, while HINOTORI was in an equatorial orbit.

For DE-2 data, we did not produce models of Te and Ne, because DE-2 was in an eccentric orbit in addition to a polar orbit (0 to ± 90 degrees), which made it difficult to model realistically. On the other hand, HINOTORI was in a circular, low latitude orbit of 31 degrees inclination. Due to the difficulty of model construction using DE-2 data, we only compare the results with IRI (International Reference Ionosphere[23]). It is generally accepted that IRI reproduces Ne below NmF2 reasonably well.

2.1. *HINOTORI-Te behavior*

The Japanese Sun Observation satellite was put into equatorial orbit at the height of 600 km in February 1981. Although the satellite was dedicated to study solar physics, two unique plasma probes that were developed in Japan were accommodated. The two probes are a resonance rectification probes (see Fig. 1) for electron temperature, Te[24,25] and an impedance probe for electron density, Ne[26] and had been flown in a number of sounding

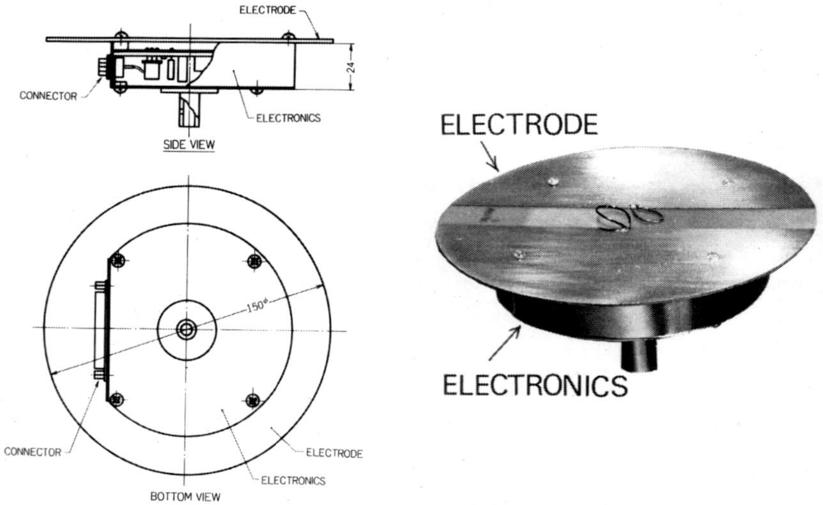

Fig. 1. Resonance rectification probe, which gives accurate Te measurements.

rockets and satellites, and their performance was well established. Data were accumulated for 16 months, until the satellite operation was terminated in June 1982 due to a battery problem. The instruments gave accurate and reliable data. Especially the resonance rectification probe provided very accurate Te.

Here we show examples of Te behavior as well as for Ne for one large earthquake, which occurred on November 22, 1981 with magnitude of 6.6, depth of 37 km, and epicenter at 14.09 N and 124.35 E.[27]

Te varies for local time, seasons, latitude, longitude, and solar activity. Figure 2 shows the local time variation of Te, with geomagnetic latitude as horizontal axis for equinox (March, September), northern summer (May, June, and July) and northern winter (November, December, and January).

For all three seasons (northern summer, northern winter, and equinox), Te takes the minimum value at ~4 LT, rises steeply and shows a peak at ~9 LT (named "morning overshoot"). After the morning overshoot, Te reduces toward noon, starts increasing at ~15 LT, and shows the second peak at ~17 LT (named "afternoon overshoot"). Finally the peak disappears at around 20 LT. The first peak is not so sensitive to geomagnetic latitude, comparing to the second peak. The second peak shows a clear latitudinal variation; the peak is not remarkable over the equator, but becomes clearer at higher latitudes.

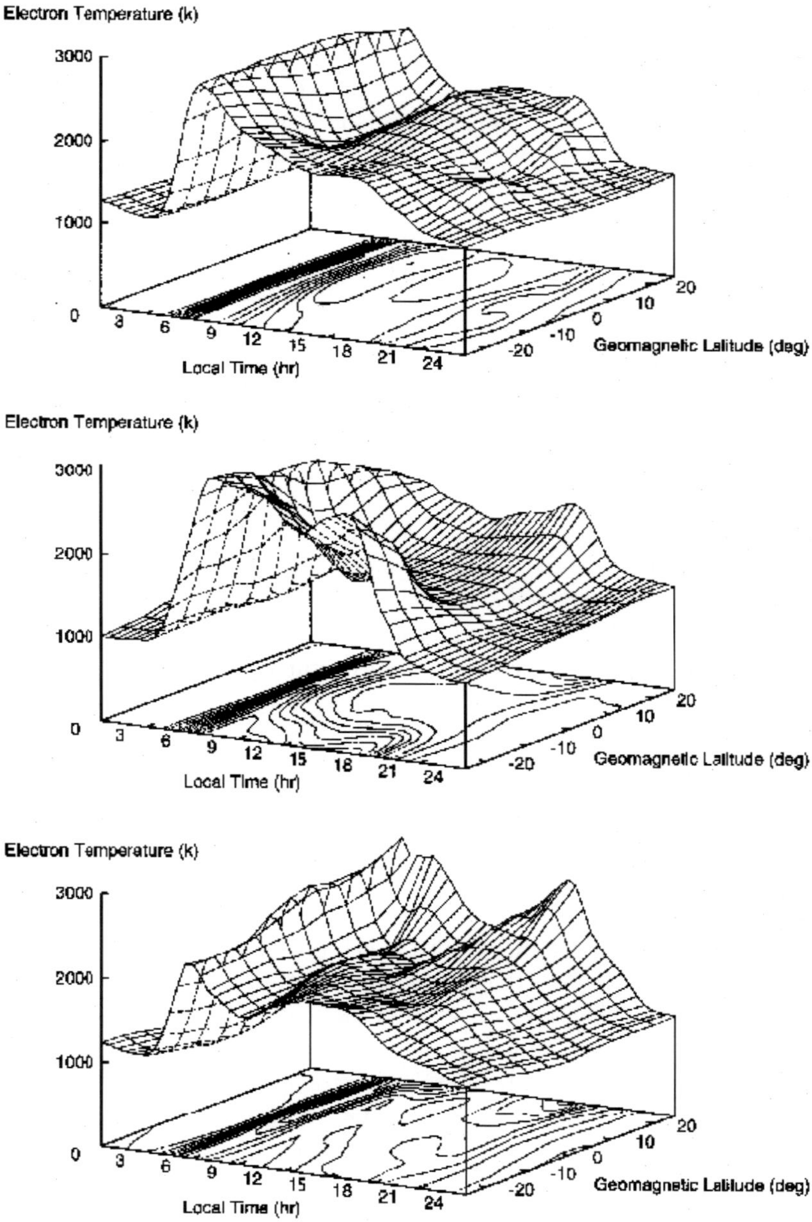

Fig. 2. Local time and seasonal dependence of Te at 600 km height with respect to geomagnetic latitude for the equinox (upper panel). Northern summer (middle panel), and Northern winter (bottom panel).

The peak at ∼9 LT is explained as follows. Thermal electrons have already been created from photoelectrons before the neutral density increase, which results in less energy loss due to electron-neutral collisions. In Fig. 2 the maximum Te morning overshoot is higher in winter than in summer, which is a result of higher Ne in the summer than in the winter hemisphere. The difference of Ne between two hemispheres is mainly caused by the summer-to-winter neutral wind. On the other hand the afternoon overshoot is explained mainly by non-local heating, although the neutral wind effect is also working. That is; photoelectrons which travel along the magnetic line of force from the place where they are produced at lower altitude, are heated at the height of 600 km. Photoelectrons from other hemisphere which travel along the magnetic filed line can also contribute. It is noted that even for the morning overshoot, contribution from the lower ionosphere in the same hemisphere as well as other hemisphere exist.[28]

It is noted here that Banks and Kockarts[29] show that Te in the afternoon overshoot at the height of 600 km is more sensitive than any other plasma parameter such as Ne, and ion temperature (Ti).

We found that largest Te reduction, which is shown in Fig. 2, occurs mainly in the afternoon overshoot. One example of Te reduction of the satellite orbit is shown in the upper panel of Fig. 3 for November, 20 1981, 2 days before the EQ occurs. The lower panel of the figure shows Ne, which was measured simultaneously. Black and blue circles in each panel plot Te/Ne values from the model and observations, respectively. Te profile shows the large reduction at afternoon overshoot. The model illustrates that Te starts to elevate at 16:30 and shows the peak before 18 LT (afternoon overshoot), and reduces at 19:06. Whilst Te values of the orbit keeps constant value during the period when Te is expected to elevate. As shown in the upper panel of Fig. 3, the observed Te matches the model values some time during or after the afternoon overshoot. The earliest local time when the observed Te and model value match is, of course, when the afternoon overshoot starts. We found that Te deviates for longer in LT as the earthquake day approaches and as the satellite orbit is closer to the epicenter. It seems that deviation of Te shows the maximum 1–2 days after the earthquake. This might give us further clues about the mechanism of the effects of earthquakes on the ionosphere.

As one can see in the lower panel, the observed Ne follows the model value or takes slightly higher values than the model at the beginning. Later it starts to be lower than the model values. This is one of the important

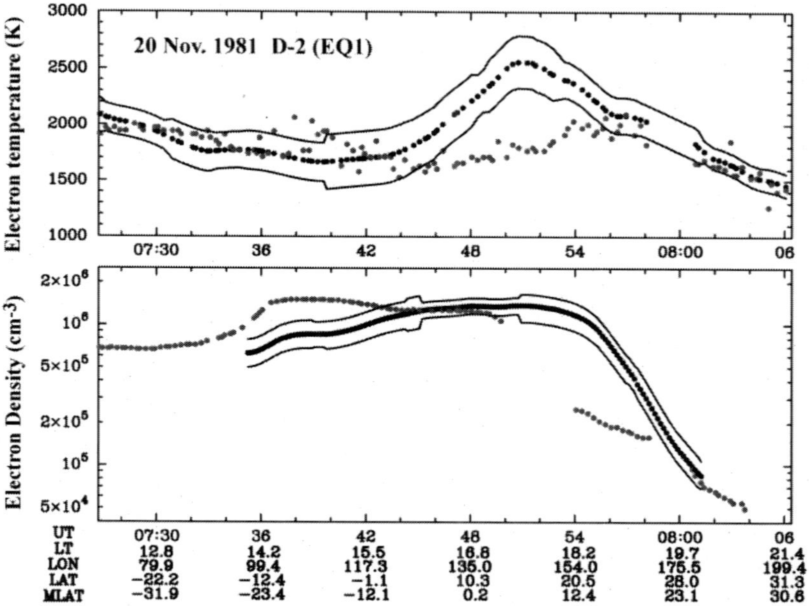

Fig. 3. An example showing the depression of Te (upper panel) in the afternoon overshoot on the November 20, 1981, 2 days prior to EQ. Black and blue dots show the model and observation for Te (upper panel) and Ne (lower panel) respectively. Two thin lines show the upper and lower values of standard deviation.

clues when we try to explain the reduction of Te in the afternoon overshoot. The feature is also consistent with DE-2 satellite data.

In addition to the earthquake of November 22, 1981, we have intensively studied two large earthquakes, which occurred on January 11, 1982 (M7.4) and January 24, 1982 (M6.6). We found that the features, which we described above, are common for all three earthquakes, both prior to and after the earthquake. The variation of Te in the afternoon overshoot also occurs during geomagnetic storms, which is the case at the beginning of November 1981. Storm time Te increases or reduces depends on longitude and latitude. The variation of Te during the magnetic disturbance occurs all over the world, whilst Te deviation during the earthquake activity is limited to the epicenter region and shows systematic gradual change.

Figure 4 shows the spatial distribution of Δ Te for the November 1981 earthquake by plotting each satellite orbit in geographic longitude-latitude coordinate before and after the earthquake. It is noted that the region of large Te deviation ranges from more than 60 degrees to the west and 40 degrees to the east from the epicenter respectively. It is also noted that

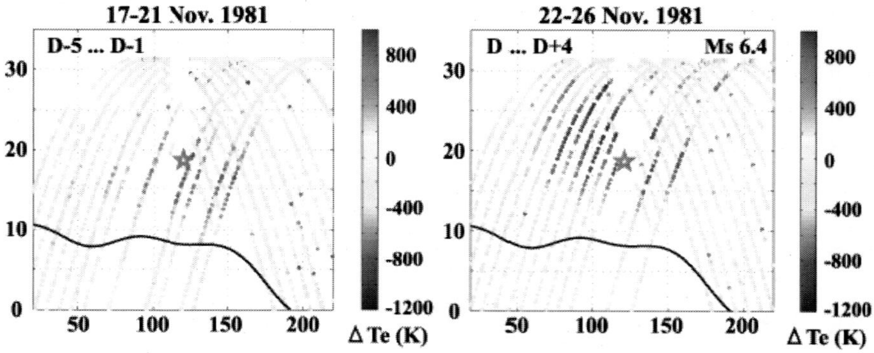

Fig. 4. Spatial distributions of Te during EQ (Epicenter; 14.09 N and 124.35 E) 5–1 days (D-5 $\cdots$ D-1) before (left) and 0–4 days (D $\cdots$ D + 4) after the earthquake day (right). D indicates the earthquake day. The star in each panel shows the epicenter locations. Note that Te deviation spreads more than 50 degrees in longitude toward both west and east from epicenter.

along the latitude of the epicenter less depression of afternoon overshoot appears to exist. Oxygen ion density, which was observed by DE-2 for another earthquake, shows the minimum deviation over the epicenter as we describe later. These two features might be another clue to the mechanism of the effect of earthquake on the ionosphere.

The disturbed region extends up to 60 degrees longitude to the East as well as to the west. The similar spatial distributions are drawn for other 2 earthquakes before and after the earthquakes, which shows more clearly symmetric deviation in longitude with respect to the epicenter. If we compare the disturbed region among three earthquakes, the EQ, which occurred on January 11, 1982 is wider than the other two.

It is noted that the largest deviation of Te with respect to latitude does not match the epicenter, because electrons easily move along the geomagnetic line of force.

Figure 5 shows the deviation of Te (Δ Te) from the model against day for November 1981. The longitude and latitude ranges are limited to 20–220 degrees and 0–32 degrees respectively. The deviation of Te which is larger than 500 K has statistical significance. Dst and Kp indices, which indicate magnetic disturbance, are also plotted.

Magnetic disturbance occurs on November 23, 1981 with the lowest Dst values of -75 nT. In the lower panel, color scale in Δ Te panel indicates that the large deviation of Te was observed both during the morning and afternoon. The deviation of Te from the model starts to increase on

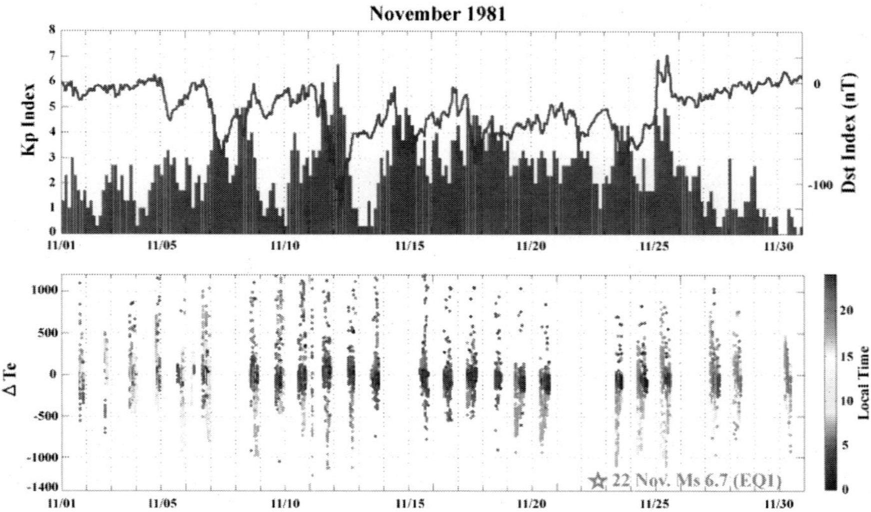

Fig. 5. Deviation of Te observation from model values for November 1981. Top panel shows Dst (Blue) and Kp indices (Black). Bottom panel shows the Te deviation with respect to date. Depending on the local time when the data were obtained, the data points are colored from dark blue (early morning) to deep red (midnight) evening. A red star shows the day when earthquake occurred. Please note that orange color data points show the largest deviation of electron temperature.[27]

November 18, 1981. Two days data gap exists, and no measurement is available on the earthquake day. It appears that the largest deviation of Te in the afternoon overshoot is found around November 23, 1981. Negative deviations of Te from the model are seen in the morning overshoot from November 6 to November 13, which might be attributed to the magnetic disturbance. It is noted that deviation of Te from the model scatters in both positive and negative values during the morning overshoot period during the magnetic disturbance. It is also noted that daytime Te varies in anti-phase with Ne during the magnetic disturbance.[22] Conversely, during the earthquake disturbance deviation of Te is only biased to negative value, and shows a gradual steady reduction. The anti-phase relation between Te and Ne is not found.

It is concluded, however, that the afternoon overshoot Te is not influenced by geomagnetic disturbances. Our result suggested that there is a high possibility to find precursor effects at least near the Philippines and Taiwan. However we need further study to confirm our results. The current research situation has led us to propose constellation of micro-satellites as a first step to study earthquake precursors.

2.2. *DE-2 data-behaviour of O^+ and Ne*

The DE-2 satellite was in the orbit from June 1981. Therefore two satellites (DE-2 and HINOTORI) were simultaneously in orbit for 6 months. Although we tried to conduct joint research by using satellite data sets from DE-2 and HINOTORI, we found that it is very difficult to study earthquake effects using both data, mainly because DE-2 is in a highly eccentric orbit, and the number of the orbit is not enough. Accordingly to get data at the same location is almost impossible. This orbit factor is discussed in detail later.

One example, which shows the behavior of ionosphere O^+ density influenced by an earthquake, which occurred on the October 15, 1981 (Lat; −33.13; Lon, 73.07, Ms; 7.5). Figure 6 shows the O^+ density along the DE-2 satellite orbit, starting from 5 days before the earthquake until 5 days after the earthquake. As the observation approaches the earthquake day, the O^+ density shows a minimum over the epicenter as if the equatorial ionization trough shifted from the geomagnetic equator to the epicenter. This tendency is more evident when the observation longitude is closer to the epicenter. The longitudes where the disturbance is found extend wider than 40–50 degrees. A similar feature is seen in Ne, but O^+ density shows it more clearly.

Ne is much higher than the IRI values away from the epicenter latitude. Above the epicenter O^+ density seems very close to the value from IRI. This might give us another clue to the mechanism linking ground and upper atmosphere phenomena.

Sometimes (not always) fluctuation of O^+ density of about 1% are found, being superposed on the background O^+. This supports the fact that widening of Doppler frequency sometimes appears before an earthquake in Taiwan. Details will be published in the near future.

3. Possible Candidate to Disturb Ionosphere

We discuss three candidates, which might contribute to the disturbance of the ionosphere. First is a change of neutral density such as atomic oxygen. In fact the ratio of H^+/O^+ increases before the earthquake,[30] which might suggest a variation of neutral gas. It is clear that H_2 is emitted by the reaction between SiO_2 and H_2O.[31] Earlier Wakita *et al.*[32] reported the emission of Hydrogen from faults. However it seems impossible for Hydrogen to reach the upper atmosphere within several days. Therefore we are forced

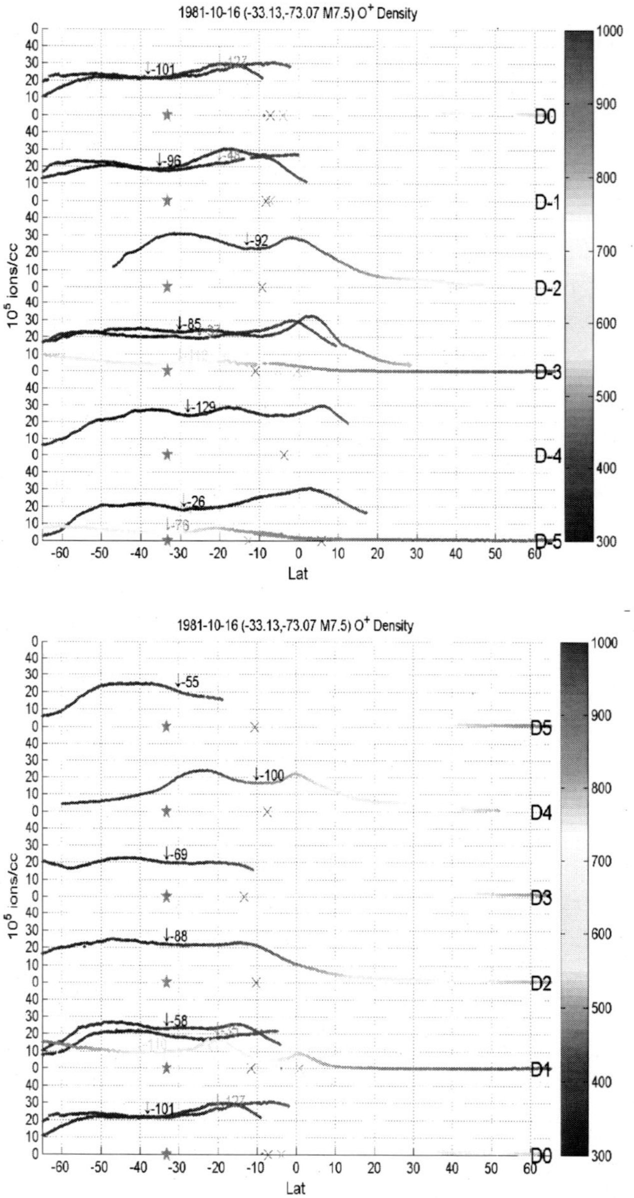

Fig. 6. Behavior of O^+ density along DE-2 Orbits five days prior (-D5, -D4, -D3, -D2, -D1) and after (D5, D4, D3, D2, D1) the earthquake which occurred on the October 16, 1981. D0 is the day when the earthquake occurred. Red star in each plot is the epicenter. Cross shows the latitude of geomagnetic equator. Color shows the satellite altitude.

to abandon this mechanism. Regarding the Te result from HINOTORI satellite, there is a possibility that the neutral wind might change Ne at 600 km and then Te. However we can not simply explain the disappearance of the afternoon overshoot by the neutral wind at 600 km. Neural winds push up or push down the plasma depending on the angle between wind and geomagnetic line of force. It is generally accepted that when Ne increases, Te reduces during the daytime. However HINOTORI Ne does not change during the initial phase of disappearance of the afternoon overshoot. We currently do not favor the role of neutral wind at 600 km as an explanation.

The most probable candidate is the electric field. Many papers that report Ne, NmF2 or TEC favor the electric field, because the effect also appears at the conjugate point and the effect appears about 5 days before the earthquake. These facts make us presume the existence of electric field that very quickly influences the ionosphere. In order to confirm this speculation electric field or plasma drift should be measured. Regarding the height of the original electric field associated with the earthquake, it might be difficult to assume that the electric field which is generated underground directly penetrates to the ionosphere. Therefore our next question is on the origin of the electric field. Although we do not want to speculate on the generation mechanism of the electric field at the moment, when information is scarce to make conclusions, we close this section by showing two plausible mechanisms below.

The strong electric field might possibly be generated at the heights where clouds formed prior to and after earthquakes. The westward electric field which shows a symmetric feature both side of the epicenter suggests the existence of a narrow region of very intense electric field near the epicenter along the north-south direction; negative at east side and positive at west side of the epicenter. The east/west width of the strong electric field region might be less than 100 km, so that, this small-scale electric field cannot reach 100 km which connect F region through magnetic line of force.

One of the candidate locations of the electric field is the thunder cloud height. High potential, which is among thunder clouds, might be mapped onto the E region ionosphere. The HINOTORI result shows that a westward electric field exists in one direction. In order to explain the single direction electric field, the distance of the clouds which is highly charged (west and east clouds are positive and negative, respectively) might be less than 100 km. Therefore if we look at the electric field at heights higher than 100 km, the electric field might become one direction.

Another mechanism is that a dynamo electric field might be modified by the neutral wind. Hagan and Forbes[33] investigated mesospheric and lower thermosphere migrating and non migrating tidal components that propagate upward from the troposphere, where they are excited by latent heat release associated with deep tropical convection. Pulinets *et al.*[7] proposes the ionization produced by the radon emanating from the Earth's crust. As a result, hydrated ions are produced, which finally release the latent heat of evaporation. Recently Immel *et al.*[34] showed that a newly discovered 1000-km scale longitudinal variation in ionospheric density might be explained by consideration of the dynamo interaction of the tides with the lower ionosphere (E–layer) in the daytime. England *et al.*[35] published a paper that seems to support this idea.

4. The Need for Constellation of Satellites

The examples we showed in the previous sections are cases where the ionosphere is modified by earthquakes. To study the earthquake effects, we need to survey the ionosphere globally to investigate whether abnormal features we find are local or global. High latitudes, above 50 degrees, might be difficult to study because high latitudes are very often highly disturbed. However, even for low/mid latitude cases, there are several cases which we cannot see any precursor effects even if the USGS list registers the earthquake. On the contrary, cases exist that Te reduces in the afternoon overshoot about 5 days before the earthquake even if we cannot find earthquakes in the USGS list. Although we are not able to yet explain this case, a case exists that if an earthquake occurs at nearly the same location one month later, the effect continues during this period. We are trying to use as much data as possible from past satellites as well as ionosonde data to explain these cases. However, we find it difficult to deduce the morphology and hence find the mechanism. Two reasons can be raised for this difficulty. First there are not many satellites in the past, which provide reliable and routine data. So far we believe that only the Langmuir probe on board DE-2[36] provides reliable values, because the effort to erase contamination effect was made by heating the electrode as well as by applying negative voltage to the electrode. We confirmed the accuracy and reliability of the DE-2 Langmuir probe data by comparing with the HINOTORI satellite. Secondly in order to make the model, the number of input parameters, such as height, longitude, latitude, solar activity, and season should be as small as possible. Constant altitude of the satellite makes the model construction much easier.

However there is no satellite with a circular orbit except HINOTORI in the past. As we mentioned above, to study the earthquake effect, we should study the global area, however the setting of ground based instruments in remote area requires effort and is time consuming and it is limited to land surfaces.

We now have the possibility to detect precursors from space and therefore we should pursue the satellite mission, and to try to confirm its possible existence. We can use micro- (10–100 kg) and mini-satellites (100–500 kg).

4.1. *Instruments candidates*

It is essential that each satellite should accommodate the same two fundamental instruments (electron temperature probe and impedance probe). The same instruments fabricated in the same organization should be onboard to obtain consistent data from 6 satellites. Two examples of mounting the two instruments are shown in Fig. 7. It is noted again that it needs special attention to get accurate electron temperature. Even electron temperatures which have been measured recently are not accurate. Langmuir probe data from recently launched satellite show an electrode contamination problem.

Data from another satellite seem to give higher temperature due to insufficient amplifier gain. Apart from these two instruments, to find the mechanism of phenomena associated with earthquake, the following instruments might be needed:

a. Electric field probe (DC to AC)
   There is a good possibility that the anomalous feature is triggered by the electric field associated with earthquakes. It is especially important to know the electric field. The measurement accuracy should be better than 0.2 mV/m to measure a deviation of 1 mV/m.
b. Plasma drift meter (University of Texas type)
   By measuring plasma drift $Vd = E \times B/B^2$, one can get the electric field which is perpendicular to the magnetic field and the information from the drift meter can be a backup of the electric field probe.
c. Ion mass spectrometer
   Recently one paper has been published[30] which used DMSP and DEMETER satellite data. Several hours before the Sumatra earthquake, a ratio H^+/O^+ increased at about 800 km height. The results seem to be acceptable. However the amount of data is still not enough to be

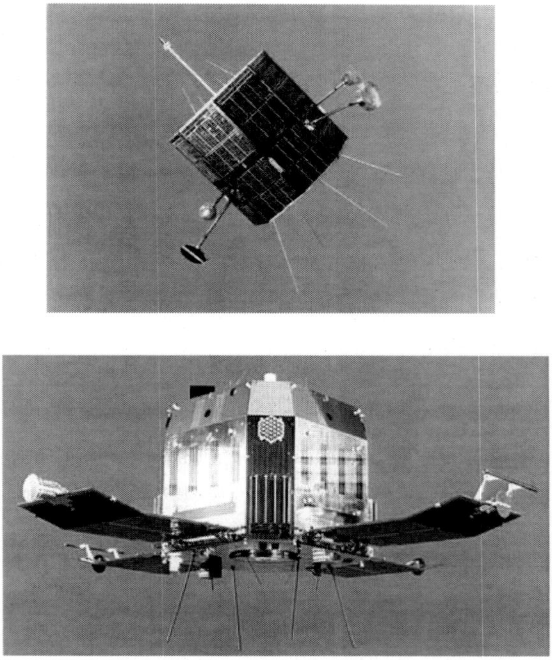

Fig. 7. Example of mounting two probes on satellite. The top picture shows electron temperature probe and electron density probe mounted at the end of the booms of TAIYO satellite (top). The bottom picture shows 4 electron temperature probes and two impedance probes (electron density) at the ends of 4 solar cell paddles of OHZORA satellite (From JAXA web page).

convincing. We need to confirm their results. Ion mass to be measured is H^+, and O^+.

d. Neutral mass spectrometer
Regarding the ration of H^+/O^+, we cannot exclude the possibility that modifications of neutral particle cause changes of the ratio.

e. Topside sounder
Pulinets[37] reports the results from topside sounders onboard intercosmos-19, and ISS-b. Both foF2 and height profile of electron density in the topside ionosphere change right before and after earthquakes. There is a possibility that height profile is modified by electric field associated with earthquakes. Therefore result from a topside sounder can be discussed together with electric field measurements. A topside sounder is surely one of the strong candidates to be accommodated. However the topside sounder emits intense RF power,

and disturbs ambient plasma as well as, potentially, the satellite itself. It is highly recommended to launch only one micro-satellite that carries a topside sounder.

f. Energetic particle analyzer
There are also several papers which report precipitation of energetic particles triggered by VLF waves associated with earthquakes[38].

g. Photometers
A report also exist that intensity of the airglow intensified. This report is reasonable, because if the plasma density increases, airglow emission in the ionosphere increase. To mount a photometer can support measurement of Ne made by the impedance probe. The narrow band photometer of narrow FOV might be able to give us 3D picture of ionosphere electron density as Formosat-3 (Taiwan satellite) conducted. The instrument can provide the change of equatorial ionization trough and cleft.[39] Through one orbit scan, we might be able to get height profile of O^+ density. 2D photometer at several wave lengths (135.6 nm, 630 nm, and 555.7 nm) might provide 2D maps at night on the atmospheric glow at several heights. Emission from atomic oxygen is a function of oxygen ion. Measurement of 557.7 nm (emission from OH) gives us the structure at the height of about 97 km, whilst 630 nm gives the structure at 200–300 km. 2D structure might indicate the wave structure of neutral wind at around 100 km. There is a possibility that electric fields associated with earthquakes are produced as a modulation of neutral wind in the Dynamo region as we mentioned before.[35]

h. VLF wave receiver (magnetic field)
Considering the data obtained from DEMETER,[40] and our data analysis of VLF measured by DE-2 satellite, it is very difficult to identify the earthquake related noise because much VLF noise exists in the ionosphere. One is in the high latitude; the second one is very often detected over geomagnetic latitude. The third one seems to be related to the Equator Temperature and wind anomaly.[41] It seems to be difficult task to identify the epicenter. We might need to find the direction. Even in this case we might encounter difficulty in identifying the source. However the direction finding might be able to give useful information on ray tracing VLF into the ionosphere, after the epicenter is identified.

The instruments, which are listed above, are the instruments which are needed for ionospheric study, and the satellites for the earthquake study are not special. The big difference between conventional science missions and this earthquake satellite mission is that researchers should examine the data with the eyes that earthquake effects should exist. The researchers, first, should conduct science to understand non-earthquake natural phenomena, and after that they should study the peculiar features associated with earthquakes.

4.2. *Constellation of micro-satellites*

To find precursor effects, the measurement of two plasma parameters is essential. Therefore we propose to put two probes in micro-satellites. The total weigh of the satellite might be about 50 kg. A spinning satellite such as TAIYO, which has solar cells at the wall, might be simplest. Two probes can be mounted at the end of a boom. TAIYO was spin stabilized with a rolling wheel controlled by a geomagnetic attitude control system. Four plasma probes were extended perpendicular to the spin axis by 0.5 m metallic booms. Power at average rate of 15 W was provided by 6000 silicon n-p solar cells. The objectives of the satellite were to study the ionosphere systematically by simultaneously observing solar ionizing radiations (Lyman-α and X-rays), the ultraviolet albedo of the earth, positive ion composition, and plasma parameters such as electron and ion densities and temperature in the ionosphere. The size of the satellite was 75 cm in diameter and 65 cm in height. Total mass of the satellite was 86 kg. The orbit is 249 km $\times$ 3129 km with an inclination of 31.54 degrees. The launch was in 1975. Modern technology will surely make it possible to reduce the weight down to 50 kg with more fine instruments. Another type of satellite might be gravity gradient, with no or very slow satellite spin so that the gravity gradient is not disturbed. However there is a concern that either the electron temperature probe or the impedance probe is in the satellite wake for a long time during one satellite orbit if the satellite does not spin. If this is the case, we need 4 electron temperature instruments, and two impedance probes to get accurate data outside the satellite wake during a whole orbit. Even with this set of instrumentation, the weight might be less than 10 kg. Apart from science instrumentations, we need instruments to provide attitude information, especially the ram direction. A sun sensor and magnetometer might be good candidates during daytime, mainly perpendicular to gravity gradient direction. The moving direction during nighttime can be guessed from daytime information assuming the satellite has slow spin.

Fig. 8. DEMETER, one example of a mini-satellite (from DEMETER web page).

4.3. *Needs of mini-satellites*

In order to understand the physics on the interaction between lithosphere, atmosphere and ionosphere, two parameters are not enough. We need at least a DEMETER (see Fig. 8) class satellite to accommodate the instruments described above, except a topside sounder. We expect that some of the countries can launch one mini-satellite. Good examples of mini-satellites are Japanese satellite "TAIYO", "HINOTORI" (circular orbit of 600 km at 31 degrees inclination, 188 kg, controlled by using geomagnetic field, http://www.isas.ac.jp/e/enterp/mission/hinotori.shtml), and the French satellite "DEMETER" (Sun synchronous circular orbit at 715 km, 130 kg, 3 axis control, http://smsc.cnes.fr/DEMETER/GP_satellite.htm)).

4.4. *Orbits of the satellites*

Inclination of the satellite might be maximum 50 degrees, for two reasons; (1) at higher latitude the plasma is highly disturbed most of the time and finding earthquake effects becomes extremely difficult, (2) the electric field associated with earthquake only drives horizontally. Inclination of 6 micro-satellites should be in low inclination of 30 degrees, and each satellite accommodates two plasma probes. The orbit should be circular at the height of 550–600 km. Orbits lower than 550 km reduce the mission life of the satellite due to the stronger atmospheric drag effect. One satellite scans in longitude 22.5 degrees every 90 minutes. If we have 6 satellites, the satellite shifts 22.5 degrees in 15 minutes. Therefore, the same latitude location can be scanned by one of the satellite every 16 satellite revolutions that means 16×90 minutes = 1440 minutes (1 day). That is, almost same location can be scanned by one of the satellites every day.

4.5. *Key issues to bear in mind for mission design*

There are several key issues which we should discuss during mission planning between the countries and in order to have a cost/time effective mission.

1. All receiving facilities should be the same, if a country newly constructs a receiving station and tracking station. If the country already has a receiving station, exchange of information is surely required. The key point is to get the same quality of data, including satellite orbit information.
2. Electronics parts of the satellite should be commercially available in order to reduce the cost of the satellite. Problems which might be caused as a result of using commercially available parts can be reduced by careful environment test and thermal vacuum test.
3. Two common instruments, resonance rectification probe and impedance probe, should be accommodated in all satellites and the instruments should be tested on ground and verified so that the instruments give the same result. Institute of Space and Astronautically Sciences, JAXA has a space plasma chamber of diameter 2 m and the length of 5m. The chamber, which is specially designed for ionosphere instruments, can produce Ne of $10^4 - 10^6$ els/cc, and Te of 500–3000 K.
4. Format for information of the satellite orbit as well as format of science data should be common to all satellites so that data can be easily handled among the participating countries.
5. The software program for data reduction of Te and Ne should be provided from Japan to save total time and effort. Each country does not need to check the program to derive final value of Te and Ne. For example, to get Ne, geomagnetic field at the satellite location is needed. Therefore in the program to calculate Ne, a subroutine to calculate geomagnetic field is needed.
6. Orbit information of each satellites and science data should be archived in each country, and should be easily accessed from member countries.

In conventional science missions, researchers do data analysis by themselves and most are interested in event studies. On the other hand, the earthquake mission needs first to collect data from all satellites, to validate data and construct models. These tasks, especially, need a lot of time and effort, and therefore a software engineer group, which can work with the researchers, should be employed.

The countries that are interested in this mission can contribute in various ways. Those are manufacturing of satellite(s), launch of the satellites, generating software, and data analysis. Although several countries already show interest in earthquake satellites, they are still not realized. An international collaborative team should be organized to promote the mission and encourage groups, which are trying to launch the satellites.

4.6. *Outreach and education*

To especially trigger the curiosity of high school students on science, the earthquake subject might be a good target and at the same time the student can learn the importance of using space in daily life. Therefore the mission might be able to provide a good tool for education. Toward achieving this goal, distribution of the data to high school students should be considered. Data distribution can be easily done through the internet. Alternatively, they can construct a set of RF receiver which provides intensive instruction for many systems for high school students. Therefore one small transmitter should be accommodated in the satellite for educational purpose. Part of the signal is transmitted from each satellite through a low power transmitter.

5. Coordination of Various Fields

As we discussed above, if we try to verify that the electric field is generated in the troposphere, we need the vertical profile of neutral gas temperature, at least up to 60 km. This might be possible using a radio occultation satellite. A constellation of 6 Taiwan satellites is producing atmospheric temperature profile up to 60 km.

We also need cloud pictures at different heights. This is available from meteorological satellites. Measurement of ground temperature by infrared camera reports the change of temperature along faults.[42–44]

In the upper ionosphere, we have already listed the necessary instruments. It is essential to study the ionosphere from various aspects.

The research on the effects of earthquakes on the ionosphere should not be limited to the ionospheric community. We need coordination among different fields.

6. Situation of Earthquake Satellite Planning in Several Countries

There are several satellites being discussed in several countries. The countries which are discussing earthquake study satellites are; Russia (Kompass-3, and Kanopus-Vulkan), Mexico (UNAMSAT-3), Ukraine (Poperedjiennia, Ionosat), USA (Quakefinder), Italy (Esparia), and China (CSES). China officially announced the launch of earthquake satellites on the occasion of DEMETER workshop, which was held in Toulouse in 2007. Recently an Australian group proposed Lightning–1 and Lightning–2. Other satellite missions are still not realized. Exchange of information among countries is urgently needed and an international satellite group for earthquake study should be formed to accelerate the realization of the mission.

7. Concluding Remarks

The electron temperature probe that was developed for pure science now has a new application. Based on our firm confidence of the reliable measurements, we have studied three EQ events, which occurred around an equator ionization anomaly. All 3 events show common features. Continuous systematic reduction of Te in the afternoon overshoot can be found prior to big earthquakes. O^+ density also shows clear features (similar feature can be seen in Ne as well).

Whether we can find smaller earthquake or high latitude earthquake in the future depends on the accuracy of the model, which can be made with repeated satellite observations. The features which have been described here strongly suggests the effect of electric field associated with earthquakes, is playing a fundamental role in the region. The electric field is the order of 1 mV/m or less and it should have very slow time variation of the order of 10 days starting from about 5 days before earthquakes. Although the generation of the electric field is puzzling, it might be atmospheric in origin, which might be first triggered by a process such as radon emission.

Our result suggests that even a small satellite, which carries two simple reliable plasma probes, can play an unexpectedly significant role in the study of precursor phenomena associated with earthquakes. Orbits of the satellites are very important. International collaboration among countries that are suffering from earthquake disasters should be established urgently to launch micro- or mini-satellites from these countries.

Finally our findings might also encourage ionosphere researchers to study lithosphere-atmosphere-ionosphere coupling, and even open new perspectives in ionosphere research, which has a long history of more than 80 years.

Acknowledgments

We would like to express our sincere thanks to Dr. M. Duldig of Ice, Ocean, Atmosphere & Climate Program, Australian Antarctic Division, Department of the Environment, Water, Heritage and the Arts, Australia for his careful reading of the manuscript. The work has been completed while one of the authors (K.-I. Oyama) is visiting the Graduate Institute of Space Science, National Central University, Taiwan.

References

1. S. Pulinets and K. Boyarchuk, Ionospheric Precursors of Earthquakes, Springer, New York, 2004.
2. J. Y. Liu, Y. I. Chen, Y. J. Chuo and H. F. Tsai, Variation of ionospheric total electron content during the Chi-Chi earthquake, *Geophys. Res. Lett.* **28** (2001) 1383–1386.
3. M. Devi, A. K. Barbara and A. Depueva, Association of total electron content (TEC) and foF2 variations with earthquake events at the anomaly crest region, *Ann. Geophys.* **47**(1) (2004) 83–91.
4. A. A. Silina, E. V. Liperovskaya, V. A. Liperrovsky and C.-V. Meister, Ionospheric phenomena before strong earthquakes, Natural Hazards and Earth system Sciences, **1** (2001) 113–118.
5. T. Ondoh, Seismo-ionospheric phenomena, *Adv. Space Res.* **26**(8) (2000) 1267–1272.
6. I. E. Zakharenkova, I. I. Shagimuratov, A. Krankowski and A. F. Lagovsky, Precursory Phenomena observed in the total electron content measurements before great Hokkaido earthquake of September 25, 2003 ($M = 8.3$), *Stud. Geophys. Geod.* **51** (2007) 267–278.
7. S. A. Pulinets, P. Biagi, V. Tramutoli, A. D. Legenka and V. Kh. Depuev, Irpinia earthquake 23 November 1980-Lessons from nature reviled by joint data analysis, Annals of Geophysics **50** (2007) 61–78.
8. H. Rishbeth, Do Earthquake Precursors Really Exist? Eos, **88**(29) (2007) 296.

9. K.-I. Oyama, A systematic investigation of several phenomena associated with contaminated Langmuir probes, *Planet. Space Sci.* **24** (1976) 183–190.
10. Y. Z. Su, K.-I. Oyama, G. J. Bailey, T. Takahashi and H. Oya, Longitudinal variation of the topside ionosphere at low latitude; satellite measurements and mathematical modeling, *J. Geophys. Res.* **101** (1996) 17191–17205.
11. Y. Z. Su, G. J. Bailey, K.-I. Oyama and N. Balan, A modelling study of the longitudinal variations in the north-south asymmetries of the ionospheric equatorial anomaly, *J. Atmos. Terr. Phys.* **59** (1997) 1299–1310.
12. Y. Z. Su, G. J. Bailey and K.-I. Oyama, Annual and seasonal variations in the low latitude topside ionosphere, *Ann. Geophysicae* **16** (1998) 974–985.
13. K.-I. Oyama, K. Schlegel and S. Watanabe, Temperature structure of plasma bubbles in the low latitude ionosphere around 600 km altitude, *Planet. Space Sci.* **36**(6) (1988) 553–567.
14. K.-I. Oyama, N. Balan, S. Watanabe, T. Takahashi, F. Isoda, G. J. Bailey and H. Oya, Morning overshoot of Te enhanced by downward plasma drift in the equatorial topside ionosphere, *J. Geomag. Geoelectr.* **48** (1996) 959–966.
15. K.-I. Oyama, M. A. Abdu, N. Balan, G. J. Bailey, S. Watanabe, T. Takahashi, de E. R. Paula, I. S. Batista, H. Oya and F. Isoda, High electron temperature associated with the prereversal enhancement and the equatorial ionosphere, *J. Geophys. Res.* **102** (1997) 417–424.
16. N. Balan, K.-I. Oyama, G. J. Bailey, S. Fukao, S. Watanabe and M. A. Abdu, A plasma temperature anomaly in the equatorial ionosphere, *J. Geophys. Res.* **102** (1997) 7485–7492.
17. G. J. Bailey, Y. Z. Su and K.-I. Oyama, Yearly variations in the low-latitude topside ionosphere, *Ann. Geophysicae* **18** (2000) 789–795.
18. S. Watanabe and K.-I. Oyama, Effects of neutral wind on the electron temperature at a height of 600 km in the low latitude region, *Ann. Geophysique* **14** (1996) 290–296.
19. K.-I. Oyama and S. Watanabe, Effects of zonal and meridional neutral winds on the electron density and temperature at the height of 600 km, JAXA RR-03-002, Japan Space Exploration Agency, Kanagawa, Japan, 2004.
20. K.-I. Oyama, P. Marinov, I. Kutiev and S. Watanabe, Low latitude Te model at 600 km based on HINOTORI satellite data, *Adv. Space. Res.* **34** (2004) 2004–2009.
21. Y. Kakinami, S. Watanabe and K.-I. Oyama, An empirical model of electron density in low latitude at 600 km obtained by Hinotori satellite, *Adv. Space Res.* **41** (2008) 1494–1498, doi: 10.1016/j.asr.2007.09.031.
22. K.-I. Oyama, D. R. Lakshmi, I. Kutiev and M. A. Abdu, Low latitude Ne and Te variations at 600 km during 1 March 1982 storm from HINOTORI satellite, *Earth Planets Space* **57** (2005) 871–878.
23. D. Bilitza, International Reference Ionosphere 2000, Radio Science, **36**(2) (2001) 261–275.
24. K. Hirao and K.-I. Oyama, An improved type of electron temperature probe, *J. Geomag. Geoelectr.* 22 (1970) 393–402.

25. K.-I. Oyama, T. Abe, K. Schlegel, A. Nagy, J. Kim and K. Marubashi, Electron temperature measurement in Martian ionosphere, *Earth and Planet. Sci.* **51** (1999) 1309–1317.
26. M.-Y. Yamamoto, T. Ono, R. T. Tsunoda, M. F. Larsen, S. Fukao and M. Yamamoto, Structures in sporadic — E observed with an impedance probe during the SEEK campaign: comparisons with neutral wind and radar echo observations, *Geophys. Res. Lett.* **25** (1998) 1781–1784.
27. K.-I. Oyama, Y. Kakinami, J. Y. Liu, H. Kamogawa and T. Kodama, Reduction of electron temperature in low latitude ionosphere at 600 km before and after large earthquakes, *J. Geophys. Res.* (2008) doi:10.1029/2008JA013367.
28. Da A. V. Rosa, The theoretical time dependent thermal behavior of the ionosphere electron gas, *J. Geophys. Res.* **71**(17) (1996) 4107–4120.
29. P. M. Banks and G. Kockarts, Aeronomy, Academic Press, New York, 1973.
30. L. G. Bankov, M. Parrot, R. A. Heelis, J.-J. Berthelier, P. G. Marinov and A. K. Vassileva, Satellite observations of ionospheric effects caused by Earth's seismic activity, accepted for *Adv. Space Res.* (2008).
31. I. Kita, S. Matsuo and H. Hiroshi, H_2 generation by reaction between H_20 and crushed rock: An experimental study on H_2 degassing from the active fault zone, *J. Geophys. Res.* **87** (1982) 10789–10796.
32. H. Wakita, Y. Nakamura, I. Kita, N. Fujii and K. Notsu, Hydrogen release: New indicator of fault activity, Science, **210** (1980) 188–190.
33. M. E. Hagan and J. M. Forbes, Migrating and no migrating diurnal tides in the middle and upper atmosphere excited by tropical latent heat release, *J. Geophys. Res.* 107(D24), (2002) 4754, doi: 10.1029/2001JD001236.
34. T. J. Immel, E. Sagawa, S. L. England, S. B. Henderson, M. E. Hagan, S. B. Mende, H. U. Frey, C. M. Swenson and L. J. Paxton, Control of equatorial ionospheric morphology by atmospheric tides, *Geophys. Res. Lett.* **33** (2006) L15108, doi: 10.1029/2006GL026161.
35. S. L. England, S. Maus, T. J. Immel and S. B. Mende, Longitudinal variation of the E-region electric fields caused by atmospheric tides, *Geophys. Res. Lett.* **33**(L22205) (2006) 1–5.
36. J. P. Krebbiel, L. H. Brace, R. F. Theis, W. H. Pinks and R. B. Kaplan, The Dynamic Explorer Langmuir probe instrument, *Space Sci. Instr.* **5** (1981) 493–502.
37. S. A. Pulinets, Strong earthquake prediction possibility with the help of topside sounding from satellite, *Adv. Space Res.* **21** (1998) 455–458.
38. Iu. I. Gal'Perin, V. A. Gladyshev, N. V. Dzhordzhio, V. I. Larkina and M. M. Mogilevskii, Precipitation of trapped energetic particles in the magnetosphere over the epicenter of an impending earthquake, Kosmicheskie Issledovaniia **30** (1992) 89–106.
39. M. L. Hsu, Study of ionosphere by using 135.6 nm nightglow, PhD thesis, National Central University, 2008.
40. M. Parrot, J. J. Berthelier, J. P. Lebreton, J. A. Sauvaud, O. Santolik and J. Blecki, Examples of unusual ionospheric observations made by the

DEMETER satellite over seismic regions, *Physics and Chemistry of the Earth* **321** (2006) 486–494.

41. R. Raghavarao, R. Suhashini, H. G. Mayr, W. R. Hoegy and L. E. Wharton, Equatorial spread F (ESF) and vertical winds, *J. Atmos. Solar-Terr. Phys.* **61** (1999) 607–617.
42. D. Ouzounov and F. Freund, Mid-infrared emission prior to strong earthquakes analysed by remote sensing data, *Adv. Space Res.* **33**(3) (2004) 268–273.
43. S. A. Pulinets, D. Ouzounov, A. V. Karelin, K. A. Boyarchuk and L. A. Pokhmelnykh, The physical nature of thermal anomalies observed before strong earthquakes, *Phys. and Chemistry of the Earth* **31** (2006) 143–153.
44. D. Ouzounov, N. Bryant, T. Logan, S. Pulinets and P. Taylor, Satellite thermal IR phenomena associated with some of the major earthquakes in 1999–2003, Physics and Chemistry of the Earth, **31** (2006) 154–163.

Advances in Geosciences
Vol. 21: Solar & Terrestrial Science (2008)
Ed. Marc Duldig

SOLAR TRANSITION REGION IN THE QUIET SUN AND ACTIVE REGIONS

H. TIAN*
School of earth and space sciences,
Peking University, Beijing, China
Max-Planck-Institut für Sonnensystemforschung,
Katlenburg-Lindau, Germany
**tianhui924@gmail.com*
tian@mps.mpg.de;tian@pku.edu.cn

W. CURDT
Max-Planck-Institut für Sonnensystemforschung,
Katlenburg-Lindau, Germany

J.-S. HE
Max-Planck-Institut für Sonnensystemforschung,
Katlenburg-Lindau, Germany

The solar transition region (TR), in which above the photosphere the temperature increases rapidly and the density drops dramatically, is believed to play an important role in coronal heating and solar wind acceleration. Long-lasting upflows are present in the upper TR and interpreted as signatures of mass supply to large coronal loops in the quiet Sun. Coronal bright points (BPs) are local heating phenomena and we found a different Doppler-shift pattern at TR and coronal temperatures in one BP, which might be related to the twisted loop system. The dominant energy loss in the lower TR is the Ly$-\alpha$ emission. It has been found that most Ly$-\alpha$ radiance profiles are stronger in the blue peak, an asymmetry opposite to higher order Lyman lines. This asymmetry is stronger when the downflow in the middle TR is stronger, indicating that the TR flows play an important role in the line formation process. The peak separation of Ly$-\alpha$ is found to be larger in coronal holes than in the quiet Sun, reflecting the different magnetic structures and radiation fields between the two regions. The Lyman line profiles are found to be not reversed in sunspot plume and umbra regions, while they are obviously reversed in the surrounding plage region. At TR temperatures, the densities of the sunspot plume and umbra are a factor of 10 lower than of the plage, indicating that the sunspot plasma emitting at TR temperatures is higher and possibly more extended in height above sunspots than above the plage region.

1. Introduction

The solar transition region (TR), in which above the photosphere the temperature increases rapidly and the density drops dramatically, is believed to play an important role in coronal heating and solar wind acceleration. Most of the TR emission comes from the VUV (vacuum ultraviolet) range of the electromagnetic spectra. And thus spectroscopic diagnostics in the ultraviolet spectral range can provide ample information on the structures and properties of the TR.

It was believed that the network pattern is the dominant emission structure in TR images. The magnetic structures in the network are a combination of diverging funnels and low-lying cool loops.[1] The funnels can be connected to the solar wind, or just legs of large coronal loops.[2] Observational results and theoretical models of the TR were reviewed by Refs. [3] and [4].

The spectral range of the SUMER (Solar Ultraviolet Measurement of Emitted Radiation) instrument[5,6] onboard SOHO (Solar and Heliospheric Observatory) includes hundreds of emission lines formed in the TR.[7,8] Through combined analyses between the data obtained by SUMER and other instruments, we have studied the magnetic structures and plasma properties of the TR in the quiet Sun and in active regions. In this paper we summarize the main results of our recent works. For the sake of simplicity, we divide the TR into three parts according to the temperature: lower TR (8000 K to 30000 K), middle TR (30000 K to 600000 K), and upper TR (600000 K to 1000000 K).

2. Flows in the TR

Since three decades ago, it has been found that emission lines formed in the middle TR are on average redshifted by several km/s.[9–12] The magnitude of the redshift is positively correlated with the line radiance,[12] and thus much larger in network than in internetwork regions.

With increasing temperature, the average Doppler shift of TR lines turns from red shift into blue shift.[10,11] Patches of blue shift on the Dopplergrams of upper TR lines like Ne VIII are found everywhere in coronal holes.[13–16] These blue shifts are interpreted as signatures of the nascent solar wind and believed to be associated with magnetic funnels which originate from supergranular boundaries.[17]

In the quiet Sun, magnetic funnel structures might also exist.[19,24] In Ref. [19], we have shown a quiet-Sun funnel structure reconstructed from photospheric magnetic field observation. The typical funnel structure is: several small funnels originating from different supergranular boundaries expand with height, and finally merge into a single wide open field region. One may speculate that coronal funnel structures in coronal holes should be of a similar form, yet the merging height might be different.

However, most of the quiet-Sun funnels do not associate with significant Ne VIII blue shift. In fact, large blue shifts are found mainly in network junctions and associated with legs of large coronal loops. Thus they are more likely to be signatures of mass supply to coronal loops rather than solar wind outflows.

The relationship between the red shift of cool lines and blue shift found at higher temperatures is not well understood. In coronal holes, the blue and red shifts were regarded as the upflow and downflow after magnetic reconnection between open field lines in coronal funnels and their side loops.[17,18]

In the quiet Sun, the scenario of continuous reconnection might also be the case, if the magnetic polarities of side loops are opposite to those of funnel-like loop legs,[19,21,36] thus enabling reconnection. The outflows produced by reconnections around a funnel tend to converge towards the center of the funnel. In contrast, the hot plasma trapped in low-lying loops are pulled down when they cool, and the downflows are stronger at the boundary of the network where side loops are accumulated. Thus, the bi-directional flows are likely to be detected as the not-fully-cospatial blue shift of Ne VIII and red shift of emission lines formed in the middle TR.[16,22] An alternative mechanism, as suggested in Ref. [23], is the heating and cooling process: Cool plasma (in photosphere and chromosphere) might continuously enter any loop leg through a certain process from outside, but then flow up and speed up after heating occurred, leading to a strong blue shift of lines formed in the upper TR. Due to the onset of possible (radiative) cooling effects, the flow might again decelerate above a certain height (perhaps in the lower corona) and finally turn downwards and accelerate under gravity, which may lead to emission by the dense plasma at lower temperatures and then cause the red shift of middle-TR lines.

In summary, we conclude that the dominant flows at upper-TR temperatures in quiet-Sun coronal loops are long-lasting upflows rather than siphon flows. once the mass is supplied into coronal loops and goes upward to the loop apexes, it may fall downward when cooling is switched on. Sometimes the coronal loops might transiently open due to magnetic reconnection and thus can release mass into the ambient corona or even into the solar wind. The steadiness of the observed shifts suggests that all these processes should occur continuously and persistently, indicating the existence of "coronal circulation".[25]

3. TR Structures in Coronal Holes and in the Quiet Sun

Figure 1 illustrates different magnetic structures in coronal holes and in the quiet Sun. In CHs (coronal holes) the dominant magnetic structures are coronal funnels which are connected to the solar wind. Most of the magnetic loops reside in the lower part of the solar atmosphere. There are almost

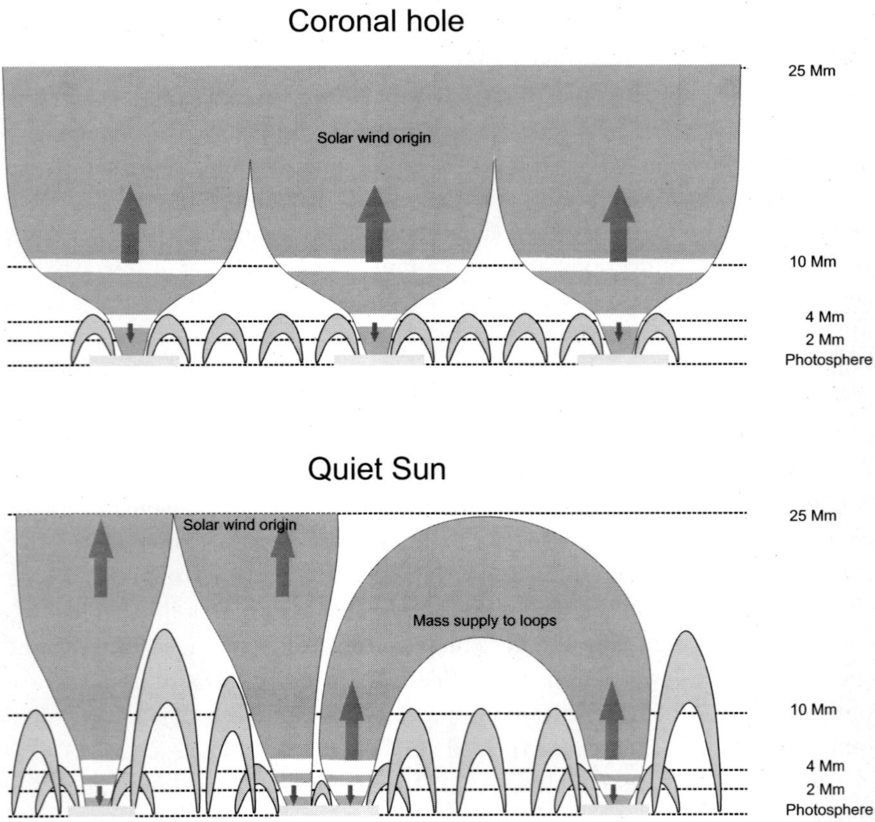

Fig. 1. Schematic presentation of the magnetic structures of the transition region in coronal holes (upper) and quiet Sun (lower). The top and bottom of the TR are indicated by the two white bars in each funnel or loop leg. The TR downflows and upflows in upper layers are marked in red and blue, respectively. The yellow bars indicate the supergranular boundaries. The figure is adapted from Ref. [31].

no large loops.[26–28] Thus, in CHs magnetic funnels can expand strongly through the TR, as revealed by the temperature variation of TR features in Ref. [26]. While magnetic loops of different sizes are crowded in the quiet Sun, so that large coronal loops can not expand drastically with height and the dense cool loops only permit a weaker expansion of the loop legs.

Through correlation analyses between 3-D extrapolated magnetic field and EUV spectroscopic observations, we can roughly estimate the emission height of TR lines. This method is based on the concept of magnetoconvection, with the magnetic field being frozen in the plasma flow. If we regard the emission height of Ne VIII as the top of the TR, we can study the spatial extension of the TR emission. Based on the results in Refs. [17], [29] and [30], the height extension of the TR is approximately 4–10 Mm in coronal holes and only 2–4 Mm in the quiet Sun. So the TR is higher and more extended in coronal holes than in the quiet Sun. However, because of the strong non-uniformity and dynamics, these height ranges are only a rough estimate and can vary with time and location.

4. Cool and Hot Components of a Coronal Bright Point

Coronal bright points (BPs) are small-scale bipolar features representing local heating in the TR and lower corona. The energization of BPs may result from the interaction between two magnetic fragments of opposite polarities,[32] magnetic reconnection along separator field lines,[33] or current sheets induced by photospheric motions.[34]

The height (or temperature) dependence of BPs properties has not been well studied. In the study of Ref. [35], we combined the spectra obtained by SUMER and EIS (EUV Imaging Spectrometer), and investigated the different emission morphology and Doppler pattern of a BP at lower and higher temperatures. We found that the transition from the cool to the hot component of the BP occurs at a temperature of about $\log(T/K) = 5.7$. From Fig. 2 we can see a totally different boundary between up- and downflows in the BP (almost perpendicular) for lines with lower and higher temperatures. The different boundary might be a result of a syphon flow along a loop system which twists or spirals at its upper segment.

The different emission morphology and Doppler pattern between the cool and hot components of the BP may also imply a different powering mechanism of the two components. Reference [36] proposed a two-stage heating process, in which magnetoconvection-driven reconnection occurs in and supplies energy to the cool BPs, whereupon the increased energy supply

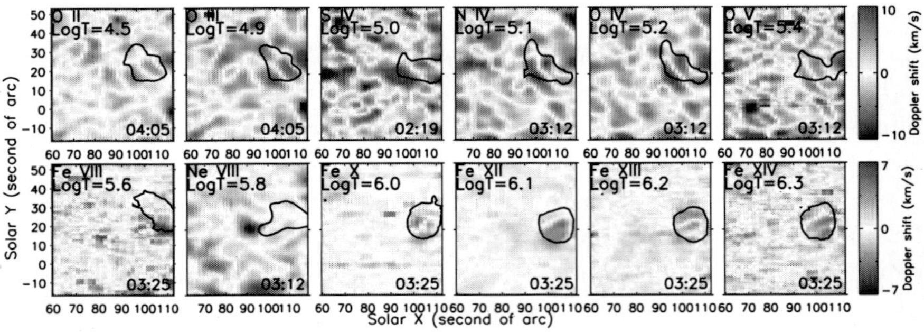

Fig. 2. Dopplergrams of the EUV lines. The approximate time when the BP was scanned is shown in the lower right corner of each map. The black contours outline the positions of the BP as seen in different wavelengths. The figure is adapted from Ref. [35].

leads to an expansion of the loop system, which interacts with the overlying coronal magnetic flux through fast separator reconnection and produces hot BPs. From magnetograms taken before and after the observation periods of SUMER and EIS, we found indeed a signature of flux cancelation. We also found an agreement between the separators in the magnetic skeleton and the orientation of the hot component of the BP. The flux cancelation and separator reconnection are likely to power the cool and hot components of the BP, respectively. Thus, our observations seem to support the two-stage powering mechanism.

5. Ly$-\alpha$ and Ly$-\beta$ Radiance Profiles in the Quiet Sun and in Coronal Holes

Hydrogen is the most abundant element in the solar system and its resonance lines, especially the Ly$-\alpha$ line, play a key role in the overall radiative energy transport of the solar atmosphere. In fact, the Ly$-\alpha$ line is the strongest emission line in the VUV spectral range and the energy loss[37] through its emission is the most important radiative loss in the lower TR. Also, the spectral irradiance at the center of the solar Ly$-\alpha$ line profile — in previous observations hampered by geocoronal absorption — is the main excitation source responsible for the atomic hydrogen resonant scattering in cool cometary and planetary material.[38] Moreover, the Lyman line profiles are very useful to diagnose the fine structures in solar prominences[39,40] and non-thermal effects in flares.[41,42]

Despite its importance, in the past 12 years high-quality observations by the SUMER spectrograph could only be completed for the higher Lyman lines,[43,44] but not for Ly$-\alpha$, since its high radiance would saturate the detectors. In June 2008, for the first time, an unconventional method (partly closing the aperture door) was applied to reduce the incoming photon flux to a level of about 20% and thus acquire undisturbed solar Ly$-\alpha$ profiles.

The initial results have been presented in Refs. [45] and [46]. We found that most Ly$-\alpha$ profiles are strongly reversed at the center, indicating a very strong radiative transfer effect. As shown in Fig. 3, most Ly$-\alpha$ profiles are asymmetric with a stronger blue peak, which is opposite to profiles of higher order Lyman lines. This result is unexpected and detailed modeling

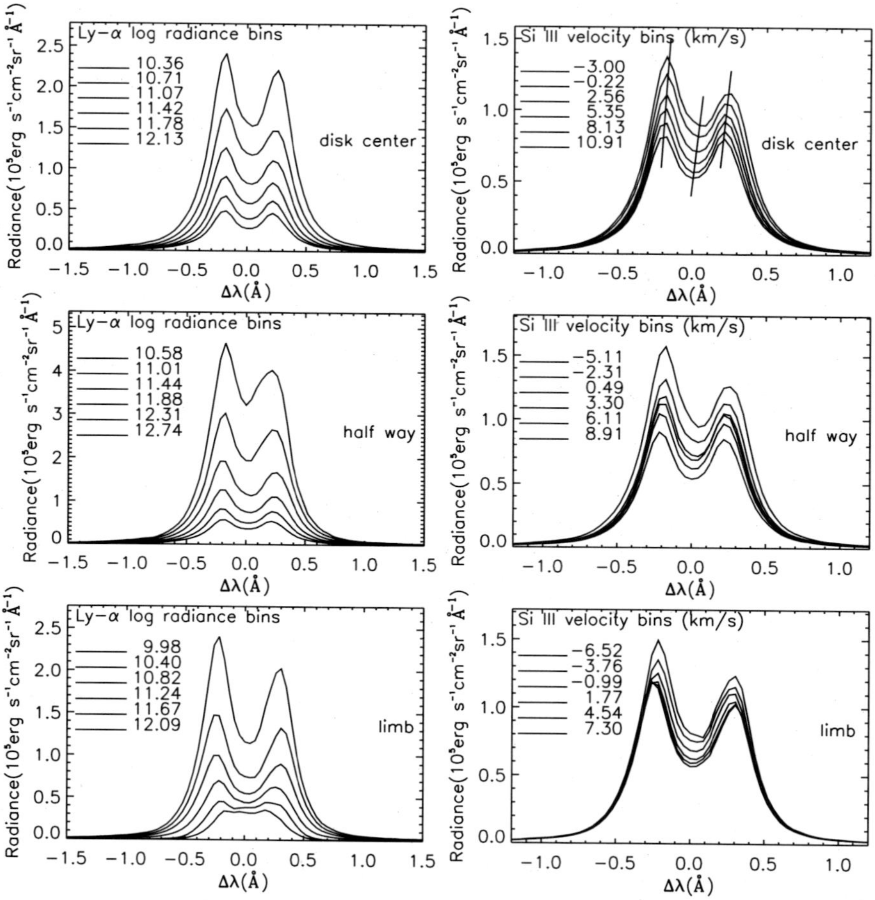

Fig. 3. Ly$-\alpha$ profiles in different radiance bins (left) and Doppler-shift bins (right). The figure is adapted from Ref. [45].

work is needed to understand it. Moreover, we found a clear correspondence between the asymmetry and the redshift of middle-TR lines like Si III: the asymmetry tends to be more prominent if larger downflows are present, indicating that downflows might play a fundamental role in the process of line profile formation. Thus, static TR models have to be revised. Also interplanetary models of hydrogen scattering may need to be reconsidered in view of the new direct and undisturbed measurement of the irradiance at the line center.

The spatial distribution of Ly$-\alpha$ profiles was studied in Ref. [46] and we found that profiles tend to be less reversed in the network than in internetwork regions. In the network, magnetic loops of different sizes and funnels are crowded in the chromosphere and TR, and the Lyman line emission originates from the outskirts of these structures. While in the internetwork region, only cool low-lying loops are present and most of the Lyman line emission sources are located at a much lower height as compared to the network. Thus the opacity in the internetwork is enhanced and the Lyman radiation penetrating the upper layers will be more strongly absorbed.

The peak separation of Ly$-\alpha$ profiles was also investigated in Ref. [47] and it turned out to be larger in coronal holes than in the quiet Sun, which might be related to the different magnetic structures as shown in Fig. 1. Another possibility is that the radiation field is much reduced in coronal holes and thus there are more hydrogen atoms in the corona, leading to more absorption of the profiles.

Another interesting phenomenon is that the asymmetries of the Ly$-\alpha$ profiles show no clear difference, while the Ly$-\beta$ profiles are significantly different in coronal holes and in the quiet Sun. Most of the Ly$-\beta$ profiles have a stronger red peak in the quiet Sun, while in coronal holes most[47] or a significant portion of[44] Ly$-\beta$ profiles are stronger in the blue peak. Thinking about the different situation of flows in the two regions, we speculate that this difference might be related to the solar wind outflow in coronal holes. In order to understand this result, a combination of new observations and NLTE modeling is needed.

6. Properties of the TR above Sunspots

Sunspots and plages are frequently formed in active regions. Spectral lines formed in the upper-TR often show significantly enhanced radiation at locations overlying sunspot umbrae. These plume-like emission features are usually termed sunspot plumes. A sunspot plume might be the common footpoints of several long-reaching loops. It usually has one end point anchored in the umbra and the other can reach far from the sunspot. We performed a detailed study of the sunspot reference spectra obtained by SUMER, and found that the TR above sunspots has some distinctive properties compared to the surrounding plage regions.[49]

The lower order hydrogen Lyman line profiles in plage regions are obviously reversed and stronger in the red peak, similar to those in the quiet Sun. This similarity might be due to the similar magnetic structures in the two regions. In both regions, the Lyman line emission originates by a large fraction from the chromosphere, and is strongly absorbed in the upper chromosphere and lower TR. On the contrary, the Lyman line profiles are found to be almost not reversed in sunspot umbrae and plumes,[48,49] which might indicate that the ratio between the chromospheric and TR contributions to the Lyman line radiation is reduced so that there is less absorption. This might also indicate that the TR is more extended above sunspots. Our density diagnostics suggests that the densities of the umbra and plume are a factor of 10 lower than of the plage, at TR temperatures. This result might suggest that the TR above sunspots is higher than in the TR above plage regions. Our scenario indicates that the sunspot TR temperature should be much lower than the surrounding temperature at the same heights, similar to the temperature structure of sunspots proposed by Ref. [50].

The DEM (Differential Emission Measure) analysis is helpful to study the thermal structure of the solar atmosphere. Ref. [51] performed the first DEM analysis for sunspot plumes by using CDS (Coronal Diagnostic Spectrometer) spectra. Since the CDS spectra include very few low-temperature lines, their DEM curves were not well determined at the low-temperature part. In Ref. [49], SUMER spectra were used for the first time to derive DEM curves of the sunspot umbra, penumbra, plume, and plage region. We found that the DEM curve of the plume is clearly different from those of other regions. It peaks at a lower temperature of around $\log(T/K) = 5.45$, which exceeds the DEM of other regions by one to two orders of magnitude at these temperatures. The low-temperature part of the DEM curves has been well established in our study. We found that at $\log(T/K) \leq 5.0$, the slope of the DEM curve is similar in the four regions, indicating that the thermal structure in this temperature range is similar everywhere in and around the sunspots.

The reason why the plume emission is so strong at upper-TR temperatures is very puzzling. We think that it might be the result of a large filling factor. The strongly enhanced emission at TR temperatures and the reduced continuum may explain the fact that many normally weak lines become prominent in sunspot plumes.

Acknowledgments

The main results mentioned in this paper were obtained through collaborations with Prof. C.-Y. Tu, Prof. E. Marsch, Dr. L. Teriaca, Dr. L.-D. Xia, Dr. E. Landi, Dr. U. Schühle, Dr. P. Lemaire, Prof. B. N. Dwivedi, and Prof. J.-C. Vial. Hui Tian is very grateful to Prof. S. S. Hasan for the invitation and Prof. S. K. Solanki for the financial support, to participate in the AOGS 2009 meeting. Hui Tian also thanks the long-lasting support from the graduate school of Peking University (especially from Mr. Feng He) and from the IMPRS graduate school (especially from Dr. D. Schmidt) run jointly by the Max Planck Society and the Universities of Göttingen and Braunschweig. The support by China Scholarship Council for Hui Tian's stay at Max-Planck-Institut für Sonnensystemforschung from September 2007 to September 2008 is also acknowledged.

References

1. J. F. Dowdy Jr., D. Rabin and R. L. Moore, *Sol. Phys.* **105** (1986) 35.
2. H. Peter, *A&A* **374** (2001) 1108.
3. J. T. Mariska, *The solar transition region*, (Cambridge University Press, Cambridge, UK, 1993).
4. K. Wilhelm, E. Marsch, B. N. Dwivedi and U. Feldman, *Space Sci. Rev.* **133** (2007) 103.
5. K. Wilhelm, W. Curdt, E. Marsch *et al.*, *Sol. Phys.* **162** (1995) 189.
6. P. Lemaire, K. Wilhelm, W. Curdt *et al.*, *Sol. Phys.* **170** (1997) 105.
7. W. Curdt, P. Brekke, U. Feldman *et al.*, *A&A* **375** (2001) 591.
8. W. Curdt, E. Landi and U. Feldman, *A&A* **427** (2004) 1045.
9. J. Chae, H. S. Yun and A. I. Poland, *ApJS* **114** (1998) 151.
10. H. Peter and P. G. Judge, *ApJ* **522** (1999) 1148.
11. L.-D. Xia, E. Marsch and K. Wilhelm, *A&A* **424** (2004) 1025.
12. W. Curdt, H. Tian, B. N. Dwivedi and E. Marsch, *A&A* **491** (2008) L13.
13. D. M. Hassler, I. E. Dammasch, P. Lemaire *et al.*, *Science* **283** (1999) 810.
14. I. E. Dammasch, K. Wilhelm, W. Curdt and D. M. Hassler, *A&A* **346** (1999) 285.
15. L.-D. Xia, E. Marsch and W. Curdt, *A&A* **399** (2003) L5.
16. T. Aiouaz, H. Peter and P. Lemaire, *A&A* **435** (2005) 713.

17. C.-Y. Tu, C. Zhou, E. Marsch *et al.*, *Science* **308** (2005) 519.
18. J.-S. He, C.-Y. Tu and E. Marsch, *Sol. Phys.* **250** (2008) 147.
19. H. Tian, C.-Y. Tu, E. Marsch, J.-S. He and G.-Q. Zhou, *A&A* **478** (2008) 915.
20. S. W. McIntosh, A. R. Davey, D. M. Hassler *et al.*, *ApJ* **654** (2007) 650.
21. J.-S. He, C.-Y. Tu, H. Tian and E. Marsch, *Advances in Space Research*, in press (2009).
22. H. Tian, C.-Y. Tu, L.-D. Xia *et al.*, *A&A* **489** (2008) 1297.
23. H. Tian, E. Marsch, W. Curdt, J.-S. He, *ApJ*, **704** (2009) 883.
24. J.-S. He, C.-Y. Tu, E. Marsch, *A&A* **468** (2007) 307.
25. E. Marsch, H. Tian, J. Sun, W. Curdt and T. Wiegelmann, *ApJ* **684** (2008) 1262.
26. H. Tian, E. Marsch, C.-Y. Tu, L.-D. Xia and J.-S. He, *A&A* **482** (2008) 267.
27. T. Wiegelmann and S. K. Solanki, *Sol. Phys.* **225** (2004) 227.
28. J. Zhang, J. Ma and H. Wang, *ApJ* **649** (2006) 464.
29. H. Tian, L.-D. Xia, J.-S. He, B. Tan and S. Yao, *ChJAA* **8** (2008) 732.
30. C.-Y. Tu, C. Zhou, E. Marsch *et al.*, *ApJ* **624** (2005) L133.
31. H. Tian, C.-Y. Tu, E. Marsch, J.-S. He, C. Zhou and L. Zhao, *Proceedings of Solar Wind 12*, in press, (2009).
32. E. R. Priest, C. E. Parnell and S. F. Martin, *ApJ* **427** (1994) 459.
33. D. W. Longcope, *ApJ* **507** (1998) 433.
34. J. Büchner, *Space Sci. Rev.* **122** (2006) 149.
35. H. Tian, W. Curdt, E. Marsch and J.-S. He, *ApJ* **681** (2008) L121.
36. S. W. McIntosh, *ApJ* **670** (2007) 1401.
37. J. M. Fontenla, E. J. Reichmann and E. Tandberg-Hanssen, *ApJ* **329** (1988) 464.
38. C. Emerich, P. Lemaire, J.-C. Vial *et al.*, *Icarus* **178** (2005) 429.
39. J.-C. Vial, H. Ebadi and A. Ajabshirizadeh, *Sol. Phys.* **246** (2007) 327.
40. S. Gunar, P. Heinzel, U. Anzer and B. Schmieder, *A&A* **490** (2008) 307.
41. C. Fang, N. Feautrier and J.-C. Henoux, *A&A* **297** (1995) 854.
42. Z. Xu, C. Fang and W.-Q. Gan, *ChJAA* **5** (2005) 519.
43. H. P. Warren, J. T. Mariska and K. Wilhelm, *ApJS* **119** (1998) 105.
44. L.-D. Xia, in *Ph.D. Thesis* (Göttingen: Georg-August-Univ., 2003).
45. W. Curdt, H. Tian, L. Teriaca, U. Schühle and P. Lemaire, *A&A* **492** (2008) L9.
46. H. Tian, W. Curdt, E. Marsch and U. Schühle, *A&A* **504** (2009) 239.
47. H. Tian, L. Teriaca, W. Curdt and J.-C. Vial, *ApJ* **703** (2009) L152.
48. W. Curdt, B. N. Dwivedi and U. Feldman, *J. Astrophys. Astr.* **21** (2000) 397.
49. H. Tian, W. Curdt, L. Teriaca, E. Landi and E. Marsch, *A&A* **505** (2009) 307.
50. K. R. Nicolas, O. Kjeldseth-Moe, J.-D. F. Bartoe and G. E. Brueckner, *Sol. Phys.* **81** (1982) 253.
51. J. W. Brosius and E. Landi, *ApJ* **632** (2005) 1196.

Advances in Geosciences
Vol. 21: Solar & Terrestrial Science (2008)
Ed. Marc Duldig

ACTIVE LATITUDES FOR SUNSPOT OCCURRENCES IN THE SOLAR CYCLE 23

V. V. ZHARKOVA
Mathematics Department, University of Bradford,
Bradford, West Yorkshire, BD7 1DP, UK
v.v.zharkova@brad.ac.uk

S. I. ZHARKOV
Mullard Space Science Laboratory, University College London,
Holmbury St. Mary, Dorking, RH5 6NT, UK
sz2@mssl.ucl.ac.uk

In this paper we carry out the principal component analysis of the sunspot and background magnetic fields in the cycle 23. There are two main latitude periodical components of the opposite polarities derived in sunspot magnetic flux which behave rather off-phase in the Northern and Southern hemispheres. These waves are shown to be modulated by the two primary waves of the background magnetic field which have the opposite polarities and travel slightly off-phase in each hemisphere. In the cycle 23 the maxima in sunspot components correspond to the minima in the background magnetic field components. It indicates the presence of two waves in the background magnetic field travelling in each hemisphere, possibly, in the opposite directions and with the opposite polarities), which affect the sunspot appearances in these hemispheres.

1. Introduction

The solar activity is well accepted to be governed by the dynamo waves traveling from one pole to another[1] while producing flux tubes with strong magnetic field (sunspots/active regions) and regenerating their magnetic fields with the help of mean kinetic helicity from the mechanical turbulent convective motion of plasma at the solar convective zone (SCZ). The dynamo is believed to operate at the levels of SCZ, or beneath of the so-called overshoot layer. In the basic dynamo theory accounting for a generation of magnetic field in the Sun, which is observed on the solar surface as background (poloidal) and sunspot/active region (toroidal) magnetic fields, there are two major factors affecting their interaction: α and Ω effects, where Ω is the angular velocity and α is the factor linked to the mean kinetic helicity of the turbulent motion.[2]

The Ω effect is associated with the gradient of angular $G = \partial\Omega/\partial r$ which appears from the helioseismic observations to be positive in the deep layers of the SCZ in the low latitudes where flux tubes with strong tororidal magnetic field, or sunspots, are formed.[3,4] Although recently there are reports about more complicated variations of this parameter with the solar radius during different phases of the solar cycle.[5] Then due to turbulent motion during the flux tube migration to the solar surface, the α effect becomes responsible for the regeneration of poloidal (background) magnetic field from the toroidal one via the mean kinetic helicity ($\vec{v}\nabla\vec{v}$) of a turbulent motion.[6]

Although the locations of magnetic field generation at SCZ and close under-photospheric layers and the magnitudes and signs of α and Ω effects are still unknown. A large number of the dynamo models solve the induction equation for constant and varying magnetic diffusivity in different conditions of the solar interior. This allows to provide a plausible interpretation of the observational features of the solar activity, e.g. butterfly diagrams and the relation between the toroidal and poloidal magnetic field, in a particular time interval (see Jouve *et al.*[7] and references therein).

However, there are still open questions remaining about the consistent relationship between these two fields throughout the whole solar cycle and the manner in which the dynamo waves travel along latitudes and longitudes. Recently, from the simultaneous observations of the background and sunspot magnetic fields Zharkov *et al.*[8] confirmed the two dynamo wave periods in sunspot areas and magnetic fields in each hemisphere: about 11 and 2.5 years. They show that these periods are defined by the variations of the solar background magnetic field (BMF, thereafter) (outside active regions and sunspots), which reveal the BMF anti-symmetric and symmetric components. The former lasting for about 11 years is the stronger one separated in each hemisphere by the polarities opposite to the leading polarities of sunspots. The latter has a much smaller magnitude and the same polarities in each hemisphere every 2–2.5 years changing to the opposite one in the next 2–2.5 years.

Therefore, this symmetric BMF component can either coincide with or be opposite to the leading polarity of sunspots, causing the excess (when opposite) or deficit (when coinciding) in the number of sunspots appearing on the solar surfaces in the different hemispheres. This is observed as a strong North-South asymmetry in sunspot areas covered in each hemisphere and resulting magnetic fields contained inside sunspots in these hemispheres.

Recent helioseismic observations of the solar differential rotation reveal a significant radial gradient $\partial\Omega/\partial r$ around the SCZ and the latitudinal gradient of the angular velocity $\partial\Omega/\partial\theta$ near the solar surface.[4] These observations were utilized in the dynamo models considering these biennial periods ($\sim$ 2 and 11 years) of the solar activity being caused by the dynamo mechanisms: the lower frequency one (11 years) operating at the base of SCZ because of the large-scale radial shear and the higher frequency one generated in the subsurface regions by the latitudinal shear.[9,10]

The two-level model gives a plausible explanation of the temporal appearance of the solar activity with two basic periods, e.g. low and high frequency However, it does not yet explain the interaction between the two types of shear, their distribution and relationships in latitude between the background (poloidal) and sunspot (toroidal) magnetic fields.

In the current paper we continue the investigation of the relationships between these two kinds of field present in the Sun started in our previous paper,[8] in order to derive the eigenvectors with highest eigenvalues of the observed periodicities in both the background and sunspot magnetic fields and to establish the links between the latitude variations of these fields in relation to the tilts and charge separation in sunspot groups.

2. Data Description

2.1. *Solar feature catalogue for sunspots*

The searchable Solar Feature Catalogues (SFCs) with 12 year coverage (1996–2008) are developed from digitized solar images by using the automated pattern recognition techniques.[11] These techniques were applied to detection of sunspots, active regions, filaments and line-of-sight magnetic neutral lines in the automatically standardized full disk solar images in Ca II K1, Ca II K3 and Hα lines taken at the Paris-Meudon Observatory and white light images and magnetograms from SOHO/MDI. The results were verified with the NOAA and Meudon synoptic maps and statistical data that revealed accuracy of 96% for sunspots[12] and 92% for active regions[13] have been obtained.

The sunspots are extracted from the SOHO/MDI images by using the automated edge-detecting technique and verified by the presence of a strong magnetic field.[12] The parameters extracted include heliographic area, heliographic diameter, number of umbras, Carrington and planar coordinates of the sunspot center of gravity, maximum magnetic field, minimum field, excess flux, total (absolute) flux, maximum umbra field, minimum umbra field, excess umbra flux, total (absolute) umbra flux and various observational parameters populated into a relational database of Solar Feature Catalogues on the web site *http://solar.inf.brad.ac.uk/*.

2.2. *Wilcox solar observatory (WSO) data*

For large-scale background magnetic field variations, low resolution magnetograms obtained daily by using the Zeeman splitting of the 525.02 nm Fe I spectral line with the WSO's Babcock solar magnetograph[14] for the line-of-sight component of the solar background magnetic field.

The observations of BMF were carried out from the beginning of the Carrington Rotation 1,642, up to the year 2005.16, or CR 2027. The data covers the solar activity cycles 21, 22, and 23.[14,15] In the current study, we consider the averaged magnetic field in the cycles 21–23 only within the 30 μ-hemispheres in the sine of heliographic latitude (from the latitudes of 75.2° North to 75.2° South) averaged in heliographic longitude within the latitudinal strip of 5°.

3. Results and Discussion

3.1. *Quasi-3D butterfly diagram for sunspots*

Latitudinal variations of sunspot excess magnetic flux, a sum of magnetic fluxes within areas covered by sunspots, calculated in the narrow latitude strips for the whole longitude range in different Carrington rotations are presented in Fig. 1. It is similar to the classic Maunder's butterfly diagram[16] with the third dimension associated with the magnitude of a magnetic flux within sunspots at different times in different latitudinal strips summarized for all longitudes (μ-spheres).

The butterfly diagram for sunspot flux, similar to that for areas, reveals a well-defined fine structure in latitude for each phase of the cycle appearing in a shape of vertical threads, or strips. These strips at given time intervals are brighter (or have larger fluxes according to the colour bar on the right hand side) in one hemisphere than in the other one. Then in the next time period the strip with larger areas shifts to other latitude or even to the opposite hemisphere. This points out that there are some periodic oscillations in the appearance of the solar activity.

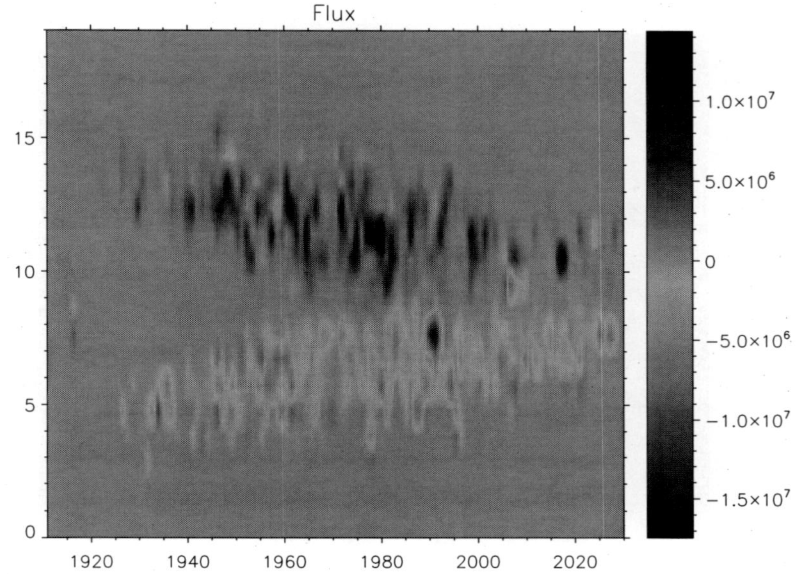

Fig. 1. The butterfly diagram in the cycle 23 for total sunspot excess magnetic flux summarized for all the longitudes within the narrow latitude strips of 5 degrees. The areas are measured in squared degrees and magnetic field in Gauss. CR number of 1910 corresponds to the start of 1996 and 2040 to the spring of 2005.

This is confirmed by the self-correlation of the sunspot excess magnetic flux residuals of the means, calculated with the running 1 year and 4 year averaging filters (e.g. 1–4y) applied to exclude the major anti-symmetric component varying in 11 years) at the given latitudinal strip presented in Fig. 2 and calculated for the sums of the magnetic field magnitudes inside sunspots taken with their signs and averaged with one year filter. The self-correlation of the excess magnetic fluxes reveal well defined period of ~2.5 years with a negative coefficient confirming the change of sign with the defined period of 2.5 years that supports the model simulation by Benevolenskaya.[9]

3.2. *Latitudinal variations of sunspot activity*

3.2.1. *Sunspot group tilt and charge separation*

The latitude distributions of tilt angles presented in Fig. 3 (the left plot) for normal sunspot group (GT) tilts reveal the well known Joy's law with the group tilt (GT) increasing from the equator to high latitudes and approaching the maximum at the latitude of 32°. There is a well noticed

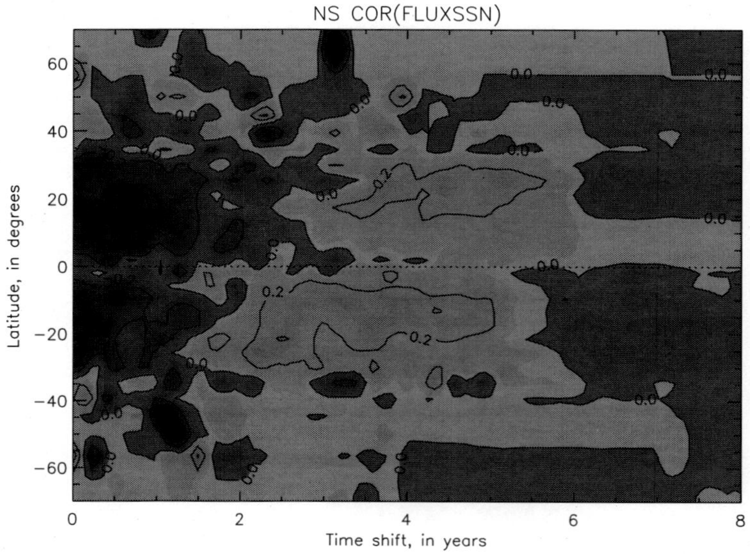

Fig. 2. The C-correlation map for the time lags in years between the 1–4y residuals of the SMF and sunspot EMF in the Northern and Southern hemispheres. Numbers at the contours show the C-correlation coefficients.

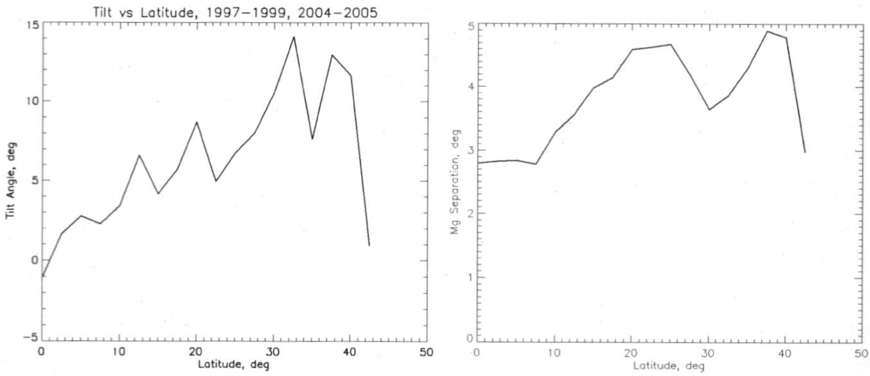

Fig. 3. Left plot: the normal tilts distributions versus latitudes for sunspot groups (Joy's law). Right plot: the polarity separation versus latitudes in sunspot groups in the cycle 23.

periodical structure in the latitudinal tilt distributions with a period of about 8°, with the maxima at 5°, 13°, 20° and 32° followed by a sharp drop of the group tilts after the latitude of 40°. The magnetic polarity separation Fig. 3 (the right plot) also increases from the equator towards the poles reaching the main maximum of 4.9° at the latitude of about 35° with the other one of 4.7° at the latitude of about 25°. Similarly to the tilts themselves, the polarity separation sharply decreases after the latitude of 35°. These maxima in the magnetic separation occur at the latitudes where there are the minima in the tilt distributions in latitude as noted by Zharkova and Zharkov.[17] Therefore, the changes of magnetic polarity separation are in the anti-phase with those in tilts.

The periodicities can be a reflection of some physical processes affecting the sunspot groups (rotations, twists or dynamo waves) which is minimal (leading to the tilt maxima) at the latitudes of 32°, 23°, 13° and 3° with the magnitudes increasing towards the top of the latitudinal zone where the butterfly diagram is formed. This periodicity in latitude can to be validated by the independent mathematical method analyzing the principal components.

3.3. *Principal component analysis*

In order to derive the main periods present in the observational data of solar activity: sunspot areas and sunspot magnetic fields at given latitudes affected by combination of various physical processes (dynamo waves, meridional flows, torsional oscillations etc.), let us apply the Principal Component Analysis (PCA) to the observational data.

PCA is the orthogonal linear transformation allowing a vector space of N observations $\vec{X}_N$ each containing M points to be transformed to a new coordinate system reducing the multi-dimensional data $\mathrm{X}(\mathrm{M} \times \mathrm{N})$ to lower dimensions for analysis producing eigenvectors with given eigenvalues. The eigenvalues and eigenvectors $\vec{V}$ are paired; the eigenvectors $\vec{V}$ with the highest eigenvalues are the principal components of the original matrix X which has the greatest variance by any projection of the data lies on the first coordinate.[18]

3.3.1. *The main latitude components of the sunspot magnetic field*

There are the two principal components of the sunspot excess magnetic flux plotted in Fig. 4 as a function of latitude (the upper plot) and time (the lower plot). It is evident that the main components are rather different in the Northern and Southern hemispheres. This points to a presence of the two dynamo waves traveling in each hemisphere (possibly, in the opposite directions and having the opposite polarities), which travel different routes

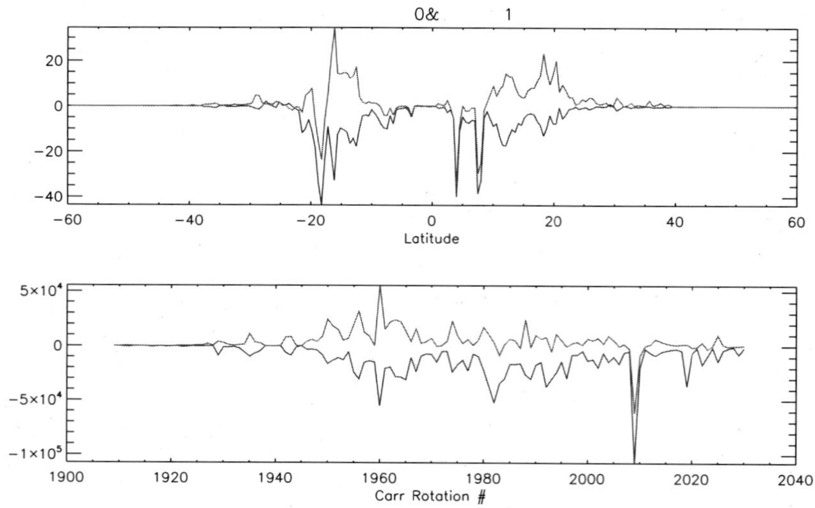

Fig. 4. The two main eigenvalues of the SMF distribution in latitudes (upper plot) and versus time of the cycle 23 (lower plot).

that, in turn, can explain the observed N-S asymmetry in the sunspot areas and magnetic flux.[8]

Also the largest eigenvalues are in the sunspots located in the latitude positions intermediate to those reported for the maximum of tilts in the Joe's law in Fig. 3 (left plot) and to the minimal magnetic polarity separation as per Fig. 3 (right plot). This points to higher twists of magnetic flux occurring at the latitudes with higher tilts compared to those where the tilts have minimums.

The two principal components in temporal variations of the sunspot magnetic flux plotted in the bottom plot of Fig. 4 also point to a pair of travelling waves, which have opposite magnetic polarities. Their maximum value was reached at the Autumn 2003 when the series of Halloween flaring events occurred.

3.3.2. *The main latitude components of BMF variations*

In order to understand the PCA component in latitudinal distributions of the sunspot magnetic flux, let us again explore its links with the background magnetic field. We have carried out the PCA analysis of the background magnetic field for the whole set of 21–23 cycles and detected the coupled 2 main components for the whole set and 2 coupled components for each cycle,

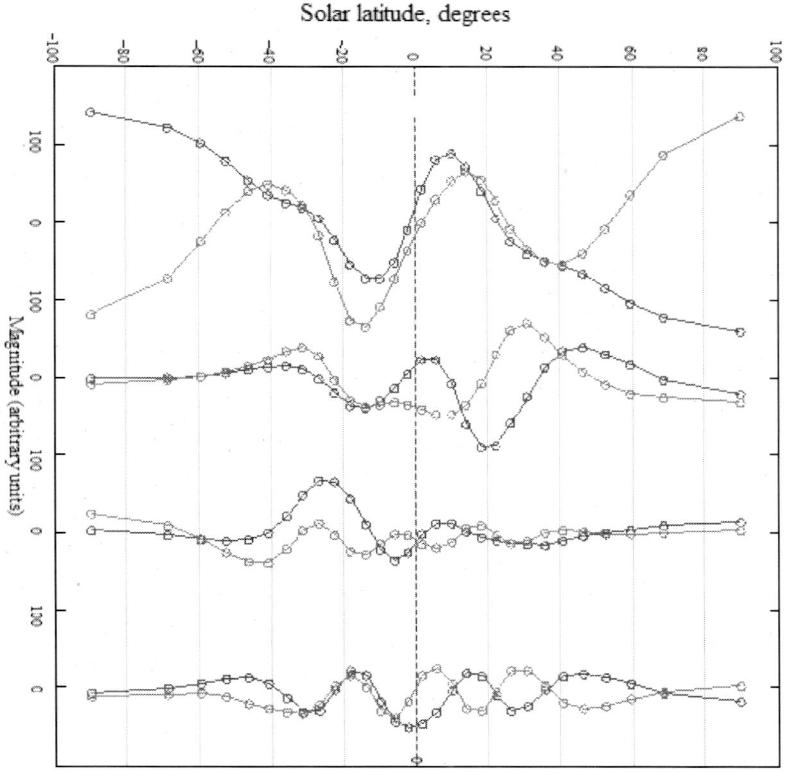

Fig. 5. The 2 main eigenvalues of the latitudinal BMF distribution in the total set of cycles 21–23 (the upper set of blue-red curves) and in the cycles 21, 22 and 23 (the next sets of the blue-red curves from the top to the bottom, respectively).

which are plotted in Fig. 5. It can be noted that the two main components of BMF derived for the whole dataset resemble the two waves with the maxima around ±10 degrees, which have the opposite magnetic polarities and travel from one hemisphere to another. These two components are likely to be the same anti-symmetric BMF components discussed in Sec. 3.1 which define magnetic field variations in the 11 year cycle.

In addition, there are secondary two components in each cycle which are unique to a given cycle, e.g. in the cycles 21 and 23 there are the eigenvalue components increased in the Northern hemisphere while in the cycle 22 — in the Southern one. These components are likely to modulate the north-south asymmetries in sunspot areas and magnetic fluxes reported for the cycles 21–23.[8,19] The absolute eigenvalues of the vectors are decreasing from the cycle 21 to 23 (compare the three bottom sets of curves in Fig. 5 that also reflects a decrease of the solar activity index in these cycles.

The variations of the two main components of BMF in the cycle 23 (the bottom two curves) have the maxima in the latitudes complimentary to (between) those where are the maximums in the sunspot excess magnetic flux. This points out that the occurrence of sunspots is strongly modulated by and is complementary to the fluctuations of BMF not only in time, but also in latitude that needs to be reflected in the future dynamo models.

4. Conclusions

In the current paper we continue the investigation of relationships between the background and sunspot magnetic fluxes as function of latitudes started in our previous paper[8] by carrying out the PCA analysis, in order to derive the eigenfunctions of the observed periodicities in both the background and sunspot magnetic fields.

We derived the two main latitude components in sunspot magnetic flux which are rather different in the Northern and Southern hemispheres. This points to a presence of two waves travelling in each hemisphere (possibly, in the opposite directions or having the opposite polarities). These waves are modulated by the two primary waves of the background magnetic field where the maxima in sunspot the components correspond to the minima in the BMF components.

These BMF waves have the opposite polarities and travel slightly off-phase in each hemisphere. Their existence can explain the observed asymmetries in sunspot area and excess magnetic fluxes observed in the cycle 23.[8] More research is required into the nature of this off-phase relation and evolution of these waves over the cycles and this will be reported in a forthcoming paper.

References

1. E. N. Parker, *Astrophys. J.* **122** (1955) 293.
2. M. Steenbeck, F. Krause and K.-H. Rädler, *Zeitschrift Naturforschung Teil A* **21** (1966) 369.
3. S. Basu, J. Christensen-Dalsgaard and M. J. Thompson, *Astron. Astrophys.* **321** (1997) 634.
4. J. Schou, H. M. Antia, S. Basu, R. S. Bogart, R. I. Bush and S. M. Chitre, J. Christensen-Dalsgaard, M. P. di Mauro, W. A. Dziembowski, A. Eff-Darwich, D. O. Gough, D. A. Haber, J. T. Hoeksema, R. Howe,

S. G. Korzennik, A. G. Kosovichev, R. M. Larsen, F. P. Pijpers, P. H. Scherrer, T. Sekii, T. D. Tarbell, A. M. Title, M. J. Thompson and J. Toomre, *Astrophys. J.* **505** (1998) 390.
5. R. Howe, M. Rempel, J. Christensen-Dalsgaard, F. Hill, R. Komm, R. M. Larsen, J. Schou and M. J. Thompson, *Astrophys. J.* **649** (2006) 1155.
6. F. Krause and K. H. Raedler, *Mean-field magnetohydrodynamics and dynamo theory* 1980.
7. L. Jouve, A. S. Brun, R. Arlt, A. Brandenburg, A. Dikpati, A. Bonanno, P. J. Käpylä, D. Moss, M. Rempel, P. Gilman, M. J. Korpi and A. G. Kosovichev, *Astron. Astrophys.* **483** (2008) 949.
8. S. Zharkov, E. Gavryuseva and V. Zharkova, *Solar Phys.* **248** (2008) 339.
9. E. E. Benevolenskaya, *Astrophys. J. Lett.* **509** (1998) L49.
10. E. E. Benevolenskaya, *Solar Phys.* **216** (2003) 325.
11. V. V. Zharkova, J. Aboudarham, S. Zharkov, S. S. Ipson and A. K. Benkhalil and N. Fuller, *Solar Phys.* **228** (2005) 361.
12. S. Zharkov, V. V. Zharkova, S. S. Ipson, *Solar Phys.* **228** (2005) 377.
13. A. Benkhalil, V. V. Zharkova, S. Zharkov and S. Ipson, *Solar Phys.* **235** (2006) 87.
14. J. T. Hoeksema, Structure and evolution of the large scale solar and heliospheric magnetic fields, PhD thesis, AA(Stanford Univ., CA.), (AA(Stanford Univ., CA.), 1984).
15. P. H. Scherrer, J. M. Wilcox, L. Svalgaard, T. L. Jr. Duvall, P. H. Dittmer and E. K. Gustafson, *Solar Phys.* **54** (1977) 353.
16. W. Mrs. Maunder, *Mon. Not. Roy. Astron. Soc.* **64** (1904) 226.
17. V. V. Zharkova, S. I. Zharkov, *Advances in Space Research* **41** (2008) 881.
18. I. T. Jolliffe, *Principal Component Analysis, 2nd edition.* (Berlin: Springer Series in Statistics, 2002).
19. M. Temmer, A. Veronig and A. Hanslmeier, *Astron. Astrophys.* **390** (2002) 707.

Advances in Geosciences
Vol. 21: Solar & Terrestrial Science (2008)
Ed. Marc Duldig

MAGNETIC HELICITY INJECTION AND CHIRALITY CHANGING IN PENUMBRAL FILAMENTS

J. T. SU*
National Astronomical Observatories,
Chinese Academy of Sciences 20A Datun Road,
Chaoyang District, Beijing, 100012, China
sjt@bao.ac.cn

Key Laboratory of Solar Activity,
National Astronomical Observatories,
Chinese Academy of Sciences

We separately sunspot calculate the magnetic helicity transports of positive and negative polarities in active region (AR) NOAA 10930. We find the rate of magnetic helicity transport of the positive polarity changed from negative to positive after a major X3.4 flare. This rate mainly came from the motions of a small positive sunspot. We also find 2 days before the flare, the penumbral filaments in the magnetic polarity inversion line (PIL) region first unwound, then they re-twisted in opposite direction. It is a chirality change in this AR at the photosphere level in response to the transport of magnetic helicity to the corona.

1. Introduction

Magnetic helicity quantifies how much the magnetic field is distorted and stressed beyond its lowest energy state (potential field). It is almost preserved on a timescale less than the global diffusion time scale. Because of this conservation property, it has been suggested that Coronal Mass Ejections (CMEs) are needed to avoid the endless helicity accumulation in the corona.[1–3] Zhang *et al.* (2006)[4] even conjectured there is an upper bound on the total magnetic helicity that a force-free field can contain. Thus, the accumulation of magnetic helicity in excess of this upper bound would initiate a non-equilibrium situation, resulting in a CME expulsion .

Certainly, we much expect there is such an upper bound on the magnetic helicity accumulation in the corona, which would be a powerful

*Work supported by grant NSFC projects (nos. 10878016 and 10778723) and and National Basic Research Program of China under grant 2006CB806301.

tool used for pace-weather prediction. However, a theoretical work by Amari *et al.* (2003)[5] suggested that a large enough helicity is a necessary but not sufficient condition for an ejection to occur, and this was supported by the observational work of Nindos & Andrews (2004).[6] Whether or not that the helicity accumulation in the corona is very important to the origin and initiation of CMEs is still hotly in debate.[2,7–12] However, after all, more recently LaBonte *et al.* (2007)[13] found a necessary condition for the occurrence of an X-flare is that the peak helicity flux has a magnitude $> 6 \times 10^{36}$ $\mathrm{Mx}^2\ \mathrm{s}^{-1}$.

In the frame of the emergence of a twisted flux tube into the solar atmosphere, the simulations[14–16] show at the late stage of emergence there is a saturation of the helicity injected into the atmosphere. At the early stage of emergence, strong shear flows are driven on the surface due to vigorous flux emergence caused by Ω-loops that compose the upper half of the flux tube, contributing to an efficient injection of helicity. On the other hand, after the axis emerges into the surface, the flux emergence becomes weak because the lower half of the flux tube is composed of U-loops that are relatively hard to lift compared to the Ω-loops. The weakening of flux emergence leads to saturation of the helicity injected into the atmosphere.[16] The transport of magnetic helicity in AR 10930 seemed to behave like this way.[17,18] Therefore, it is worthwhile paying more attention to the saturation of helicity transport for the early warning of space-weather.

Very significant advances have been made in the last decade to estimate the magnitude of H in the corona and in the interplanetary space (see Demoulin (2007) for a review).[19] In theory, Berger & Field (1984)[20] first derived an expression for relative helicity flux that can be applied to an individual solar AR occupying an area $\mathbf{S}$ of the photosphere,

$$dH_m/dt = -2 \int_s [(\mathbf{A}_p \cdot \mathbf{v})\mathbf{B} - (\mathbf{A}_p \cdot \mathbf{B})\mathbf{v}] \cdot d\mathbf{S}, \tag{1}$$

where $\mathbf{A}_p$ is the vector potential of a reference potential field that reproduces the normal component of the magnetic field $\mathbf{B}$ on $\mathbf{S}$, and $\mathbf{v}$ is the plasma velocity at the surface $\mathbf{S}$ through which the helicity is passing. The first term in Eq. (1) is usually associated with magnetic flux emergence, whereas the second is associated with shearing/twisting motions.

The method is widely used to estimate dH_m/dt is called Local Correlation Tracking (LCT) developed by Chae *et al.* (2001).[21] Since the measured velocity is within the plane of the magnetogram, Chae *et al.*[21] believed that LCT permits an estimate only of the shearing/twisting term in Eq. (1). Démoulin and Berger (2003)[22] made a further simplification of dH_m/dt by defining the velocity $\mathbf{u} = \mathbf{v}_t - \mathbf{v}_n \mathbf{B}_t/\mathbf{B}_n$, which is tangential to the surface S. Then, the photospheric helicity rate simplifies to:

$$dH_m/dt = -2\int_s [(\mathbf{A}_p \cdot \mathbf{u})] B_n dS. \tag{2}$$

If the magnetic field is simply advected (without distortion or reconnection) across the photosphere, Démoulin and Berger (2003)[23] have shown that $\mathbf{u}$ is the velocity of the footpoint of the magnetic flux tubes on the photosphere. Therefore, the full helicity rate across the photosphere can be estimated by the LCT method. However, as pointed out by Pariat *et al.* (2005),[14] the integrand $G_A = -2(\mathbf{A}_p \cdot \mathbf{u})B_n$in Eq. (2) can exhibit spurious signals. Thus, they defined a new estimate of the integrand called G_θ, by which the spurious signals in the calculation of dH_m/dt are greatly reduced. However, for long-term calculations of total accumulated helicity, both proxies give practically identical results.[23] So far both the MHD simulations of Cheung *et al.* (2005, 2008)[15,24] and the observations[25–28] have confirmed that the emergence term in Eq. (1) contributes to most of the helcity flux during the early stages of emergence, while the shear term becomes more important later when the magnetic field becomes more vertical.

In the following we will show the calculation results for dH_m/dt in the AR NOAA 10930 over the whole field of view (FOV) of Fig. 1a using the LCT technique developed by Chae *et al.* (2001).[21] The rate of the helicity injection through the photosphere is estimated via Eq. (2). In the previous work of Su *et al.* (2009),[29] they have shown as one of solar magnetic features, penumbral filaments are also with chirality. In this work, we find the penumbral filaments can change their chirality, such as from right- to left-handedness, which may be a response to the negative helicity transport by a sunspot rotating in the AR NOAA 10930.

1.1. *Observations and data reduction*

AR NOAA 10930 consisted of a major stable sunspot (MSS) and a minor rotating sunspot (MRS). Min and Chae (2009) suggested that the rotation of this small sunspot may be closely related to the dynamic development of emerging twisted magnetic fields.[30] Correspondingly, an inverse sigmoid-loop formed in the corona and a X3.4 flare occurred on December 13 were thought to be a result of the continuous counterclockwise rotation

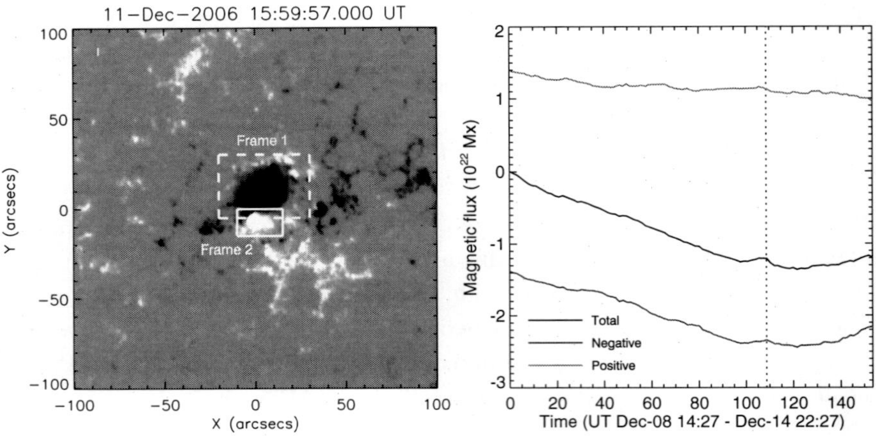

(a) (b)

Fig. 1. Left: MDI magnetogram for AR NOAA 10930 on 2006 December 11. The FOV is $200'' \times 200''$. Two frames are marked for a late illustration. Right: Temporal development of magnetic fluxes over the FOV for negative, positive polarity and total of them. Dotted line marks the peak time of X3.4 flare.

of the MRS.[31–33] Moreover, there was a channel structure formed in this AR. Wang *et al.* (2008)[34] found the channel was associated with new flux emergence in the middle of existing penumbra connecting the sunspot.

We use 96 min full-disk longitudinal magnetograms taken by the Michelson Doppler Imager on board the Solar and Heliospheric Observatory (MDI/SOHO) to calculate dH_m/dt during 2006 December 8-14. We carry out the calculations using the Lev1.8 data, which were recalibrated in December, 2008. The other data set used are G-band 4305 Å images (GBIs), and vector magnetograms, which were taken by the SOT instruments onboard Hinode.[35–39] The spatial resolutions are about $0.218''$ for GBIs with a time cadence of 2 min, and $0.32''$ for vector magnetograms, respectively. All of the GBIs used in this study have been calibrated using the standard Solar Software IDL routines.

The SOT/SP magnetograms are derived from the inversion of the full Stokes profiles based on the assumption of the Milne-Eddington atmosphere model, and are corrected for the projection effect.[40] We use an automated ambiguity-resolution code based on the minimum energy algorithm[41] to correct the magnetograms. The method is promising although it is time consuming. The noise level is determined by selecting one quiet region within the magnetogram to calculate 2σ standard deviation of the average field strength.

1.2. *Magnetic helicity transport in NOAA 10930*

Figure 1a shows a MDI magnetogram on 2006 December 11, which provides the basic magnetic structures of the AR. Figure 1b shows temporal development of negative (blue) and positive (red) magnetic fluxes and the total unsigned (black) over the FOV. We can find that the negative flux increased in magnitude before the flare and decreased after the flare, while the positive flux kept nearly unchanged throughout the observing period. Thus, it is natural that the total flux had a similar evolving trend with that negative flux. We follow Tian and Alexander (2009)[42] to separately calculated the contribution of positive and negative polarities to the helicity injected into the corona. For a comparison, we also calculate the total helicity flux of positive and negative polarities.

Figure 2a shows temporal development of dH_m/dt deduced from MDI during 2006 December 8–14. From this plot, we can see that before the flare, dH_m/dt of the positive polarity (red) was predominantly negative while after the flare, it first decreased toward zero in magnitude, then increased toward the positive. On the other hand, that of the negative polarity (blue) had a little different evolving trend. It was negative only before the first 64 hours (before 6:00 UT on December 11). However, after that time, it was more inclined to be positive. In the period of the flare (marked by the shadow area), there was a sharp variation for all dH_m/dt curves. This variation may be caused by the distortions of the profile of the spectral line being used for the magnetic measurements during the flare.[43] However, Zhang *et al.* (2009)[44] believed that it is the real change and depends on the properties of the flare and of its host active region. The total

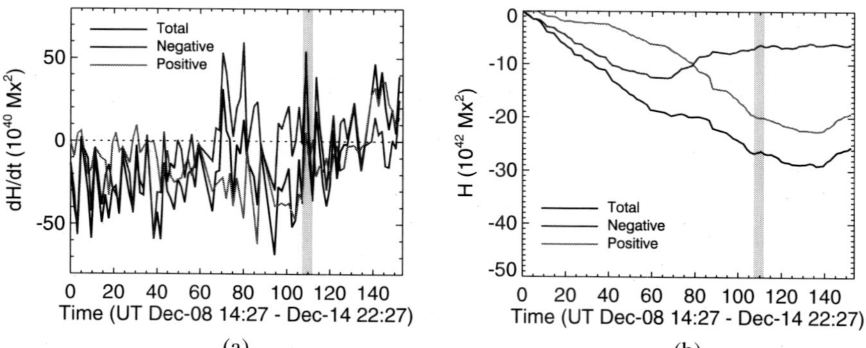

Fig. 2. (a) dH_m/dt in the period of 2006 December 8–14. (b) Accumulation of magnetic helicity. The shadow area indicates the phrase where an X-class flare occurred.

transport rate (black) had a similar evolving trend with that of the positive polarity.

In Fig. 2b, we can find all the magnetic helicity accumulations of positive polarity (red), negative polarity (blue), and total of them (black) were negative. The total helicity accumulation and that of the positive polarity decreased in magnitude after the flare in response to the sign changes of dH_m/dt in Fig. 2a. That of the negative polarity also showed such change, but it appeared earlier than the above two. It is very interesting that Magara and Tsuneta (2007)[17] have found the helicity accumulation in this AR first increases rapidly, then it become saturated in the late phase of the X3.4 flare. They used a model of Priest and Forbes (2000)[45] to calculate dH_m/dt:

$$dH_m/dt = v_s\Phi^2/L, \tag{3}$$

where v_s is the shear velocity of the footpoints of magnetic flux tubes, L and are the length and flux of an uniformly sheared arcade, respectively. Therefore, dH_m/dt calculated by this formula is only associated with shearing motions of sunspots. Our calculating results with the LCT method are similar to theirs.

Some discussions on the LCT method are necessary. In case that there are significant magnetic flux emergences, the method may not work properly since it is based on the comparison of the same features between two different frames. By comparing a time series of the derived velocity vectors, we find this method can trace the main velocity properties of the MRS: rotating and eastwards moving. However, despite of that, we must accept the fact that LCT can not trace all emergence velocity properties.

1.3. *Photosphere response to helicity transport*

Figure 3 shows the evolutions of the average values of $\alpha_z = (\nabla \times \mathbf{B})_z/B_z$ of the two sunspots during 2006 December 9–13. The SP/SOT vector magnetograms with very high-resolution (0.32″) are used to calculate the α_z maps on each day. We can see the twist of the MRS was ~3 times larger than that of the MSS.

The average values of α_z of the two sunspots behaved a large variation with time, especially in the MRS. On the whole, they increased in magnitude from 14:00 UT of December 09, while decreased after 06:00 UT of December 11 and declined up to the minimums before the

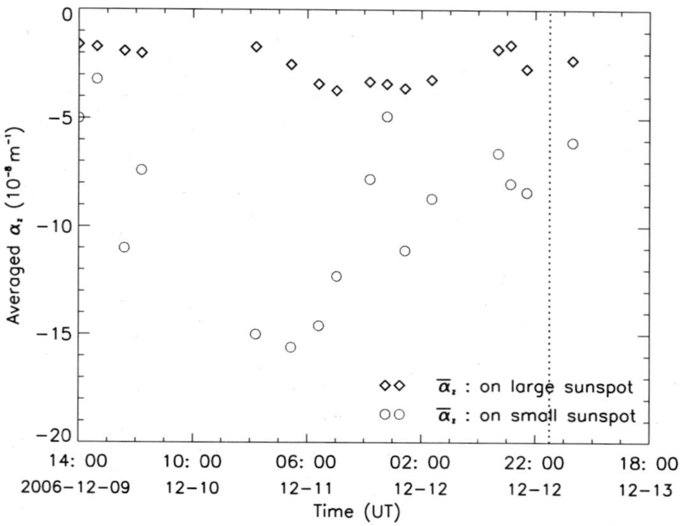

Fig. 3. Evolutions of the averaged α_z of the two sunspots. The data are chosen within the two Frames 1 and 2 shown in Fig. 1a, respectively. The standard deviation of the average value was $\sim 10^{-11}\mathrm{m}^{-1}$. The dotted-line indicates the onset of the X3.4 flare.

onset of the X3.4 flare. It is very interesting that the evolutions of the averaged α_z showed the same behaviour as the variations of the magnetic shear.[18]

1.4. *Negative (positive) helicity increased (decreased) in PIL region*

In the above section, we have found the averaged values of α_z of the two sunspots increased in magnitude before 06:00 UT of December 11. In this section, we will analyse the chirality change of the penumbral filaments in the PIL region in response to the changes of α_z.

In Fig. 4, the penumbral filaments to be studied are marked by two dashed-rectangles in (a) and (b), respectively. The filament in the first rectangle shows a clear right-handed twist (positive helicity), while the one in the second rectangle does not show so clear twist as the first one. However, about 2 hours later, a filament with left-handed twist (negative helicity) is found formed there (see Fig. 8). The maps of local twist α_z related to the filaments evolution are shown with contour lines in Fig. 5. The background is the continuum intensity and its FOV has been marked with a solid-line square in Fig. 4.

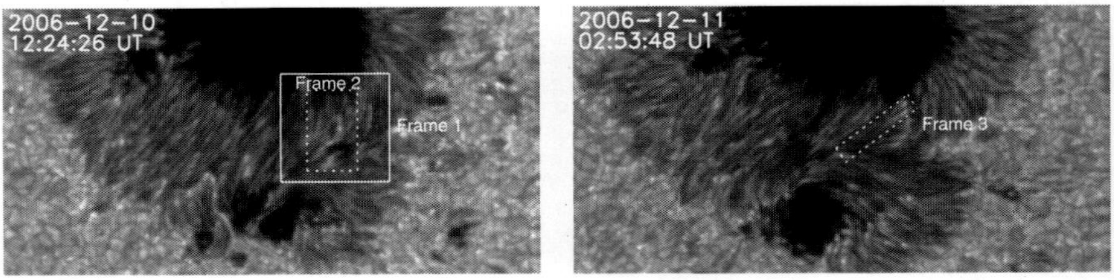

(a) (b)

Fig. 4. GBIs for the southern half of the active NOAA 10930 on 2006 December 10 and 11, respectively. The FOV is $76'' \times 38''$. A solid-line square and two dashed-line rectangles show three regions to be studies.

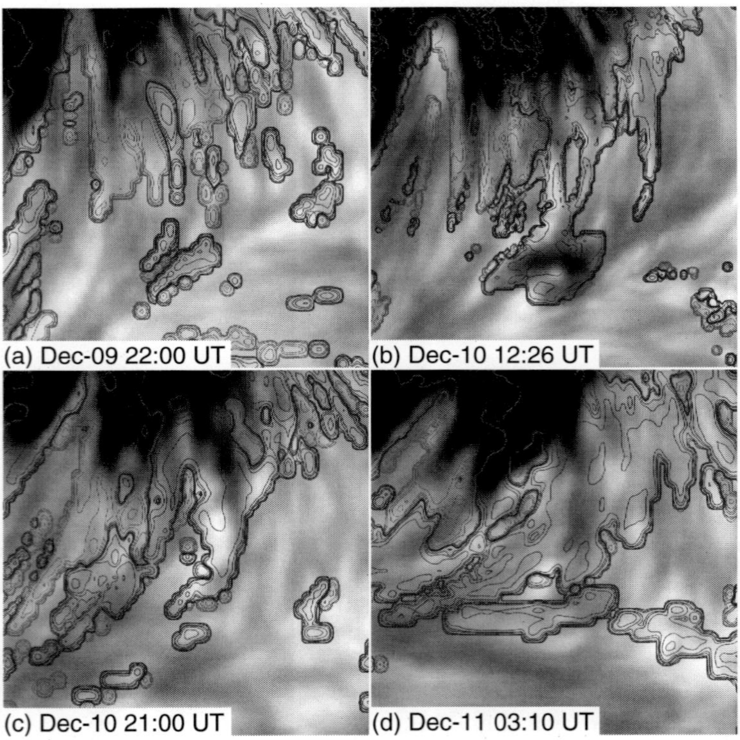

Fig. 5. Temporal development of the local twist α_z. The background is the continuum intensity image. Red(blue) represents positive (negative) value. The contours are $\pm 0.5,1,2,6 \times 10^{-7}\mathrm{m}^{-1}$. Its FOV($15.4'' \times 15.4''$) is marked with a solid-line square in Fig. 4.

Table 1. Summations $\sum \alpha_z^+$, $\sum \alpha_z^-$, and $\sum \alpha_z$ over the FOV of Fig. 5.

Sum (m^{-1})	Dec-09 22:00 UT	Dec-10 12:26 UT	Dec-10 21:00 UT	Dec-11 03:10 UT
$\sum \alpha_z^+$	0.00100	0.00103	0.00065	0.00043
$\sum \alpha_z^-$	−0.00107	−0.00176	−0.00137	−0.00228
$\sum \alpha_z$	−0.00011	−0.00073	−0.00073	−0.00185

Figure 5 shows the regions of the positive twist (red) shrank with time while those of the negative twist (blue) enlarged. Another feature is those small blue patches in the first panel seemed to consolidate to a large patch. Table 1 lists the summations $\sum \alpha_z^+$, $\sum \alpha_z^-$, and $\sum \alpha_z$ over the FOV of Fig. 5. $\sum \alpha_z^+$ decreases from 0.0010 m^{-1} to 0.0004 m^{-1} while $\sum \alpha_z^-$ increases from −0.00107 m^{-1} to −0.0023 m^{-1}. Thus, the total twist $\sum \alpha_z$ increased from −0.00011 m^{-1} to −0.00185 m^{-1}. In the following section, we will see these changes for the helicity well correspond to the chirality changes of the filaments there from right- to left-handed.

1.5. *Chirality change of penumbral filaments*

Figure 6 are the GBIs which shows temporal development of the penumbral filaments in the PIL region. The filaments interested are marked with two horizontal arrows in the first panel and two vertical arrows in the last

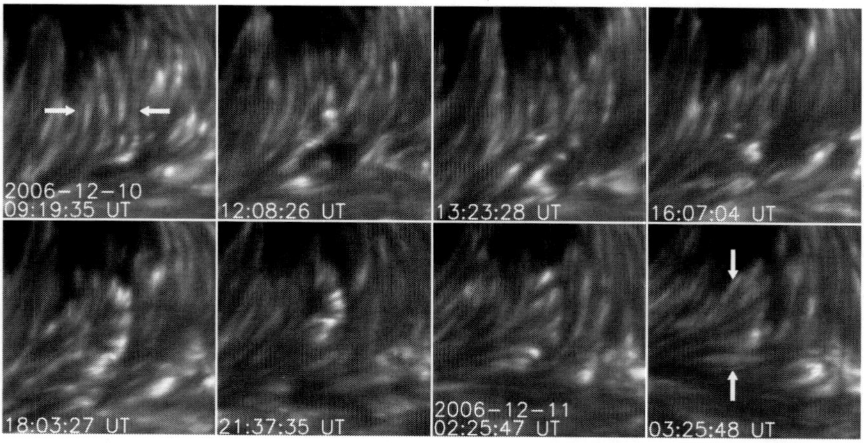

Fig. 6. Temporal development of the penumbral filaments in the PIL region (GBIs). Its FOV (15.4″ × 15.4″) is marked with a solid-line square in Fig. 4.

panel. At first, their chirality was right-handed at 09:19 UT of December 10. About three hours later, they evolved into a thicker one with the same chirality. At 13:23 UT, another thick filament nearby emerged, then two thick filaments had a trend to twist together at 16:07 UT. More complicate evolutions occurred at 18:03 UT through 02:25 UT of December 11. At last, the new filaments with left-handed chirality formed there and they overlain on the PIL.

To study the evolutions of the filaments in detail, we show two more figures in Figs. 7 and 8. Figure 7 shows the time sequence of GBIs on December 10. It is found that the thick filament consists of two slender ones, marked with f1 and f2, and they twist together in the outer penumbra (downwards). At 12:48 UT, the lower part of the filament obviously starts to expand. At 13:15 UT, the individual profiles of f1 and f2 can be discerned due to their counter-clockwise twisting. At 14:03 UT, f1 and

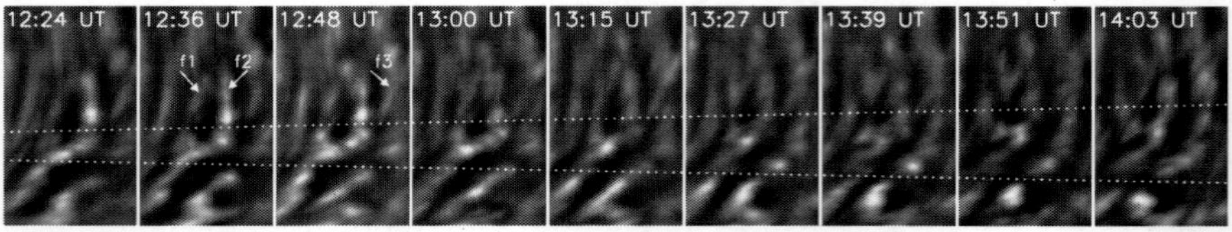

Fig. 7. Time sequence of GBIs for a rolling filament (consisting of f1 and f2). f3 marks an emerging filament. Its FOV is marked with a dashed-rectangle in Fig. 4a. The FOV is $6.6'' \times 11''$.

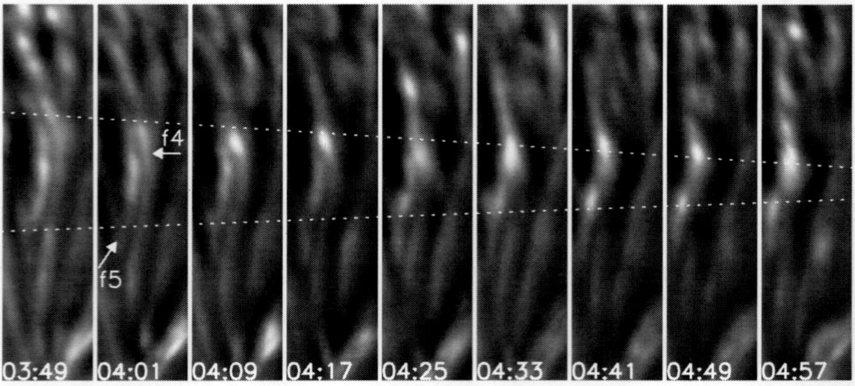

Fig. 8. Time sequence of GBIs for the generation of a filament with left-handed chirality. It developed from Filaments f4 and f5. Its FOV is marked with a dashed-rectangle in Fig. 4b. The FOV is $3.3'' \times 13.2''$.

f2 seem to go in different ways. f2 keeps its identity while f1 only leaves some remanet at the top. Similarly, a neighboring filament f3 to the right also undergoes such rolling motions. It quite resembles a multi-wire cable.[46]

We can trace the shifts of two separating dark features located above and below the branching point of f1 and f2 at 12:36 UT. Two dotted-lines thread them at each frame. The top/lower one moved upward/downward at a speed of $\sim$0.31/0.24 km/s. That two dark features separated means the filament untwisted and expanded.

Figure 8 shows that at 03:49 UT of December 11, filaments f4 and f5 are paralleled with each other and each consisted of two branches (two smaller filaments). In the frame from left to right, the right branch of f5 and the left branch of f4 become brighter (maybe also more twisted). At 04:17 UT, they twist together at the top. The new filament has a trend to shrink, which might be responsible for two kinds of subsequent filament activities. One is the brightening patch obviously moving downwards during the period of 04:09-04:57 UT. The other is one additional brightening occurred at the tip of the filament at 04:25 UT and more brightenings are found in the whole top-half of the filament at 04:57 UT, which might be caused by magnetic reconnections.

Similar to the above work, we find two converging features in this figure in response to the filament twisting. Two dotted-lines thread them again at each frame. The top/lower feature moves downward/upward at a speed of $\sim$0.44/0.33 km/s, separately. This gradual convergence indicates the filament twisting and shrinking.

Figure 9 shows two topologically different cases of twisted tubes. Each one has both θ- and z-component of the magnetic field. The tube in Fig. 9a is right-handed with a positive helicity and that in Fig. 9b left-handed with a negative helicity. The configurations in (a) and (b) are similar to those of the filaments in Figs. 7 and 8, respectively. The chirality change for the filaments should be caused by the change of θ-component of the magnetic field. Moreover, we have talked about the brightenings in the top-half of the filament at 04:57 UT, which might be caused by magnetic reconnections. However, we do not find any magnetic cancellation features at those sites by checking the NFI/SOT longitudinal magnetograms. It is expected that two flux tubes (like f4 and f5 in Fig. 8) with the same chirality and z-component of the magnetic field could coalesce to form one tube when they approach and squeeze each other. It is the occurrence of θ-reconnection in them.

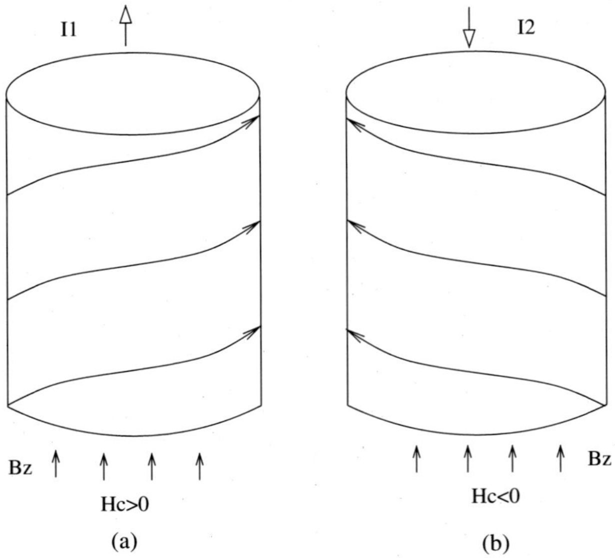

Fig. 9. Sketch of two twisted flux tubes used to determine the chirality of penumbral filaments. (a) is the right-handed chirality with a positive helicity and (b) the left-handed with a negative helicity.

2. Conclusions

In this paper, we separately calculate the rates of magnetic helicity transport positive and negative polarities in the AR NOAA 10930, and the accumulations of helicity injection over 7 days using MDI 96 min line-of-sight magnetograms and a LCT technique. It is found the rate of magnetic helicity transport of the positive polarity kept negative, however, after the X3.4 flare, it became positive. The total transport rate of two polarities had a similar evolution with the above rate. This result is consistent with the finding of Magara and Tsuneta (2008)[17] that the magnetic helicity in this active region first increases rapidly, while it become saturated in the late phase of the flare.

Two days before the flare (the first 64 hours in Figs. 1b and 2), we find (1) the positive flux kept nearly unchanged while the negative kept increased; (2) the rate of helicity transport of positive polarity kept increase in magnitude while that of negative polarity kept decrease; (3) the averaged α_z of the positive polarity sunspot seemed to increase in magnitude, but that for the negative sunspot was unclear (for the data loss). Combining the above facts, we can find that the negative flux increased and it contributed less helicity transport, whereas the positive flux kept unchanged and it contributed larger helicity transport. In this period, the penumbral filaments in the PIL region first unwound and then they twisted again in opposite direction. We conjecture that the chirality change of the filaments may be caused by the counterclockwise rotations of the positive-polarity sunspot instead of flux emergence. So the chirality change in this work may be not associated with the magnetic channel that Wang *et al.* (2008)[34] described.

Acknowledgments

This work is supported by the AJFC projects (nos. 10878016 and 10778723), the NSFC projects (nos. 10473016, 10273002, 10843010, 10673016, 60673158, 10733020 and 10803012), and National Basic Research Program of China under grant 2006CB806301. Hinode is a Japanese mission developed and launched by ISAS/JAXA, with NAOJ as domestic partner and NASA and STFC (UK) as international partners. It is operated by these agencies in cooperation with ESA and the NSC (Norway).

References

1. D. M. Rust, *Geophys. Res. Lett.* **21** (1994) 241.
2. B. C. Low, *Sol. Phys.* **167** (1996) 217.
3. M. Zhang, *Annu. Rev. Astron. Astrophys.* **43** (2005) 103.
4. M. Zhang, N. Flyer and B. C. Low, *ApJ* **644** (2006) 575.
5. T. Amari, J. F. Luciani, J. J. Aly, Z. Mikic and J. Linker, *ApJ* **595** (2003) 1231.
6. A. Nindos and M. D. Andrews, *ApJ* **616** (2004) L175.
7. D. M. Rust, *J. Geophys. Res.* **106** (2001) 25075.
8. K. Kusano, T. Maeshiro, T. Yokoyama and T. Sakurai, *ApJ* **610**, (2004) 537.
9. S. K. Antiochos, C. R. DeVore and J. A. Klimchuk, *ApJ* **510** (1999) 485.
10. S. D. Spicer, P. MacNiece, S. Antiochos and J. M. Finn, *Magnetic Fields and Helicity in the Sun and Heliosphere, 25th meeting of the IAU, Joint Discussion* 3 (2003).
11. A. D. Phillips, P. J. MacNeice and S. K. Antiochos, *ApJ* **624** (2005) L129.
12. J. Wang, G. Zhou and J. Zhang, *ApJ* **615** (2004) 1021.
13. B. J. LaBonte, M. K. Georgoulis and D. M. Rust, *ApJ* **671** (2007) 955.
14. E. Pariat, P. Demoulin and M. A. Berger, *A&A* **439** (2005) 1191.
15. M. Cheung, M. Schussler and F. Moreno-Insertis, *Proceedings of the International Scientific Conference on Chromospheric and Coronal Magnetic Fields (ESA SP-596).* In D.E. Innes, A. Lagg, S.A. Solanki (Eds.), ESA SP-596, p.54.1 (2005).

16. T. Magara and D. W. Longcope, *ApJ* **586** (2003) 630.
17. T. Magara and S. Tsuneta, *PASJ* **60** (2008) 1181.
18. T. Magara, *ApJ* **702** (2009) 386.
19. P. Démoulin, *Adv. Space Res.* **39** (2007) 1674.
20. M. A. Berger and G. B. Field, *J. Fluid Mech.* **147** (1984) 133.
21. J. Chae, H. Wang, J. Qiu, P. R. Goode, L. Strous and H. S. Yun, *ApJ* **560** (2001) 476.
22. P. Demoulin and M. A. Berger, *Sol.Phys.* **215** (2003) 203.
23. J. Chae, *Adv. Space Res.* **39** (2007) 1700.
24. M. C. M. Cheung, M. Schussler, T. D. Tarbell and A. M. Title, *ApJ* **687** (2008) 1373.
25. L. Tian, D. Alexander and R. Nightingale, *ApJ* **684** (2008) 747.
26. L. Tian and D. Alexander, *ApJ* **673** (2008) 532.
27. J. Liu and H. Zhang, *Sol.Phys* **234** (2006) 21.
28. H. Jeong & J. Chae, *ApJ* **671** (2007) 1022.
29. J. T. Su, T. Sakurai, Y. Suematsu, M. Hagino and Y. Liu, *ApJ* **697** (2009) L103.
30. S. Min and J. Chae, *Sol. Phys.* **258** (2009) 203.
31. Y.N. Su, L. Golub, A. van Ballegooijen *et al.*, *PASJ* **59** (2007) 785.
32. J. Zhang, L. Li and Q. Song, *ApJL* **662** (2007) 35.
33. X. L. Yan, Z. Q. Qu, C. L. Xu, Z. K. Xue and D. F. Kong, *Research in Astronomy and Astrophysics* **9** (2009) 596.
34. H. Wang, J. Jing, C. Tan, T. Wiegelmann and M. Kubo, *ApJ*, **687** (2008) 658.
35. T. Kosugi, K. Matsuzaki, T. Sakao *et al.*, *Sol. Phys.* **243** (2007) 3.
36. S. Tsuneta, K. Ichimoto, Y. Katsukawa *et al.*, *Sol. Phys.* **249** (2008) 167.
37. Y. Suematsu, S. Tsuneta, K. Ichimoto *et al. Sol. Phys.* **249** (2008) 197.
38. K. Ichimoto, B. Lites, D. Elmore *et al.*, *Sol. Phys.* **249** (2008) 233.
39. T. Shimizu, S. Nagata, S. Tsuneta *et al. Sol. Phys.* **249** (2008) 221.
40. G. A. Gary and M. J. Hagyard, *Sol. Phys.* **126** (1990) 21.
41. K. D. Leka, G. Barnes, A. D. Crouch, T. R. Metcalf, G. A. Gary, J. Jing and Y. Liu, *Sol. Phys.* **139** (2009).
42. L. R. Tian and D. Alexander, *ApJ* **695** (2009) 1012.
43. J. Qiu and D. E. Gary, *ApJ* **599** (2003) 615.
44. Y. Zhang, B. Tan and Y. Yan, *ApJ* **704** (2009) 1622.
45. E. Priest and T. Forbes, *Magnetic reconnection: MHD theory and applications, (New York: Cambridge University Press)* (2000).
46. M. Ryutova, T. Berger and A. Title, *ApJ* **676** (2008) 1356.

Advances in Geosciences
Vol. 21: Solar & Terrestrial Science (2008)
Ed. Marc Duldig

ON THE OBSERVATIONS OF MULTIPLE MHD OSCILLATIONS IN THE SOLAR LOOPS

A. K. SRIVASTAVA
Aryabhatta Research Institute of Observational Sciences (ARIES),
Manora Peak, Nainital-263 129, India
aks.astro.itbhu@gmail.com
www.aries.res.in

Observations of multiple oscillations of various magnetohydrodynamic (MHD) modes may play crucial role in the diagnostics of the local plasma conditions of solar loops. Magnetically structured coronal loops anchored into the photosphere exhibit various kinds of MHD oscillations such as fast sausage, kink, slow acoustic oscillations etc. Exploiting various harmonics of such observed MHD oscillations may be a tracer of important local plasma conditions and structuring, e.g. scale heights and signature of stratification, equilibrium condition, characteristic speed, density contrast, loop expansion factor/magnetic field divergence etc, which can shed new light on the dynamics of the solar loops. In this review, I briefly summarize the first observational signatures of multiple MHD oscillations in various types of solar loops and discuss their significance in diagnosing the solar atmosphere.

1. Introduction

The combination of complex magnetic field and plasma generates MHD waves and oscillations in the solar atmosphere, which may be the major candidates for the heating and dynamics of the solar atmosphere. The idea of exploiting such observed MHD modes as a diagnostic tool for determining the local plasma properties of the solar atmosphere was firstly suggested by Roberts *et al.*[1] However, the application of MHD seismology has suffered by the lack of high-resolution observations of MHD oscillations till last decade. This scenario has changed drastically due to recent space-based observations by the Solar and Heliospheric Observatory (SOHO), the Transition Region and Coronal Explorer (TRACE), and high resolution spectra/images from the Hinode and STEREO spacecrafts. The new techniques for the back end instruments of ground-based solar telescopes (e.g. adaptive optics of Swedish Solar Telescope; Fast Spectrometers of ROSA etc) are also now providing the ubiquitous presence of such MHD waves and oscillations in the lower solar atmosphere. The newly adapted instruments from ground and space are now showing well the ubiquitous presence of various MHD modes in different solar structures, and thus providing more accurate inputs in form of the observed various wave parameters for the theory of MHD seismology to study the local plasma conditions and dynamics of the solar atmosphere.

Magnetically structured coronal loops anchored into the photosphere exhibit various kinds of MHD oscillations such as fast sausage, kink, torsional, slow acoustic oscillations etc. Exploiting such observed oscillations as a diagnostic tool for deducing physical conditions of solar plasma, is known as a potential tool of MHD seismology. The fast kink wave is most frequently observed mode as it can be directly detected by periodic spatial displacement of coronal loop axis (e.g. Nakariakov *et al.*[2]; Aschwanden *et al.*[3]; Wang & Solanki[4]; Verwichte *et al.*[5]; O'Shea *et al.*[6]; De Moortel & Brady[7]; Van Doorsselaere *et al.*[8]; Erdélyi & Taroyan[9]; Ruderman *et al.*,[10] and references cited there). Pandey & Srivastava[11] have also reported the first observational signature of kink oscillations in the stellar loops. On the other hand, the fast MHD sausage wave causes the variation of pressure and magnetic field in a coronal loop and therefore can be observed as intensity oscillations,[12] intensity and Doppler shift oscillations[9] or periodic modulation of coronal radio emission[13,14] from various types of solar loops. The fundamental mode of slow acoustic oscillations are also now well observed in flaring and non-flaring loops both (e.g. Wang *et al.*[15,16]; Srivastava & Dwivedi[17] and references cited therein) which causes the significant modulation of plasma emissions near the loop apex. These observations provide a basis for the estimation of coronal plasma properties (e.g. Nakariakov & Ofman[18]; Van Doorsselaere *et al.*[19] The most fascinating trend is now developing in terms of the oscillations of higher harmonics of these MHD modes in the solar loops along with the first harmonics, which may provide the crucial informations about the local plasma dynamics of the solar loops (e.g. O'Shea *et al.*[6]; Srivastava *et al.*[12]; Srivastava & Dwivedi[17] and references cited there).

In this paper, I briefly summarize the first observations of multiple oscillations of various MHD modes in the solar atmosphere in light of my recent observational findings, and discuss their role in diagnosing the local plasma conditions and dynamics.

2. Observations of Multiple MHD Oscillations

2.1. *The multiple kink oscillations in cool flaring loops*

O'Shea *et al.*[6] have observed the first and second harmonics of fast kink mode of magnetoacoustic waves in the cool postflare loop system. Inspite of its incompressibility, the kink mode causes the weak perturbations in the density near the loop boundary.[20] Therefore, it leads an intensity oscillation in the EUV light curve of O V 629 Å emitted from the cool flaring TR loop system as observed by SOHO/CDS. The details of the observations and data reduction is given in O'Shea *et al.*.[6] The observations of various harmonics of fast kink oscillations as obtained by O'Shea *et al.*[6] are summarized in the Table 1 for the two cases : (i) Stationary flare in AR 10820 as observed in the dataset 33813 of SOHO/CDS on 10th Nov. 2005, (ii) Propagating disturbances (PDs) in AR 10820 as observed in the dataset 33821 of SOHO/CDS on 11th Nov 2005, O'Shea *et al.*[6] have firstly found the signature of first and second harmonics of kink oscillations in the cool, flaring TR loops associated with AR 10820. By using the observed oscillation periods of kink modes, valuable seismological informations of this flaring region associated with AR 10820 can be obtained. For the stationary flare, the period ratio is $P_1/P_2 \sim 2.4$, which probably indicates the presence of an expanding TR loop system associated with the observed flare. Using Eq. (96) of Verth & Erdélyi,[21] we estimate the loop expansion factor $\Gamma = 1.52$, which is the proxy of loop expansion and magnetic field divergence. The higher value of loop expansion factor indicates the significant divergence/expansion of magnetic field lines in these cool TR loops in AR 10820. In the present case, magnetic field expansion also dominates over the density stratification in these loops as the period ratio becomes greater than 2.0,[22] which indicates that probably the magnetic field is governing the kink mode oscillation periods in the observed loops

Table 1. Observations of Kink Harmonics in AR 10820 [adapted from O'Shea *et al.*[6]].

Observations of AR 10820 on 10th-11th Nov 2005 CDS/SOHO	First kink harmonics ω_1	Second kink harmonics ω_2 by
(i) Stationary flare of data set 33813 (10th Nov 2005)	Frequency = 2.10 mHz Probability = 99.2% Amplitude = 43.8%	Frequency = 5.04 mHz Probability = 98% Amplitude = 26.2%
(ii) PD of the data set 33821 (11th Nov 2005)	Frequency = 2.54 mHz Probability = 100% Amplitude = 11.5%	Frequency = 6.10 mHz Probability = 99% Amplitude = 6.1%

as expected. The period ratio shift >2.0 is also noticed in the case of propagating disturbances (PDs) of the data set 33821 as observed on 11th Nov 2005 in AR 10820, which is consistent with the results of stationary flare. Using Eq. (37) of Andries *et al.*[22] and the observed loop length ($2L$) of 2.02×10^8 m,[6] the antinode shift of the second harmonics (or first overtone) of 5.04 mHz observed in the stationary flaring loop is estimated as $\sim$15.44 Mm towards its loop apex. It should be noted that the average loop length has been estimated by O'Shea *et al.*[6] during the evolution of propagating disturbances (PDs) on 10th Nov 2005 between 21:30 UT and 21:52 UT. It seems that the loop system is also presented during the evolution of flare between 19:50 UT and 21:52 UT of the same day, although brightening could not allow to measure it. We assume that approximately same shift may occur but towards the loop footpoint when effect of density stratification will be considered dominant in the loop.[22] Therefore, by considering the approximately same shift of antinode of second harmonics as 15.44 Mm and assuming also the effect of density stratification, the approximate density scale height of such loop can be measured by using Eq. (36) of Andries *et al.*,[22] which will be $\sim$18.50 Mm. The density stratification does not seem dominant effect compared to magnetic field stratification in this case as period ratio is >2.0, although the consideration of same magnitude of antinode shift in both the cases allows us to measure the expected magnitude of scale height. Hence, $2L \sim 2.02 \times 10^8$ m, $H \sim 18.50$ Mm may cause a shift of $\sim$15.44 Mm in the antinode of second harmonics or first overtone towards loop footpoint. The similar but opposite shift may be caused by expansion ($\Gamma = 1.52$) in the observed loop. This scale height ($\sim$18.5 Mm) may be super-hydrostatic compared to the hydrostatic scale height of $\sim$13 Mm for the temperature 0.25 MK at which these loops are maintained in AR 10820 and visible in TR O V 629 Å line. The exact theoretical explanation of this super-hydrostatic scale height is still open. However, to conduct more accurate and realistic magneto-seismology of solar loops, the magnetic field divergence should be taken into account.[21] The theory of kink harmonics and related MHD seismology of solar loops have been reported by Verth and Erdélyi[21] and also discussed indetail by Andries *et al.*[22]

2.2. *The multiple slow acoustic oscillations in non-flaring coronal loop*

Srivastava & Dwivedi[17] have studied the intensity oscillations near the apex of a non-flaring coronal loop to find the signature of MHD oscillations.

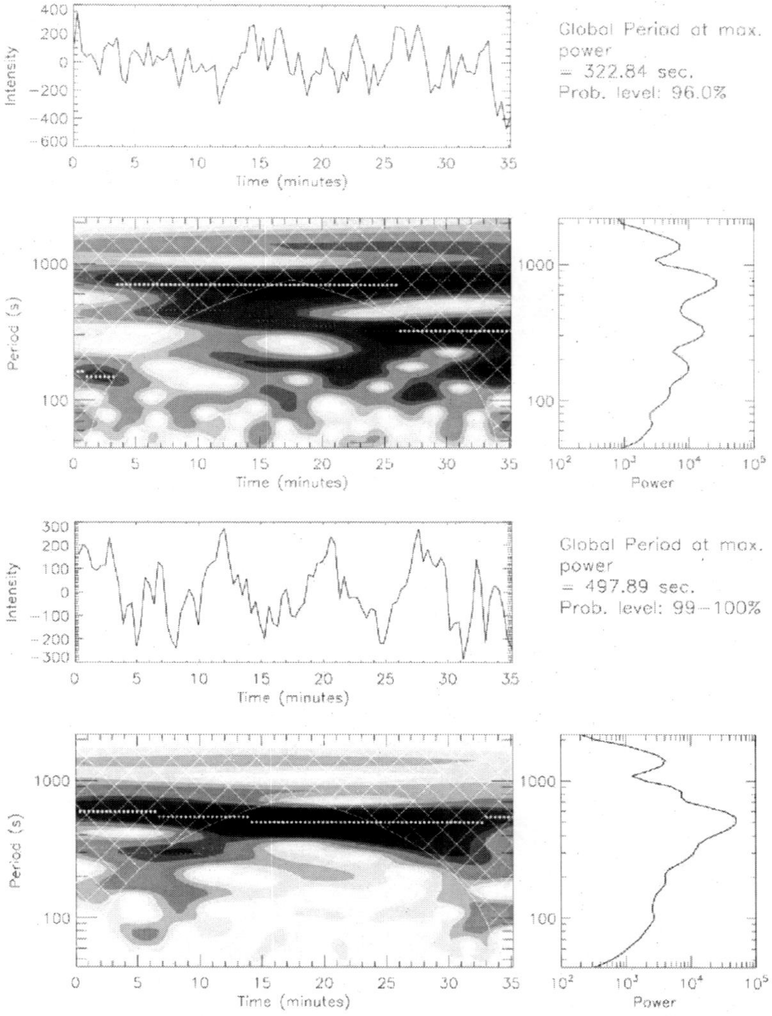

Fig. 1. The wavelet spectra for time series of Fe XII 195.12 Å for the second harmonics near the apex of non-flaring loop(left panel), and the first harmonics near its footpoint (right panel): In each wavelet, the top panel shows the variation of intensity, the wavelet power spectrum is given in the bottom-left panel, and global wavelet power in the bottom-right panel. [Figures are adapted from Srivastava & Dwivedi.[17]] Using 40" slot, the observations of this non-flaring loop system were acquired on 8 May 2008 between 09:35 UT and 10:10 UT by Hinode/EIS.The details of the observations, data reduction, power spectral analyses using wavelet technique are described in Srivastava & Dwivedi.[17]

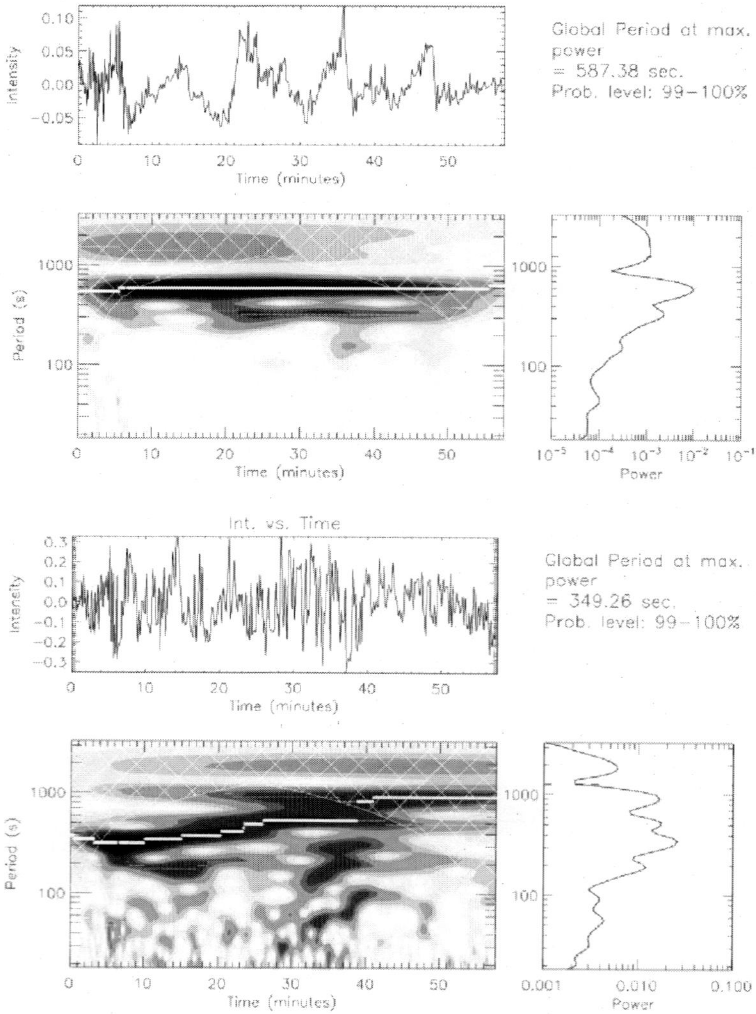

Fig. 2. The wavelet spectra of the time series of H-alpha for the first harmonics near the apex of cool post flare loop(top−left panel), and the second harmonics near its footpoint (top−right panel): In each wavelet, the top panel shows the variation of intensity, the wavelet power spectrum is given in the bottom-left panel, and global wavelet power in the bottom-right panel [Figures are adapted from Srivastava *et al.*.[12]] The cool post flare loop system is shown in the bottom most panel, which is observed from 15-cm Solar Tower Telescope at ARIES on 02 May 2001.The details of the observations, data reduction, power spectral analyses using wavelet technique are described in Srivastava *et al.*[12]

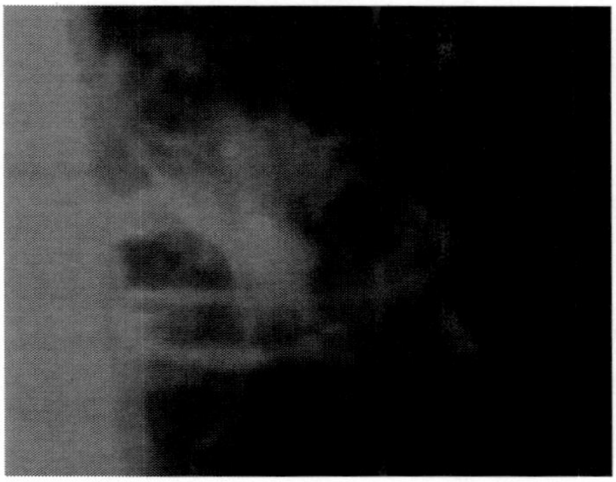

Fig. 2. (*Continued*)

They analyse the time series of the strongest Fe XII 195.12 Å image data, observed by 40" SLOT of the EUV Imaging Spectrometer (EIS) onboard the Hinode spacecraft. Using a standard wavelet tool, they produce power spectra of intensity oscillations near the apex of a non-flaring coronal loop system, and detect intensity oscillations of a period of ~322 s with a probability of 96%. This oscillation period of ~322 s is found to be in good agreement with theory of the second harmonics (or first overtones) of standing slow acoustic oscillations.[23] They detect, for the first time, the observational signature of multiple (first and second) harmonics of slow acoustic oscillations in such non-flaring coronal loop. Such oscillations have been observed in the past in hot and flaring coronal loops only, but have been predicted recently to exist in comparatively cooler and non-flaring coronal loops as well.[24] They find the periodicity 497*s* with the probability 99−100% near the clearly visible western footpoint of the loop, and interpret these oscillations to be likely associated with the first harmonics (or fundamental mode) of slow acoustic oscillations. The details of this unique observational result are given in Srivastava & Dwivedi.[17]

In this observation, the period ratio $P_1/P_2 = 1.54$ is significantly smaller than 2.0 which may be associated with the density stratification in such loops.[25] Using the period ratio shift and seismology theory of McEwan *et al.*[25] [Eq. 24], the estimated scale height ~10 Mm is found to be very low compared to the hydrostatic scale height at the inner coronal temperature. This may be the signature of the departure of such loops from the hydrostatic equilibrium. The period ratio shift may also be due to the magnetic field divergence in the coronal loop system. However, it is ruled out due to the fact that the slow wave is longitudinal compressive wave and weakly depends on the magnetic field in a plasma of low-beta. Hence, the magnetic stratification may not have a strong effect on the period of slow waves which is mainly affected by the temperature structure along the loop. Finally, we confirmed the fact that the density stratification dominates over the magnetic field divergence in the studied EUV loop system as the period ratio is less than 2.0.

2.3. *The multiple sausage oscillations in cool post flare loop*

Recently, Srivastava *et al.*[12] have firstly observed the multiple sausage oscillations of the period $P_1 = 597$ s (first harmonics or fundamental mode) and $P_2 = 349$ s (second harmonics or first overtones) in the cool post-flare loop of length $L = 97$ Mm and width $2a = 6.0$ Mm with their H-alpha observations, and studied the seismological properties of the selected loop. They have found that the density contrast is very high ~600 in that loop, and therefore, they concluded that the observed global sausage modes can exist only in the overdense cool post flare loops only. This prediction supports the previous findings of Nakariakov *et al.*[14] and Aschwanden *et al.*,[26] while this is the first case applied to the cool H-alpha post flare loop system. The postflare loops usually have a very high density contrast of the order of 100–1000. The studied loops are cool post flare ones which are overdense and the estimated density ratio (600) falls in the expected limit. Therefore, the trapped sausage mode may occur in such loops, although wave leakage can not be ruled out as a prominent dissipative mechanism. Using the period ratio and theory of both Andries *et al.*[27] and McEwan *et al.*,[25] the scale height is estimated as ~17 Mm which is much greater than the scale height at the formation temperature of H-alpha line in the lower solar atmosphere.The super-hydrostatic scale heights are well observed in the coronal loops, however, it is unique and the first observational signature of super-hydrostatic scale height (17 Mm) in the overdense and cool post flare loop system. This may be due to its departure from hydrostatic equilibrium condition, although the exact theoretical explanations are still open for this super-hydrostatic scale height estimated for cool post flare loop system from the period ratio observation. Srivastava *et al.*[12] have also estimated Alfvénic speed ~300 km/s outside the loop, and also the damping time of such wave modes as 20 min for the selected loop. The estimation of such crucial seismological parameters indicates the potential capabilities to diagnose the observed multiple sausage oscillations in determining loop dynamics and plasma conditions.

The ratio between the periods of the fundamental and those of the first harmonics of sausage waves is proportional to $P_1/P_2 \sim 1.68$, which is significantly smaller than 2.0. Similar phenomena were observed for fast kink oscillations,[5] slow acoustic oscillations[17] for the coronal loops. The deviation of P_1/P_2 from 2.0 in homogeneous loops is very small due to the wave dispersion,[25] although the longitudinal density stratification may cause this significant period ratio shift from 2.0 for longitudinal tubular modes (e.g. sausage, slow-acoustic modes).[8,17,27] Since the oscillations are sausage MHD modes, the resonant absorption is ruled out as a dissipation mechanism. Therefore the wave leakage in the surroundings may be the most plausible cause for the wave damping. There is again a signature of the domination of density stratification over the magnetic field divergence in this loop system.

3. Discussion and Conclusions

Magnetically structured coronal loops anchored into the photosphere exhibit various kinds of MHD oscillations such as fast sausage, kink, torsional, slow acoustic oscillations etc. Exploiting various harmonics of such observed oscillations may be a tracer of important local plasma conditions, e.g. scale heights and signature of stratification, equilibrium conditions, characteristic speeds, density contrasts, loop expansion factor/ magnetic field divergence etc, which may provide new informations on the plasma conditions and dynamics of the solar loops.

This short paper summarizes the recent first observational signatures of multiple MHD oscillations of different MHD modes in various types of solar loops, and their diagnostic capabilities. The observations of multiple kink oscillations in the cool, TR flaring loops are found to be a tracer of the magnetic field divergence and loop expansion which dominates over the effect of density stratification.The density scale height is found to be $\sim$18 Mm in the cool TR loops exhibiting the kink oscillations. The estimated density scale height is found to be a super-hydrostatic scale height and hence may have important implications on the dynamics of these cool, TR, flaring loops. The observations of multiple slow acoustic oscillations in the non-flaring coronal loop system are found to be a tracer of the non-hydrostatic and density stratified plasma environment in which the plasma flow and slow standing waves are simultaneously presented,[17] and the magnetic field stratification and loop expansion are not a well dominating candidate to control the plasma dynamics related with these slow modes.

Similarly, the observations of multiple sausage oscillations in the cool post flare loops are found to be a tracer of overdense, non-hydrostatic, density stratified atmosphere. There, the magnetic field divergence/loop expansion is again not a probable candidate to control the oscillation dynamics of tubular sausage mode coupled with the magnetic field. Furthermore, the super-hydrostatic scale height is observed for the first time in these loops complementing that they are well known for the coronal loops. Although this may not be universal phenomenon, the oscillation dynamics may be governed by the domination of density stratifications in the loops which support various harmonics of longitudinal tubular modes (e.g. sausage mode which perturbs both the magnetic field and density, and slow acoustic mode which perturbs density only). The magnetic field divergence/flux tube expansion may not be well dominant in such loops even after its presence in all kinds of solar flux tubes (e.g. Srivastava *et al.*[12]; Srivastava & Dwivedi[17]). The loops which support kink harmonics, may have well dominant effect of magnetic field divergence/flux tube expansion over the density stratifications as observed by O'Shea *et al.*[6] However, this case study should be tested with the forthcoming high resolution space based observations.

In conclusion, the observations of multiple MHD oscillations provide the wide range of diagnostics capabilities of local plasma conditions in various types of solar loops. The forthcoming high resolution observations of multiple MHD oscillations from both ground and space and the sophisticated new theory of MHD seismology of solar loops will be very important to reveal the mystery of the complex plasma motions/oscillations and its physical conditions in the various solar loops.

Acknowledgments

AKS acknowledges financial support from the Department of Science and Technology (DST), Ministry of Science and Technology of India and the Russian foundation for Basic Research (RFBR), Russia in form of project grant. AKS thanks both the referees for their constructive comments which improved the manuscript considerably.

References

1. B. Roberts, P. M. Edwin and A. O. Benz, *Astrophys. J.* **279** (1984) 857.
2. V. M. Nakariakov, L. Ofman, E. E. Deluca, B. Roberts and J. M. Davila, *Science* **285** (1999) 862.

3. M. J. Aschwanden, L. Fletcher, C. J. Schrijver and D. Alexander, *Astrophys. J.* **520** (1999) 880.
4. T. J. Wang and S. K. Solanki, *Astron. Astroph.* **421** (2004) L33.
5. E. Verwichte, V. M. Nakariakov, L. Ofman and E. E. Deluca, *Solar Phys.* **223** (2004) 77.
6. E. O'Shea, A. K. Srivastava, J. G. Doyle and D. Banerjee, *Astron. Astroph.* **473** (2007) L13.
7. I. De Moortel and C. S. Brady, *Astrophys. J.* **664** (2007) 1210.
8. T. Van Doorsselaere, V. M. Nakariakov and E. Verwichte, *Astron. Astroph.* **473** (2007) 959.
9. R. Erdélyi and Y. Taroyan, *Astron. Astroph.* **489** (2008) L49.
10. M. S. Ruderman, G. Verth and R. Erdélyi, *Astrophys. J.* **686** (2008) 694.
11. J. C. Pandey and A. K. Srivastava, *Astrophys J.* **697** (2009) L153.
12. A. K. Srivastava, T. V. Zaqarashvili, W. Uddin, B. N. Dwivedi and P. Kumar, *Mon. Not. R. Astr. Soc.* **388** (2008) 1899.
13. M. J. Aschwanden, *Solar Phys.* **111** (1987) 113.
14. V. M. Nakariakov, V. F. Melnikov and V. E. Reznikova, *Astron. Astroph.* **412** (2003) L7.
15. T. Wang, S. K. Solanki, W. Curdt, D. E. Innes and I. E. Dammasch, *Astrophys. J.* **574** (2002) L101.
16. T. J. Wang, S. K. Solanki, D. E. Innes and W. Curdt, *Astron. Astroph.* **435** (2005) 753.
17. A. K. Srivastava and B. N. Dwivedi, *New Astronomy* **15** (2010) 8.
18. V. M. Nakariakov and L. Ofman, *Astron. Astroph.* **372** (2001) L53.
19. T. Van Doorsselaere, V. M. Nakariakov, P. R. Young and E. Verwichte, *Astron. Astroph.* **487** (2008) L17.
20. F. C. Cooper, V. M. Nakariakov and D. Tsiklauri, *Astron. Astroph.* **397** (2003) 765.
21. G. Verth and R. Erdélyi, *Astron. Astroph.* **486** (2008) 1015.
22. J. Andries, T. van Doorsselaere, B. Roberts, G. Verth, E. Verwichte and R. Erdélyi, *Space Science Reviews* **76** (2009).
23. V. M. Nakariakov, D. Tsiklauri, A. Kelly, T. D. Arber and M. J. Aschwanden, *Astron. Astroph.* **414** (2004) L25.
24. Y. Taroyan and S. Bradshaw, *Astron. Astroph.* **481** (2008) 247.
25. M. P. McEwan, G. R. Donnelly, A. J. Díaz and B. Roberts, *Astron. Astroph.* **460** (2006) 893.
26. M. J. Aschwanden, V. M. Nakariakov and V. F. Melnikov, *Astrophys. J.* **600** (2004) 458.
27. J. Andries, M. Goossens, J. V. Hollweg, I. Arregui and T. Van Doorsselaere, *Astron. Astroph.* **430** (2005) 1109.

Advances in Geosciences
Vol. 21: Solar & Terrestrial Science (2008)
Ed. Marc Duldig

MODELLING THE CORONAL MAGNETIC FIELD USING HINODE (AND FUTURE) DATA

M. S. WHEATLAND* and S. A. GILCHRIST
School of Physics, Sydney Institute for Astrophysics,
University of Sydney, NSW 2006, Australia
**m.wheatland@physics.usyd.edu.au*
www.physics.usyd.edu.au/~wheat

S. RÉGNIER
School of Mathematics and Statistics,
University of St Andrews, St Andrews,
Fife KY 16 9SS, United Kingdom
stephane@mcs.st-andrews.ac.uk

There is considerable interest in accurate modelling of the solar coronal magnetic field using photospheric vector magnetograms as boundary data, and the nonlinear force-free model is often used. However, recent studies using Hinode data have demonstrated that this modelling fails in basic ways, with the failure attributable to the departure of the inferred photospheric magnetic field from a force-free state. The solar boundary data are inconsistent with the model, which leads to inconsistencies in calculated force-free solutions. A method for constructing a self-consistent nonlinear force-free solution is described, which identifies a force-free solution that is close to the observed boundary data. Steps towards developing more sophisticated magneto-hydrostatic modelling — taking into account pressure and gravitational forces at the level of the solar boundary data — are also outlined.

1. Background

1.1. *Coronal magnetic field modelling*

Solar flares are large magnetic explosions in the Sun's corona, powered by intense active region magnetic fields around sunspots.[1] Large flares influence our local space weather, including producing energetic particle storms which can damage satellite electronics. It has been estimated that a space weather 'superstorm' at the next solar maximum could produce an economic loss of up to \$70 billion[2] due to lost revenue and infrastructure. Space weather effects provide a motivation for developing accurate modelling of solar coronal magnetic fields, with the aim of better understanding and predicting large-scale solar activity.

Spectro-polarimetric measurements of certain photospheric lines permit a determination of the vector magnetic field at the photosphere. The resulting vector magnetograms are inferences rather than direct measurements,[3] and involve a number of uncertainties, but they provide the most detailed available information about active region magnetic fields. A new generation of instruments, including the spectropolarimeter (SP) of the Solar Optical Telescope (SOT) on the Hinode satellite,[4] is producing state-of-the-art data. In principle vector magnetograph data provides a set of boundary values for modelling the overlying coronal field, but in practice basic difficulties prevent the construction of reliable models from the data.[5,6]

1.2. *The nonlinear force-free model*

The nonlinear force-free model, involving a static balance of purely magnetic forces, is often used to describe the coronal magnetic field in the low density solar corona.[7–9] A nonlinear magnetic field $\mathbf{B}$ satisfies

$$\mathbf{J} \times \mathbf{B} = 0 \quad \text{and} \quad \nabla \cdot \mathbf{B} = 0, \tag{1}$$

where $\mathbf{J} = \mu_0^{-1} \nabla \times \mathbf{B}$ is the electric current density. The first of Eqs. (1) states that $\mathbf{J}$ is everywhere parallel to the magnetic field, and hence the equations may be reformulated as:

$$\nabla \times \mathbf{B} = \alpha \mathbf{B} \tag{2}$$

$$\mathbf{B} \cdot \nabla \alpha = 0, \tag{3}$$

introducing the force-free parameter α. Eq. (3) states that that α is constant along magnetic field lines.

The boundary conditions for the problem comprise a specification of the normal component of $\mathbf{B}$ in the boundary (denoted B_n), together with a specification of α over one polarity (sign) of B_n.[7,10–12] Equivalently, the boundary condition on α may be replaced by a specification of the normal component J_n of the electric current density over one polarity of B_n. We label the two choices of polarity as P (corresponding to $B_n > 0$), and N (corresponding to $B_n < 0$). It is only necessary to specify the force-free parameter (or J_n) on one polarity because of the constancy of α along field lines in the model.

In application to vector magnetograph data, boundary values for α may be constructed via finite differencing of the vector magnetogram values $B_x^{\rm vm}$

and $B_y^{\rm vm}$ parallel to the surface, according to

$$\alpha^{\rm vm} = \frac{1}{B_z^{\rm vm}} \left(\frac{\partial B_y^{\rm vm}}{\partial x} - \frac{\partial B_x^{\rm vm}}{\partial y} \right). \tag{4}$$

In the modelling described here solar curvature is ignored, and the boundary field values are assumed to lie in the plane $z = 0$. If Eq. (4) is applied over the observed vector magnetogram region it provides two possible choices for boundary values on α, corresponding to the P and N choices.

The nonlinear force-free equations are difficult to solve in the general 3-D case. A variety of iterative numerical methods have been developed,[13] and demonstrated to work on theoretical test cases.[14] Some of the methods use the specification of the full vector field in the boundary as boundary conditions, rather than the boundary conditions outlined above. This is formally an over-prescription, but should not introduce a problem if the boundary values are consistent with the force-free model.

The current-field iteration, or Grad-Rubin,[10] method provides one numerical approach to solving the nonlinear force-free model. This method involves solution, at iteration k, of the scheme:

$$\mathbf{B}^{[k]} \cdot \nabla \alpha^{[k+1]} = 0, \tag{5}$$

$$\nabla \times \mathbf{B}^{[k+1]} = \alpha^{[k+1]} \mathbf{B}^{[k]}. \tag{6}$$

If $\mathbf{B}^{[k]}$ is known then Eq. (5) is a linear equation for $\alpha^{[k+1]}$, allowing propagation of boundary values of the force-free parameter into the volume. The right-hand side of Eq. (6) is then known, and it is also a linear equation (Ampère's law) for $\mathbf{B}^{[k+1]}$, which may be solved in the volume subject to the boundary conditions on the field.

A fixed point of the iteration, $\mathbf{B} = \mathbf{B}^{[k+1]} = \mathbf{B}^{[k]}$, is a solution of the nonlinear Eqs. (2) and (3). There are a number of specific implementations of this method, involving different methods of solution of the equations, and different ways of imposing the boundary conditions.[12,15,16] Typically the scheme is started with from the potential field $\mathbf{B}^{[0]} = \mathbf{B}_{\rm pot}$ satisfying the boundary values of B_z. All Grad-Rubin methods use the mathematically correct boundary conditions on α specified above.

A fast current-field iteration implementation was presented in Ref. [16], namely a method with a computational time which scales as N^4, for a grid with N^3 points. Eq. (5) is solved by field line tracing such that $\nabla \cdot \mathbf{J}^{[k]} = 0$ is enforced, and Eq. (6) is solved using 2-D Fourier Transforms and a vector potential, thereby ensuring $\nabla \cdot \mathbf{B}^{[k]} = 0$.

1.3. *The failure of the nonlinear force-free model*

Two recent workshops have applied nonlinear force-free modelling to Hinode SOT/SP data.[5,6] A workshop in 2007 used data for NOAA active region AR 10930, observed on 12–13 December 2006, and a workshop in 2008 used data for AR 10953, observed on 30 April 2007. The workshops found that the force-free methods fail in basic ways in application to the solar data. Different methods produce solutions which are inconsistent with one another. For example, the estimates of magnetic free energy for AR 10930 varied between 1% and 32% of the energy $E_{\rm p}$ of the potential field with the same boundary values $B_z^{\rm vm}$, for solutions using different methods of solution of the force-free model but based on the same 'preprocessed' boundary data (this procedure is discussed below).[5] The solutions produced by individual methods are also not *self-consistent*. For example, for AR 10953, the P and N solutions produced by the fast current-field iteration method[16] had free magnetic energies equal to 3% and 18% of $E_{\rm p}$ respectively.[6]

The left-hand panel of Fig. 1 illustrates the inconsistency. This shows the field lines for the P solution (dark lines) and the N solution (light lines)

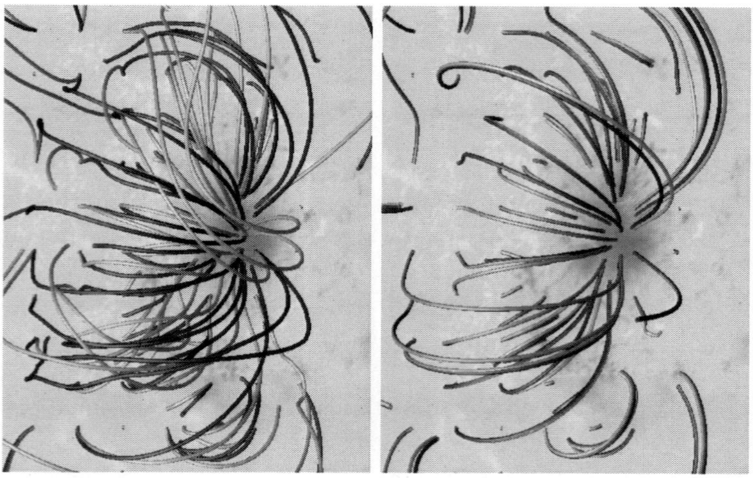

Fig. 1. Nonlinear force-free magnetic field solutions for AR 10953, constructed using the fast current-field iteration method applied to Hinode SOT/SP data. The field lines for the solutions are shown for the central part of the active region, in a view looking down on the computational domain. Left panel: The P (dark field lines) and N (light field lines) solutions obtained from the original data — these solutions use the boundary values for the force-free parameter on the positive and negative polarity respectively, and illustrate the inconsistency of the boundary data with the force-free model. Right panel: The P and N solutions after 10 self-consistency cycles.

for AR 10953, for the central part of the active region, in a view looking directly down on the computational domain. The field lines are qualitatively very dissimilar, and the light field lines are clearly more twisted, reflecting the fact that this solution has larger values of $|\alpha|$, and a correspondingly greater free energy. Despite this basic inconsistency, nonlinear force-free modelling continues to be applied to solar boundary data. Caveat emptor![a]

The failure of the model may be attributed to the departure of the vector magnetogram field values from a force-free state. Errors in the field determination may contribute, but it is generally believed that the photospheric field is subject to non-magnetic forces.[17] The net force and torque on the coronal magnetic field may be evaluated via surface integrals of the boundary field values, which provide necessary conditions for consistency of the boundary data with the force-free model.[18] These integrals are typically significantly non-zero for vector magnetogram data.

One approach to the problem is 'preprocessing,' whereby the boundary data are modified so as to satisfy the necessary conditions.[19] Preprocessing of solar-like boundary data has shown to lead to improved nonlinear force-free solutions,[20] and preprocessing was applied to the data at the workshops in 2007 and 2008 in the modelling of ARs 10930 and 10953.[5,6] However, preprocessed solar photospheric vector magnetic field data *remains inconsistent with the force-free model*, as shown by the inconsistency of the P and N solutions in the left panel of Fig. 1. The conditions being enforced are necessary but not sufficient, and there are an infinite number of such conditions.[11] Although preprocessing enforces some of these conditions, it does not enforce others, as shown explicitly in Ref. [6] for the AR 10953 data (see Fig. 5). A further problem with preprocessing is that it typically involves smoothing the data, which is undesirable. The Hinode data are distinguished by its high resolution, and this should be exploited in modelling.

2. Self-Consistent Nonlinear Force-Free Modelling

An alternative approach involves identifying a force-free solution which is, in some sense, close to the observed (non-force-free) boundary data. A specific method, in the context of the current-field iteration approach to solving the nonlinear force-free model, is presented in Ref. [21], and is demonstrated in application to Hinode SOT/SP data for AR 10953.

---

[a] Let the buyer beware!

2.1. The self-consistency method

There are three steps in the 'self-consistency' method.[21] First, the P and N solutions are constructed, from the (unpreprocessed) vector magnetogram boundary data. The vector magnetogram-derived boundary values of the force-free parameter are denoted by $\alpha_0 \pm \sigma_0$, where σ_0 are uncertainties calculated from the corresponding uncertainties in the inferred magnetic field values.[22] The P solution provides a mapping, along field lines, of boundary values at points in the region P to points in N, and thereby defines new possible boundary values $\alpha_1 \pm \sigma_1$ of the force-free parameter at points in N. Similarly, the N solution maps boundary values at points N to points in P, and so defines new possible boundary values $\alpha_1 \pm \sigma_1$ at points in N. Hence the two solutions define a complete new set of boundary values $\alpha_1 \pm \sigma_1$ (over both polarities).

Step two involves deciding on a new boundary value for α at each boundary point based on two possible choices $\alpha_0 \pm \sigma_0$ and $\alpha_1 \pm \sigma_1$. For this purpose Bayesian probability is applied.[23] The assumption of Gaussian errors and an application of Bayes's theorem lead to a most probable value, and associated uncertainty[21]:

$$\alpha_2 = \frac{\alpha_0/\sigma_0^2 + \alpha_1/\sigma_1^2}{1/\sigma_0^2 + 1/\sigma_1^2} \quad \text{and} \quad \sigma_2 = \left(\frac{1}{\sigma_0^2} + \frac{1}{\sigma_1^2}\right)^{-1/2}. \tag{7}$$

In other words, the new value is an uncertainty-weighted average of the two possible values. The new set of boundary values over all boundary points is still, in general, inconsistent with the force-free model, but the values are expected to be closer to consistency.

Step three involves iterating steps one and two. Two new force-free solutions are constructed for the P and N choices using the boundary values $\alpha_2 \pm \sigma_2$, via two separate applications of current-field iteration. Together, the two new solutions define new mappings between the P and N regions, and hence a complete new set of boundary conditions $\alpha_3 \pm \sigma_3$. Eqs. (7) are applied to $\alpha_2 \pm \sigma_2$ and $\alpha_3 \pm \sigma_3$ to decide on most probable values $\alpha_4 \pm \sigma_4$ at each boindary point, and so on.

Each iteration of the procedure [construction of the P and N solutions, and then application of Eqs. (7)] is referred to as a 'self-consistency cycle.' It is expected that the procedure converges to a self-consistent solution, i.e. a set of boundary values of the force-free parameter for which the P and N solutions are the same.

2.2. *Application to Hinode data*

In Ref. [21] the achievement of self-consistency is demonstrated using the Hinode SOT/SP data for AR 10953. The right-hand panel of Fig. 1 illustrates the final, self-consistent solution, after 10 self-consistency cycles. The field lines for the P and N solutions, shown by dark and light lines respectively, agree closely. The energies for the two solutions differ by less than 0.03%.

The self-consistency procedure changes the boundary values. The changes are quite significant for the AR 10953 case, representing an rms change of about 120 G in B_x and about 100 G in B_y. This reflects the fairly gross discrepancies between the initial P and N solutions seen in the left panel of Fig. 1. The changes are comparable to the changes introduced by preprocessing for the preprocessed data used at the workshop (preprocessing changes the B_z values also, whereas the B_z values are preserved in the self-consistency procedure).

It should be noted that the application to AR 10953 demonstrated in Ref. [21] is at the level of a 'proof of concept' and could be improved in many ways. Uncertainties were not incorporated into the procedure (all points were assumed to have equal uncertainties). Also, the Hinode SOT/SP data used was embedded in a larger SOHO/MDI field of view for which only B_z values were available, and α was assigned to zero for this region. The rms changes introduced by the procedure are expected to be smaller once these limitations are removed. The calculation will be repeated including uncertainties and avoiding embedding in the near future.

3. Magneto-Hydrostatic Modelling

Whilst the self-consistency procedure outlined above provides one approach to constructing a coronal magnetic field model, and addresses the specific difficulty preventing the construction of a nonlinear force-free model, it is desirable to have a physical solution to the problem, i.e. an approach incorporating non-magnetic forces.

It is likely that the magnetic field at the photospheric level is subject to magnetic forces as well as pressure gradients, and dynamical forces (associated with flows). All of these forces are expected to be significant at the photospheric level and in at least a narrow layer above that level. However, here we consider the next simplest model, i.e. a static model incorporating pressure and magnetic forces, but neglecting gravity, referred to as the magneto-hydrostatic (MHS) model. The MHS equations may be written:

$$\mathbf{J} \times \mathbf{B} = \nabla p \quad \text{and} \quad \nabla \cdot \mathbf{B} = 0, \tag{8}$$

where p is the gas pressure, the new dependent variable in the model.

Other forms of the equations are also useful. The first of Eqs. (8) implies

$$\mathbf{B} \cdot \nabla p = 0, \tag{9}$$

$$\mathbf{J} \cdot \nabla p = 0, \tag{10}$$

i.e. p is constant along field lines of $\mathbf{B}$ and stream lines of $\mathbf{J}$. It is convenient to separate $\mathbf{J}$ into components parallel and perpendicular to $\mathbf{B}$:

$$\mathbf{J} = \mathbf{J}_{\|} + \mathbf{J}_{\perp}, \tag{11}$$

with the parallel component represented in terms of a scalar function α:

$$\mathbf{J}_{\|} = \frac{1}{\mu_0} \alpha \mathbf{B}. \tag{12}$$

Setting $\nabla \cdot \mathbf{J} = 0$ leads to

$$\mathbf{B} \cdot \nabla \alpha + \mu_0 \nabla \cdot \mathbf{J}_{\perp} = 0, \tag{13}$$

and taking the cross product of $\mathbf{B}$ with the first of Eqs. (8) gives

$$\mathbf{J}_{\perp} = \frac{1}{B^2} \mathbf{B} \times \nabla p. \tag{14}$$

The boundary conditions for the problem comprise a specification of B_n together with a specification of J_n and p over one polarity of B_n.[10] The need to specify J_n and p over only one polarity follows from Eqs. (9) and (10).

3.1. *The Grad-Rubin method*

There have been relatively few attempts to solve the 3-D MHS equations from boundary conditions. One approach to solving the model in the solar context was presented in Refs. [24] and [25], and subsequently developed in spherical co-ordinates.[26,27] That approach is a generalisation of the optimisation method for calculating nonlinear force-free fields.[28,29] Here we consider a different method, a generalisation of current-field iteration, based on a qualitative description given in Grad and Rubin in Ref. [10].

The Grad-Rubin iteration scheme for the MHS problem may be summarised as:

$$\mathbf{B}^{[k]} \cdot \nabla p^{[k+1]} = 0 \tag{15}$$

$$\mathbf{J}_{\perp}^{[k+1]} = \mathbf{B}^{[k]} \times \nabla p^{[k+1]} / (B^{[k]})^2, \tag{16}$$

$$\mathbf{B}^{[k]} \cdot \nabla \alpha^{[k+1]} = -\mu_0 \nabla \cdot \mathbf{J}_{\perp}^{[k+1]}, \tag{17}$$

$$\nabla \times \mathbf{B}^{[k+1]} = \alpha^{[k+1]} \mathbf{B}^{[k]} + \mu_0 \mathbf{J}_{\perp}^{[k+1]}. \tag{18}$$

If $\mathbf{B}^{[k]}$ is known then Eq. (15) is a linear equation for $p^{[k+1]}$, allowing propagation of boundary values of pressure into the volume. Eq. (16) then allows the calculation of the perpendicular component of the current density $\mathbf{J}_{\perp}^{[k+1]}$ in the volume and on the boundary. Given those values, Eq. (17) is a linear equation for $\alpha^{[k+1]}$ allowing propagation of boundary values of the scalar function $\alpha^{[k+1]}$ into the volume. [The boundary values of $\alpha^{[k+1]}$ need to be constructed from the boundary values of J_n and the calculated boundary values of $\mathbf{J}_{\perp}^{[k+1]}$ at the given iteration, according to Eqs. (11) and (12).] The right-hand side of Eq. (18) is then known in the volume, and this equation is a linear equation for $\mathbf{B}^{[k+1]}$ (Ampère's law), which may be solved in the volume subject to the boundary conditions on the field. The scheme may be started from the potential field $\mathbf{B}^{[0]} = \mathbf{B}_{\rm pot}$ satisfying the boundary values of B_z. A fixed point of the iteration is a solution of the nonlinear problem.

A code implementing a method of solution of the iteration scheme (15)–(18) is currently being developed and tested. The code uses techniques similar to those used in the fast current-field iteration code for the nonlinear force-free problem.[16] Full details of the code and methods, and initial results for a simple test case will be given in a later publication.

3.2. *Prospects for application to data*

In principle spectro-polarimetric data provides not just magnetic, but also thermodynamic information, so that it is possible to infer plasma parameters from the observations.[30,31] Typically the thermodynamic parameters are ignored, in standard 'inversion' procedures. However, methods have been developed for reliable extraction of additional information, including the pressure profile in the atmosphere.[3,32] In principle this data provides additional boundary information for coronal field modelling.

In practice it may be necessary to introduce gravity into the magneto-hydrostatic modelling, to make the atmospheric model more realistic. This requires an additional dependent variable, density, but with the assumption of an equation of state the density is simply related to pressure. The Grad-Rubin scheme may also be applied in this case. However, prospects for modelling from the data remain to be investigated in detail.

4. Summary

In principle, photospheric vector magnetogram data provides boundary data for coronal magnetic field modelling, and state-of-the-art data are being returned by new instruments including the SOT/SP on the Hinode spacecraft. In practice, however, modelling has proved difficult.

The nonlinear force-free model is appropriate in the corona, but it does not describe the field at the photospheric level, corresponding to the solar data. The model is inconsistent with the data, and this has lead to a failure of nonlinear force-free modelling.[5,6] Preprocessing the data to make it more compatible with the model[19] does not solve the inconsistency problem.

An approach to identifying a 'self-consistent' nonlinear force-free solution, which has boundary values close to the observed data, is described.[21] This provides one possible avenue for successful modelling. An alternative is to consider more sophisticated physical models, incorporating non-magnetic forces. A method for solving the magneto-hydrostatic (MHS) equations using Grad-Rubin iteration[10] is briefly described. A code implementing this method is currently being developed.

Acknowledgments

Stephane Régnier acknowledges the support of a Royal Society Short Visit Grant during a visit to the University of Sydney.

References

1. E. R. Priest and T. G. Forbes, *Astron. Astrophys. Rev.* **10** (2002) 313.
2. S. Odenwald, J. Green and W. Taylor, *Adv. Sp. Res.* **38** (2006) 280.
3. J. C. Del Toro Iniesta and B. Ruiz Cobo, *Solar Phys.* **164** (1996) 169.
4. S. Tsuneta, K. Ichimoto, Y. Katsukawa, S. Nagata, M. Otsubo, T. Shimizu, Y. Suematsu, M. Nakagiri, M. Noguchi, T. Tarbell, A. Title, R. Shine,

W. Rosenberg, C. Hoffmann, B. Jurcevich, G. Kushner, M. Levay, B. Lites, D. Elmore, T. Matsushita, N. Kawaguchi, H. Saito, I. Mikami, L. D. Hill and J. K. Owens, *Solar Phys.* **249** (2008) 167.
5. C. J. Schrijver, M. L. DeRosa, T. Metcalf, G. Barnes, B. Lites, T. Tarbell, J. McTiernan, G. Valori, T. Wiegelmann, M. S. Wheatland, T. Amari, G. Aulanier, P. Demoulin, M. Fuhrmann, K. Kusano, S. Régnier and J. K. Thalmann, *Astrophys. J.* **675** (2008) 1637.
6. M. L. DeRosa, C. J. Schrijver, G. Barnes, K. D. Leka, B. W. Lites, M. J. Aschwanden, T. Amari, A. Canou, J. M. McTiernan, S. Régnier, J. K. Thalmann, G. Valori, M. S. Wheatland, T. Wiegelmann, M. C. M. Cheung, P. A. Conlon, M. Fuhrmann, B. Inhester and T. Tadesse, *Astrophys. J.* **696** (2009) 1780.
7. T. Sakurai, *Solar Phys.* **69** (1981) 343.
8. A. N. McClymont, L. Jiao and Z. Mikic, *Solar Phys.* **174** (1997) 191.
9. G. A. Gary, *Solar Phys.* **203** (2001) 71.
10. H. Grad and H. Rubin, in *Proc. 2nd Int. Conf. on Peaceful Uses of Atomic Energy*, Vol. 31, United Nations, Geneva (1958), p. 190.
11. J. J. Aly, *Solar Phys.* **120** (1989) 19.
12. T. Amari, T. Z. Boulmezaoud and J. J. Aly, *Astron. Astrophys.* **446** (2006) 691.
13. T. Wiegelmann, *J. Geophys. Res. A* **113** (2008) 3.
14. C. J. Schrijver, M. L. Derosa, T. R. Metcalf, Y. Liu, J. McTiernan, S. Régnier, M. S. Wheatland and T. Wiegelmann, *Solar Phys.* **235** (2006) 161.
15. T. Amari, J. J. Aly, J. F. Luciani, T. Z. Boulmezaoud and Z. Mikic, *Solar Phys.* **174** (1997) 129.
16. M. S. Wheatland, *Solar Phys.* **245** (2007) 251.
17. T. R. Metcalf, L. Jiao, A. N. McClymont, R. C. Canfield and H. Uitenbroek, *Solar Phys.* **247** (1995) 269.
18. M. M. Molodenskii, *Sov. Astron.* **12** (1969) 585.
19. T. Wiegelmann, B. Inhester and T. Sakurai, *Solar Phys.* **233** (2006) 215.
20. T. R. Metcalf, M. L. Derosa, C. J. Schrijver, G. Barnes, A. A. van Ballegooijen, T. Wiegelmann, M. S. Wheatland, G. Valori and J. M. McTtiernan, *Solar Phys.* **247** (2008) 269.
21. M. S. Wheatland and S. Régnier, *Astrophys. J.* **700** (2009) L88.
22. K. D. Leka and A. Skumanich, *Solar Phys.* **188** (1999) 3.
23. E. T. Jaynes, *Probability Theory*, ed. G. Larry Bretthorst, Cambridge University Press, Cambridge, UK (2003).
24. T. Wiegelmann and B. Inhester, *Solar Phys.* **214** (2003) 287.
25. T. Wiegelmann and T. Neukirch, *Astron. Astroph.* **457** (2006) 1053.
26. T. Wiegelmann, T. Neukirch, P. Ruan and B. Inhester, *Astron. Astroph.* **475** (2007) 701.
27. P. Ruan, T. Wiegelmann, B. Inhester, T. Neukirch, S. K. Solanki and L. Feng, *Astron. Astroph.* **481** (2008) 827.
28. M. S. Wheatland, P. A. Sturrock and G. Roumeliotis, *Astrophys. J.* **540** (2000) 1150.

29. T. Wiegelmann, *Solar Phys.* **219** (2004) 87.
30. J. Jefferies, B. W. Lites and A. Skumanich, *Astrophys. J.* **343** (1989) 920.
31. E. Landi Degl'Innocenti and M. Landolfi, *Polarization in Spectral Lines*, Dordrecht, Kluwer (2004).
32. B. Ruiz Cobo and J. C. del Toro Iniesta, *Astrophys. J.* **398** (1992) 375.

Advances in Geosciences
Vol. 21: Solar & Terrestrial Science (2008)
Ed. Marc Duldig

SOLAR MASS EJECTION IMAGER (SMEI) AND INTERPLANETARY SCINTILLATION (IPS) 3D-RECONSTRUCTIONS OF THE INNER HELIOSPHERE

B. V. JACKSON*, P. P. HICK, A. BUFFINGTON,
M. M. BISI† and J. M. CLOVER
Center for Astrophysics and Space Sciences,
University of California, San Diego
9500 Gilman Dr. #0424, La Jolla, CA 92093-0424, U.S.A.
**bvjackson@ucsd.edu*
http://smei.ucsd.edu/ or http://ips.ucsd.edu/
†mmbisi@ucsd.edu

M. TOKUMARU
Solar-Terrestrial Environment Laboratory, Nagoya University
Furo-cho, Chikusa-ku, Nagoya 464-8601, Japan
http://stesun5.stelab.nagoya-u.ac.jp/index-e.html
tokumaru@stelab.nagoya-u.ac.jp

The Solar Mass Ejection Imager (SMEI) direct white-light data as well as higher-level data products are available on our UCSD Website at http://smei.ucsd.edu/ from first light in early February 2003 to nearly the present day. These analyses provide densities in the inner heliosphere, show many familiar CMEs in three dimensions (3D) during this interval, and provide animations and individual images of them. This 3D analysis is enhanced by use of interplanetary scintillation (IPS) velocity observations to help provide the overall form of the structures reconstructed. Our time-dependent 3D reconstruction technique is discussed, and the different ways we test and validate these 3D results. These checks include both internal consistency checks, and comparisons with *in situ* measurements at various near-Earth spacecraft, at Ulysses, at the STEREO spacecraft, and from magnetic field data at Mars.

1. Introduction

The Solar Mass Ejection Imager (SMEI)[1–2] was launched 6 January 2003 on the Air Force Space Test Program satellite *Coriolis*. The instrument consists of three baffled cameras whose $3° \times 60°$ fields of view are aligned in the long

Fig. 1. SMEI is depicted on the Coriolis spacecraft in its terminator polar orbit at 840 km with an orbital inclination of 98°. SMEI looks away from the Earth at 30° above the local horizontal to avoid sunlight reflected from the Earth and from the *Windsat* radiometer antenna situated at the top of the Corilois spacecraft. The fields of view of the three cameras (each shown as shaded wedges extending from the satellite in this artists impression) together cover nearly 160° of sky and, as the instrument orbits Earth, sweep out nearly all the sky around it.

dimension to achieve a combined ~160° wide field of view that scans most of the sky every 102-minute orbit (Fig. 1). The cameras view the heliosphere in Thomson-scattered light with ~0.5° angular resolution. Approximately 4500 four-second exposure CCD-camera data frames per orbit as in Fig. 2a are combined into a map of the sky as shown in Fig. 2b. Calibration of the SMEI data to an absolute level of ~4% using known star brightness has been accomplished.[3] Using a combination of three dimensional (3D) reconstruction and image-analysis techniques, SMEI observations quantify the extents of transient structures and their interactions with each other and with slow-moving ambient solar wind. These techniques have mapped the 3D extents of the 28 October 2003 halo coronal mass ejection (CME) and

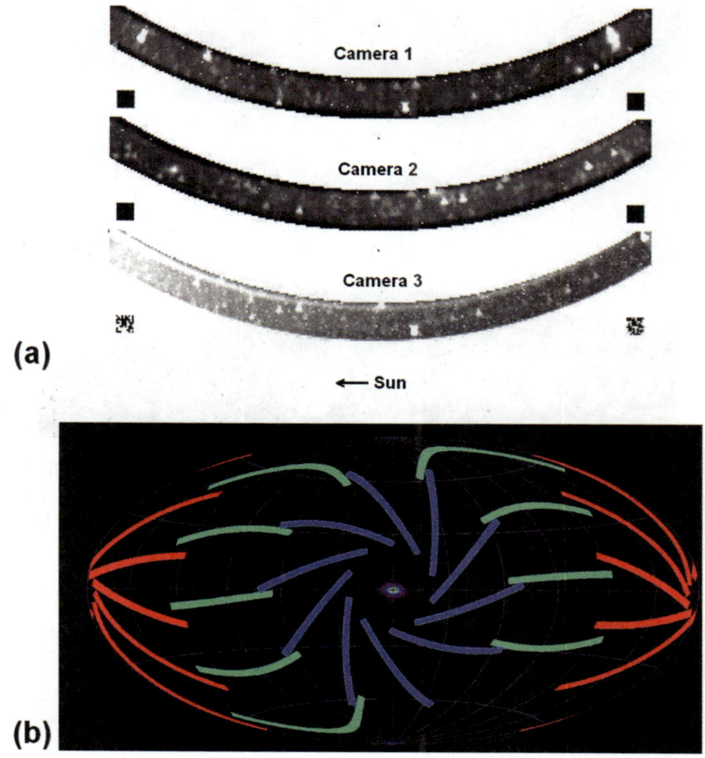

Fig. 2. (a) $3^\circ \times 60^\circ$ image frames from each of the three SMEI cameras. Camera 1 (top) views farthest from the Sun; camera 3 (bottom) closest, with the Sun towards the left in each frame. (b) Full-sky Hammer-Aitoff projection showing the placement of a sample of SMEI camera image frames with the Sun centered in the projection (in October). About 4500 frames per orbit fill in the sky in a clockwise direction every 102 minutes. Camera 1 is shown in red, Camera 2 is in green, and Camera 3 is in blue.

its interplanetary counterpart (ICME) in low resolution and determined its mass.[4] Other studies that have included interplanetary scintillation (IPS) observations[5] have mapped the 3D extent of the high velocity structure surrounding the 28 October 2003 ICME and determined the energy present in its outward-flowing material. These same techniques, using a combination of SMEI and IPS data from the Solar-Terrestrial Environment Laboratory (STELab) (Fig. 3) radio antennas,[6] have also been used to map both the energy and mass of a sequence of halo CMEs that erupted from the Sun on 27–28 May 2003.[7] IPS analyses alone have been used to determine the structure location, mass, and solar wind speeds from ICMEs[5,8] and,

Fig. 3. STELab IPS radio array (one of four now operating) near Mt. Fuji. The arrays used singly measure scintillation intensity (or g-value) for about 50 radio sources each day. Scintillation signals cross-correlated between arrays give a robust IPS velocity determination.

by combining both mass and speed, Jackson *et al.*[9] demonstrated that IPS plasma ram pressure measurements from Mars Global Surveyor (MGS) magnetometer observations compared satisfactorily with these analyses.

The University of California, San Diego (UCSD) time-dependent tomographic analysis that allows solar wind temporal variations to be mapped in 3D is discussed in this article. This current analysis fits data in essentially the same manner for SMEI, IPS, or a combination of these data sets in the 3D-reconstructions. Several of the different tools and error analysis techniques we have developed to certify the 3D reconstruction results are shown. We give examples of these analyses for specific events from both data sets, and using combinations of them. These analyses are also compared with *in-situ* measurements and other remote-sensing observations for specific events. The analysis programs using IPS data from STELab or from SMEI now reside at the Community Coordinated Modeling Center (CCMC) at the Goddard Space Flight Center for use by others, and so this article serves as an explanation for, and the caveats present in using these analysis routines at the CCMC.

This article uses the term 'CME' in reference to a sequence of events in the low corona as viewed in the SOlar and Heliospheric

Observatory (SOHO) Large-Angle Spectroscopic COronagraph (LASCO)[10] or the twin Solar-TErrestial RElations Observatory (STEREO)[11] near-Sun coronagraph observations, but in keeping with the now common terminology for these, describes these events in the interplanetary medium as 'ICMEs'; the switch occurring when this material is viewed by SMEI and in IPS observations and measured *in situ* near Earth.

Section 2 describes the time-dependent tomographic analysis routines developed by our group at the University of California, San Diego (UCSD) for fitting SMEI brightness, and STELab IPS velocity and g-level data. Section 3 presents examples of these analyses for the 27–28 May 2003 CME/ICME, and shows some of the measurement techniques we have developed to certify that these analyses are accurate. Section 4 gives examples of these results with *in situ* measurements and remote-sensing observations. We conclude in Section 5.

2. 3D-Reconstruction Analysis

Computational aspects of the UCSD 3D-reconstruction program have been discussed in other articles over the past decade.[12–15] These analyses and references therein describe early versions of the analysis that use the assumption that the heliosphere corotates. In more recent articles,[4,5,7,9,16,17] the assumption that heliospheric structure corotates has been relaxed. In this latter modification of the technique, described more fully here, line-of-sight segments and their 3D weights are projected back in space and time to a solar wind inner boundary (a source surface) that is set at a height below the closest approach of all lines of sight to the Sun (usually 15 R_s). Each line of sight is followed out to typically 2AU from the Earth and projected to the source surface consistent with the velocity and interactions present in the solar wind model.

In current analyses the inversion process updates boundary conditions for the kinematic 3D solar wind model to better fit observations using a least-squares fitting procedure. This procedure minimizes the differences between modeled and observed brightness, or modeled and observed IPS g-value and velocity, or a combination of these. As explained elsewhere,[7,18] unlike IPS g-level, it is not possible to distinguish the slowly-varying Thomson-scattering signal due to the ambient solar wind from the large and very bright zodiacal light signal. Because of this, the mean value for this ambient is modeled using the average *in-situ* density measured from the mean solar wind value at 1 AU throughout the period of the observation.

A least-squares fitting program developed specifically for this type of analysis inverts the weighted, projected model values on the two-dimensional (2D) inner boundary source surface at the different time steps to provide solar wind model outflow parameters. The model values are directly inverted on the source surface at the discrete source surface times to give new solar wind parameter values; these are converged for each data set.

In the least-squares fitting process, ratios of modeled-to-observed values and a modeled-to-observed chi-squared are monitored to indicate a rate of convergence for this interval. Velocity and density corrections to the 3D model are made separately. First, the inversion changes are made to previous velocity boundary conditions on the inner boundary surface. Second, the 3D solar wind model is updated and new projected locations of each line-of-sight point on the inner-boundary surface are determined. Third, inversion changes are made to previous density boundary conditions on the inner boundary surface. Finally, the 3D model is again updated with all the newest boundary values.

The Carrington maps of velocity and density at the inner boundary are smoothed each iteration using a 2D Gaussian spatial filter that incorporates equal-solar-surface areas, and a Gaussian temporal filter. Locations where insufficient information is available to change the model values are left blank in the final result. For the analysis presented here, the blank areas include sections of heliospheric volume on the opposite side of the Sun from Earth that cannot be observed or reconstructed by SMEI or IPS at the digital resolutions used.

The reconstruction program generally converges to an unchanging model within a few iterations, but is set to operate for 9 iterations to guarantee convergence.[12] For a typical rotation and the digital resolutions of the current SMEI data sets, a set of density and velocity iterations generally takes about fifteen minutes on a 3.0 GHz Intel® Pentium IV™ computer. Normally those IPS-velocity observations and SMEI-brightness lines of sight throughout the period that do not fit within a three-sigma limit of the mean ratio change ascribed at that location by the model (typically ∼1% of the SMEI brightness or IPS velocity lines-of-sight) are removed from the data set. This allows a further safeguard for the removal of lines of sight that are outliers which do not fit the model values. We sometimes find that the overall chi-squared fit, for g-level or brightness or velocity, is reduced when this criterion is imposed. The program then operates for 9 more iterations (18 in total). The solutions are insensitive to the initial model values, and after a few iterations any residual signature of the initial values has disappeared. Other tests[12] show that the 3D reconstruction of a set of artificial observations using a known 3D input reproduces the input.

In the following subsections we go into more detail about the specific numerical algorithms that are used in the SMEI brightness and IPS time-dependent 3D reconstructions. The data used in the 3D reconstructions are often noisy; this necessitates using different smoothing techniques to mitigate the effects of the noise, and to smooth the intermediate steps in the iterative procedure.

2.1. *IPS and Thomson-scattering remote sensing*

This explanation of the algorithm was given originally in Hick and Jackson,[15] and is adapted for the current analysis in the time-dependent form described here. The derivation presented here closely follows this previous work but is altered after the present Eq. 11 to include time-dependent tomography. The present formulae retain IPS g^2, but now also include the Thomson-scattering measured brightness B.

The observational data used in the IPS tomographic reconstruction are the IPS g-level (or 'disturbance factor') and IPS velocity $V_{\rm IPS}$ observations. Each observation represents a line-of-sight integration through the solar wind in the inner heliosphere. To date we have used g-level data from the IPS arrays in Cambridge (UK), Ooty (India), and Nagoya (Japan); $V_{\rm IPS}$ data have been available from Nagoya and Ooty. The purpose of the reconstruction is to create a model 3D heliosphere (i.e. a density and velocity distribution) that reproduces these observations as well as possible. The reconstruction uses an iterative scheme to minimize the differences between actually observed and calculated model values. Since the reconstruction only involves the solar wind density n and the (radial) solar wind outflow velocity V, the observed quantities g and $V_{\rm IPS}$ need to be related to these two solar wind parameters. The g-level is related to the scintillation index m:

$$g = m/\langle m\rangle, \tag{1}$$

where m is the instantaneous observed scintillation index for an IPS source; $\langle m\rangle$ is the expected "quiet" scintillation index, based on an average of

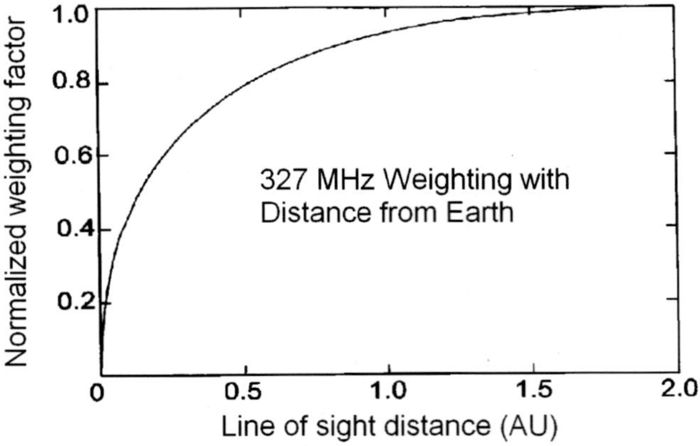

Fig. 4. IPS 327 MHz line of sight weighting with distance from Earth as described in Jackson *et al.*[12]

past source observations as a function of solar elongation, and g depends only weakly on elongation (or heliocentric distance). The scintillation index m is related to the small-scale density variations along the line of sight s by:

$$m^2 = \int_0^\infty ds\, W\rho(s)\delta n(s)^2. \tag{2}$$

The "weight" function $W\rho(s)$

$$W\rho(s) = 2\pi \int_0^\infty q dq \sin^2(q^2 \lambda_O s/(4\pi)) \exp(-\Theta_O^2 q^2 s^2/2) q^{-3} \tag{3}$$

as formulated by Young[19] and refined by Kojima *et al.*,[12] depends on observing wavelength λ_O, the angular size of the radio source Θ_O, and the turbulence power spectrum with wave vector q. The same Eq. 3 weak-scattering weighting factor along the line of sight is assumed for fast and slow solar wind, and different solar wind structure. This weighting is given for STELab data assuming a nominal angular source size of 0.1 arc sec (Fig. 4). The small-scale density fluctuations $\delta n(s)$ along the line of sight cause the scintillation. $\delta n(s)$ depends not only on macroscopic properties (solar wind speed, density, magnetic field), but also on microscopic properties associated with turbulence in the solar wind. However, empirical evidence[20] suggests that changes in $\delta n(s)$ scale with changes in values of the bulk electron density. Quantitatively we model this behavior by expressing $\delta n(s)$ in terms of the heliocentric distance r and the solar wind electron density n, or equivalently, the normalized density $\hat{n}$ defined as $\hat{n} = (r/r_0)^2 n$:

$$\delta n(r, \hat{n}) = \delta n_0 (r_0/r)^{2-\beta r} (\hat{n}/n_0)^{\beta n}. \tag{4}$$

A value for r_0 of 1 AU normalizes results to observations at this distance from the Sun.[14] The tomography is calculated using values for the powers $\beta n \approx 0.5$[2,3] and $\beta r \approx 0$ that are derived by iteratively fitting remote-sensing results to *in-situ* solar wind velocity and density measurements.[12,20] The mean scintillation index $\langle m \rangle$ (and hence the g-level) depends on parameters $\hat{n}^{\text{mean}}$ and δn_0^{mean} describing the density and density fluctuation in the "average" background solar wind.

The IPS velocity is defined in terms of the same weight function and small-scale density fluctuations,

$$V_{\text{IPS}} = \int_0^\infty ds\, W\rho(s) \delta n(s)^2 V_\perp(s) \Big/ \int_0^\infty ds\, W\rho(s) \delta n(s)^2. \tag{5}$$

Here, $V_\perp$ is the component of the solar wind velocity perpendicular to the line of sight at a distance s from the observer. The most accurate calculation of V_{IPS} uses the cross correlation of the IPS signals at different IPS stations.[6,14] The above expression is consistent with this more accurate calculation to within $\sim$10 kms^{-1} while significantly reducing the required computational resources.[14]

We introduce a few simplifying definitions to generate the equations for g and V_{IPS} that are implemented in our numerical algorithm. First, write the background solar wind density $\hat{n}^{\text{mean}}$ (needed to calculate $\langle m \rangle$) as:

$$\hat{n}^{\text{mean}}(s) = n_{\text{mean}} \bar{n}_{\text{mean}}(s). \tag{6}$$

The constant n_{mean} sets the absolute density scale. $\bar{n}_{\text{mean}}$ is a dimensionless function defining the shape of the mean background solar wind density distribution. Also define η as the normalized density $\hat{n}$ to absorb the constants $\delta n_0, \delta n_0^{\text{mean}}$, and n_{mean}:

$$\eta = (\delta n_0 / \delta n_0^{\text{mean}})^{1/\beta n} \hat{n} / n_{\text{mean}}. \tag{7}$$

This "modified normalized density" will be useful later in the discussion of the kinematic solar wind model (Section 2.3.1). Define the "γ function":

$$\gamma = \eta^{2\beta n}. \tag{8}$$

Finally, define a new weight function $W\sigma$ that absorbs the dependence on heliocentric distance:

$$W\sigma(s, \beta r) = W\rho(s) r(s)^{-2(2-\beta r)}. \tag{9}$$

With these definitions g-level and IPS speed V_{IPS} become:

$$g^2 = \int_0^\infty ds\, W\sigma(s, \beta r)\gamma(s) \Big/ \int_0^\infty ds\, W\sigma(s, \beta r)(\bar{n}_{\mathrm{mean}}(s))^{2\beta n}, \text{ and} \tag{10}$$

$$V_{\mathrm{IPS}} = \int_0^\infty ds\, W\sigma(s, \beta r)\gamma(s) V(s) \sin\chi(s) \Big/ \int_0^\infty ds\, W\sigma(s, \beta r)\gamma(s) \tag{11}$$

where we put $V_\perp(s) = V(s)\sin\chi(s)$, and assume $V(s)$ is radial. $\chi(s)$ is the angle between the direction to the Sun and to the Earth from the position at distance s along the line of sight. In Eq. (10) we usually set $n_{\mathrm{mean}} = 1$; i.e. we assume a background solar wind density with a $1/r^2$ drop-off.

In the analysis of brightness from Thomson scattering as given in Jackson and Hick,[18] and adapted for this description:

$$B = \int_0^\infty ds\, n(s) Wt(s), \tag{12}$$

where $n(s)$ is the bulk electron density per cubic centimeter, at distance s in cm, along the line of sight; $Wt(s)$, the scattered intensity per electron, serves as a brightness 'weight factor' for the density. For the large distances from the Sun viewed by SMEI:

$$Wt(s) = 1/2\, \sigma_T F_S(s)(r_0/r(s))^2(2 - \sin^2\chi(s)), \tag{13}$$

where σ_T is the Thomson-scattering cross section, $F_S(s)$ is the flux received from the solar disk at a distance r_0, $r(s)$ is the distance of the electron from the Sun, and $\chi(s)$ is the angle between the incident solar radiation from the Sun and the direction of scattering towards the spacecraft instrument viewing the sky.[21] To evaluate Eq. (13) both $r(s)$ and $\chi(s)$ are evaluated as functions of the distance R of the observer from the Sun, elongation ε of the line of sight, and distance s along the line of sight. Using R, ε, and s as independent variables $Wt(s)$ scales as R^{-2}. For a fixed R and ε, the function $Wt(s)$ has a maximum at $s = R\cos\varepsilon$, and is symmetric around this point. For $\varepsilon < 90°$ the maximum is at the point of closest approach of the line of sight to the Sun. Figure 5 shows the weight function at $\varepsilon = 16°$, $31°$, and $90°$ for $R = 1$ AU.

For tomography based on Thomson scattering brightness rather than g-level, Eq. (12) replaces Eq. (10). Note the structural similarity between

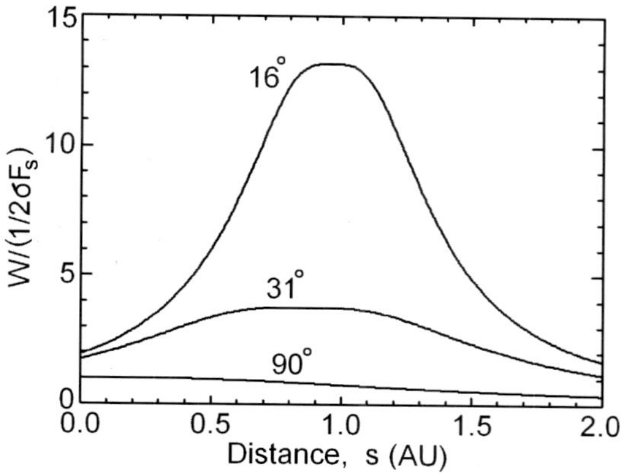

Fig. 5. Thomson-scattering line of sight weighting at three different elongations with distance from the Earth as described in the text (from Jackson and Hick[18]).

these equations. The weight ($W\sigma$ and Wt) captures the line-of-sight geometry and the microscopic scattering physics and γ (implicitly, Eq. 10), and n (explicitly, Eq. 12) describes the dependency on bulk solar wind density. This similarity allows us to apply the same algorithm to both IPS and Thomson scattering tomography.

There are three undetermined functions in these equations: γ (or equivalently in the case of IPS, the normalized density $\hat{n}$), the radial solar wind outflow speed V, and $\bar{n}_{\mathrm{mean}}$, the shape of the background normalized solar wind density. In the case where both Thomson-scattering and IPS is used in the same tomographic routine, γ is assumed directly related to n by use of Eqs. (7) and (8), and the equivalence of γ to $\hat{n}$. The first two of the unknown functions, γ (or n) and V, are our primary interest in the time-dependent 3D reconstructions described in the following sections. The reconstruction problem can be formulated as follows: for a given shape $\bar{n}_{\mathrm{mean}}$ (specified over the heliospheric volume of interest), and for a given set of g-level and V_{IPS} observations, find the functions γ (or equivalently n) and V that best fit these observations.

Several points can be made about Eq. (10) and Eq. (11):

1. g^2 is a more 'natural' observational quantity than g itself. Both g^2 and V_{IPS} are described in terms of very similar integrals, with $\gamma(s)$ and $V(s)\sin\chi(s)$ specifying the contribution of a line-of-sight segment at distance s from the observer to the observed g-level and IPS speed, respectively.

2. The numerator and denominator for g and $V_{\rm IPS}$ have the same dependence on heliocentric distance; thus $V_{\rm IPS}$ and also g (except with unusual radially structured solar wind) will be nearly independent of solar elongation.
3. The influence of both $n_{\rm mean}$ and βn has been reduced to determining the density scale (Eq. 7) together with the unknown constants δn_0 and $\delta n_0^{\rm mean}$. The solution for γ and V depends explicitly on βr, which controls the weighting for the line-of-sight contribution $\gamma(s)$ and $V(s)\sin\chi(s)$.
4. For a given shape $\bar{n}_{\rm mean}$, if the pair of functions $\hat{n}$, V is the solution for a specific $n_{\rm mean}$, then the pair $\alpha\hat{n}, V$ is the solution for $\alpha n_{\rm mean}$ for any constant α.

The last point implies that for a given set of g-level and $V_{\rm IPS}$ observations, the solution for the normalized density $\hat{n}$ is determined only up to a constant, i.e. the absolute density scale of the solution cannot be determined from the tomographic reconstruction itself. This is not surprising: the g-level, our proxy for the solar wind density, is only a relative statement about the state of the solar wind along the line of sight to the IPS source as compared with 'average' conditions (Eq. 1). What these 'average' conditions are must be established using external information. For instance, the density scale can be calibrated against solar wind densities observed *in situ* at 1 AU near Earth. This external calibration defines the relationship between γ and n in Eqs. (7) and (8), and hence also implicitly defines the constants δn_0 and $\delta n_0^{\rm mean}$.

Unlike the g-level, Thomson-scattering (Eq. 12) changes in brightness determine the absolute scale of electron density variations $\Delta n(s)$. However, in our tomographic algorithm, the total density is needed in the solar wind model. There is a difficulty with Thomson-scattering brightness in determining the ambient density because only a tiny percentage ($\leq$1%) of the total brightness of the heliosphere arises from sunlight scattering off ambient solar wind. The far larger fraction is due to the zodiacal light, scattering from interplanetary dust. If we divide the solar wind density into two components:

$$n(s) = n_{\rm quiet}(s) + \Delta n(s) \qquad (14)$$

then the differential Thomson scattering measurements provide information about the variable part $\Delta n(s)$. Because of this, the mean value for the ambient $n_{\rm quiet}(s)$ needs to be established externally and determined over s, the distance from the observer. This is usually accomplished using *in situ* density measured from the mean solar wind value at 1 AU throughout the Carrington rotation (CR) period of the observation, and mapped everywhere in the heliosphere assuming a radial density fall-off. Finally it is assumed that, at the distance of 1 AU, solar wind speeds provide a constant mass flux throughout a CR regardless of heliographic latitude or longitude. With this modification to the analysis, Thomson scattering intensity can replace g-levels in the tomographic reconstruction. With this caveat, the Thomson-scattering equations and the SMEI results derived from them are determined in a far more straightforward manner and with far fewer assumptions than in the case of the IPS analysis.

2.2. *The numerical algorithm*

The reconstruction task has been reduced to finding the functions γ (or n) and V for a given set of g-level, $V_{\rm IPS}$, or B observations satisfying Eqs. (10), (11), or (12). The IPS normalized density $\hat{n}$ follows from γ using Eqs. (7) and (8), or in the case where Thomson scattering brightness is measured, Eq. (14) and a radial fall-off of density currently provides $\hat{n}$ for the determination of $V_{\rm IPS}$.

2.2.1. *Notation: use of subscripts*

In the following, subscripts i, j, k, and l are used when a quantity refers to the 3D heliospheric grid (next subsection) in computing the heliospheric γ-function and velocity, and quantities at the source surface (when only i, j, and l appear). The subscript μ is used to identify a line of sight while ν refers to a segment at a certain distance from the observer along the line of sight.

2.2.2. *Remote-sensing observations*

The reconstruction is based on a set of $N_{\rm obs}$ line-of-sight observations for g-level, IPS velocity $V_{\rm IPS}$, and Thomson-scattering brightness B. Because there are far more lines of sight from SMEI observations than from the IPS, the heliospheric resolutions obtained from the analyses with the greatest numbers of lines of sight are presumed to dominate, and for this reason there are two different programs used in the reconstructions; one with spatial and temporal resolutions specific for the highest numbers of SMEI lines of sight, and another program that operates at the resolutions of the IPS data set:

$$(g^2)^{\text{obs}}_{\mu}, V^{\text{obs}}_{\text{IPS},\mu}, B^{\text{obs}}_{\mu} \quad (\mu = 1, N_{\text{obs}}) \tag{15}$$

"Model values" are calculated from the latest iterative 3D model of heliospheric γ-function and velocity V, and are compared with the actual observations. The purpose of the reconstruction is to create a model 3D heliosphere that matches these observations as closely as possible. The model values are given by

$$(g^2)^{\text{mdl}}_{\mu}, V^{\text{mdl}}_{\text{IPS},\mu}, B^{\text{mdl}}_{\mu} \quad (\mu = 1, N_{\text{obs}}) \tag{16}$$

Each line of sight is subdivided into N_{los} segments of length ds_{μ}. The segment length is expressed as a constant in units of the Sun-Earth distance. Since for Earth-based observations this distance varies with time, this means that the segment length is slightly different for each line of sight. The distance from observer to the center of each line segment is given by:

$$s_{\nu,\mu} = (\nu - 0.5)ds_{\mu} \quad (v = 1, N_{\text{los}}; \mu = 1, N_{\text{obs}}) \tag{17}$$

Values of line-of-sight segments are normalized to the spatial longitudinal resolution (λ_R) of the reconstructions in degrees such that $ds_{\mu} = \lambda_R \times r/200$ for $\lambda_R \leq 10°$ and $N_{\text{los}} = 2r/ds_{\mu}$, so that for an observer at Earth ($r = 1\,\text{AU}$) each line of sight extends out about 2 AU.

2.2.3. *The solar wind plasma*

The grid used in the reconstruction is regular in heliographic longitude, heliographic latitude, heliocentric distance, and time, and is fixed relative to the Sun (i.e. rotates with the sidereal solar rotation rate, P_{sid}). The range of longitudes covered by the grid (360°, i.e. a whole solar rotation), is expressed in terms of the 'Carrington variable', c. One unit in a Carrington variable covers 360° in heliographic longitude. The integer part, int(c), is a conventional CR number, and effectively sets the time period of interest. The fraction is related to the heliographic longitude, $\lambda = 360° \times \{1 - (c - \text{int}(c))\}$.

The range of heliographic longitudes for the grid is set by a start and end "Carrington variable": c_{beg} and c_{end}. Note that the grid does not need to start at 0°, i.e. at the exact start of a new CR. Associated with the variables c_{beg} and c_{end} are the times, t_{beg} and t_{end} at which the corresponding heliographic longitude crossed the center of the solar disk as seen from Earth (or, more general, "the observer"). These times determine which observations are used for the reconstruction. E.g. all g-level, V_{IPS}, and brightness observations inside the time interval $[t_{\mathrm{beg}} - t_{\mathrm{marg}}, t_{\mathrm{end}} + t_{\mathrm{marg}}]$ are used. The "safety margin" on each end of the time interval is used to include all observations that can impact the time period of interest.

The latitude grid covers the full range $-90°$ to $+90°$. The radial grid covers the range from the "source surface" at R_s to some outer boundary at R_{max}.

$$\begin{aligned}
c_i &= c_{\mathrm{beg}} + (i-1) \times dC \quad dC = (c_{\mathrm{end}} - c_{\mathrm{beg}})/(N_c - 1) \ \ (i = 1, Nc)\\
l_j &= -\pi/2 + (j-1) \times dL \quad dL = \pi/(N_l - 1) \ \ (j = 1, N_l)\\
r_k &= R_s + (k-1) \times dR \quad dR = (R_{\mathrm{max}} - R_s)/(N_r - 1) \ \ (k = 1, N_r)\\
t_l &= (t_{\mathrm{beg}} - t_{\mathrm{marg}}) + (l-1) \times dT \quad dT = (t_{\mathrm{end}} - t_{\mathrm{beg}} + 2t_{\mathrm{marg}})/\\
&\qquad\qquad\qquad\qquad\qquad\qquad\qquad (N_t - 1) \ \ (l = 1, N_t)
\end{aligned} \tag{18}$$

Typical values for the SMEI grid parameters are $R_s = 15$ solar radii, $R_{\mathrm{max}} = 3\,\mathrm{AU}$, $dR = 0.1\,\mathrm{AU}$, $dC = 1/54$ (i.e. $6.67°$ in heliographic longitude), $dL = 6.67°$, and $dT = 0.5$.

The tomography reconstructs the 3D heliospheric γ-function (and its associated normalized density $\hat{n}$, or n) and velocity V in grid points of Eq. (18):

$$\gamma_{i,j,k,l} \text{ or } n_{i,j,k,l}, V_{i,j,k,l} (i = 1, N_c; j = 1, N_l; k = 1, N_r; l = 1, N_t) \tag{19}$$

The lower boundary of the heliospheric grid ($k = 1$), the "source surface", plays a crucial role in the tomographic reconstruction. The γ-function or n and velocity at the source surface are simply the above equation evaluated only for $k = 1$, but with the other indices evaluated over the full range of values. Specifying the content of these two maps at every time t_l initializes the reconstruction; each iteration is completed by updating them.

2.2.4. *Discretization of the line-of-sight integrals*

The calculation of model g-level and IPS velocity V_{IPS}, observations requires an integration through the model heliosphere along the same directions (lines of sight) as the actual observations (Eqs. 10, 11, and 12). In discrete form this becomes:

$$(g^2)^{\mathrm{mdl}}_{\mu} = \sum_{\nu=1}^{N_{\mathrm{los}}} H_{\nu,\mu}\gamma_{\nu,\mu} \quad \text{or} \quad (n)^{\mathrm{mdl}}_{\mu} = \sum_{\nu=1}^{N_{\mathrm{los}}} H_{\nu,\mu}\gamma_{\nu,\mu} \tag{20}$$

with weight factors

$$H_{\nu,\mu} = \sigma_{\nu,\mu} / \sum_{\nu=1}^{N_{\mathrm{los}}} \sigma_{\nu,\mu}, \tag{21}$$

and

$$V^{\mathrm{mdl}}_{\mathrm{IPS},\mu} = \sum_{\nu=1}^{N_{\mathrm{los}}} K_{\nu,\mu} V_{\nu,\mu} \sin\chi_{\nu,\mu} \tag{22}$$

with weight factors

$$K_{\nu,\mu} = \sigma_{\nu,\mu}\gamma_{\nu,\mu} / \sum_{\nu=1}^{N_{\mathrm{los}}} \sigma_{\nu,\mu}\gamma_{\nu,\mu}, \tag{23}$$

and where

$$\sigma_{\nu,\mu} = \sigma(s_{\nu,\mu}) = \rho(s_{\nu,\mu}) r_{\nu,\mu}^{-2(2-\beta r)}. \tag{24}$$

The μ-dependence of $s_{\nu,\mu}$ enters through the μ-dependence of ds_{μ} (Eq. 17). The function $\rho(s)$ does not depend on the line-of-sight orientation (i.e. the elongation), but does depend on the distance along the line of sight.

The heliospheric γ or n function and velocity in all line-of-sight segments

$$\gamma_{\nu,\mu}, \text{ or } n_{\nu,\mu}, V_{\nu,\mu} \quad (\nu = 1, N_{\mathrm{los}}; \mu = 1, N_{\mathrm{obs}}) \tag{25}$$

are obtained by linear interpolation on the 4D γ or n function and velocity (Eq. 19) at the center positions (Eq. 17) of all line-of-sight segments.

2.3. *The iterative process*

The iterative process is started by specifying γ or n function values and velocity V at the 3D (spatial, time) source surface. Using these source surface values the 4D γ function values and velocities (Eq. 19) in the heliospheric grid (Eq. 18) are obtained by applying a solar wind model for the propagation of mass from the source surface out into the heliosphere.

We assume radial outflow and apply simple kinematic arguments to conserve mass and mass flux (Section 2.3.1). At this stage also "traceback" information is accumulated which connects each heliospheric grid point to its "source location" at the source surface.

Model line-of-sight observations (Eq. 16) are calculated by integrating along the appropriate directions through the 4D heliosphere, and include taking into account the light travel time of photons originating at different distances from the observer. These model values are compared with the actual observations (Eq. 15). This comparison provides the main convergence criterion for the iterative process. The observed-to-modeled ratios for all lines of sight will be used to determine the source surface update, completing the iteration.

All line-of-sight segments are projected back to the source surface using the "traceback" information, carrying along the observed-to-modeled ratio of the line of sight they belong to. At the source surface all the segments of all lines of sight are assigned to the nearest grid point. The γ-function or n and velocity in the grid point are then updated by combining observed-to-modeled ratios of all line of segments assigned to the grid point.

We now follow the main steps in this process in detail.

2.3.1. *The kinematic solar wind model*

Given are the γ function values and velocities on the source surface at heliocentric distance $r_0 = R_s$. From these we need to reconstruct the γ function values and velocities at "higher levels", i.e. heliocentric distances r_k $(k = 2, N_l)$. The problem is solved by induction.

Given the γ values and velocities at level k are available, the γ values and velocity at level $k + 1$ need to be determined. The connection between the levels is established using simple kinematic arguments based on conservation laws. Currently we use conservation of mass and mass flux (though other choices, such as conservation of momentum, are easily implemented). Each grid point i, j, l on level k (Eq. 19) represents a parcel of mass with a normalized density $\tilde{n}_{ij,k,l}$ (related to $\gamma_{i,j,k,l}$ through Eqs. 7 and 8) and velocity $V_{i,j,k,l}$. The parcel of mass is assumed to travel radially outward at the local speed. When it arrives at level $k + 1$ at a later time it has a normalized density $\hat{n}_{i,j,k,l}$ and velocity $\hat{V}_{i,j,k,l}$. The conservation laws for the parcels of material (mass and mass flux, respectively) are:

$$\tilde{n}_{i,j,k,l} = \hat{n}_{i,j,k,l}; \quad \tilde{n}_{i,j,k,l} V_{i,j,k,l} = \hat{n}_{i,j,k,l} \hat{V}_{i,j,k,l} \tag{26}$$

In the corotating heliographic coordinate system the parcel will have moved to a larger Carrington variable (smaller heliographic longitude). The parcel will arrive at level $k+1$ at Carrington variable:

$$\hat{c}_{i,j,k,l} = c_{i,j,k,l} + dR/V_{i,j,k,l} P_{\rm sid} \tag{27}$$

at time

$$\hat{t}_{i,j,k,l} = t_{i,j,k,l} + dR/V_{i,j,k,l} \tag{28}$$

where $P_{\rm sid}$ is the sidereal rotation period of the Sun where $dR = r_{k+1} - r_k$. Note that the parcel only shifts in longitude and time, not in heliographic latitude using the current conservation laws. The position $\hat{c}_{i,j,k,l}$ and time $\hat{t}_{i,j,k,l}$ place the parcel somewhere in between grid points (in space and time) at level $k+1$. Here we will only discuss longitude (and its increment i), but in fact a similar interpolation is also made in time. Let the new grid points be $(i_{\rm near}, j)$ and $(i_{\rm far}, j)$, where $|i_{\rm near} - i_{\rm far}| = 1$. Let grid point $(i_{\rm near}, j)$ be closest to $\hat{c}_{i,j,k,l}$, and define the difference in Carrington variable $\delta\hat{c}_{i,j,k,l} = |\hat{c}_{i,j,k,l} - \hat{c}_{i{\rm near},j,k,l}|$. At level k+1 each parcel of mass is split up in two fractions, which are assigned to the neighboring grid points $(i_{\rm near}, j)$ and $(i_{\rm far}, j)$. A fraction $f_{\rm near} = 1 - \delta c_{i,j,k,l}$ is assigned to $(i_{\rm near}, j)$, and the remaining fraction $f_{\rm far} = \delta c_{i,j,k,l}$ is assigned to $(i_{\rm far}, j)$. The normalized density and velocity at each grid point (i, j, l), at level $k+1$ are obtained by averaging all nearest parcel fraction values weighted according to the fraction, in space and time, assigned to a given grid point.

The displacements $\hat{c}_{i,j,k,l} - c_{i,j,k,l}$ and $\hat{t}_{i,j,k,l} - t_{i,j,k,l}$ from Eq. (27) and Eq. (28) respectively are used to construct a "traceback matrix" S^c and S^t that connects each heliospheric grid point in space and time to its origin at the source surface (i.e. the point on the source surface which lies on the same stream line as the grid point), and for the Carrington variable this is:

$$c^{\rm source}_{i,j,k,l} = c_{i,j,k,l} + S^c_{i,j,k,l} \tag{29}$$

For radial outward flow (currently assumed), the value of $S^t_{i,j,k,l}$ is simply the time difference between the Carrington variable at the upper grid point and the Carrington variable at the source surface. This 'traceback matrix' is needed in the final phase of the iteration to project the line-of-sight observations back to the source surface (Section 2.3.3).

2.3.2. *Convergence criterion and rejection of outliers*

Improving the model from one iteration to the next is based on a comparison of model observations (Eq. 16) and actual observations (Eq. 15). Error estimates for g-levels, IPS velocities $V_{\rm IPS}$, and brightness B are defined as:

$$\sigma_g^2 = N_{\rm obs}^{-1} \sum_{\mu=1}^{N_{\rm obs}} (\tau_\mu^g - 1)^2 \tag{30}$$

$$\sigma_{V_{\rm IPS}}^2 = N_{\rm obs}^{-1} \sum_{\mu=1}^{N_{\rm obs}} (\tau_\mu^{V_{\rm IPS}} - 1)^2 \tag{31}$$

$$\sigma_B^2 = N_{\rm obs}^{-1} \sum_{\mu=1}^{N_{\rm obs}} (\tau_\mu^B - 1)^2 \tag{32}$$

and where we define the ratios of observed and model values:

$$\tau_\mu^g = (g^2)_\mu^{\rm obs}/(g^2)_\mu^{\rm mdl} \tag{33}$$

$$\tau_\mu^{V_{\rm IPS}} = V_{{\rm IPS},\mu}^{\rm obs}/V_{{\rm IPS},\mu}^{\rm mdl} \tag{34}$$

$$\tau_\mu^B = B_\mu^{\rm obs}/B_\mu^{\rm mdl} \tag{35}$$

These quantities (Eqs. 30–32) give estimates of the relative deviation of model values and actual observations, and are the best convergence criteria available. These should move closer to zero from iteration to iteration.

The relative differences of model and actual observations for individual lines of sight (normalized to the "average deviation" for all sources; Eqs. 30–32)

$$\delta g_\mu = (\tau_\mu^g - 1)/\sigma_g \tag{36}$$

$$\delta V_{{\rm IPS},\mu} = (\tau_\mu^{V_{\rm IPS}} - 1)/\sigma_{V_{\rm IPS}} \tag{37}$$

$$\delta B_\mu = (\tau_\mu^B - 1)/\sigma_B \tag{38}$$

are used as a criterion to flag individual observations as bad. If after a specified number of iterations the relative difference for an observation is above a certain threshold (typically set to 3 for g-level, $V_{\rm IPS}$, and B observations) this is used to justify throwing out the data point. The iterative process is then restarted with these "outlier" data points removed.

2.3.3. *Projection to the source surface*

To finish the iteration, the results from Section 2.3.2. need to be translated to the source surface. Let the heliographic coordinates of the line-of-sight segments (Carrington variable, heliographic latitude, heliocentric distance, and time respectively) be:

$$c_{v,\mu}, l_{v,\mu}, r_{v,\mu}, t_{v,\mu}. \tag{39}$$

This location is 'traced back' to the source surface using the "traceback matrix" S (from Eqs. 28 and 29). The traceback value at the line-of-sight segment $S_{v,\mu}$ is calculated from a linear interpolation on neighboring heliospheric grid points. The source location is then defined by:

$$c_{\nu,\mu}^{\text{source}} = c_{\nu,\mu} + S_{\nu,\mu}^{c}; \quad l_{\nu,\mu}^{\text{source}} = l_{\nu,\mu}; \quad r_{\nu,\mu}^{\text{source}} = R_s; \quad t_{\nu,\mu}^{\text{source}} = t_{\nu,\mu} + S_{\nu,\mu}^{t}. \tag{40}$$

2.3.4. *Source surface updates*

The projected locations (Eq. 40) of all line-of-sight segments will be scattered across the source surface. Figure 6 shows samples of these

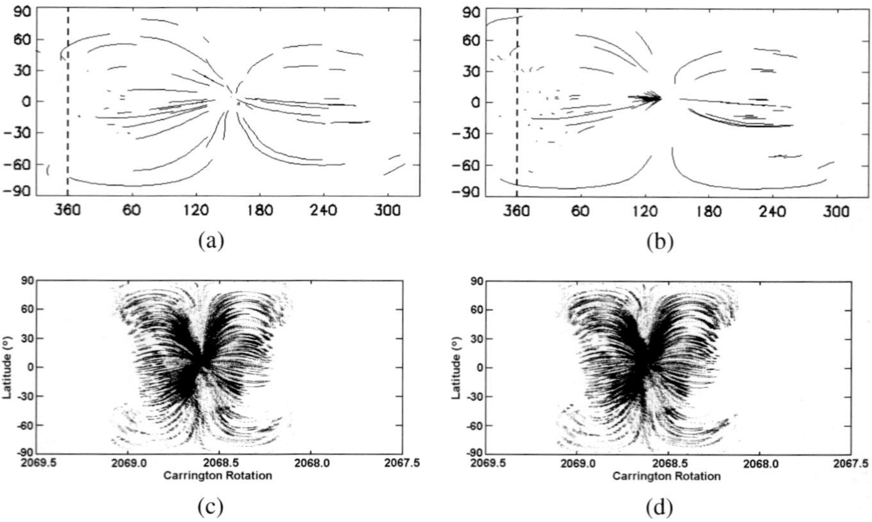

Fig. 6. Samples from the time-dependent analysis of line-of-sight segments projected onto a Carrington map. Lines of sight begun in one map often complete in another (a) and (b) STELab velocity line-of-sight projections on 14 and 15 July 2000 shown for one 360° Carrington rotation. (c) and (d) SMEI line-of-sight projections for two half-day intervals in the beginning of April, 2008.

projected line-of-sight segments. We show adjacent periods of the IPS 3D reconstructions in Fig. 6a and 6b, and two adjacent periods from the SMEI 3D reconstructions in Fig. 6c and 6d. For each grid element (i, j, l) at the source surface, all segments located within one half grid spacing and time of the grid element are collected:

$$\begin{aligned} c_{i,j,l} - dC/2 \le c^{\text{source}}_{\nu,\mu} \le c_{i,j,l} + dC/2; \quad l_{i,j,l} - dL/2 \le l^{\text{source}}_{\nu,\mu} \\ \le l_{i,j,l} + dL/2; \quad \text{and} \quad t_{i,j,l} - dT/2 \le t^{\text{source}}_{\nu,\mu} \le t_{i,j,l} + dT/2. \end{aligned} \tag{41}$$

The ratios in Eqs. 33–35 for these line-of-sight segments are then used to update the source surface γ function or n and velocity. The group of segments near bin (i, j, l) is defined by a group of pairs:

$$(\nu_a, \mu_a), \quad a = 1, N^{\text{segments}}_{i,j,l} \tag{42}$$

The ratios for this group of segments are combined in a weighted mean to obtain a correction factor to the source surface values:

$$\begin{aligned} \gamma^{\text{new}}_{i,j,l}/\gamma^{\text{old}}_{i,j,l} &= \sum_{a=1}^{N^{\text{segments}}_{i,j,l}} H_{\nu_a,\mu_a} \tau^g_{\mu_a} \Big/ \sum_{a=1}^{N^{\text{segments}}_{i,j,l}} H_{\nu_a,\mu_a} \\ n^{\text{new}}_{i,j,l}/n^{\text{old}}_{i,j,l} &= \sum_{a=1}^{N^{\text{segments}}_{i,j,l}} H_{\nu_a,\mu_a} \tau^n_{\mu_a} \Big/ \sum_{a=1}^{N^{\text{segments}}_{i,j,l}} H_{\nu_a,\mu_a} \\ V^{\text{new}}_{i,j,l}/V^{\text{old}}_{i,j,l} &= \sum_{a=1}^{N^{\text{segments}}_{i,j,l}} K_{\nu_a,\mu_a} \tau^V_{\mu_a} \Big/ \sum_{a=1}^{N^{\text{segments}}_{i,j,l}} K_{\nu_a,\mu_a} \end{aligned} \tag{43}$$

Thus, for each segment included in the sum the correction factor τ is weighted proportional to the weight it had in the calculation of the model observation.

Before continuing with the next iteration the new source-surface values are smoothed by applying a circular spatial (relative to a grid point on the source surface), and a temporal Gaussian filter as mentioned earlier across the entire source surface[4,5,7]. The purpose of this is mainly to 'dampen' the solution, and thereby improve the stability of the iteration process.

3. Analysis Examples

SOHO/LASCO C3 coronagraph images of 27 and 28 May 2003 show several halo CMEs with considerable structure and mass combined with background coronal features. The first of these events moves out slowly such that the latter, having started about 18 hours later but having over twice the outward expansion rate, has caught up with the first on their way to Earth.

The 27–28 May 2003 CME sequence was one of the first recognized in direct SMEI images as a halo event observed in the interplanetary medium.[2,22] The SMEI direct images contain numerous artifacts; to remove these effectively for observing ICMEs takes considerable effort (see, e.g. Jackson *et al.*[2,7]) for a more detailed explanation of how the removal of many of the artifacts present in the SMEI data is accomplished in order to prepare them for the 3D reconstruction analysis). Figure 7 shows a sky map presentation from the 27–28 May 2003 CME sequence. Figure 7a shows where the lines of sight within ±1.5 hours of the time period of the Hammer-Aitoff (full sky) map presentation are placed to produce the reconstruction coverage. 7b and 7c present a sequence of 3D-reconstructed fisheye sky maps as in Jackson *et al.*[7] and show the progression of the 27–28 May 2003 ICMEs outward in 2D sky maps. The sky maps are derived from the 3D reconstructed density volumes that have been fit to the observed SMEI data as described in Section 2. The times given in Fig. 7 are instantaneous times interpolated to the appropriate value from the (a) reconstruction analyses. The brightness includes the r^{-2} heliospheric density fall-off, and thus provides brightness as observed in direct images. The sky maps from the tomographic modeling have a $3.3\,\mathrm{e^-\,cm^{-3}}$ base at 1 AU removed from them that is also modeled with an r^{-2} heliospheric density fall-off (see text Section 2.1.), and show excursions from a mean of zero, in order to match direct images from SMEI, where all but the changing brightness has been removed. Since the fit to a heliospheric solar wind model uses time series from multiple sky maps, and reduces signals that do not participate in the outward progression of the solar wind, gaps or bad spots in the data are filled in to provide continuous inner-heliospheric coverage. For most of the time period covered by the reconstructed sky maps, the area blanked out near the center of the direct sky maps of Fig. 7 is located in approximately the same place. This region is back-filled somewhat from the lines of sight that exist at larger elongations, and more at some position angles than others to a circle set at 18° from the Sun. Since few non-heliospheric artifacts such as zodiacal light and aurora remain in these maps constructed from volumes from the edited time series, the image contrast of

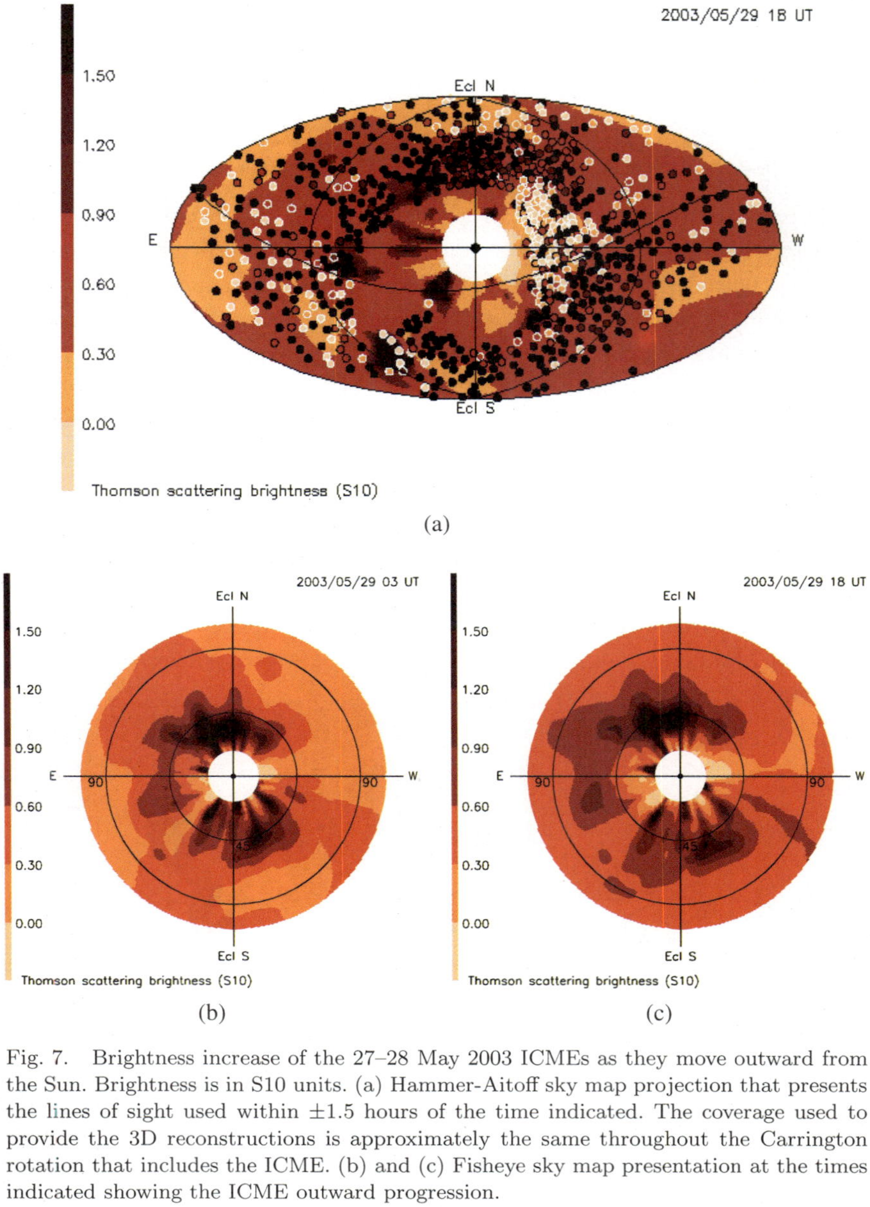

(a)

(b) (c)

Fig. 7. Brightness increase of the 27–28 May 2003 ICMEs as they move outward from the Sun. Brightness is in S10 units. (a) Hammer-Aitoff sky map projection that presents the lines of sight used within ±1.5 hours of the time indicated. The coverage used to provide the 3D reconstructions is approximately the same throughout the Carrington rotation that includes the ICME. (b) and (c) Fisheye sky map presentation at the times indicated showing the ICME outward progression.

the map can be enhanced above that of direct sky maps, thus highlighting more detail.

4. Remote-Observer Views and *in-situ* Comparisons

The 3D-reconstructed volumes can also be viewed as a remote observer would view them or be used to determine a mass within a specific region of the reconstruction volume as in Fig. 8. Perhaps more pertinent are planar cuts through the volumetric data and *in-situ* comparisons. Figure 9 gives one such example for the 27–28 May 2003 ICME sequence in the SMEI data, and also shows the 3D-reconstruction of the associated velocity during this period. Regions that are accessed by fewer than ten lines of sight per resolution element in the SMEI analysis are left blank, and these locations are carried over to the much lower resolution volume elements in the velocity data unless they too have fewer than the required number, at which location they are also left blank. Both 3D-reconstructed velocity and density are combined to determine ram pressure for the ICME highlighted in this example, and are shown as the primary example of these measurements available at Mars and compared with Mars Global Surveyor magnetic field measurements[9].

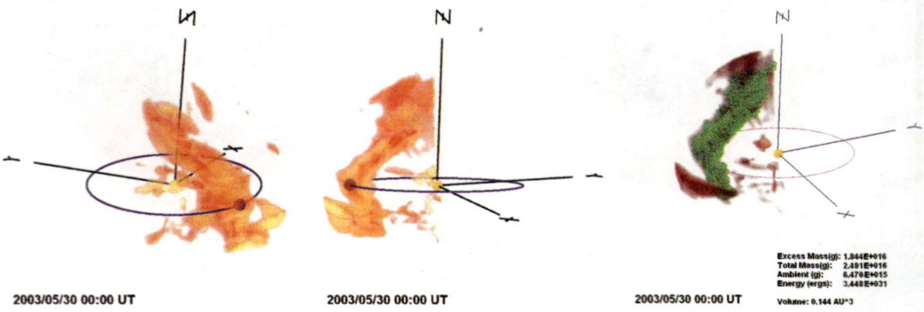

(a) (b) (c)

Fig. 8. 3D mass determination for the 27-28 May 2003 halo CME sequence as the associated ICMEs reach Earth vicinity at 00:00 UT 30 May 2003. Electron density is contoured upward from $15\,\mathrm{cm}^{-3}$. (a) The ICME observed from about 45° East of the Sun-Earth line and 20° above the ecliptic plane, just as the event sequence begins to engulf the Earth. (b) The ICME observed from about 70° West of the Sun-Earth line and 5° above the ecliptic plane. (c) The total event now highlighted and filled with cubes from approximately this same perspective, has a volume of $0.144\,\mathrm{AU}^3$ above this contour level, a total mass of 2.49×10^{16} g, and is comprised of several halo CMEs. (see Jackson *et al.*[7]).

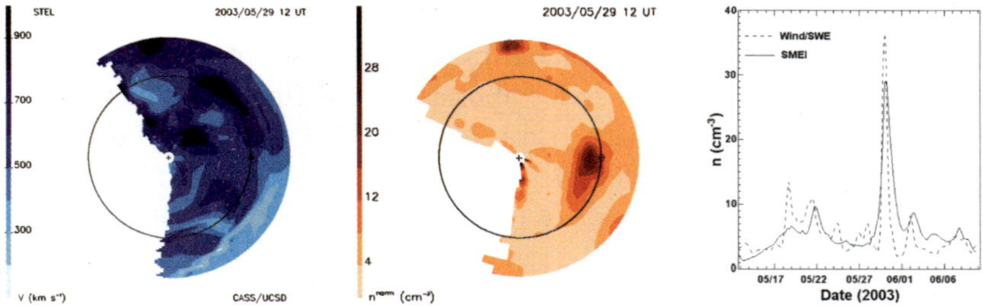

(a) (b) (c)

Fig. 9. 3D reconstruction of the 27–28 May 2003 halo ICME sequence of events as it reaches Earth displayed as an ecliptic-plane cut of the reconstruction at the time indicated, as viewed down from the North. The Sun is at the center marked with a +, an ellipse marks the Earth's orbital path in the ecliptic, and the Earth is shown by ⊕. (a) A velocity cut at the time indicated. There is high velocity to the West of the Sun-Earth line associated with the CME sequence. (b) A density cut at the same time as in (a). The density scale is given to the left. To best display the structure, an r^{-2} density fall-off has been removed to scale densities to 1 AU. The main structure at Earth is associated with the halo CME sequence observed by LASCO on 27–28 May 2003, and shows that the density enhancement of the ICME hitting Earth in this event is more extensive to the East of the Earth than to the West (from Jackson *et al.*[7]). (c) Time-series plot of the density at Earth from the reconstruction and from Wind proton observations for the whole Carrington rotation that includes the ICME. The latter *in situ* observations are combined into 12-hour averages matching the temporal and spatial resolutions of the SMEI reconstruction. The correlation coefficient within 6 days of the ICME passage is 0.86[7].

5. Conclusions

We have described the mathematical analyses in the current UCSD 3D-reconstruction algorithm, and given examples of some of the recent measurements and presentations that they provide. The time-dependent analysis programs using IPS data from STELab and Thomson-scattering brightness data from SMEI now reside at the CCMC for use by others. This article serves as an explanation for these analyses, and some of the caveats present in using them. The actual Fortran programs used in these analyses that are here described mathematically are available from both the CCMC and the UCSD Website at: *http://cass.ucsd.edu/solar/resources.html#resources*.

Acknowledgments

We thank M. Kojima for his encouragement throughout the length of this project and for originally making available IPS data for these analyses under the auspices of a joint CASS/UCSD-STELab cooperative agreement. SMEI was designed and constructed by a team of scientists and engineers from the US Air Force Research Laboratory, the University of California at San Diego, Boston College, Boston University, and the University of Birmingham, UK. The authors also wish to thank the Wind/SWE group for use of their data. The work of B.V. Jackson, A. Buffington, P.P. Hick, and J.M. Clover was supported at the University of California at San Diego by NSF grants ATM-0331513 and ATM- 0852246 and NASA grants NAG5-134543 and NNG05GG45G. In addition, M.M. Bisi and B.V. Jackson were supported in part by AFOSR grant FA9550-06-1-0107.

References

1. C. J. Eyles, G. M. Simnett, M. P. Cooke, B. V. Jackson, A. Buffington, P. P. Hick, N. R. Waltham, J. M. King, P. A. Anderson and P. E. Holladay, The Solar Mass Ejection Imager (SMEI), *Solar Phys.* **217** (2003) 319.
2. B. V. Jackson, A. Buffington, P. P. Hick, R. C. Altrock, S. Figueroa, P. Holladay, J. C. Johnston, S. W. Kahler, J. Mozer, S. Price, R. R. Radick, R. Sagalyn, D. Sinclair, G. M. Simnett, C. J. Eyles, M. P. Cooke, S. J. Tappin, T. Kuchar, D. Mizumo, D. F. Webb, P. Anderson, S. L. Keil, R. Gold and N. R. Waltham, The Solar Mass Ejection Imager (SMEI) mission, *Solar Phys.* **225** (2004) 177.
3. A. Buffington, J. S. Morrill, P. P. Hick, R. A. Howard, B. V. Jackson and D. F. Webb, Analyses of the comparative responses of SMEI and LASCO, *Proc. SPIE* **6689** 66890B (2007) 1–6, doi:10.1117/12.734658.
4. B. V. Jackson, A. Buffington, P. P. Hick, X. Wang and D. Webb, Preliminary three-dimensional analysis of the heliospheric response to the 28 October 2003 CME using SMEI white-light observations, *J. Geophys. Res.* **111**(A4) (2006) A04S91.
5. B. V. Jackson, P. P. Hick, A. Buffington, M. M. Bisi, M. Kojima, and M. Tokumaru, Comparison of the extent and mass of CME events in the interplanetary medium using IPS and SMEI Thomson scattering observations, in *Astronomical and Astrophysical Transactions* edited by I. Chashei and V. Shishov, **26**(6) (2007) 477–487, doi:10.1080/10556790701612221.
6. M. Kojima and T. Kakinuma, Solar–cycle evolution of solar-wind speed structure between 1973 and 1985 observed with the interplanetary scintillation method, *J. Geophys. Res.* **92** (1987) 7269.
7. B. V. Jackson, M. M. Bisi, P. P. Hick, A. Buffington, J. M. Clover and W. Sun, Solar Mass Ejection Imager (SMEI) 3D reconstruction of the 27-28 May 2003 CME sequence, *J. Geophys Res.* **113** A00A15 (2008) doi:10.1029/2008JA013224.

8. M. Tokumaru, M. Kojima, K. Fujiki, M. Yamashita and B. V. Jackson, The source and propagation of the interplanetary disturbance associated with the full-halo coronal mass ejection on 2003 October 28, *J. Geophys. Res.* **112** A05106 (2007) doi:10.1029/2006JA012043.
9. B. V. Jackson, J. A. Boyer, P. P. Hick, A. Buffington, M. M. Bisi and D. H. Crider, Analysis of solar wind events using interplanetary scintillation (IPS) remote sensing 3D reconstructions and their comparison at Mars, *Solar Phys.* **241** (2007) 385.
10. G. E. Brueckner, R. A. Howard, M. J. Koomen, C. M. Korendyke, D. J. Michels, J. D. Moses, D. G. Socker, K. P. Dere, P. L. Lamy, A. Llebaria, M. V. Bout, R. Schwenn, G. M. Simnett, D. K. Bedford and C. J. Eyles, The Large Angle Spectroscopic Coronagraph (LASCO): visible light coronal imaging and spectroscopy, *Solar Phys.* **162** (1995) 357.
11. M. L. Kaiser, The STEREO mission: an overview, *Adv. Space Res.* **36** (2005) 1483.
12. B. V. Jackson, P. L. Hick, M. Kojima and A. Yokobe, Heliospheric tomography using interplanetary scintillation observations, 1. Combined Nagoya and Cambridge observations, *J. Geophys. Res.* **103** (1998) 12049.
13. K. Asai, M. Kojima, M. Tokumaru, A. Yokobe, B. V. Jackson, P. L. Hick and P. K. Manoharan, Heliospheric tomography using interplanetary scintillation observations, 3. Correlation between speed and electron density fluctuations in the solar wind, *J. Geophys. Res.* **103** (1998) 1991.
14. M. Kojima, M. Tokumaru, H. Watanabe, A. Yokobe, K. Asai, B. V. Jackson and P. L. Hick, Heliospheric tomography using interplanetary scintillation observations, 2. Latitude and heliocentric distance dependence of solar wind structure at 0.1-1 AU, *J. Geophys. Res.* **103** (1998) 1981.
15. P. P. Hick and B. V Jackson, Heliospheric tomography: an algorithm for the reconstruction of the 3D solar wind from remote sensing observations, *Proc. SPIE* **5171** (2004) 287–297.
16. B. V. Jackson, A. Buffington and P. P. Hick, A heliospheric imager for Solar Orbiter, in *Proc. of Solar Encounter: The First Solar Orbiter Workshop*, Puerto de la Cruz, Tenerife, Spain, 14–18 May 2001 (*ESA SP*) **493** (2001) 251.
17. M. M. Bisi, B. V. Jackson, P. P. Hick, A. Buffington and J. M. Clover, 3D reconstructions of the early November 2004 CDAW geomagnetic storms: preliminary analysis of STELab IPS speed and SMEI density, *J. Geophys Res. — Space Physics Special Edition — Geomagnetic Storms of Solar Cycle 23* **113** (2008) A00A11.
18. B. V. Jackson and P. P. Hick, Corotational tomography of heliospheric features using global Thomson scattering data, *Solar Phys.* **211** (2002) 344.
19. A. T. Young, Interpretation of interplanetary scintillation, *Astrophys J.* **168** (1971) 543.
20. B. V. Jackson, P. P. Hick, A. Buffington, M. Kojima, M. Tokumaru, K. Fujiki, T. Ohmi and M. Yamashita, Time-dependent tomography of heliospheric features using interplanetary scintillation (IPS) remote sensing observations, *Solar Wind 10 Conference Proceedings*, Pisa, June 17–21 (2002), pp. 31–34.

21. D. E. Billings, A guide to the solar corona, Academic, New York, (1966) 150.
22. S. J. Tappin, A. Buffington, M. P. Cooke, C. J. Eyles, P. P. Hick, P. E. Holladay, B. V. Jackson, J. C. Johnston, T. Kuchar, D. Mizuno, J. B. Mozer, S. Price, R. R. Radick, G. M. Simnett, D. Sinclair, N. R. Waltham and D. F. Webb, Tracking a major interplanetary disturbance with SMEI, *Geophys. Res. Lett.* **31** (2004) L02802.

Advances in Geosciences
Vol. 21: Solar & Terrestrial Science (2008)
Ed. Marc Duldig

INTERPLANETARY SHOCK GENERATED IONOSPHERIC TRAVELING CONVECTION VORTEX NEAR LOCAL NOON

X.-Y. ZHOU
Jet Propulsion Laboratory,
California Institute of Technology,
4800 Oak Grove Drive, Pasadena,
California 91109, USA
xiaoyan.zhou@jpl.nasa.gov

D. G. SIBECK
Goddard Space Flight Center,
Greenbelt, Maryland 20771, USA

O. AMM
Finnish Meteorological Institute,
Geophysical Research Division,
P.O. Box 503, FIN - 00101 Helsinki, Finland

J. M. RUOHONIEMI
Applied Physics Laboratory,
Johns Hopkins University, Johns Hopkins Rd,
Laurel, Maryland 20723, USA

Theoretical studies have shown that a pair of ionospheric vortices with opposite polarities can be created by interplanetary shock impingement. Field-aligned currents run into one vortex and out of the other, which couple the magnetosphere and ionosphere. Here we report such a vortex event that occurred on November 22, 1997. The ionospheric vortex was centered at ~67° magnetic latitude (MLAT) and ~13 magnetic local time (MLT), and traveled westward at a speed of $\sim 7\,\mathrm{km\,s^{-1}}$ with a lifetime of only 4–6 min. The ionospheric Hall currents suddenly increased from near zero to $\sim 350\,\mathrm{mA\,m^{1}}$ when the vortex occurred. A downward field-aligned current of $\sim 2\,\mu\mathrm{A\,m^{-2}}$ flowed into the center of the vortex based on the magnetograms. The SuperDARN radar provided supplementary observations that indicated a possible existence of another vortex that might have centered at higher latitude of ~73° MLAT on the auroral oval and at ~11 MLT. Observations shown in this paper are consistent with the theoretical prediction of shock generated vortices. In addition, we speculate that shock-generated detached proton auroral spots may be caused by the presence of downward field-aligned currents into the center of a vortex as well as electromagnetic ion cyclotron waves.

1. Introduction

The local noon sector (09–14 local time) of the magnetosphere and ionosphere responds to solar wind variations almost instantaneously. One of such occasions is the dayside auroral intensification caused by the arrival of fast forward interplanetary shocks and pressure pulses.[1–5] Other evidences include ionospheric travelling convection vortices (TCVs) (e.g. Araki,[6] Russell and Ginskey[7]). This paper reports an interplanetary shock caused TVC on closed magnetic field lines in the local noon sector. We also discuss in what circumstance the vortex may be related to dayside proton auroral spots. These auroral spots are located below the auroral oval and some are detached from the oval.

Convection vortices are intense swirls of ionospheric plasma motion. The convection has an opposite sign of the Hall current that can be detected by ground-based magnetometers. A convection vortex seen by one magnetometer has signatures of a bipolar variation in the H component and of a single positive (or negative) excursion in the D component. If these signatures are detected in a temporal sequence by several longitudinally (or latitudinally) separated magnetometers, the convection vortex is traveling longitudinally (or latitudinally) as well, therefore, is named a traveling convection vortex (TCV).

While it is still under study, causes of the ionospheric TCVs were found in the solar wind. Previous studies showed that TCVs can be driven by the solar wind pressure and interplanetary magnetic field (IMF) variations,[8,9] magnetic cavities,[10,11] or by local plasma nonuniformities in the magnetosphere.[12] These TCVs mainly occur on the auroral oval, the boundary between open and closed field lines. The peak occurrence is near 10 magnetic local time (MLT).[2–13] Lifetime of the TCVs is typically ∼5–15 min. They have amplitudes less than 150 nT and predominantly travel antisunward at speeds of 3–10 km s^{-1}.

For northward IMF conditions using east-west chain at subauroral latitudes around 55° geomagnetic, Russell and Ginskey[7] found that in contrast to TCVs discussed above, the ionospheric vortices generated by shock compressions appeared to extend deep into the magnetosphere and move tailward simultaneously on both dawn and dusk sides when shocks propagated down tail. Engebretson *et al.*[14] found that in the southern and northern polar caps vortices generated at the sudden commencement (SC) propagated poleward at a very high speed (20–30 km s^{-1}) and had a very short lifetime (∼1 min). The vortices centered at ∼67° magnetic latitude (MLAT), ∼10 MLT and ∼71° MLAT, ∼08 MLT were explained as a result of the imposition of a pair of field-aligned currents (FACs). Moretto *et al.*[15] studied a weak interplanetary shock generated effect and found that two vortices were generated in the polar cap at ~09 and 15 MLT near 80° MLAT. The lifetime was only ~4 min during which no vortex traveling was noticeable. They commented that TCV is usually a quiet-time high-latitude phenomenon, so ionospheric vortices generated by interplanetary shocks may not be TCV.

2. Theoretical Prediction

Theories predict that a pair of ionospheric vortices of opposite polarizations can be created during the magnetosphere compression. Both enhanced magnetic reconnection on the magnetopause and the coupling with compressional waves can generate transverse Alfvén waves that carry field-aligned currents into the ionosphere.[6,16–20] Due to shock compressions, a sharp gradient in the magnetospheric density is generated in the region between the compressed outer magnetosphere and the uncompressed inner magnetosphere. FACs are generated in this area because of the plasma nonuniformity[21–23] that can be identified as a sharp gradient in the Alfvén speed.[24] An intense shock can move the dayside magnetopause inside geosynchronous orbit. So the FACs can be generated on the L-shells below the auroral oval. Since the compression is the most intense near local noon, but gradually weakens elsewhere, so a pair of FAC as shown in Fig. 1 is

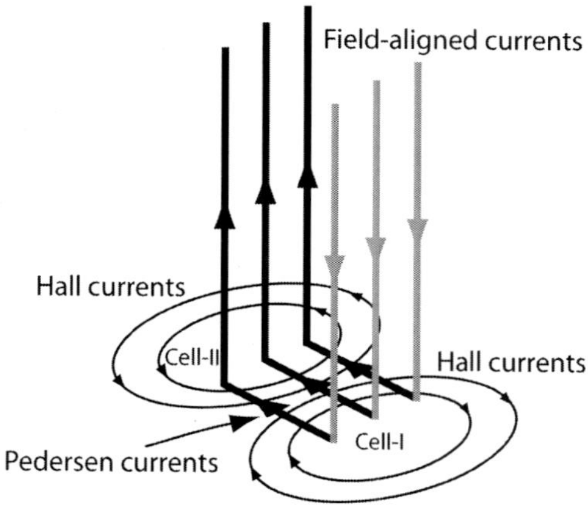

Fig. 1. A schematic of local noon ionospheric twin vortices and a pair of field-aligned currents generated by an interplanetary shock impingement (after Kivelson and Southwood[17]).

expected to occur near local noon. As illustrated in Fig. 1, the FACs flow downward into one vortex (Cell-I) upward out of the ionosphere from the other vortex (Cell-II) and are closed via Pedersen currents in the ionosphere.

This paper studies a shock event on November 22, 1997 to examine the scenario in Fig. 1. We used solar wind observations from the Wind spacecraft for the shock identification, the ground-based magnetograms from the IMAGE chain and SuperDARN convection maps for the ionospheric TCVs. It is found that the magnetospheric and ionospheric response to the shock compression is well consistent with this scenario.

3. Observations

3.1. *Wind and polar observations*

While located at GSM $(\mathrm{x,y,z}) = (181, -5, 32)\ \mathrm{R}_E$, Wind detected a fast forward interplanetary shock at 0912 UT on November 22, 1997, which is shown by the vertical dashed line in Fig. 2. At the shock, the IMF magnitude, the solar wind thermal velocity, the plasma density and the solar wind speed increased simultaneously. The solar wind ram pressure ($P_{ram} = \rho V^2$) increased from ~3 to 13 nPa. The IMF B_z was about −4 nT and turned to further southward. The IMF B_y had an abrupt increase from 4 to 16 nT. The enhancement in southward IMF B_z increased the likelihood of the dayside reconnection, while the enhancement in IMF B_y should cause newly reconnected magnetic field lines connected to the

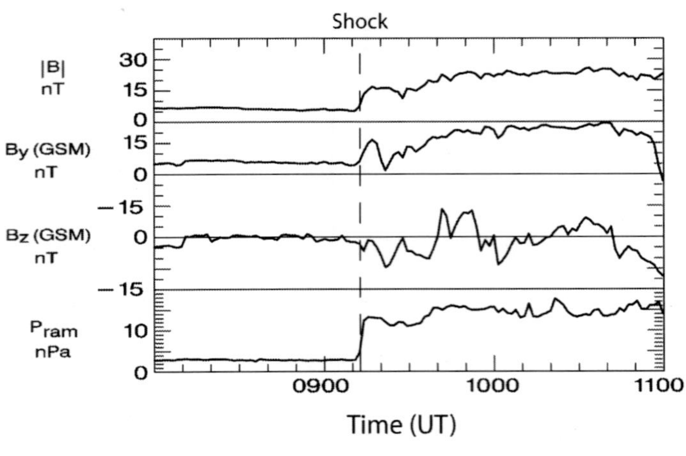

Fig. 2. The November 22, 1997 interplanetary shock event observed by the Wind spacecraft.

northern ionosphere to move dawnward through local noon. Using the shock speed, the calculated shock arrival time at the nose of the magnetopause is ∼0948:45 UT. A SC occurred at 0949 UT.

3.2. *IMAGE ground-based magnetometer observations*

When the interplanetary shock arrived at the subsolar magnetopause, the IMAGE magnetometer chain was located at 12–14 MLT and latitudes of 66°–74° MLAT. Figure 3 presents station locations of the IMAGE chain and magnetogram stack plots. In panel a, solid black circles and arrows

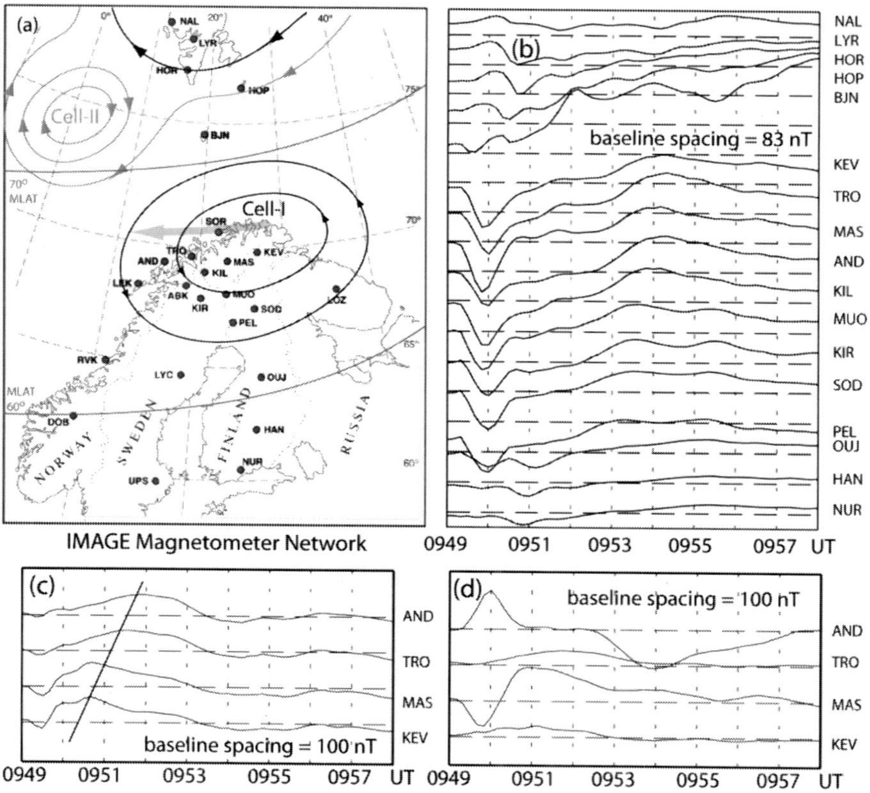

Fig. 3. Geomagnetic perturbations measured by the IMAGE ground-based magnetometers from 0949 to 0958 UT on November 22, 1997. (a) A map of IMAGE stations. Arrows are the plasma flow directions. The geomagnetic latitudes on 60° and 70° are shown by two curves crossing the panel. (b) The X-component (northward) stack plot. (c) The Y-component (eastward) stack plot. (d) The Z-component (downward) stack plot. The data time resolution is 10 sec resolution.

indicate a vortex (Cell-I) detected by the magnetometers. Another vortex (Cell-II) is illustrated by grey lines and arrows (see more explanation in the Section 3.3 and Fig. 5). Figure 3a shows that the IMAGE chain was right beneath Cell-I. Geomagnetic components, X, Y and Z, point to north, east, and downward, respectively. The time resolution is 10 seconds. All IMAGE stations recorded sudden disturbances at ~0949:10 UT, corresponding to the shock arrival. The X-component (Fig. 3b) perturbations were positive at the higher latitude stations (HOP-NAL), but negative at the lower latitude stations (BJN-NUR), indicating a westward current above the Scandinavia mainland and eastward current above Svalbard. This means a low latitude eastward plasma flow and high latitude westward flow. The Y component of the IMAGE/KEV-AND east-west chain is shown in Fig. 3c to identify the vortex motion. The peak time of positive deviations were connected by a straight line which slope indicates an azimuthal velocity. The peak in the Y component reached KEV at ~0950:50 UT, then AND at 0951:50 UT, which indicated a westward motion at $\sim 7\,\mathrm{km\,s^{-1}}$. The opposite variation of the Z components at MAS and AND (Fig. 3d) implied a northward flowing current between the two stations. The magnetic field reversals and the decays showed not only that the vortex moved westward, but also that its shape changed.

To affirm this spatial variation effect, we have calculated the equivalent current system (see next figure and discussion). The calculation is based on a 2D spherical elementary current systems (SECS) upward continuation technique[25] assuming uniform conductance. Figure 4 shows six patterns of the equivalent current. Within 30 sec from 0949:10 to 0949:40 UT, the equivalent current near 70°N increased from ~0 to $200\,\mathrm{mA\,m^{-1}}$. The bottom row shows that Cell-I traveled westward rapidly and moved out of the area. The center of the vortex moved into the IMAGE coverage at 0950:00 UT at (72°N, 26°E). By 0951:20 UT, it had moved to (70°N, 12°E). So the center traveling speed was $\sim 7\,\mathrm{km\,s^{-1}}$ that is consistent with the speed obtained from the Y-component (see Fig. 3c). At 0953:30 UT, the vortex disappeared as the flow below 70°N turned to north-west, which is consistent with the SuperDARN observation (see Fig. 4e). The maximum current density at 0950:00 UT was $\sim 330\,\mathrm{mA\,m^{-1}}$.

3.3. *SuperDARN observations*

Figure 5 presents plasma flow patterns derived from SuperDARN radar line-of-sight observations from 0944 to 1000 UT on Nov 22, 1997. Vectors

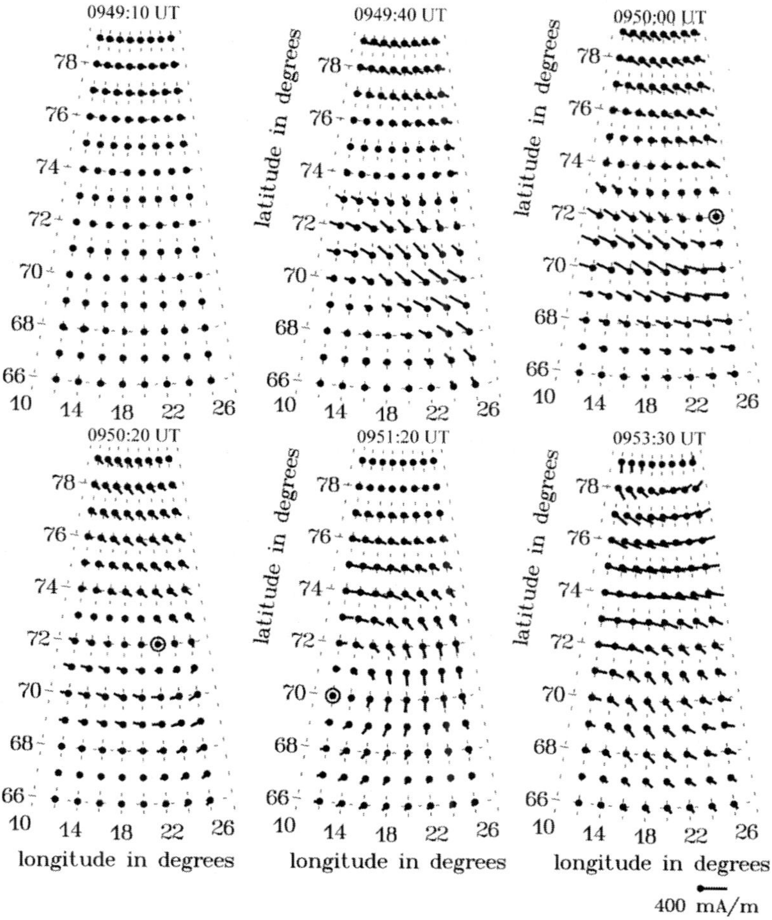

Fig. 4. Ionospheric equivalent current obtained from the IMAGE magnetometer chain in 0949:10-0953:30 UT on November 22, 1997. The latitudes and longitudes are in geographic coordinates. Circled dots show the vortex center. The current intensity is presented by the vector length, and the current direction points away from the dot. Note the plasma flow direction is opposite to the current direction.

are plotted where observations were available for this event. Each vector points in the direction (away from the dot) of the solved velocity, while the color scale and tail length indicate the speed value.

Following the shock arrival, the potential pattern experienced a significant perturbation. The extension of the dusk cell across the noon meridian acquired a mesoscale protrusion (0946–0948 UT), which is seen to be associated with the emergence of a vortex (Cell-II). This cell might

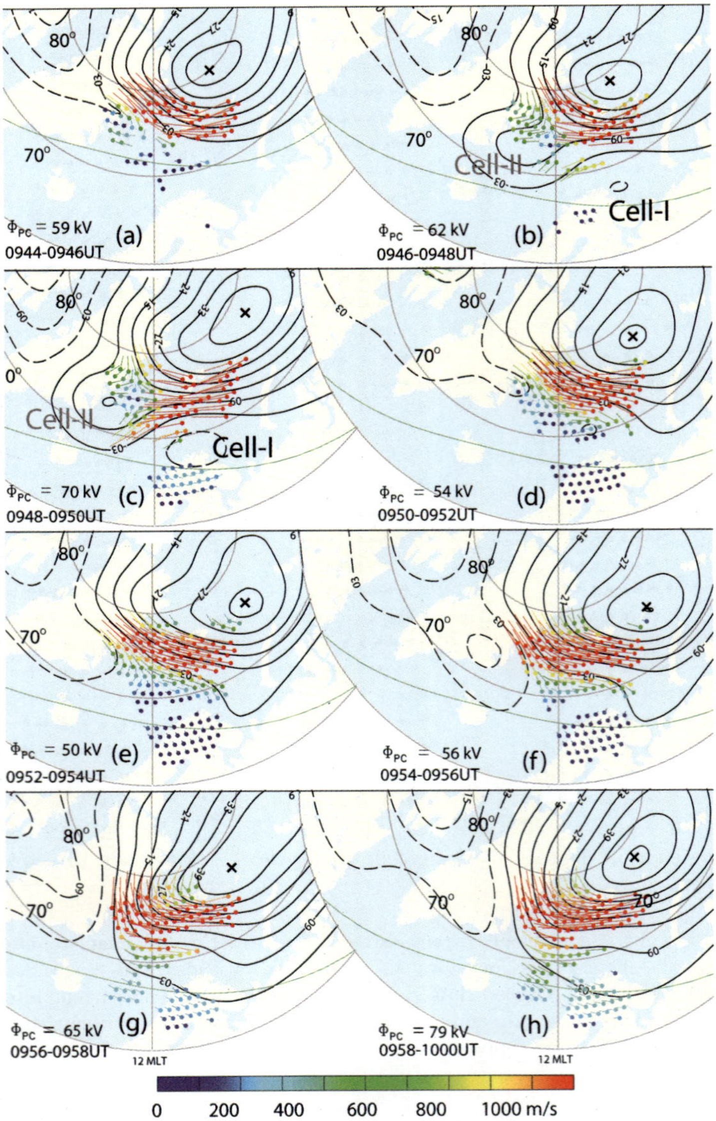

Fig. 5. SuperDARN northern hemisphere convection patterns during the shock arrival from 0944 to 1000 UT. Self-consistent velocity vectors are shown at locations where line-of-sight velocity measurements were obtained. Equipotential contours are separated by 6 kV increments to indicate the electrostatic potential Φ. Negative (positive) potential contours are displayed as solid (dashed) lines. Magnetic local noon is directed down to be convenient for the comparison with the IMAGE map and data. Each vector shows the velocity magnitude (color bar on the bottom) and direction (pointing away from the dot by the tail).

be located at 70°–75° MLAT within 10–12 MLT. A stagnation point (the blue dot at ~76° MLAT and 1140 MLT) separated this cell from the main convection cell in the dusk sector. Since this vortex was near the edge of the radar coverage, the observation cannot decisively confirm its existence. But it was depicted by the statistical fitting results (in Panels b and c). So it is possible that Cell-II existed. At lower magnetic latitude around 68° in 12–13 MLT, there was another vortex (Cell-I) indicated by eastward convection over Scandinavia and westward convection above 70° latitudes (Panels b and c). This convection signature was consistent with the IMAGE X-component perturbation (Fig. 3b). Panels c to e showed, through the reversal of the eastward flows below 70° latitude, that Cell-I disappeared after 0952 UT and hence had a lifetime less than 6 min. By the time at 0954 UT, the convection had evidently turned to westward bellow 70° MLAT, consistent with the calculated equivalent currents.

By 1000 UT, the westward flow at Scandinavia merged with the large scale dusk cell. DMSP-F11 observations (not shown in this paper) indicated that the open/closed boundary between the mantle and LLBL lay near 76.5° MLAT and 1342 MLT at 0918 UT, while DMSP-F13 encountered the LLBL from 0942 to 0943 UT over locations ranging from 78.3° MLAT and 1318 MLT to 75.9° MLAT and 10 MLT (P. Newell, personal communication, 2005). Consequently, the center of Cell-I was located well within the closed magnetic field line region, as was the center of the potential Cell-II.

It should be noted that it is not easy to obtain the snapshots of vortices from SuperDARN because the ionospheric backscatter usually does not suffice to cover such a large area. Radar observations of short-lived and localized events are rare. Super DARN plasma flow velocities are generally derived from fits to line-of-sight velocities observed by several radars and a statistical model. In this case, with only one radar available, the eastward components of the velocities in Fig. 5 were determined by the statistical model. However, the consistence of Cell-I among Figs. 3–5 confirmed that the SuperDARN modeling result is reliable for this event.

4. Discussions

This paper has reported an ionospheric TCV near local noon at subauroral latitudes. The shock-caused vortex occurred within 30 sec after the shock arrival with a center at ~67° MLAT and ~13 MLT (the Cell-I). The SuperDARN data show it is likely there was another vortex (Cell-II) at ~73° MLAT and ~11 MLT. Such double vortex structure can be explained by a theoretical scenario in Fig. 1. The lifetime of the vortex was only 4–6 min. Based on the IMAGE magnetometer data, the vortex traveled westward at a speed of $\sim$7 km s^{-1}.

The pattern and magnitudes of the Hall currents generated by the shock impingement were calculated and shown in Fig. 4. The current can also be estimated from the magnetic disturbances, $J_H = \frac{2}{\mu_0}\Delta B_h$, where ΔB_h is the variation of geomagnetic H component. Both methods showed that the maximum Hall current density is about 350 mA m^{-1}. The density of the FACs can be calculated using $J_{//}= D/\alpha$. Here $D = (\nabla x J_{eq})_z$, i.e. the Z component of the curl equivalent current, and $\alpha = \Sigma_H/\Sigma_P$. When the conductance is uniform ($\alpha = 1$), then, the FAC at (26°E, 72°N) at 0950:00 UT is $\sim$2 μA m^{-2}. For a quiet solar wind and magnetospheric condition, the measured TCV-related FAC was found to be $\sim$1.0 μA m^{-2}.[20] In general, field-aligned currents in the dayside auroral zone were found to be from $\sim$0.06 to $\sim$3.3 μA m^{-2}.[26,27]

The TCV event on November 22 has two special features. One is the location at subauroral latitudes near local noon. The other is the very short lifetime. There are two proposed models for FAC generation near local noon. First, the shock impingement generates a compressional MHD wave that propagates isotropically in the magnetosphere. Meanwhile a FAC system on the magnetopause is required to couple to the ionosphere. The FAC is carried by the field-aligned Alfvén mode.[14,15] Second, at a plasma and magnetic field uonuniformity, such as a sharp gradient in plasma density and magnetic field intensity, the MHD waves are converted from isotropic mode to transverse mode. Then, a FAC is generated at this gradient.[21,22] When shocks compress the magnetopause, this gradient can be generated at low L-shells of closed field lines.

The east coast of Greenland was near the west side of Cell-II at $\sim$0948 UT. Magnetograms from four east coast stations NRD, DMH, DNB and SCO (from north to south) were also examined (data are not shown). DNB and SCO, which are at 75° and 72° MLAT, showed a vortex signature with a bipolar variation in the X component and negative deviation in the Y component. However, the magnitude of X component variation is a factor of 2–3 larger than the Y variation. That presented a significant impulsive disturbance that propagated from the low latitude to the high latitude (J. Watermann, communication, 2004).

Shock auroras occurred during this event (not shown in this paper). There have been many studies that reported in addition to the aurora on and along the oval, electron and proton auroras on dayside and below the oval may also occur due to the shock compression.[4,5,28–30] In some occasions the detached proton auroras appeared to be bright spots.[30,31] Fuselie *et al.*[29] suggested that the electromagnetic ion cyclotron (EMIC) waves cause the particle-wave interaction and proton precipitation. Later Zhang *et al.*[32] found the EMIC waves on the magnetic field lines at which footprints proton auroras were recorded. We speculate that the reported bright proton auroral spot may have been caused by the presence of EMIC waves as well as a downward FAC into the center of a vortex, such as Cell-I.

Acknowledgments

The results reported here represent one aspect of research carried out by the California Institute of Technology Jet Propulsion Laboratory for the National Aeronautics and Space Administration.

References

1. X.-Y. Zhou and B. T. Tsurutani, *Geophys. Res. Lett.* **26** (1999) 1097.
2. V. G. Vorobjev *et al.*, *J. Geophys. Res.* **106** (2001) 28897.
3. P. E. Sandholt, H. C. Carlson and A. Egeland, *Dayside and Polar Cap Aurora,* Kluwer Academic Publishers, Astrophysics and Space Sci. Library, Dordrecht, Holland, **270** (2002).
4. K. Liou *et al.*, *Geophys. Res. Lett.* **29** (2002), doi:10.1029/2001GL014182.
5. B. Hubert, J. C. Gerard, S. A. Fuselier and S. B. Mende, *Geophys. Res. Lett.* **30** (2003) 1145, doi:10.1029/2002GL016464.
6. T. Araki, in *Solar Wind Sources of Magnetospheric Ultra-Low-Frequency Waves*, edited by M. J. Engebretson, K. Takahashi and M. Scholer, AGU, Washington, D.C. (1994) 183–200.
7. C. T. Russell and M. Ginskey, *J. Geophys. Res.* **100** (1995) 23,695.
8. E. Friis-Christensen, M. A. McHenry, C. R. Clauer and S. Vennerstrom, *Geophys. Res. Lett.* **15** (1988) 253.
9. D. G. Sibeck and G. I. Korotova, *J. Geophys. Res.* **101** (1996) 13,413.
10. R. J. Sitar *et al.*, *J. Geophys. Res.* **103**(A10) (1998) 23357–23372, doi:10.1029/98JA01916.
11. D. L. Murr and W. J. Hughes, *Geophys. Res. Lett.* **30**(7) (2003) 1354, doi:10.1029/2002GL015498.
12. M. M. Lam and A. S. Rodger, *J. Geophys. Res.* **109** (2004) A02204, doi:10.1029/2003JA010214.
13. E. Zesta, W. J. Hughes and M. Engebretson, *J. Geophys. Res.* **107**(A10) (2002) 1317, doi:10.1029/1999JA000386.
14. M. J. Engebretson *et al.*, *J. Geophys. Res.* **104**(A10) (1999) 22433–22452, doi:10.1029/1999JA900237.

15. T. Moretto, A. J. Ridley, M. J. Engebretson and O. Rasmussen, *J. Geophys. Res.* **105** (2000) 2521–2531.
16. T. Tamao, *J. Geomagn. Geoelectr.* **16** (1964) 89–114.
17. M. G. Kivelson and D. J. Southwood, *J. Geophys. Res.* **96** (1991) 1661.
18. K.-H. Glassmeier and C. Heppner, *J. Geophys. Res.* **97** (1992) 3977.
19. R. L. Lysak and D.-H. Lee, *Geophys. Res. Lett.* **19** (1992) 937–940.
20. H. Vogelsang *et al.*, *J. Geophys. Res.* **108** (2003), doi:10.1029/2002JA009407.
21. T. Tamao, *Sci. Rep. Tohoku Univ.: Geophys.* **17** (1965) 43–72.
22. D. J. Southwood, *Planet. Space Sci.* **22** (1974) 483–491.
23. D. G. Sibeck *et al.*, *J. Geophys. Res.* **108**(A2) (2003) 1095, doi:10.1029/2002JA009675.
24. S. Fujita *et al.*, *J. Geophys. Res.* **108** (2003), doi:10.1029/2002JA009407.
25. O. Amm and A. Viljanen, *Earth, Planets and Space* **51** (1999) 431.
26. K.-H. Glassmeier, *Annales Geophysicae* **5A** (1987) 115.
27. L. J. Lanzerotti *et al.*, *J. Geophys. Res.* **96** (1991) 14009.
28. M. Meurant *et al.*, *Geophys. Res. Lett.* **30** (2003) 2032, doi:10.1029/2003GL018017.
29. S. A. Fuselier *et al.*, *J. Geophys. Res.* **109** (2004), doi:10.1029/2004JA010393.
30. T. A. Yahnina, H. U. Frey, T. Bösinger and A. G. Yahnin, *J. Geophys. Res.* **113** (2008) A07209, doi:10.1029/2008JA013099.
31. Y. Zhang, L. J. Paxton, T. J. Immel, H. U. Frey and S. B. Mende, *J. Geophys. Res.* **108** (2002), doi:10.1029/2002JA009355.
32. Y. Zhang, L. J. Paxton and Y. Zheng, *J. Geophys. Res.* **113** (2008) A01212, doi:10.1029/2007JA012554.

Advances in Geosciences
Vol. 21: Solar & Terrestrial Science (2008)
Ed. Marc Duldig

IONOSPHERIC OBSERVATION USING FM-CW RADAR ARRAY

A. IKEDA*, A. YOSHIKAWA and M. G. CARDINAL
Department of Earth and Planetary Sciences,
Kyushu University, Japan
**a-ikeda@geo.kyushu-u.ac.jp*

K. YUMOTO
Space Environment Research Center,
Kyushu University, Japan

M. SHINOHARA
Kagoshima National College of Technology, Japan

K. NOZAKI
National Institute of Information and
Communications Technology, Japan

B. M. SHEVTSOV and V. V. BYCHKOV
Institute of Cosmophysical Research and Radiowaves
Propagation of the Far Eastern Branch of Russian
Academy of Sciences, Russia

Q. M. SUGON, Jr. and D. McNAMARA
Manila Observatory, Philippines

In order to understand the electro-magnetic phenomenona in space, the Space Environment Research Center (SERC) of Kyushu University is deploying the MAGnetic Data Acquisition System of the Circum-pan Pacific Magnetometer Network (MAGDAS/CPMN). In this network, we also installed FM-CW (Frequency Modulated Continuous Wave) radars at Paratunka, Russia (PTK: Magnetic Latitude = 45.8°, Magnetic Longitude = 221.6°), Sasaguri, Japan (SAS: M. Lat. = 23.2°, M. Lon. = 199.6°), and Manila, Philippines (MNL: M. Lat. = 4.19°, M. Lon. = 192.4°) to detect the ionospheric electric field variations. The FM-CW radar is a kind of HF (High Frequency) radar that can measure the ionospheric Doppler velocity, from which we can calculate the ionospheric electric fields. In the present paper we will introduce our FM-CW radar array observations and its preliminary scientific results. The results are as follows: (1) At the time of SC, dusk-to-dawn and subsequent dawn-to-dusk electric field penetrate into the low-latitude ionosphere. (2) Pi 2-associated ionospheric electric field at $L = 2.05$ is a manifestation of the plasmaspheric cavity mode. (3) Ground Pc 5 pulsation seems to be driven by ionospheric electric field at low and equatorial region.

1. Introduction

The Space Environment Research Center (SERC) of Kyushu University in Fukuoka, Japan is deploying a magnetic observation array, which is called the MAGnetic Data Acquisition System of the Circum-pan Pacific Magnetometer Network (MAGDAS/CPMN).[1–3] In this network, we also initiated the installation of FM-CW (Frequency Modulated Continuous Wave) radars to measure the ionospheric electric fields. An FM-CW radar is a kind of HF (High Frequency) radar that can detect Doppler velocity of a target for ionospheric observation.[4]

The magnetic field measured on the ground is a superposition of various internal and external sources. The external sources are the solar wind, the magnetosphere, and the ionosphere. Among these sources, FM-CW radar observation provides only the information from the ionosphere. Therefore an analysis that combines the data from the MAGDAS/CPMN and FM-CW radars is expected to give us a better understanding of magnetospheric and ionospheric phenomenona from the ground observations. For example, one benefit of combining ionospheric electric field data and magnetic field data is to allow us to identify the propagation modes of geomagnetic pulsations.

It is known that HF Doppler measurements at low- and mid-latitudes often show a good correlation with the variations of geomagnetic phenomena (e.g. Ref. [5]). For example, SCs (e.g. Refs. [6, 7]), Pc 5 pulsations (e.g. Refs. [8, 9]), Pc 3–4 pulsations (e.g. Refs. [10, 11]), and Pi 2 pulsations (e.g. Refs. [12, 13]) were reported using HF Doppler measurements. However, the comparison analysis of ionospheric response and ground magnetic field is limited. In this paper, we introduce our FM-CW radar observation and show some geomagnetic phenomenon observed simultaneously by the FM-CW radars and magnetometers of MAGDAS/CPMN.

2. Observation and Instruments

We installed FM-CW (Frequency Modulated Continuous Wave) radars at Paratunka, Russia (PTK: Magnetic Latitude = 45.8°, Magnetic Longitude

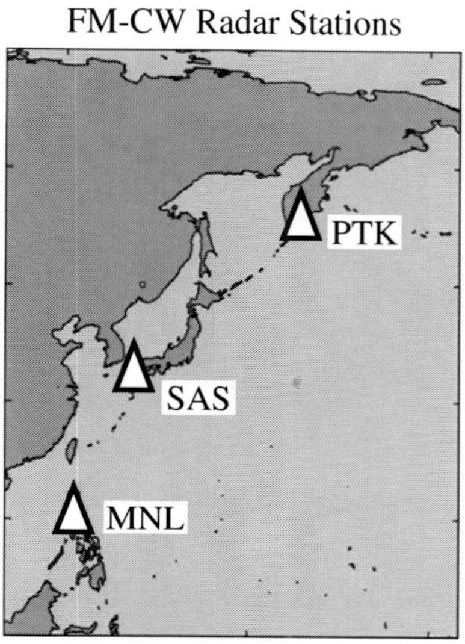

Fig. 1. Station map of FM-CW radars. The three stations are located at PTK, Russia, SAS, Japan, and MNL, Philippines.

= 221.6°, LT = UT + 10.5 hrs), Sasaguri, Japan (SAS: M. Lat. = 23.2°, M. Lon. = 199.6°, LT = UT + 9.5 hrs) and Manila, Philippines (MNL: M. Lat. = 4.19°, M. Lon. = 192.4°, LT = UT + 8.5 hrs). Figure 1 shows a station map of the currently installed FM-CW radars. They are located almost in the same longitude. The FM-CW radars at SAS and PTK were installed in 2000 and 2005, respectively. The latest installation was done on March 2009 at MNL. Figure 2 shows the FM-CW radar system at Manila, Philippines and Fig. 3 shows the tower for the radar. The FM-CW radar system consists of a receiver, a transmitter, and a control PC (from left). The PC controls the FM-CW radar and sends observational data to SERC via internet. The radar tower at MNL is a 100-foot tower with a dipole antenna mounted on top of it.

An FM-CW radar is a type of HF radar that can measure the range of target as well as its Doppler information. Our radar is an improved version of the FM-CW radar developed by Nozaki and Kikuchi.[14,15] Poole[16] and Poole and Evance[17] used FM-CW radar for ionospheric Doppler measurement for the first time. Our Doppler observation is based on the fast Fourier transform (FFT) method which is a variation of a technique developed by Barrick[18] to measure ocean waves. This technique provides

FM-CW Radar System at Manila

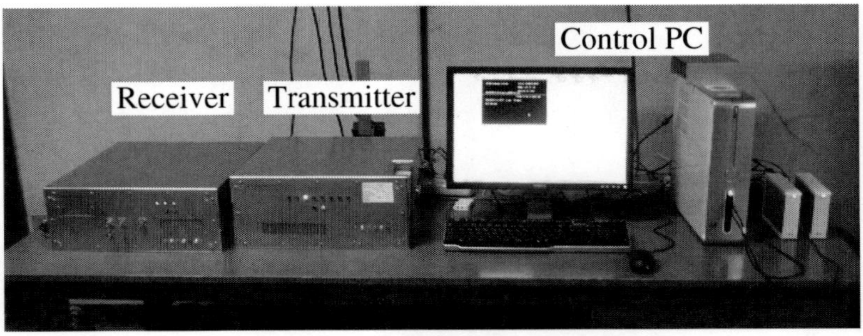

Fig. 2. FM-CW radar system at MNL. The components are a receiver, a transmitter, and a control PC (from left).

FM-CW Radar Tower at Manila

Fig. 3. A tower of FM-CW radar at MNL. The length of the tower is 30 foot. A dipole antenna is set to the tower.

us the Doppler velocity of a target and its distance. Generally FM-CW radars sweep their transmitting frequency slightly (e.g. from 3.00 MHz to 3.01 MHz) for Doppler observation and transmit signals continuously. In the FFT method, range and Doppler information can be estimated using a double FFT. The first FFT over each sweep yields the range information, and the second FFT for each range bin over a number of sweeps provides the Doppler frequency. For our observation, the sweep rate is 200 or 100 kHz/sec and sweep time is 0.1 seconds. The observed Doppler frequency Δf is represented by

$$\Delta f = f_0 \times 2V^*/c \tag{1}$$

where f_0 is the transmitting frequency and V^* is vertical drift velocity of the ionosphere describe by

$$V^* = c \times \Delta f/2f_0 \tag{2}$$

We use low frequency (e.g. 2.5 MHz) for the transmitting frequency f_0 at night and we use higher frequency (e.g. 5.0 MHz) at daytime. The data of Doppler velocity is digitized with 3-sec or 10-sec sampling. The data accuracy of the vertical drift velocity is 3.9 m/s at 3.0 MHz.

By assuming that the V^* is caused by the frozen-in effects in the ionosphere, we can estimate the east-west ward electric field (E_y). The equation of the frozen-in effect is described

$$\mathbf{E} = -\mathbf{V} \times \mathbf{B} \tag{3}$$

where **E** is east-west electric field (E_y), and **B** is the horizontal component (H component) of the magnetic field in the ionosphere, and **V** is obtained by FM-CW radars. A schematic diagram of the Doppler measurement by an FM-CW radar is shown in Fig. 4.

3. Observational Results and Discussion

3.1. *Sudden commencement*

Figure 5 shows the geomagnetic H component (H) at Kuju, Japan (KUJ: M. Lat. = 23.6°, M. Lon. = 203.2°, LT = UT + 8.7 hrs) and the ionospheric electric field E_y at SAS on 4 November, 2003. The time interval is 10 minutes from 06:22 UT to 06:32 UT corresponding to the local daytime. During this SC event, the transmitting frequency f_0 was 8.0 MHz and the observed ionospheric altitude was about 260 km (virtual height). A time resolution of E_y and H for this event is 10 seconds, and 3 seconds, respectively. The E_y at SAS shows a small variation prior to the onset of the SC (described by the vertical dashed line in Fig. 5). At 06:25:45 UT (indicated by the black arrow), the E_y started to decrease within a few of

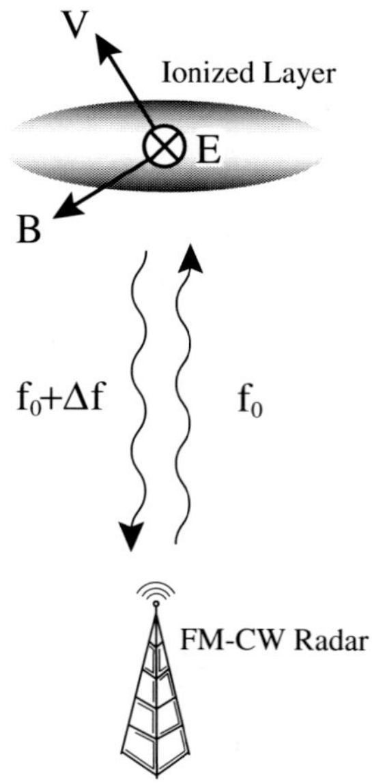

Fig. 4. Schematic diagram of the Doppler measurement of the ionosphere. When the eastward electric field penetrates into the low-latitude ionosphere, it drifts upward because of $\mathbf{E} \times \mathbf{B}$ drifts.

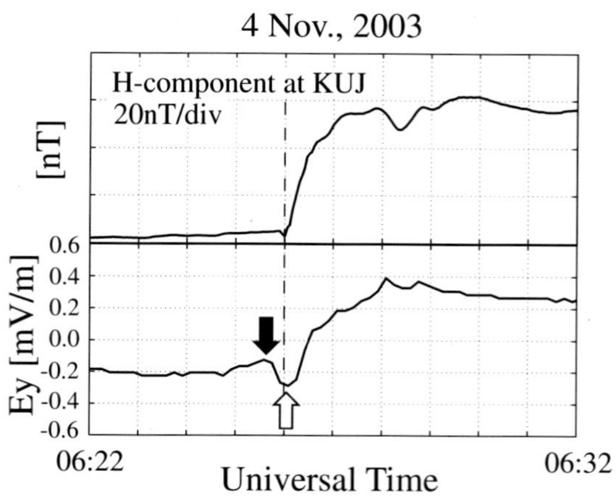

Fig. 5. SC at daytime on 4 November, 2003. The upper panel shows the ground H component at KUJ. The bottom panel shows the E_y at SAS, which is calculated from the observed ionospheric Doppler velocity. The transmitting frequency of the FM-CW radar was 8.0 MHz during this event.

tens seconds and reached the minimum value of −0.29 mV/m at 06:26:05 UT (indicated by the white arrow). The value was −0.15 mV/m (westward) and the duration was 30 seconds. This signature coincides with slightly negative variation of the H at KUJ which is the so-called PRI (preliminary reverse impulse). The PRI is well known that it often appears at the time of PI (Preliminary Impulse) in daytime low- and equatorial region.[19] The westward electric field at the time of PRI corresponds to the dusk-to-dawn direction.

After the E_y attained its minimum value (indicated by the white arrow), it increased rapidly from −0.29 to 0.39 mV/m during a period from 06:26:05 UT to 06:28:05 UT. The amplitude of this increase is 0.68 mV/m (eastward). This eastward electric field corresponded with dawn-to-dusk electric field. In addition, the timing of the eastward electric field and increasing of H (so-called Main Impulse MI) was same.

The nighttime SC event is shown in Fig. 6. The format is the same as that of Fig. 5. The time interval is from 17:06 UT to 17:16 UT on 21 January, 2005. During this SC event, the FM-CW radar observed the altitude of about 360 km (virtual height) at 2.5 MHz. At 17:11:15 UT (indicated by the black arrow), the E_y showed a sudden increase from 0.0 mV/m to 0.26 mV/m within a few of tens of seconds. This eastward electric field preceded a sudden increase of the magnetic H component. Therefore it can be identified as a dusk-to-dawn electric field corresponding with daytime PRI.

At the time of MI onset indicated by the vertical dashed line in Fig. 6, the E_y decreased with increasing H almost simultaneously at 17:11:35 UT (indicated by the black arrow). The decrease of 1.16 mV/m in E_y indicates the existence of westward electric field. The direction is dawn-to-dusk.

It has been known that the interplanetary and magnetospheric electric field can penetrate into the low-latitude ionosphere (e.g. Ref. [19]). The IEF (Interplanetary Electric Field) can continuously penetrate to the low-latitude ionosphere without significant attenuation.[20] Short life time (<3 hours) dawn-dusk IEF penetrate into the ionosphere without shielding.[21] Also penetration of electric fields has been reported in the case of SC (e.g. Refs. [6, 22]). Araki[23] concluded that the PI is caused by a dusk-to-dawn electric and the MI is caused by a dawn-to-dusk electric field. The two types of electric fields are generated in the magnetospehre by a solar wind pressure enhancement and penetrate into the polar ionosphere and low-latitude ionosphere. This is consistent with our result.

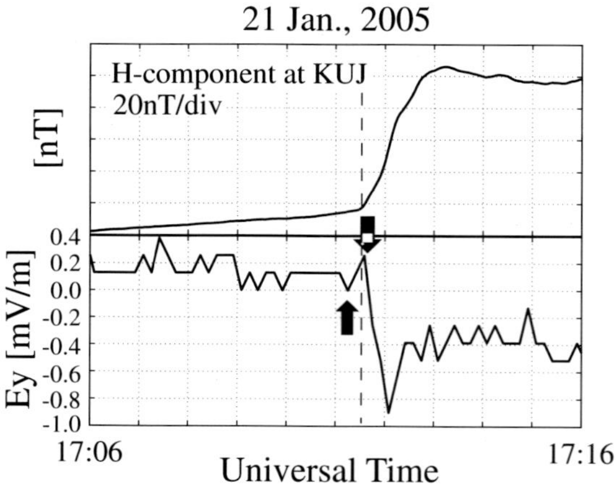

Fig. 6. SC at nighttime on 21 January 2005. The format is the same as that of Fig. 4. The transmitting frequency was 2.5 MHz during this event.

3.2. *Pi 2 pulsation*

Figure 7 shows a nighttime Pi 2 (40–150 seconds) event observed by an FM-CW radar and a MAGDAS magnetometer at PTK on 19 October 2007. During this event, the FM-CW radar observed the altitude of about 250 km (virtual height) at 3.0 MHz. The ground magnetic *H* and *D* components (*D*) obtained at PTK (L = 2.05, LT = UT + 10.5 hrs) and Ashibetsu, Japan (ASB; M. Lat. = 34.7°, M. Lon. = 209.6°, L = 1.48, LT = UT + 9.5 hrs), and E_y at PTK are plotted in Fig. 7. The E_y is 3-sec time resolution and the ground magnetic data is 1-sec time resolution for this event. The thick lines indicate *H* components and the thin lines indicates *D* components.

Figure 8 shows the band pass filtered data. We can see that Pi 2 pulsations were observed simultaneously in *H* and V^*. They started around 1644 UT and attained their peak around 16:48 UT (see the *H* component at PTK in Fig. 8). The dominant frequency was 15.4 mHz for all waves. Fig. 8 also shows the filtered *H* and *D* components at PTK and ASB, and the filtered E_y at PTK. The peak-to-peak amplitude of *H* at PTK, *D* at PTK, *H* at ASB, *D* at ASB, and E_y at PTK are 3.3 nT, 1.4 nT, 2.5 nT, 1.2 nT, and 0.43 mV/m, respectively. The ground magnetic perturbation was dominant at *H* components rather than at *D* components.

We calculated the cross correlation between E_y at PTK and geomagnetic field data. As a result, the maximum correlation coefficients were 0.95 for $E_y - H$ (PTK), 0.95 for $E_y - H$ (ASB), 0.79 for $E_y - D$

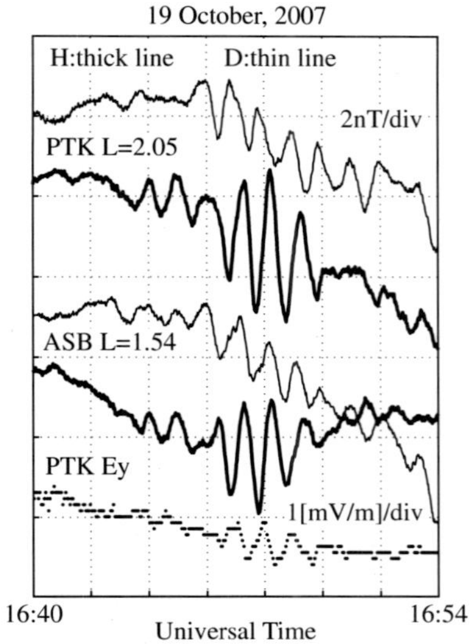

Fig. 7. Pi 2 pulsations on 19 October 2007. H and D components at PTK and ASB and E_y at PTK. The thick lines are H components and the thin lines are D components. The transmitting frequency of the FM-CW radar was 3.0 MHz.

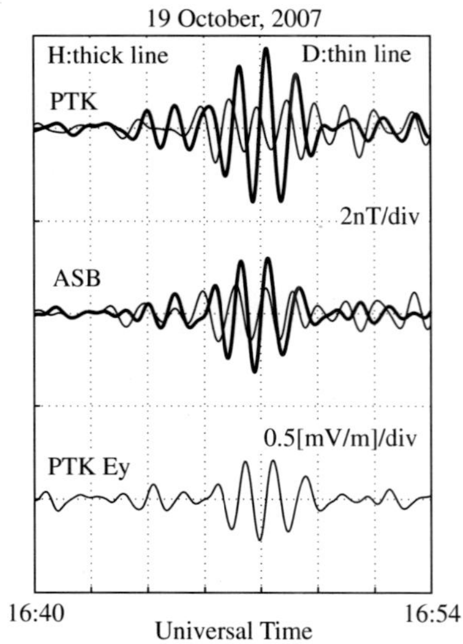

Fig. 8. Band pass filtered (40–150 seconds) H and D components at PTK and ASB and E_y at PTK on 19 October 2007. The thick lines are H components and the thin lines are D components for magnetic field data.

(PTK), 0.72 for $E_y - D$ (ASB). Thus the correlation coefficient of $E_y - H$ is higher than that of $E_y - D$. In addition, the correlation coefficient of $E_y - H$ (ASB) is higher than that of $E_y - H$ (PTK).

Pi 2 pulsations are well known as the signal for the onset of magnetic substorms (e.g. Ref. [24]). Moreover low-latitude Pi 2 pulsations are explained in terms of the plasmaspheric cavity mode (e.g. Ref. [25]). Takahashi *et al.*[26] analyzed the ground magnetic H component at KAK ($L = 1.26$) and the magnetic field variation obtained by AMPTE CCE inside the plasmasphere. They found that nighttime Pi 2 pulsations originated from a cavity-mode-type resonance excited in the inner magnetosphere. In addition, the magnetic field variations on Pi 2 pulsations in the ionosphere and on the ground given by Han *et al.*[27] were consistent with the radial cavity mode. They found that the Pi 2 waves observed at KAK and by the low-altitude satellite ($h = 638 - 849$ km) Ørsted were highly coherent and oscillated without phase lag at nighttime. They concluded that the nightside Pi 2 pulsations at low latitude are directly propagated from the magnetosphere to the ground and became the cavity-mode oscillation. Thus low-latitude Pi 2 pulsations in ground H components are explained in terms of the plasmaspheric cavity mode. On the basis of previous studies, it seems that our result also suggests that the ionospheric E_y at PTK ($L = 2.05$) is a manifestation of the plasmaspheric cavity mode because the Pi 2 pulsation in E_y shows high correlation coefficient with lower-latitude H components.

3.3. *Pc 5 pulsation*

The daytime Pc 5 pulsation (150–600 seconds) observed by an FM-CW radar at SAS are shown in Fig. 9. The H component at equatorial station YAP (M. Lat. = 0.42°, M. Lon. = 209.9°, L = 1.00, LT = UT + 9.2 hrs) and KUJ (L = 1.19, LT = UT + 8.7 hrs) and E_y at SAS are plotted in Fig. 9. A time resolution of E_y and H for this event is 10 seconds, and 3 seconds, respectively. YAP, SAS, and KUJ were located at the local dayside sectors. During this event, the FM-CW radar at SAS recorded an altitude of 300 km at 8.0 MHz. The peak-to-peak amplitude of E_y at SAS was about 1.0 mV/m. The magnetic Pc 5 pulsation at YAP was larger than that at KUJ. The peak-to-peak amplitude at YAP was more than 100 nT. This would be due to equatorial enhancement because of the high ionospheric conductivity in the equatorial region (e.g. Ref. [24]). Inspecting the vertical dashed lines in Fig. 9, the peaks of H at KUJ and YAP almost corresponds

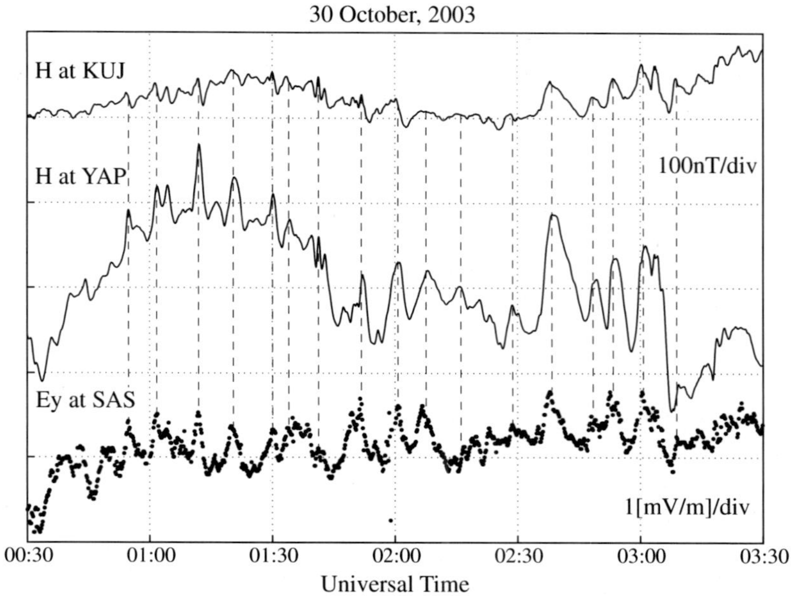

Fig. 9. Pc 5 pulsations on 31 October 2003. The ground magnetic H and D components at YAP and KUJ and E_y at SAS are plotted. The FM-CW radar observed the altitude of about 300 km at 8.0 MHz.

with that of E_y at SAS. The positive peaks of E_y correspond with positive peaks of the H.

The oscillation of E_y at SAS corresponds with the H at KUJ and YAP without significant time delay. This result suggests that the ground Pc 5 was excited by the ionospheric electric fields which drive ionospheric currents. The source of E_y for Pc 5 would be caused by the polar dawn-to-dusk electric fields which are excited by the DP 2 type current system.[28]

4. Summary

In the MAGDAS/CPMN, SERC also deploys FM-CW radar observations to understand the Magnetosphere-Ionosphere phenomenona in space. In this paper, we showed SC, Pi 2, Pc 5 events which were observed by MAGDAS/CPMN and FM-CW radars simultaneously. For these events, ionospheric perturbation correlated with ground magnetic field variation. In the future, we will study longitudinal structure of ionospheric electric fields associated with geomagnetic phenomenon by using the FM-CW radars at PTK, SAS, and MNL.

Acknowledgment

We would like to thank Mr. Ryuichi Ishihara, Mr. Kazuhiro Mori, and Mr. Toshiki Shinbaru for helping to construct the FM-CW radar at Sasaguri, Japan. The FM-CW radar at Paratunka, Russia was installed with the support of the Institute of Cosmophysical Research and Radiowaves Propagation (IKIR) of the Far Eastern Branch of Russian Academy of Sciences. The FM-CW radar at Manila, Philippines was installed with the support of the Manila Observatory. We acknowledge all supporters for the FM-CW radar observation. We also acknowledge World Data Center for Geomagnetism, Kyoto (http://wdc.kugi.kyoto-u.ac.jp/index.html) for calculating the IGRF model.

References

1. K. Yumoto and the CPMN Group, *Earth Planets Space* **53** (2001) 981.
2. K. Yumoto and the MAGDAS Group, *Solar Influence on the Heliosphere and Earth's Environment: Recent Progress and Prospects* **309** (2006) 981.
3. K. Yumoto and the MAGDAS Group, *Bull. Astr. Soc. India* **35** (2007) 511.
4. A. Ikeda, K. Yumoto, M. Shinohara, K. Nozaki, A. Yoshikawa and A. Shinbori, *Mem. Fac. Sci., Kyushu Univ., Ser. D, Earth & Planet. Sci. Vol. XXXII* **1** (2007) 1.
5. K. L. Chan, D. P. Kanellakos and O. G. Villard Jr., *J. Geophys. Res.* **67** (1962) 2066.
6. T. Kikuchi, T. Ishimine and H. Sugiuchi, *J. Geophys. Res.* **90** (1985) 4389.
7. T. Araki, S. Fujitani, M. Emoto, K. Yumoto, K. Shiokawa, T. Ichinose, H. Luehr, D. Orr, D. Milling, H. Singer, G. Rostoker, S. Tsunomura, Y. Yamada and C. F. Liu, *J. Geophys. Res.* **102**(A7) (1997) 14075.
8. J. H. Sastri, K. B. Ramesh, D. R. K. Rao and J. V. S. V. Rao, *J. Atmos. Terr. Phys.* **55** (1993) 1271.
9. T. Motoba, T. Kikuchi, T. F. Shibata and K. Yumoto, *J. Geophys. Res.* **109**(A06214) (2004) doi:10.1029/2004JA010442.
10. R. A. Marshall and F. W. Menk, *Ann. Geophys.* **17** (1999) 1397.
11. F. W. Menk, *Planet. Space Sci.* **40** (1992) 495.
12. J. Klostermeyer and J. Röttger, *Planet. Space Sci.* **24** (1976) 1065.
13. J. F. Grant and K. D. Cole, *Planet. Space Sci.* **40** (1992) 1461.
14. K. Nozaki and T. Kikuchi, *Mem. Natl Inst. Polar Res., Spec. Iss.* **47** (1987) 217.
15. K. Nozaki and T. Kikuchi, *Proc. NIPR Symp. Upper Atm. Phys.* **1** (1988) 204.

16. A. W. V. Poole, *Radio Sci.* **20** (1985) 1609.
17. A. W. V. Poole and Evans, *Radio Sci.* **20** (1985) 1617.
18. D. E. Barrick, *NOAA Technical Report ERL*, **283-WPL** (1973) 26.
19. M. C. Kelley, J. J. Makela, J. L. Chau and M. J. Nicolls, *Geophys. Res. Lett.* **30**(4) (2003) 1158, doi:10.1029/2002GL016321.
20. C. -S. Huan, S. Sazykin, J. L. Chau, N. Maruyama and M. C. Kelley, *Atmos. Sol. Terr. Phys.* **69**(10–11) (2007) 1135.
21. Y. Wei, M. Hong, W. Wan, A. Du, J. Lei, B. Zhao, W. Wang, Z. Ren, and X. Yue, *Geophys. Res. Lett.* **35**(L02102) (2008), doi:10.1029/2007GL032305.
22. Y. -N. Huang, *J. Geophys. Res.* **81** (1976) 175.
23. T. Araki, *Geophysical Monograph 81* **183** (1994) 200.
24. T. Saito and S. Matsushita, *J. Geophys. Res.* **73** (1968) 267.
25. T. K. Yeoman and D. Orr, *Planet. Space Sci.* **38** (1989) 1367.
26. K. Takahashi, S. -I. Ohtani and B. J. Anderson, *J. Geophys. Res.* **100** (1995) 21929.
27. D. S. Han, T. Iyemori, M. Nose, H. McCreadie, Y. Gao, F. Yang, S. Yamashita and P. Stauning, *J. Geophys. Res.* **109**(A10209) (2004) 1210, doi:10.1029/2004JA010576.
28. T. Motoba, T. Kikuchi, H. Luhr, T.-I. Kitamura, H. Tachihara, K. Hayashi and T. Okuzawa, *J. Geophys. Res.* **107**(A2) (2002), doi:10.1029/2001JA900156.

Advances in Geosciences
Vol. 21: Solar & Terrestrial Science (2008)
Ed. Marc Duldig

DYNAMICS OF THE EARTH'S MAGNETOSPHERE FROM SOME MODERN DATA ANALYSIS TOOLS

A. T. Y. LUI
The Johns Hopkins University Applied Physics Laboratory Laurel, MD 20723, USA
Tony.Lui@jhuapl.edu

It is well recognized that 99.9% of space is in a plasma state consisting of highly ionized gas. Many dynamical phenomena in the Plasma Universe may be prototyped within the environment of the Earth's magnetosphere. Since the dawn of space era with the launch of artificial satellites, we have the capability of making *in situ* measurements to probe the plasma dynamics in the magnetospheres of the Earth and the other planets in our solar system. In this paper, three modern data analysis tools on *in situ* measurements will be described for the pursuit of identifying the physical processes responsible for some dynamic magnetospheric phenomena. Specifically, these tools are wavelet analysis, phase-space density analysis, and Grad-Shafranov reconstruction. These enabling tools reveal clues about the underlying physical processes for the observed magnetospheric phenomena and their plausible connection with the modern research of nonlinear dynamics. Examples of how these tools are applied to advance our understanding of energetic events in the Earth's magnetosphere are presented.

1. Introduction

There are four different approaches in advancing scientific research. One is by developing state-of-the-art instruments to probe a given phenomenon. The second is to develop powerful data analysis techniques to reveal the pertinent characteristics of the physical process. The third is to capture the essentials of the physics involved with theories based on the fundamental laws. The fourth is to perform numerical modeling to test our understanding and bring closure between theory and observations. All these approaches are complimentary to each other. For example, data analysis results will help to identify the specifications of instruments needed for the next stage of exploration of a given phenomenon. There is therefore considerable synergy between instrument design and data analysis tools.

In this paper, some modern data analysis tools are described to show how their applications in the observations of magnetospheric dynamics can help in yielding insightful clues on the underlying processes in magnetospheric phenomena. The tools expounded in this paper are only a small collection that the author has utilized and not meant to be an inclusive list of all available analysis techniques. These are wavelet analysis, phase space density analysis, and Grad-Shafranov reconstruction. The wavelet analysis decomposes signals in the frequency-time domain that provides valuable information on waves. The phase space analysis discussed here allows one to probe the past history of measured plasma in certain dynamic situations. The Grad-Shafranov reconstruction extends the vision of the plasma environment from a one-dimensional path of a satellite to a two-dimensional view. It is worth noting that other useful analysis tools, especially for multi-point analysis, have been documented in Paschmann and Daly (2008).[1]

2. Wavelet Analysis

There are several ways to investigate the information content of a given signal in the frequency-time domain as illustrated in Fig. 1. For example, Shannon decomposition is a means to obtain averages of a given signal at a fixed time interval without the desire to seek its frequency component. In

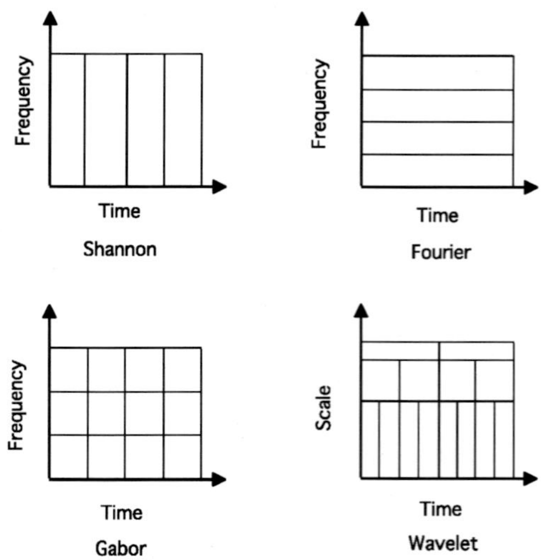

Fig. 1. Four different kinds of frequency-time decomposition.

contrast, a common practice to obtain the frequency component of a given signal without the desire of the time resolution is the Fourier transform. One could recover some aspect on the time resolution on the onsets of some frequency component of a given signal by the Gabor transform, which is basically the division of an interval into several subintervals and perform Fourier transform on each subinterval individually. This yields a uniform time resolution for all frequencies concerned. However, high frequency waves may have shorter durations than low frequency waves. A more desirable decomposition would have higher time resolution for high frequency waves and lower time resolution for low frequency waves. Furthermore, natural signals often have transient appearance, unlike the infinitely long existence of a Fourier component. A different set of basis functions than the Fourier components is more suited to model the appearance of these natural signals. Wavelet analysis is a tool aimed to decompose a given signal with these characteristics in mind. The use of the word scale instead of frequency is to emphasize the transient nature of the basis functions used in the frequency-time decomposition even though scale is basically the inverse of frequency. However, one may note that the word scale is more general than frequency in terms of its applicability to spatial scale also rather than the time scale alone.

Wavelet transform is based on a set of basis functions formed by the dilation and translation of a mother function $y(t)$. Dilation of the mother function produces long-duration low-frequency and short-duration high-frequency functions. Formally, the continuous wavelet transform of a given signal $y(t)$ is given by

$$W(a,b) = \frac{1}{\sqrt{a}} \int_{\infty}^{-\infty} s(t) y\left(\frac{t-b}{a}\right) dt. \tag{1}$$

where $a > 0$ is the scaling factor and b is the translation parameter. A commonly used mother function is the Morlet mother wavelet given by

$$y(t) = \frac{1}{\sigma\sqrt{2\pi}}[exp(-i2\pi f_0 t) - exp(-2\pi^2\sigma^2 f_0^2)]exp\left(-\frac{t^2}{2\sigma^2}\right). \tag{2}$$

where f_o is the frequency parameter which allows one to shift the frequency range for investigation and σ is the width of the Gaussian envelope.

The usefulness of wavelet transform may be demonstrated by examining the wavelet transform result from test waves, as illustrated in Fig. 2. The first test wave in the top panel of Fig. 2 is a low-frequency wave with superposition of two short-lived high-frequency waves. The wavelet

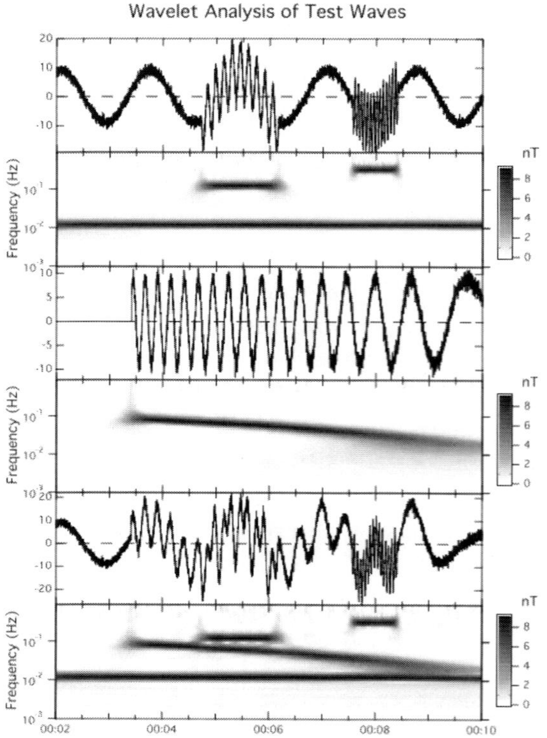

Fig. 2. Test waves to illustrate the power of the wavelet transform to extract wave components in frequency-time decomposition.

transform for this signal given in the second panel of Fig. 2 shows that all three waves can be identified. In particular, the start and end times of the high-frequency waves can be determined from this transformation result. The second test wave is a chirp signal in which the frequency is changing with time, i.e. a non-stationary signal. Non-stationary signals pose a severe challenge to Fourier transform, which is incapable of giving a proper result for this type of waves. On the other hand, the wavelet transform below this test wave shows the remarkable capability of wavelet transform to identify the nature of the signal. The final test wave is the sum of all the previous two test waves. Again, the wavelet transform result indicates the presence of all these four components in the frequency-time domain.

Current disruption is a magnetospheric phenomenon associated with highly variable magnetic field fluctuations. It is typically observed close to the neutral sheet and is perceived to be a magnetic field reconfiguration

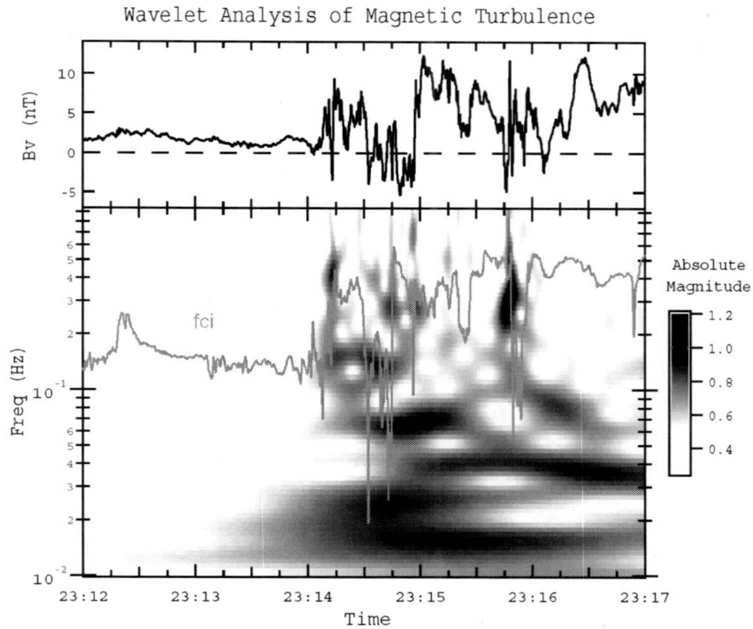

Fig. 3. Wavelet analysis of magnetic fluctuations in a current disruption event on June 1, 1985 observed by the AMPTE/CCE satellite to reveal the inverse cascade and intermittency features.

such that the highly stretched magnetic field lines in the pre-substorm-onset become relaxed to a more dipolar geometry. Wavelet analysis has been applied to investigate the nature of the magnetic field fluctuations in this phenomenon.[2] Figure 3 shows the wavelet transform result for the B_v component from a current disruption event observed by the AMPTE CCE located at $\sim$8.8 R_E downstream in the tail near the midnight magnetic local time.[3] The curve on the frequency-time spectrogram denotes the instantaneous proton-cyclotron frequency. The wavelet coefficients are normalized to the wave amplitude. The Morlet mother wavelet is employed [see Lui and Najmi[2] for details]. It may be seen that low-frequency waves around $\sim$0.03 Hz started to appear at $\sim$23:13:40 UT. The high-frequency waves around 0.2-0.6 Hz started shortly after, coincident with the onset of current disruption. Interestingly, there is a trend of the high-frequency waves descending in frequency to the lower frequency range, suggesting the possibility that the high-frequency wave energy gradually decays to the low frequency wave energy. This trend is interpreted as an inverse cascade phenomenon to contrast it with the MHD turbulence where low-frequency waves in the inertial range decays to the high-frequency waves for dissipation. This finding is the first report of inverse cascade within the magnetosphere. Recent observations from THEMIS satellites reveal similar inverse cascade feature in current disruption.[4] Another feature in current disruption revealed in wavelet analysis is the intermittent excitation of the high-frequency waves. As noted before, the high-frequency waves are excited at ∼23:14:10 UT. Another excitation of the high-frequency waves are seen at ∼23:15:40 UT having a similar inverse cascade feature. One can extend wavelet analysis to wavelet bicoherence to examine wave-wave coupling.[5]

3. Phase Space Density Analysis

Fast plasma flows are often observed in the magnetotail during intervals of activity. Figure 4 shows a fast plasma flow interval on December 10, 1996 observed by Geotail at a downstream distance of ∼25 R_E in the premidnight sector. The tailward plasma flows at ∼17:07–17:13 UT, which are mainly along the magnetic field, are accompanied by mostly southward magnetic field. This event has been reported as evidence of magnetic reconnection

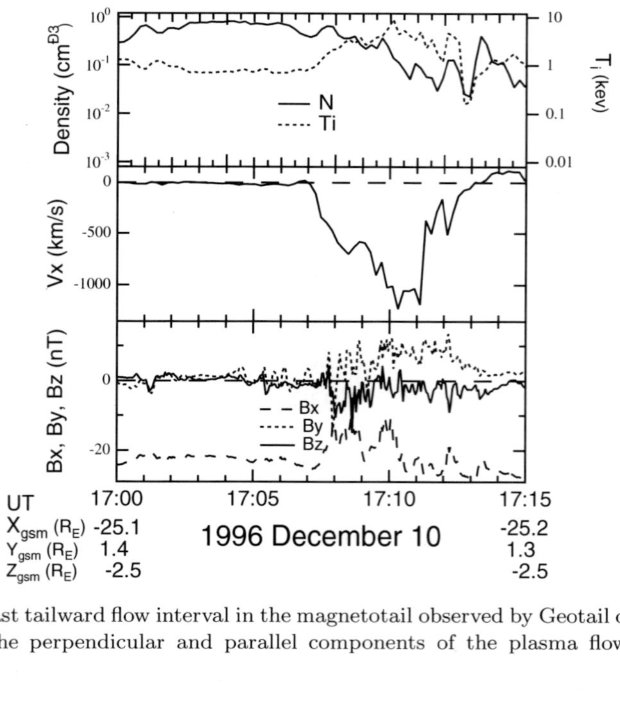

Fig. 4. A fast tailward flow interval in the magnetotail observed by Geotail on December 10, 1996. The perpendicular and parallel components of the plasma flow are shown separately.

by several authors previously.[6] Further understanding of the phenomenon requires one to ask whether these flows represent the motion of the entire population or a combined effect due to different motions of more than one population. Evidence for more than one population being observed at that time is revealed by the examination of the phase space density plots.[7,8]

Figure 5 shows the time sequence of the velocity distribution in the $V_x - V_y$ phase space plane during this activity. It can be seen that before

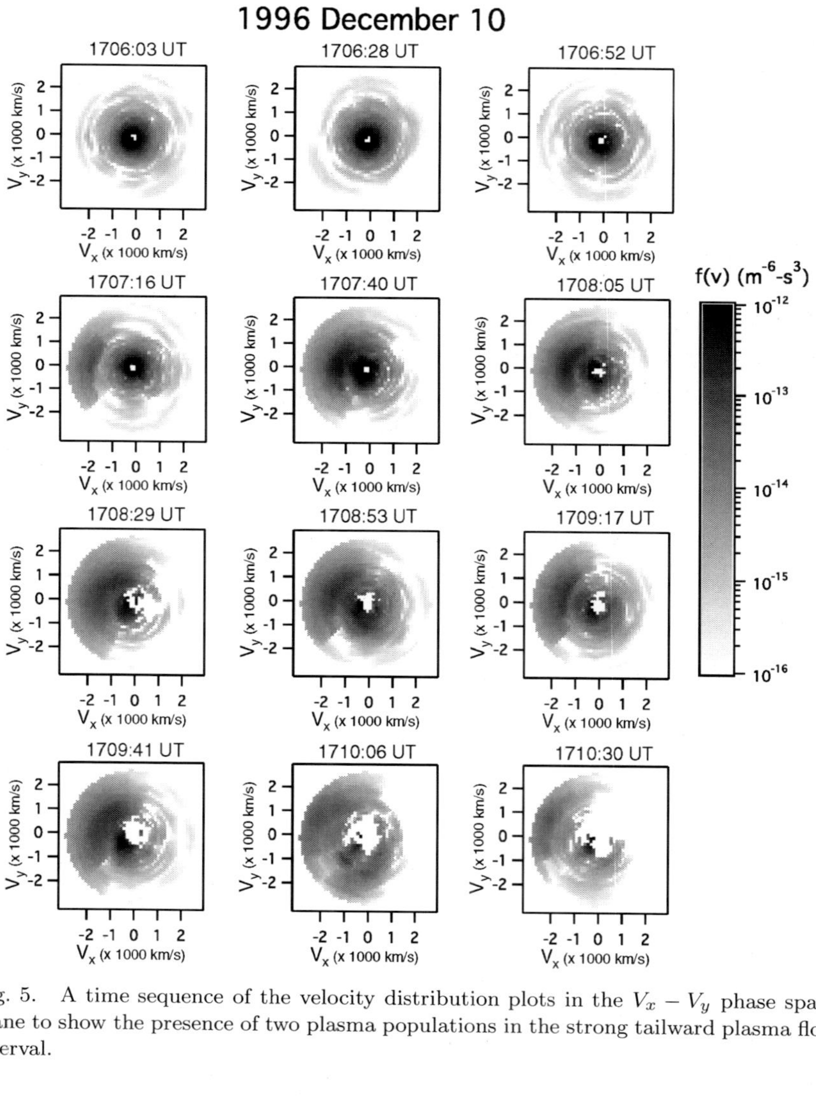

Fig. 5. A time sequence of the velocity distribution plots in the $V_x - V_y$ phase space plane to show the presence of two plasma populations in the strong tailward plasma flow interval.

the appearance of tailward flow, there is only one population. However, another population appears at ~17:07 UT when strong tailward flow is detected. This strong tailward flow population persists to the end of the interval shown while the stationary population gradually diminishes.

A careful analysis of the observed phase space density that allows the presence of multiple population shows two populations in the fast tailward flow period, as shown in Fig. 6. The solid line (moment approach) represents what is obtained by assuming one population only. The multiple-component procedure recognizes intervals when there are two populations and calculates the moments for each population separately. It can be seen that the two populations behave very differently. One population has a high-speed tailward flow while the other population stays nearly motionless. The former is referred to as the beam component while the latter is referred to as the local component. The beam component has a tailward speed up to ~2400 km/s, faster than the local Alfvén speed (~1700 km/s), and is slightly depleted in O^+ ions compared with the local component.

One may gain further insight into the phenomenon by carrying out additional analysis of the phase space density measurements.[9] Parks *et al.* (2001)[10] have also emphasized the importance of kinetic approach rather than fluid approach in analyzing magnetospheric dynamics. It is well documented that the magnetospheric plasma population has a Kappa distribution,[11] i.e.

$$f(v_\parallel, v_{\perp 1}, v_{\perp 2}) = f_0 \left(1 + \frac{(v_\parallel - v_0)^2}{ku_\parallel^2} + \frac{(v_{\perp 1} - v_1)^2 + v_{\perp 2}^2}{ku_\perp^2}\right)^{-(k+1)} \quad (3)$$

where

$$f_0 = \frac{n_0 \Gamma(\kappa + 1)}{(\kappa\pi)^{3/2}\Gamma(\kappa - 1/2) u_\parallel u_\perp^2}.$$

If we denote the magnetic field at the satellite location as B_s and that at the particle source location as B_0, then for conservation of energy and first adiabatic invariant (letting w to represent the velocity at the satellite location), we have

$$v_\parallel^2 + v_{\perp 1}^2 + v_{\perp 2}^2 = W_\parallel^2 + W_{\perp 1}^2 + W_{\perp 2}^2. \quad (4)$$

$$\frac{v_{\perp 1}^2 + v_{\perp 2}^2}{B_0} = \frac{W_{\perp 1}^2 + W_{\perp 2}^2}{B_s}. \quad (5)$$

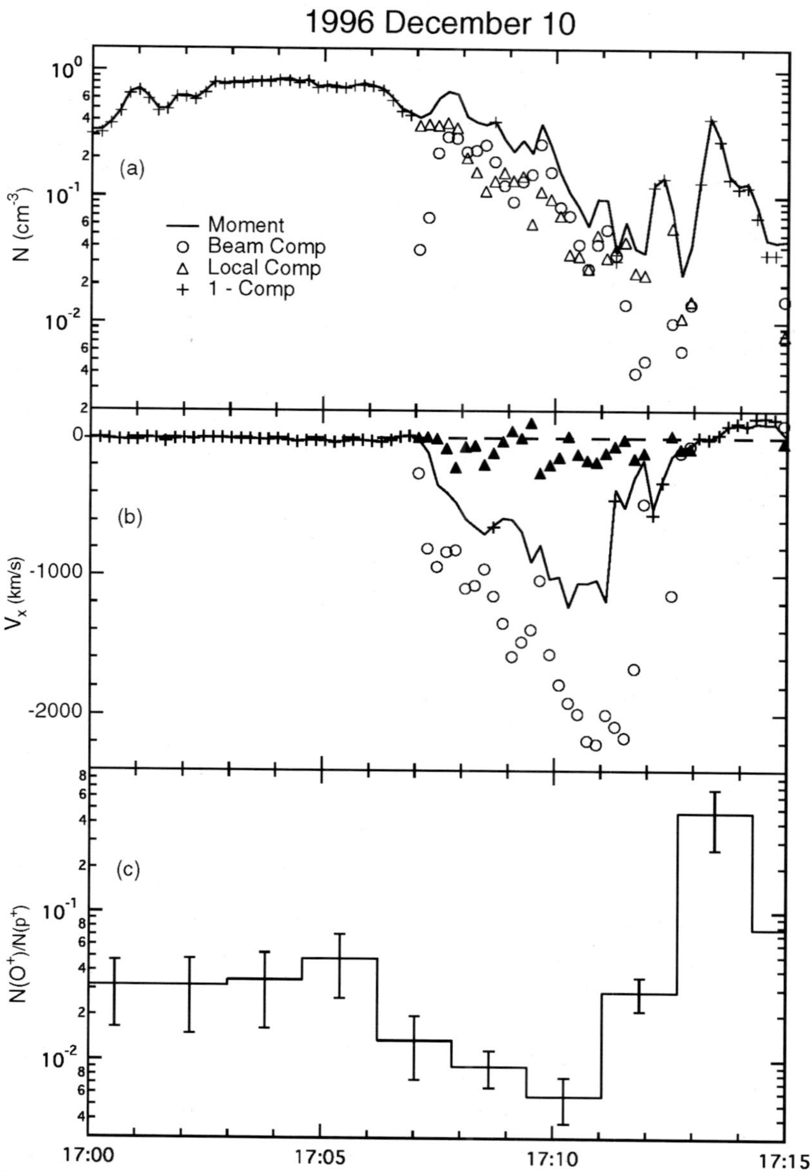

Fig. 6. A procedure separating the moments of the two populations shows that one plasma population (beam component) has high-speed flow while the other plasma population (local component) is practically stationary. The ratio of number density of O^+ ions and protons indicates that the fast flowing population is not oxygen-rich., i.e. the beam is not an oxygen ion beam.

Combining Eqs. (4) and (5) gives

$$v_{\parallel}^2 = W_{\parallel}^2 + (1-\beta)(W_{\perp 1}^2 + W_{\perp 2}^2). \tag{6}$$

$$v_{\perp 1}^2 + v_{\perp 2}^2 = \beta(W_{\perp 1}^2 + W_{\perp 2}^2). \tag{7}$$

where $\beta = B_o/B_s \leq 1$ for adiabatic compression where the magnetic field at the source is less than that at the satellite location. If we further assume that there is an energy gain through an electric potential ϕ, which can be acquired through an electric field parallel to the magnetic field or by drifting across the dawn-dusk electric field in the tail, then the velocities between the two locations for a particle of mass m and charge e are related by

$$v_{\parallel}^2 + v_{\perp 1}^2 + v_{\perp 2}^2 + 2e\varphi/m = W_{\parallel}^2 + W_{\perp 1}^2 + W_{\perp 2}^2. \tag{8}$$

leading to

$$v_{\parallel}^2 = W_{\parallel}^2 + (1-\beta)(W_{\perp 1}^2 + W_{\perp 2}^2) - 2e\varphi/m \tag{9}$$

Using Liouville's theorem to relate the two populations gives

$$f(W_{\parallel}, W_{\perp 1}^2 W_{\perp 2}^2) = f_0 \left\{ 1 + \frac{\left[\sqrt{W_{\parallel}^2 + (1-\beta)((W_{\perp 1} - v_1)^2 + W_{\perp 2}^2) - \frac{2e\varphi}{m}}\right] - v_0^2}{ku_{\parallel}^2} + \frac{\beta[(W_{\perp 1 - v_1})^2 + W_{\perp 2}^2]}{ku_{\perp}^2} \right\}^{-(k+1)}. \tag{10}$$

subject to the condition

$$W_{\parallel}^2 + (1-\beta)(W_{\perp 1}^2 + W_{\perp 2}^2) \geq 2e\varphi/m. \tag{11}$$

In other words, the velocity components and the magnetic field at the source location, as well as the electric potential experienced by the ensemble from the source location to the satellite location, can be obtained from a detailed investigation of the observed velocity distribution at the satellite location.

As an illustration of this technique, let us investigate the velocity distribution of the event shown in Figs. 4 and 6. The observed three-dimensional phase space density is fitted with the expression (10) using the nonlinear Lavenburg-Marquardt fitting procedure.[12] The following result is obtained. For the local component, $n_0 = 0.45\,\text{cm}^{-3}$, $v_{||} = 67\,\text{km/s}$, $v_\perp = 67\,\text{km/s}$, $u_{||} = 228\,\text{km/s}$, and $u_\perp = 240\,\text{km/s}$. For the beam, $n_0 = 0.014\,\text{cm}^{-3}$, $v_{||} = 1392\,\text{km/s}$, $v_\perp = 286\,\text{km/s}$, $u_{||} = 286\,\text{km/s}$, $u_\perp = 282\,\text{km/s}$, $\beta = B_0/B_s = 0.15$, $\phi = 6.3\,\text{kV}$, and $\kappa = 4.6$. The degree of freedom in the fitting of the cold (hot) population is 1770 (153) and a chi square value is 1108 (161), indicating a good fit. From this fitting procedure, one can deduce that the beam originates from a source location with a substantially lower magnetic field than that at the satellite location, consistent with the expectation that it comes from the neutral sheet location. Figure 7 shows the comparison between the observed velocity distribution and the modeled one in two phase space planes. The resemblance between the two is quite remarkable.

One can carry this nonlinear fitting procedure to the entire tailward plasma flow interval. The result is shown in Fig. 8. There is considerable continuity of the extracted parameters. The thermal speed of the source for the beam population increases with time at the beginning of the event, indicating heating with activity onset. There is a general trend of $\beta = B_0/B_s$ increasing with time, indicating that the magnetic field at the source location increasing with time, consistent with the trend of magnetic field in a dipolarization site. The time history of electric potential suggests a substantial initial acceleration lasting less than 1 min, followed by a gradual enhancement later. Overall, the phase space density analysis suggests that the observed tailward flows are due to a beam population originating from a dipolarization site, rather than a magnetic reconnection site.

The success of the technique relies on the conservation of the first adiabatic invariant for the ensemble. One may question whether this conservation can be applied to current disruption activity. Obviously, we do not have observations at the current disruption site for this particular event. However, if one could use the current disruption event on Jun 1, 1986 (shown in Fig. 3) as a representative case, one may notice that waves over a broadband frequency range are excited. In other words, the associated waves are not at a monochromatic ion cyclotron frequency. This implies that only a small portion of the population will have their first adiabatic invariant violated while most of the particles in the main population have their first adiabatic invariant conserved. Furthermore, the good agreement in fitting the observed population is another indication that this assumption holds for the main population.

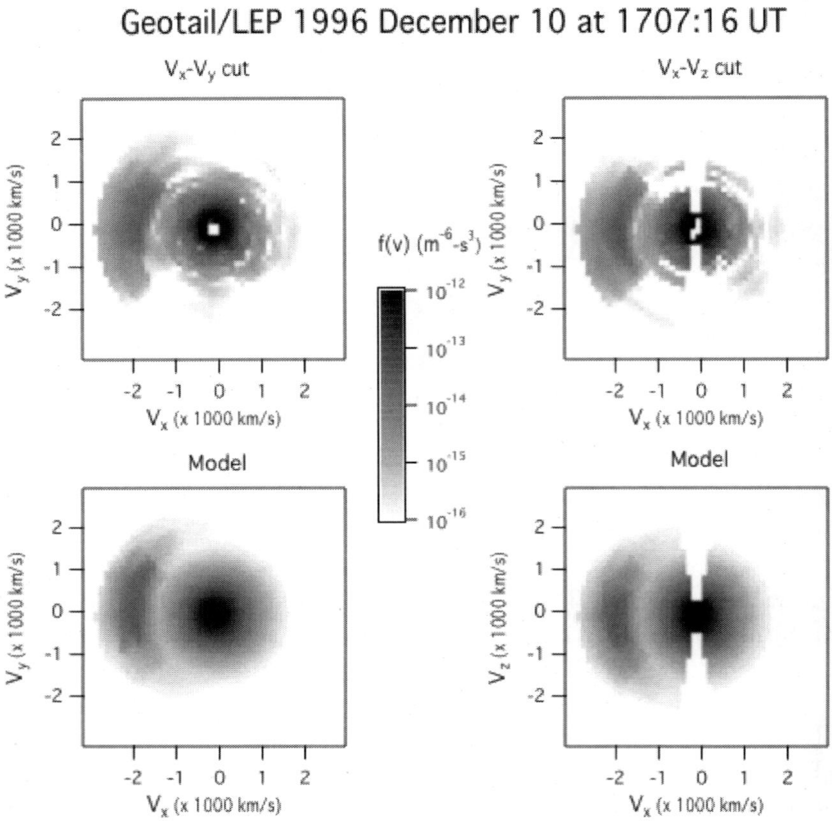

Fig. 7. Top: Observed velocity distribution of the plasma for the $V_x - V_y$ and $V_x - V_z$ planes. Bottom: Detailed modeling of the observed velocity distribution during an interval to extract information of the plasma parameters at the source location and the electric potential experienced by the source population. The nonlinear Lavenburg-Marquardt fitting procedure is used.

4. Grad-Shafranov Reconstruction

Reconstruction of plasma configuration from plasma observation from a single satellite is based on solving the Grad-Shafranov equation for a two-dimensional MHD equilibrium as a spatial initial value problem, a procedure initiated by Sonnerup and Gao (1996).[13] Many papers have been published using this technique to examine magnetopause structures.[14–20] The equation is

$$\left(\frac{\partial^2}{\partial x^2} + \frac{\partial^2}{\partial y^2}\right) A = -\mu_0 \frac{dP_t(A)}{dA} \tag{12}$$

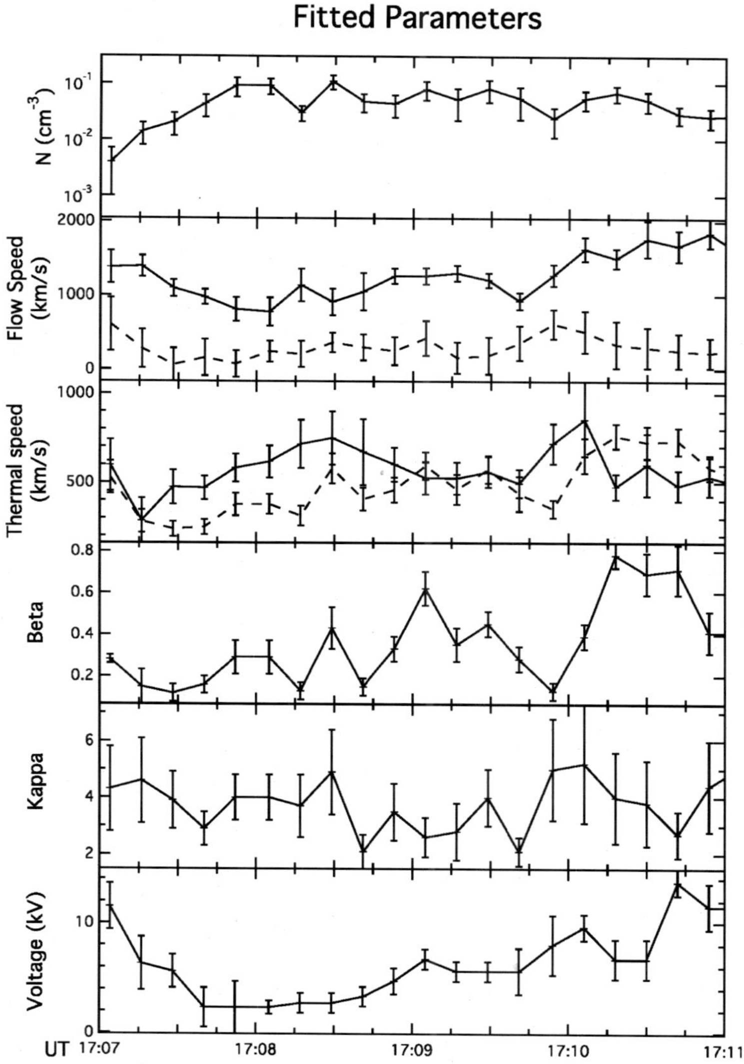

Fig. 8. Plasma parameters extracted from the phase space density analysis on the observed three-dimensional phase space densities of the beam population. The solid and dashed lines in the velocity panels refer to the beam and local populations, respectively.

where the transverse pressure is given by $P_t = p + B_z^2/2\mu_0$. The magnetic field vector **B** is related to the partial vector potential $A(x, y)$ and the axial magnetic field B_z by $\mathrm{B} = \nabla A(x, y) \times \hat{z} + B_z(A)\hat{z}$. The invariant axis ($z$-axis) is the direction along which the structure changes much more gradually in comparison with that in the plane perpendicular to this axis. In identifying this invariant axis, one needs to perform the minimum variance analysis. To examine the equilibrium condition of the structure, one also needs to perform the deHoffman-Teller transformation and verify that a suitable deHoffman-Teller frame can be obtained.[14] The suitable plane for reconstruction is found by trial and error such that the transverse pressure P_t and the B_z component along the invariant axis have a simple dependence on the partial vector potential A. The procedure determines A from integrating the time series of the B_y component along the satellite path. The Grad-Shafranov equation is then solved by designing P_t as a function of A. If the minimum variance axis shows a poor determination of the invariant axis or the deHoffman-Teller transformation shows the lack of a two-dimensional equilibrium configuration or the dependence of P_t and B_z on A is multi-value, then the structure cannot be constructed by this procedure. The above conditions are quite stringent, indicating that not all observed structures can be readily reconstructed under this scheme.

One may test the validity of such reconstruction with a theoretical model as documented in Hau and Sonnerup (1999)[14], namely, let

$$A(x, y) = ln\left\{a \sin(x) + \sqrt{1 + a^2} \cosh(y)\right\}. \tag{13}$$

where $\alpha = 0.225$, the plasma pressure by $p = e^{-2A}/(3\mu_0)$, and the axial magnetic field by $B_z = e^{-A}/\sqrt{3}$. For testing, the model structure is rotated by 5.7^o with respect to the satellite path, as done in Hau and Sonnerup.[14]

Figure 9 shows the comparison between the theoretical model and the numerical solution of the Grad-Shafranov equation from the model values along a satellite path.[21] There is hardly any noticeable difference between the two, demonstrating the accuracy of the procedure when it is applicable.

This reconstruction technique has been applied to analyze a magnetopause structure sampled by the THEMIS satellites. Figure 10 shows a magnetic flux rope observed by THEMIS D on May 20, 2007 as reported by Sibeck *et al.* (2008).[22] In the early part of this interval, THEMIS D observed a high number density, fast tailward-duskward flow and highly variable magnetic field with a southward component. These are signatures of magnetosheath. At the end of this interval, the plasma has a low number density, almost stagnant in terms of flow, and a fairly steady magnetic field with a northward component. These are signatures of magnetosphere after the crossing of the dusk magnetopause. At ~2202 UT, a feature is detected with no resemblance to the magnetosheath and magnetospheric plasma. In

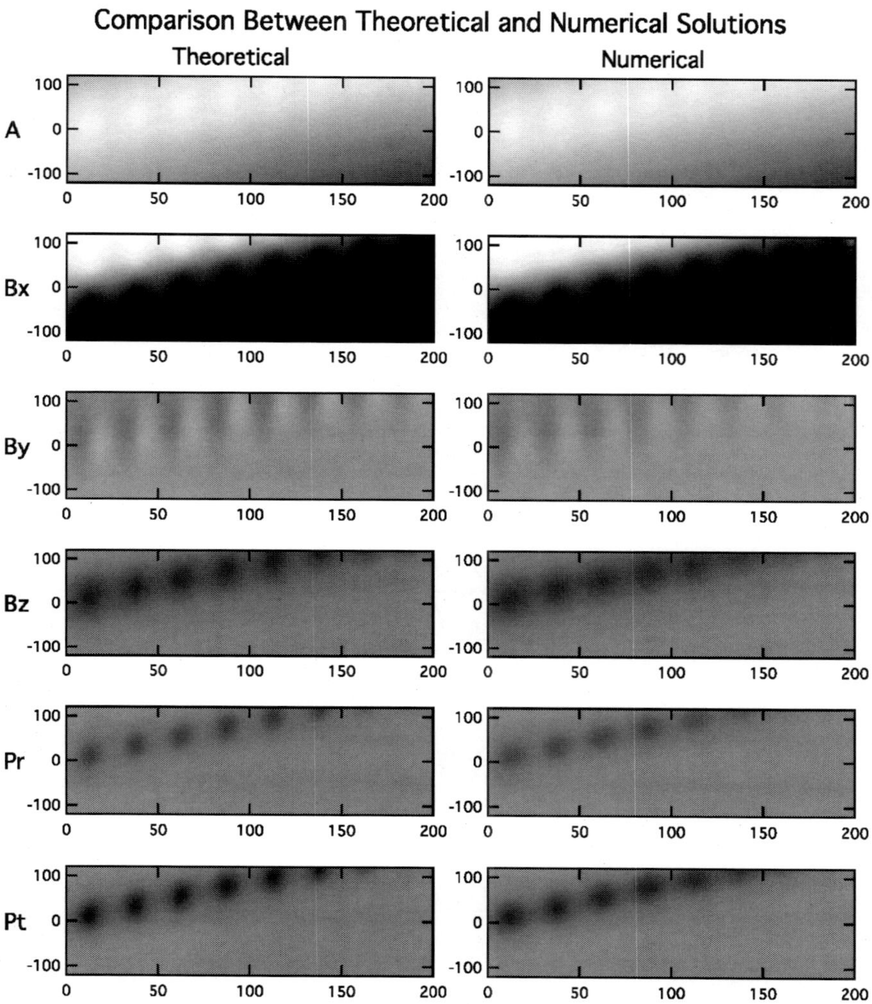

Fig. 9. A comparison between the theoretical model and the numerical solution in Grad-Shafranov reconstruction.

particular, the magnetic field is much larger than that in the two plasma environments. The bipolar signature in the B_x component and the high values of B_t and B_z suggest that this is a magnetic flux rope.

The minimum variance analysis and the deHoffman-Teller transformation are applied to this magnetic flux rope interval and the results are shown in Fig. 11. These results indicate a good determination of the invariant axis and a suitable deHoffman-Teller frame can be obtained in

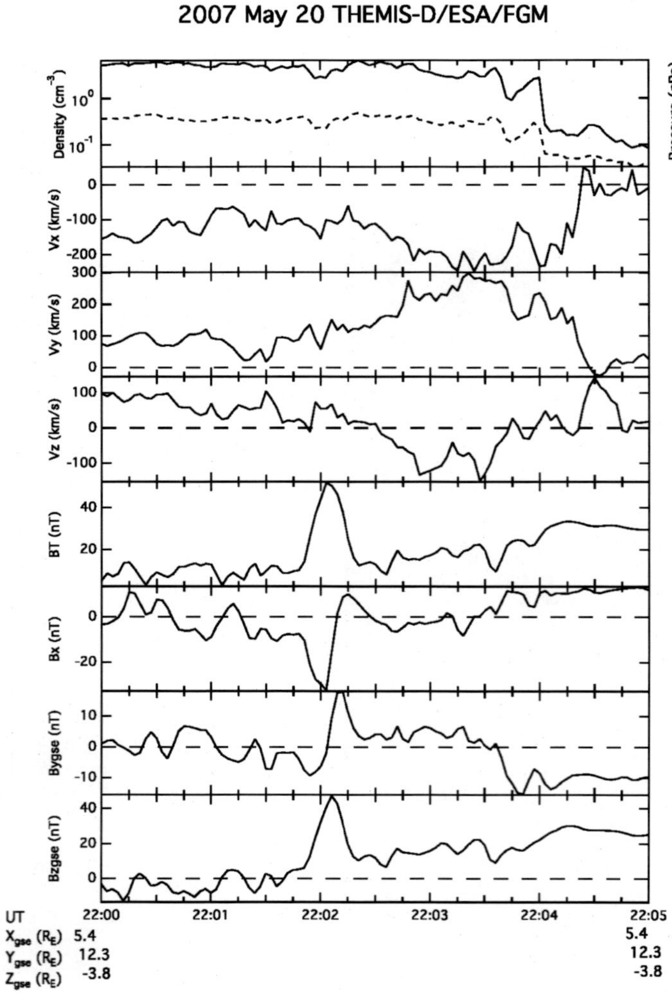

Fig. 10. Observation of a magnetic flux rope by THEMIS-D on the dusk magnetopause on May 20, 2007.

which the structure does not change appreciably, i.e. we indeed have a two-dimensional equilibrium structure. After these criteria are satisfied, the reconstruction is carried out by integrating the Grad-Shafranov equation and the parameters from the solution are transformed back to the coordinate system where the observed parameters are measured. Figure 12 shows the comparison between the observed parameters and the modeled parameters along the satellite path. It can be seen that the comparison

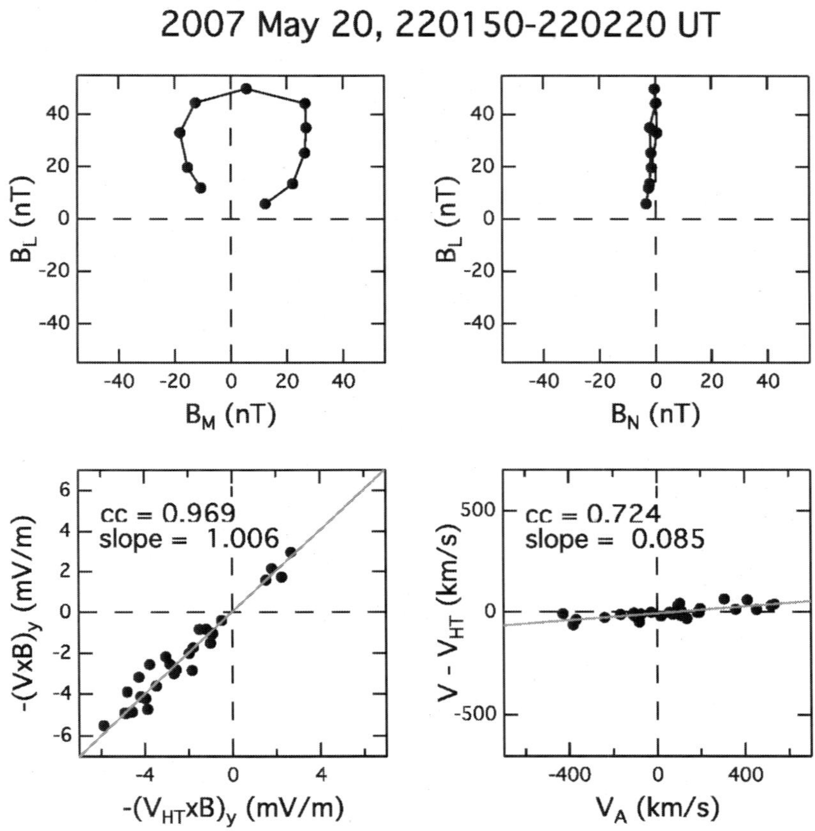

Fig. 11. The results of minimum variance analysis and deHoffman-Teller transformation applied to the magnetic flux rope interval.

is exceedingly good. A quantitative comparison of all three magnetic field components between the observed and modeled values yields a correlation coefficient of 0.998.

Two-dimensional maps of parameters based on the reconstruction are shown in Fig. 13, extending the vision of the measured parameters from the one-dimension along the satellite path. Current density can be derived from the assumption that the variation on the plane perpendicular to the invariant axis is negligible, i.e. it is basically a two-dimensional structure. Again, the current density has to be calculated on the plane perpendicular to the invariant axis. This ability of deriving the current density is extremely useful because in general one cannot obtain the three components of the current density based on magnetic field measurements from one single

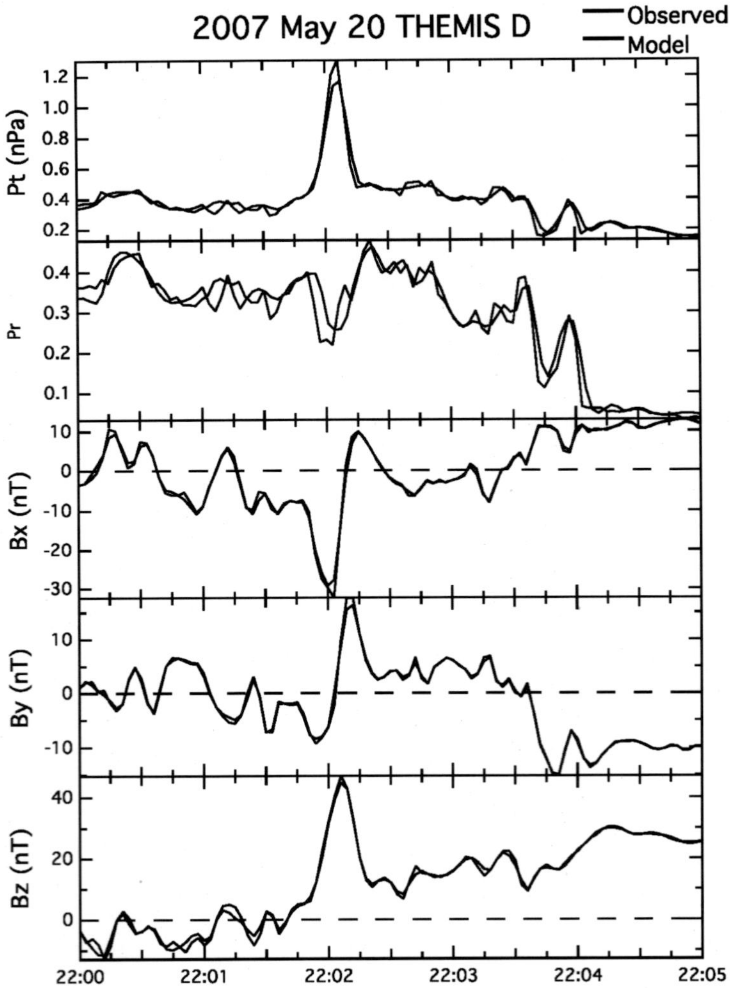

Fig. 12. Comparison between observed parameters and modeled parameters based on the Grad-Shafranov reconstruction technique. There is very good agreement between the two profiles of each of the parameters.

satellite. It can be seen that the magnetic field component and the current density component are strongest in the z-axis, as expected for a magnetic flux rope. The current density can be as high as $\sim 50\,\mathrm{nA/m^2}$. Another feature that can be noted is the depression of the plasma pressure in the core of the magnetic flux rope. This arises from the strong magnetic field component along the rope axis and the balance of the total transverse

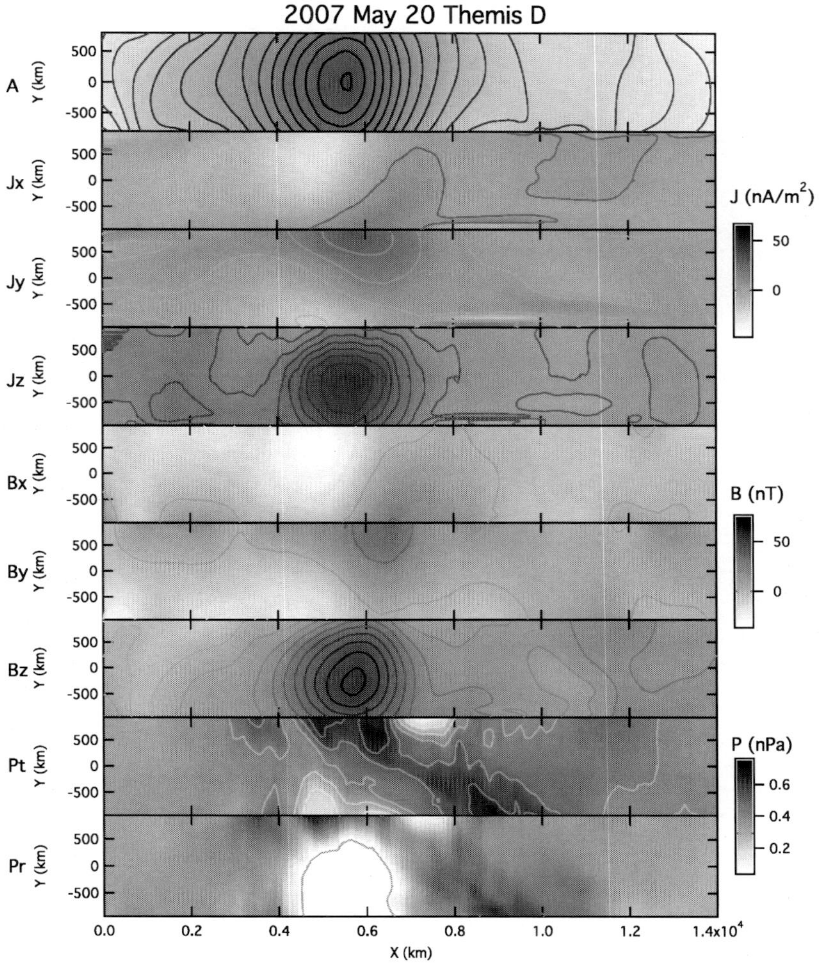

Fig. 13. Two-dimensional maps of magnetic field components, current density components, the transverse pressure, and the plasma pressure based on the Grad-Shafranov reconstruction technique.

pressure balance across the structure requires a significant depression of the plasma pressure.

Since there are five THEMIS satellites crossing the dusk magnetopause nearly simultaneously, the Grad-Shafranov reconstruction solutions of these five satellites can be combined to provide a more extended vision (in the direction perpendicular to the satellite path) of the plasma environment around the magnetic flux rope.[23] Figure 14 shows the two-dimensional maps

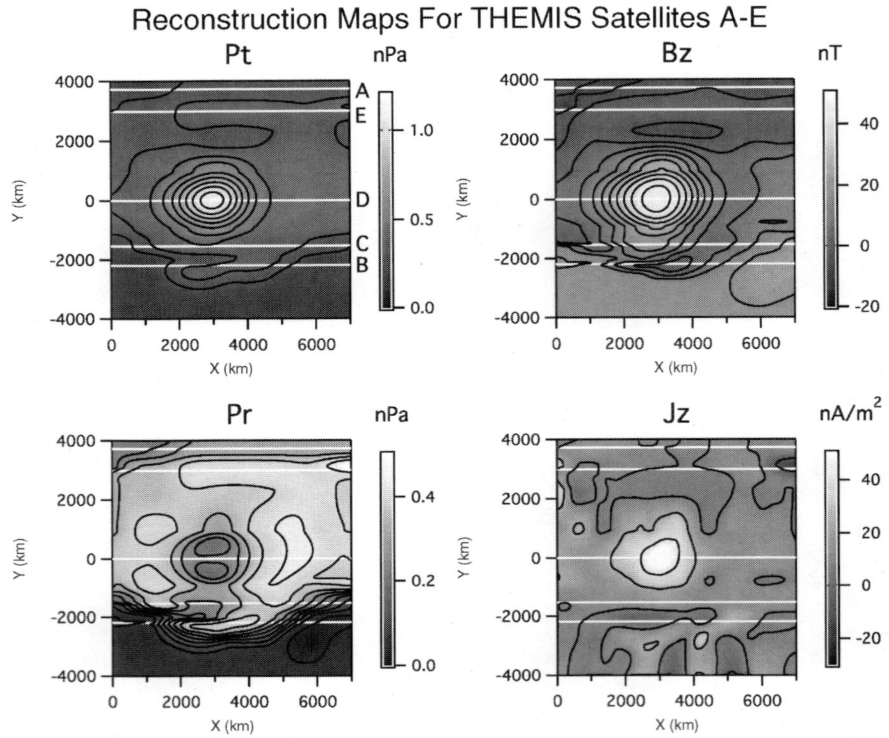

Fig. 14. Two-dimensional maps of the transverse pressure, the B_z component, the plasma pressure, and the J_z component of the magnetic flux rope intervals based on combining reconstructions from all five THEMIS satellites.

of transverse pressure, the z-component of the magnetic field, the plasma pressure, and the z-component of the current density from the combined reconstructions. The satellite paths are given in the five horizontal lines in each map. It can be seen that THEMIS D indeed passes very close to the core of the magnetic flux rope. THEMIS C passes near the edge of the magnetic flux rope. THEMIS B passes even further from the magnetic flux rope core, mostly within the magnetosphere. THEMIS A and E are quite far from the magnetic flux rope, mostly within the magnetosheath.

5. Summary and Conclusion

We have briefly described three different modern analysis tools that can be employed to reveal interesting features of magnetospheric phenomena.

The wavelet analysis allows one to decompose signals in the frequency-time domain and gain insights on inverse cascade and intermittency of the observed signal. The phase space density analysis allows one to unravel moments from more than one population measured in the velocity distribution and to probe the past history of the observed plasma in ascertaining the plasma parameters in the source region and the electric potential encountered. The Grad-Shafranov reconstruction allows one to extend the vision of plasma environment from the one-dimensional view along the satellite path to the two-dimensional view around the satellite path, as well as deriving current densities from measurements from a single satellite without assuming a planar current sheet.

Acknowledgment

This work is supported by NSF grant ATM-0630912 and NASA grant NNX09AI53G to the Johns Hopkins University Applied Physics Laboratory.

References

1. G. Paschmann and P.W. Daly, ISSI Science Report, SR-008, 17–21, (Kluwer Academic Pub. 2008)
2. A. T. Y. Lui and A.-H. Najmi, Time-frequency de-composition of signals in a current disruption event, *Geophys. Res. Lett.* **24** (1997) 3157–3160.
3. A. T. Y. Lui *et al.*, A case study of magnetotail current sheet disruption and diversion, *Geophys. Res. Lett.* **15** (1988) 721–724.
4. A. T. Y. Lui *et al.*, Inverse cascade feature in current disruption, *J. Geophys. Res.* **113** A00C06, doi:10.1029/2008JA013521 (2008a).
5. G. Consolini and A. T. Y. Lui, Symmetry breaking and nonlinear wave-wave interaction in current disruption: Possible evidence for a dynamical phase transition, *Magnetospheric Current Systems*, ed. by S. Ohtani *et al.*, *AGU Monograph 118*, (AGU, Washington, DC 2000) pp. 395–401.
6. T. Nagai *et al.*, Counterstreaming ions as evidence of magnetic reconnection in the recovery phase of substorms at the kinetic level, *Phys. Plasmas* **9**(9) (2002) 3705–3711.
7. A. T. Y. Lui *et al.*, Plasma transport from multi-component approach, *Geophys. Res. Lett.* **32** L06101, doi:10.1029/2004GL021891 (2005).
8. A. T. Y. Lui and T. Hori, Phase space density analysis of energy transport in the Earth's magnetotail, *Space Sci. Rev.* **122**(1–4) (2006) 69–80.
9. A. T. Y. Lui, Extraction of source location parameters from observed particle velocity distribution, *Geophys. Res. Lett.* **33** L21108, doi:10.1029/2006GL027922 (2006).

10. G. K. Parks *et al.*, Kinetic characterization of plasma sheet dynamics, *Space Sci. Rev.* **95** (2001) 237–255.
11. S. P. Christon *et al.*, Spectral characteristics of plasma sheet ion and electron populations during undisturbed geomagnetic conditions, *J. Geophys. Res.* **94** (1989) 13409.
12. W. H. Press *et al.*, *Numerical Recipes in Fortran 77*, (Cambridge University Press, Cambridge, UK 1992).
13. B. U. Ö Sonnerup and M. Guo, Magnetopause transects, *Geophys. Res.Lett.* **23** (1996) 3679–3682.
14. L.-N. Hau and B. U. Ö Sonnerup, Two-dimensional coherent structures in the magnetopause: Recovery of static equilibria from single-spacecraft data, *J. Geophys. Res.* **104** (1999) 6899–6917.
15. Q. Hu and B. U. Ö Sonnerup, Reconstruction of two-dimensional structures in the magnetopause: Method improvements, *J. Geophys. Res.* **108** A11011, doi:10.1029/2002JA009323 (2003).
16. W.-L. Teh and L.-N. Hau, Evidence for pearl-like magnetic island structures at dawn and dusk side magnetopause, *Earth Planets Space* **56** (2004) 681–686.
17. W.-L. Teh and L.-N. Hau, Triple crossings of a string of magnetic islands at duskside magnetopause encountered by AMPTE/IRM satellite on 8 August 1985, *J. Geophys. Res.* **112** A08207, doi:10.1029/2007JA012294 (2007).
18. H. Hasegawa *et al.*, Reconstruction of two-dimensional magnetopause structures from Cluster observations: verification of method, *Ann. Geophys* **22** (2004) 1251–1266.
19. H. Hasegawa *et al.*, Optimal reconstruction of magnetopause structures from Cluster data, *Ann. Geophys* **23** (2005) 973–982.
20. H. Hasegawa *et al.*, The structure of flux transfer events recovered from Cluster data, *Ann. Geophys* **24** (2006) 603–618.
21. A. T. Y. Lui *et al.*, Reconstruction of a magnetic flux rope from THEMIS observations, *Geophys. Res. Lett.* **35**, L17S05, doi:10.1029/2007GL032933 (2008b).
22. D. G. Sibeck *et al.*, Crater FTEs: Simulation results and THEMIS observations, *Geophys. Res.Lett.* **35** L17S06, doi:10.1029/2008GL033568 (2008).
23. A. T. Y. Lui *et al.*, Composite Grad-Shafanov reconstruction of a magnetic flux rope based on observations from five THEMIS satellites, *J. Geophys. Res.*, **113**, A00C01, doi:10.1029/2008JA013189 (2008c).

Advances in Geosciences
Vol. 21: Solar & Terrestrial Science (2008)
Ed. Marc Duldig

REMOTE SENSING OF EARTH'S PLASMASPHERE

A. K. SINGH
Department of Physics, University of Lucknow, 226 007 Lucknow
aksphys@gmail.com

Interest in the palsmasphere has recently surged with the development of new or more refined remote sensing tools and with an associated increase in awareness of region's geophysical importance. The interactions between the cool plasmasphere and hot plasma injected during storms and substorms, role of instabilities, wave induced particle scattering, coupling of plasmasphere to the nonuniform underlying ionosphere or inward displacements of the plasmapause and steepening of its density profile need more attention. From its (plasmasphere) initial detection in 1953 to the present day, remote sensing by whistlers have also provided the fact that a light ion gas (protonosphere) floats on the heavier ion gas. Whistler-mode probing, both by whistlers and transmitter signals, continues to have great potential far remote sensing of plasma motions, time variations and density structures. In the present paper, I have reviewed the latest development in ground based as well as satellite observations of plasmaspheric conditions. In short, we can say that plasmasphere has remained a challenging target and remote sensing will surely play an important role in future research.

1. Introduction

The Earth's plasmasphere is an inner part of the magneteosphere. It is located just outside the upper ionosphere located in Earth's atmosphere. It is a region of dense, cold plasma that surrounds the Earth. Although plasma is found throughout the magnetosphere, the plasmasphere usually contains the coldest plasma. The plasmasphere is the torus of plasma in the inner most region of magnetosphere, comprising the corotating of the magnetosphere. Magnetic field lines there are closed and approximately dipolar, permitting filling of the plasmasphere by plasma escaping from the Earth's ionosphere. Cold plasma with energy lower than 1–2 eV and density of 100–1,000 cm^{-3} trapped by the geomagnetic field plays the dominating role in this region. Hydrogen ions with 10–20% of helium and 5–10% of oxygen ions mainly composed the plasmasphere, which is essentially the extension of ionosphere to the higher altitudes.[1]

Based on studying the whistler propagation along the geomagnetic field, Storey[2] concluded that the electron density at an altitude of $\sim$12,000 km in the equatorial region is about 400 cm^{-3} and assumed that he measured the density in the interplanetary space (in that case particle energy was unknown). This was the first evidence for the existence of plasma far beyond normal ionosphere. The existence of plasmasphere was proved by Morgan and Allcock.[3] Shortly afterwards, the discovery of nose whistlers[4] led to the possibility of determining whistler path latitude and permitted detailed studies of plasma distribution in the plasmasphere. Many quantitative data about the structure of this part of the Earth's plasma environment were obtained.[5–8] The plasmasphere was independently identified by *in-situ* plasma measurements and ground based whistler observations in the early 1960s. Gringauz[9] used plasma observations from the Lunik moon probes to identify a steep density gradient corresponding to the outer edge of the plasmasphere. From the whistler studies, Carpenter[10] detected a sharp decrease in the electron density in the geomagnetic equator plane, initially called "a Knee". This knee was identified with the region of a sharp decrease in the ion density discovered on the Soviet satellites.[11] Carpenter[12] subsequently proposed the term "plasmasphere" and "plasmapause" by the analogy with the terms "magnetosphere" and "magnetopause", which was accepted by the scientific community. Figure 1 shows the schematic view of the Earth's plasmasphere and magnetosphere.

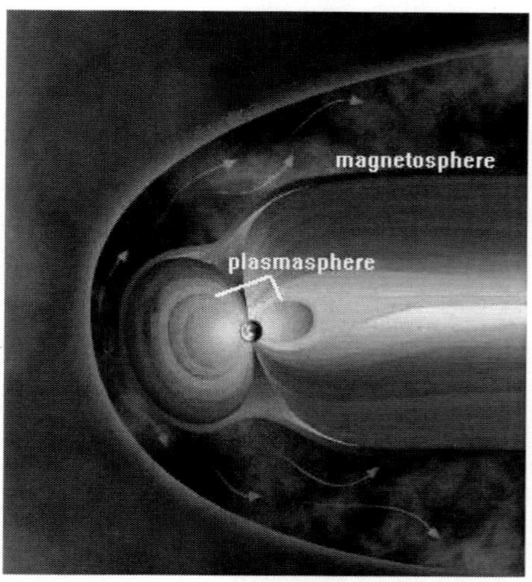

Fig. 1. Schematic of the Earth's plasmasphere and magnetosphere (http://science.nasa.gov).

In the present paper, we have reviewed the data of various plasmaspheric parameters obtained from whistlers recorded at ground stations. In fact whistler waves have been used as diagnostic tool for remote sensing of the Earth's plasmasphere and in the study of the plasma distribution in the region.

2. Whistlers and Plasmasphere

Whistlers are radio frequencies produced by lightning. These radio waves travel away from Earth and then come back again following the Earth's magnetic field lines. On the way back, they pass through the magnetosphere, the plasmasphere and the ionosphere. Whistlers provided much of the early information on the global shape of the plasmasphere, on its equatorial density profile, and on the fact that it undergoes cycles of erosion and recovery during magnetically disturbed periods. In particular, discovery of a bulge-like extension in plasmasphere radius in the dusk sector[12] stimulated early efforts to interpret the shape of the plasmasphere in terms of the inter play between the motion of plasma imposed by the Earth's rotation and the generally sunward flow in the middle magnetosphere driven by the solar wind as it impinges upon the magnetosphere.[13,14] As data were accumulated over time and at many locations, it became possible to identify major temporal variations in the density of the plasmasphere, ranging in the periods from hours to eleven years[15] One of the more pronounced and, for many years, least well understood of these is an annual variation.[16,17] The internal difficulty in understanding the inter play between the comparatively rapidly changing ionosphere and the overlying but more slowing varying plasmasphere is illustrated by the fact that mechanisms that could explain the annual variation have been found both in the properties of the ionosphere and in plasmasphere thermal structure.[18] Whistlers also made possible study of the interchange of plasma between the ionosphere and overlying plasmasphere, providing measures of the rate at which upward fluxes from the ionosphere refilled depleted overlying regions,[19] and clarifying the role of plasmasphere as a reservoir for the decaying nighttime ionosphere.[20] Many plasmaspheric parameters like electron density, total electron content of a flux tube, downward transport of flux of electrons, large scale electric fields, characteristic properties of ducts present in plasma, electron temperature, etc. can be determined from whistler studies.[17,21–23]

3. Plasmaspheric Density

Ion density distribution in the plasmasphere has been studied rather thoroughly since this parameter can be obtained during almost any measurements in the plasmasphere. The first experiments indicated that plasma density distribution strongly depends on geomagnetic activity[1]; therefore, we first consider plasma distribution under quiet geomagnetic conditions. The ion density distribution in the geomagnetic equator plane has been thoroughly studied based on the whistler observation data. The ion and electron density was also determined in the wave and plasma experiments on different satellites.[24–27] Under quiet geophysical conditions near the geomagnetic equator plane, cold plasma density decreases with increasing distance from the Earth, and a sharp outer plasmaspheric boundary is often not observed and density smoothly decreases in the cases when magnetic conditions remain quiet during several days. The density decrease rate ($d \log N_{eq}/dL$) varies from -4 to -3.[27–29] Plasma density in the plasmasphere depends mostly on local time, latitude and observation season. The density daily variations near the equator are ±10–15% for $L \sim 3$, and the density is minimal after midnight.[30] According to whistler observations, which make it possible to almost continuously monitor the equatorial plasmasphere. Figure 2 shows the variation of equatorial electron density as a function of L value for the whistlers recorded at different

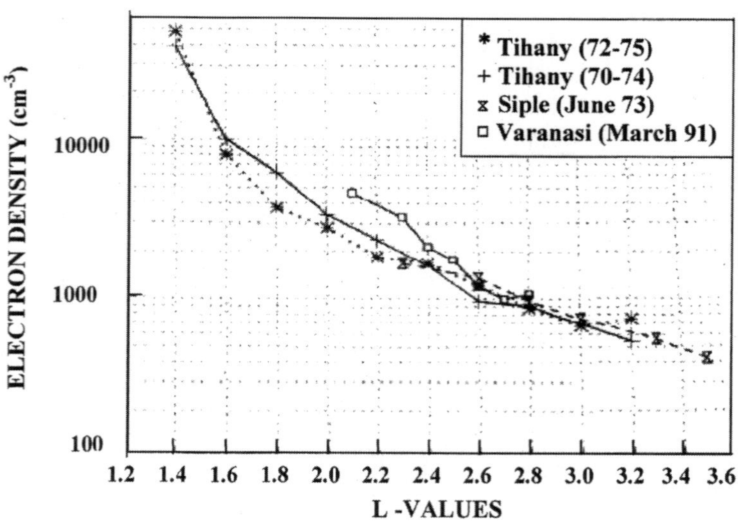

Fig. 2. Variation of plasmaspheric electron density with L-values.[21]

ground based stations.[22] The electron density sharply increases as the L value decreases.

The average electron density derived from whistlers recorded at Tihany ($L = 1.8$) and Siple ($L = 4$) are superimposed on each other, and show electron density variations between 5×10^4 electrons cm^{-3} and 10^3 electrons cm^{-3} as the L values changes from 1.4 to 3.5. In the same figure, we have presented the electron density derived from whistlers recorded at low latitude station of Varanasi ($L = 1.07$) that has the same order of magnitude and the same trend. The Varanasi data belongs to moderate magnetic activity ($K_p = 3 - 4$). Tihani data were divided into March to August, 1972–1975 (summer) and October to February, 1970–1974 (winter) periods. For $L > 2.4$, there is a lower density in summer than in the winter. Lichtenberger *et al.*,[31] analyzing whistlers recorded on the Intercosmos 24 satellite, reported N_{eq} between 520–400 electrons cm^{-3} as L changes between 3.40 and 3.65. Tarcsai *et al.*[28] have argued that, as a consequence of the difference between the magnetic and rotational axes of the Earth, the phenomenon depends not only on the magnetic latitude but also on longitude. In the absence of data from the eastern hemisphere, it is not possible to study longitudinal dependencies. Errors are introduced in the determination of N_{eq} values due to unavoidable approximations in the whistler analysis.[32] At L-values ($L < 2$), there is an increasing tendency to underestimate the electron densities whereas, for $L > 2$, the errors contribute little to the observed variations in N_{eq} and N_T (total electron content in a flux tube). At low latitudes a substantial part of a whistler path lies in the ionosphere. Park[33] gave an empirical formula to estimate the time delay due to ionospheric path, assuming that the maximum contribution comes from the F_2 — layer of ionosphere. Usually part of the sub-ionospheric path at low latitude lies in the Earth-ionosphere waveguide and a corresponding correction to the propagation delay should be taken into account. The correction would vary from event to event and can be evaluated only when source location and ionospheric exit location from the duct are precisely known.

4. Plasmaspheric Electric Fields

4.1. *Convection-corotation model*

The plasmasphere is the region in the inner magnetosphere where the corotation electric field dominates over the cross tail electric field. The cross tail electric field, directed from dawn to dusk, tends to produce equipotentials parallel to the Sun-Earth axis. The geomagnetic field, rotating with the Earth, produces a $\mathbf{v}\times\mathbf{B}$ electric field about the Earth. The outer boundary of the plasmasphere, the plasmaspause is approximately located where these two fields are equal, i.e. near an L value such that

$$E_T = (B_0/L^3)LR_E\omega \tag{1}$$

where E_T is the cross tail field, B_0 is the magnetic field strength at the Earth's surface and equator ($B_0 = 0.30\,\text{G}$), R_E is the Earth's radius, and ω is the Earth's rotational speed. For a typical plasmapause location of $L = 4R_E$, $E_T = 1\,\text{mV/m}$.[12] Magnetic field lines in the plasmasphere are closed and nearly dipolar. The toroidal surface of the plasmaphere intersects the Earth's surface at high latitudes (for $L = 4$, MLAT $= 60°$), roughly corresponding to the equatorward boundary of the auroral zone. Beyond the plasmasphere, magnetic field lines are drawn westward and stretched down the magnetotail on the night side. An early modification to the simple convection-corotation is the Stern-Volland model[34,35] which includes a semi-empirical shielding effect. Previously mentioning shielding electric fields result from polarization in the plasmasheet as it approaches the Earth, tending to shield the inner magnetosphere from the convection electric field.[36] Other than Stern-Volland model, there are some other models also which describes the effects of the electric field on plasmapause position.[37,38]

4.2. *Convection model*

Plasmaspheric studies show dynamic behaviour in response to solar activity and geomagnetic conditions. The basic convection-corotation model suggests a teardrop shape that will change in the size with changes solar/geomagnetic activity. These changes are observed along with additional storm related dynamics and structures. During periods of high solar/geomagnetic activity, the region of closed field lines is compressed closer to the Earth. At such times the plasmapause may be at $L = 2$, compared to $L = 4-6$ in typical conditions. When considered relative to K_p index, the response in plasmapause location is delayed on the order of hours. Carpenter and Park[39] found an empirical relation for the plasmapause location $L_{PP} = 5.7 - 0.47\ K_P$, valid for the midnight-dawn sector and using the maximum K_p index value over the preceding 12 hours. O'Brien and Moldwin[40] evaluated degree of correlation for relation for K_P, Dst, and AE indices for various time delays.

4.3. *Ionosphere-plasmasphere coupling*

A number of efforts have been made for modeling the coupling between the ionosphere and the Plasmasphere. Bailey *et al.*[41] describe a multi-species fluid model in which time-dependent equation of continuity, momentum and energy balance are solved along electric dipole magnetic field lines for densities, field aligned fluxes, and temperatures of different ion species and electrons. The model was applied to study the effects of the vertical $\mathbf{E} \times \mathbf{B}$ drift, natural wind, and the plasma fountain on the distribution of electrons in the equatorial topside ionosphere. Moffett *et al.*[42] describe a coupled model of the thermosphere, ionosphere and the plasmasphere. The model has been applied to study the behavior of field aligned ion fluxes in the topside ionosphere in the mid-to-low latitude regions. It is found that interhemispheric asymmetries can arise readily from the different offsets of the geomagnetic and geographic poles without any asymmetry in the inputs. A similar model was used by Guiter *et al.*[43,44] to study the distribution of O^+ ions. The model is interhemispheric and simultaneously solves the equations of continuity, momentum and energy equation for a two-ion (H^+ and O^+), quasi-neutral, current less plasma. They concluded that O^+ is responsible for the annual density variation in the plasmasphere and that it critically determines other plasmaspheric parameters.[43,44] Cole[45] determined a plasmaspheric impedance on the ionospheric current system from longitudinal pressure gradients. Comfort *et al.*[46] re-examined thermal conductivities along plasmaspheric field lines. The effects of anisotropic thermal conductivity on the temperature structure of the ionosphere-plasmasphere have been described by Khazanov *et al.*[47]

4.4. *Energetic particle — Plasmasphere coupling*

In the ionosphere when extreme ultraviolet light ionizes upper atmospheric neutral particles, photoelectrons are created. Because of their high mobility, they can escape from the ionosphere and quickly traverse the field line through the Plasmasphere and a few of them get scattered into the geomagnetic trap. Once here they are long lived and slowly deposit their energy to the thermal plasma in this region. These various stages of photoelectron evolution can all influence the cold plasma population, and all have been studied in recent years.[47,48] The ring current region is a high-energy belt, which can influence the plasmasphere structure and dynamics through the injection of a high energy plasma component into the plasmasphere. The ring current and plasmasphere can interact with each other via multiple processes such as Coulomb collisions, charge exchange, and wave-particle interactions. Coupling of the ring current with the plasmasphere due to Coulomb collisions was addressed by Fok *et al.*[49] They found that as the ring current ion energy degrades, a low-energy (<1 keV) ion population is formed in a background of thermal plasma. The energy transferred from the ring current ions to the plasmasphere results in enhanced plasma temperature at high latitude. Liemohn *et al.*[50] examined the electrodynamical coupling of the ring current with the cold plasma during the early stages of refilling, determining that its effect is quite minor.

5. Plasmasphere: After IMAGE and CLUSTER Observations

In the past several years, new techniques have been developed for observing the plasmasphere. From the ground, interpretation of magnetometer data[51] and signals from GPS satellites[52] provide prototomographic capabilities. From space, magnetospheric imaging achieves a global perspective previously only provided by models. The Imager of Magnetopause-to-Aurora Global Exploration (IMAGE) satellite[53] was launched in 2000 with two plasmasphere instruments onboard. The radio plasma instrument (RPI)[54] uses active radio wave sounding to determine remote electron density, and has yielded some needed information about the density dependence along magnetic field lines.[55] The extreme ultraviolet (EUV) imager[56] routinely obtains full global images of the Earth's plasmasphere by remote sensing solar 30.4-nm light that has been resonantly scattered by plasmaspheric He^{+} ions. Plasmaspheric imaging has indeed confirmed the existence of plumes. Based on the CLUSTER observations, Darroutzet *et al.*[57] described the density structure inside the plasmasphere. Recently Darroutzet *et al.*[58] have studied the plasmaspheric density structures and behavior of plasma in the Plasmasphere using CLUSTER and IMAGE data. Empirical models for the plasma densities in the inner magnetosphere, including plasmasphere and polar magnetosphere have been derived.[59] Pierrard *et al.*[60] have described the recent progress in physics-based models of the plasmasphere using the fluid and kinetic approaches. Recently, Darrouzet *et al.*[61] have reviewed all the recent discoveries on the plasmasphere made with CLUSTER and IMAGE observations.

6. Conclusions

The plasmasphere and upper atmosphere continue to be challenging fields of research involving global and mesoscale problems as well as microscale plasma processes. In this paper, I have reviewed the latest development in ground based as well as satellite observations of plasmaspheric conditions. Combining data from ground and/or various spacecraft as well as model simulations will help scientists to arrive at a more global picture of the state of plasmasphere. Electron content data routinely inferred from the propagation of radio signals between spacecraft or between spacecraft and the ground undoubtedly will play the important role. IMAGE and CLUSTER missions have helped to better relate plasmaspheric wave phenomena with plasmaspheric density structures, and heavy particles content of the plasmasphere, better locate the source of waves and how they propagate. The distribution of field-aligned density structures in space and time throughout plasmasphere are not yet well documented. Electric field distribution is critical in modeling the plasmasphere dynamics, need to develop higher time resolution empirical models.

Acknowledgments

AKS is thankful to Department of Science and Technology (DST), Government of India for providing financial support as a research project (File No. SR/S4/AS: 261/06). Author is thankful to referees for their critical comments and suggestions.

References

1. J. F. Lemaire and K. I. Gringauz, with contribution from D. L. Carpenter and V. S. Bassolo, The Earth's plasmasphere, Cambridge Univ. Press, Cambridge (1998).
2. L. R. Storey, An investigation of whistling atmospheric, *Phil. Trans. Roy. Soc. (London)* **246** (1953) 113–141.
3. M. G. Morgan and G. M. Allcock, Observations of whistling atmospherics at geomagnetically conjugate points, *Nature* **177** (1957) 30–31.
4. R. A. Helliwell, J. H. Crary, J. H. Pope and R. L. Smith, The "nose" whistler — a new high latitude phenomena, *J. Geophys. Res.* **61** (1956) 139–142.
5. R. L. Smith and R. A. Helliwell, Electron densities to 5 Earth radii deduced from nose whistlers, *J. Geophys. Res.* **65** (1960) 2583.

6. R. L. Smith, Properties of the outer ionosphere deduced from nose whistlers, *J. Geophys. Res.* **66** (1961) 3709–3716.
7. J. H. Pope, An estimate of electron densities in the exosphere by means of nose whistlers, *J. Geophys. Res.* **66** (1961) 67–75.
8. D. L. Carpenter, Electron density variations in the magnetosphere deduced from whistler data, *J. Geophys. Res.* **67** (1962) 3345–3360.
9. K. I. Gringauz, The structure of the ionized gas envelope of Earth from direct measurements in the USSR of local charged particle concentrations, *Planet. and Space Sci.* **11** (1963) 281–296.
10. D. L. Carpenter, Whistler evidence of a "knee" in the magnetospheric ionization density profile, *J. Geophys. Res.* **68** (1963) 1675–1682.
11. D. L. Carpenter, Whistler measurement of the equatorial profile of magnetospheric electron density, *Radio Science 1960–1963, vol. III, The ionosphere*, Ed. by G. M. Brown, Elsevier (1965) 76–91.
12. D. L. Carpenter, Whistler studies of the plasmapause in the magnetosphere. 1. Temporal variation in the position of the knee and some evidence on plasma motions near the knee, *J. Geophys. Res.* **71** (1966) 693–709.
13. A. Nishida, Formation of plasmapause, or magnetospheric plasma knee, by the combined action of magnetospheric convention and plasma escape from the tail, *J. Geophys. Res.* **71** (1966) 5669–5679.
14. N. M. Brice, Bulk motion of the magnetosphere, *J. Geophys. Res.* **72** (1967) 5193–5211.
15. C. G. Park, D. L. Carpenter and D. B. Wiggin, Electron density in the plasmasphere: whistler data on solar cycle, annual and diurnal variations, *J. Geophys. Res.* **83** (1978) 3137–3144.
16. M. A. Clilverd, A. J. Smith and N. R. Thomson, The annual variation in quiet time plasmaspheric electron density, determined from whistler mode group delays, *Planet. Space Sci.* **39** (1991) 1059–1069.
17. A. K. Singh, The study of inner magnetosphere by VLF waves, *Ph. D. Thesis.* Banaras Hindu University, India (1995).
18. P. G. Richards, T. Chang and R. H. Comfort, On the causes of the annual variation in the plasmaspheric electron density, *J. Atoms. Solar-Terr. Phys.* **62** (2000) 935–946.
19. C. G. Park, A morphological study of sustorm-associated disturbances in the ionosphere, *J. Geophys. Res.* **79** (1974) 2621–2827.
20. C. G. Park and P. M. Banks, Influence of thermal plasma flow on the mid latitude night-time F_2 layer; effects of electric fields and neutral winds inside the plasmasphere, *J. Geophys. Res.* **79** (1974) 4661–4668.
21. R. P. Singh, Whistler studies at low latitudes: a review, *Ind. J. Rad. Space Phys.* **22** (1993) 139–155.
22. R. P. Singh, A. K. Singh and D. K. Singh, Plasmaspheric parameters as derived from whistler spectrograms: a Review, *J. Atmos. Solar-Terr. Phys.* **60** (1998a) 495–508.
23. A. K. Singh and R. P. Singh, Duct Life Times at Mid Latitudes, *Periodica Polytechnica* **43** (1999) 43–52.
24. D. A. Gurnett, R. R. Anderson, F. L. Scarf *et al.*, Initial results from ISEE-1 and 2, plasma wave investigation, *Space Sci. Rev.* **23** (1979) 103–122.

25. K. I. Gringauz and V. S. Bassolo, Structure and properties of the Earth's plasmasphere (a review), *Geomagn. Aeron.* **30** (1990) 1–17.
26. A. K. Singh, Abhay K. Singh, D. K. Singh and R. P. Singh, Thermal effect on the dispersion characteristics of proton whistlers, *J. Atmos. Solar-Terr. Phys.* **60** (1998b) 551–561.
27. G. A. Kotova, V. V. Bezrukikh, M. I. Verigin *et al.*, INTERBALL1/Alpha 3 cold plasma measurements in the evening plasmasphere: quiet and disturbed magnetic conditions, *Adv. Space Res.* **30** (2002) 1831–1834.
28. G. Tarcsai, P. Szemeredy and L. Hegymegi, Average electron density profiles in the plasmasphere between $L = 1.4$ and 3.2 deduced from whistlers, *J. Atmos. Terr. Phys.* **50** (1988) 607–611.
29. D. L. Carpenter and R. R. Anderson, An ISEE/whistler model of equatorial electron density in the magnetosphere, *J. Geophys. Res.* **97** (1992) 1097–1108.
30. C. G. Park, D. L. Carpenter and D. B. Wiggin, Electron density in the plasmasphere: Whistler data on solar cycle, annual, and diurnal variations, *J. Geophys. Res.* **83** (1978) 3137–3144.
31. J. Lichtenberger, G. Tarcasi *et al.*, Whistler doublets and hyperfine structure recorded digitally by the signal analyzer and sampler on the active satellite, *J. Geophys. Res.* **96** (1991) 21149–21158.
32. G. Tarcsai, J. H. Strangeways and M. J. Rycroft, Error sources and travel time residuals in plasmaspheric whistler interpretation, *J. Atmos. Terr. Phys.* **51** (1989) 249–258.
33. C. G. Park, *Tech. Report*: Methods of determining electron concentration in the magnetosphere from nose whistlers, Stanford University, California (1972).
34. H. Volland, A semi empirical model of large-scale magnetospheric electric fields, *J. Geophys. Res.* **78** (1973) 171–180.
35. D. P. Stern, The motion of a proton in the equatorial magnetosphere, *J. Geophys. Res.* **80** (1975) 595–599.
36. J. Goldstein, B. R. Sandel, W. T. Forrester, M. F. Thomsen and M. R. Hariston, Global plasmasphere evolution 22–23 April 2001, *J. Geophys. Res.*, **110**: (2005) A12218, doi: 10.1029/2005JA011282.
37. M. V. Liemohn, A. J. Ridley, D. L. Gallagher, D. M. Ober and J. U. Kozyara, Dependence of plasmaspheric morphology on the electric field description during the recovery phase of the 17 April 2002 magnetic storm, *J. Geophys. Res.* **109** (2004) A03209, 10.1029/2003JA010304.
38. V. Pierrard, G. Khazanov, J. Cabrera and J. Lemaire, Influence of the convection electric field models on predicted plasmapause positions during the magnetic storms, *J. Geophys. Res.* **113** (2008) A08212, 1–21, doi:10.1029/2007JA012612.
39. D. L. Carpenter and C. G. Park, On what ionosphere workers should know about the plasmapause-plasmasphere, *Rev. Geophys. Space Phys.* **11** (1973) 133–154.
40. T. P. O'Brien and M. B. Moldwin, Empirical plasmapause models from magnetic indices, *Geophys. Res. Lett.* **30** (2002) 1152, doi:10.1029/2002GL016007.

41. G. J. Bailey, N. Balan and Y. Z. Su, The Sheffield University plasmasphere-ionosphere model — a review, *J. Atmos. Solar Terr. Phys.* **59** (1997) 1541–1552.
42. R. J. Moffett, G. H. Millward, S. Quegan, A. D. Aylward and T. J. Fuller-Rowel, Results from a coupled model of the thermospohere, ionosphere and plasmasphere (CTIPM), Advances in Space Research, **8** (1996) 33–39.
43. S. M. Guiter, T. E. Moore and G. V. Khazanov, Modeling of O^+ ions in the plasmasphere, *J. Geophys. Res.* **100** (1995a) 21921–21928.
44. S. M. Guiter, C. E. Rasmossen, T. I. Gombosi, J. J. Sojka and R. W. Schunk, What is the source of observed annual variations in plasmaspheric density? *J. Geophys. Res.* **100** (1995b) 8013–8020.
45. K. D. Cole, A new role of the plasmasphere in "quiet" ionospheric electrodynamics, *J. Atmos. Terr. Phys.* **57** (1995) 805–812.
46. R. H. Comfort, P. D. Craven and P. G. Richards, A modified thermal conductivity for low density plasma magnetic flux tubes, *Geophy. Res. Lett.* **22** (1995) 2457–2460.
47. G. V. Khazanov, M. V. Liemohn, J. V. Kozyara and T. E. Moore, Inner magnetospheric super thermal electron transport: photoelectron and plasmasheet electron sources, *J. Geophys. Res.* **103** (1998) 23485–23501.
48. G. V. Khazanov and M. V. Liemohn, Nonsteady state ionosphere-plasmasphere coupling of superthermal electrons, *J. Geophys. Res.* **100** (1995) 9669–9681.
49. M. C. Fok, P. D. Carven, T. E. Moore and P. G. Richards, Ring current plasmasphere coupling through Coulomb collisions. In Horwitz, J. L., Singh N., Burch, J.L. (Eds.), Cross-scale coupling in space plasma, *Geophysical Monograph, AGU* **93** (1995) 161–171.
50. M. V. Liemohn, G. V. Khazanov, P. D. Craven and J. U. Kozyara, Nonlinear kinetic modeling of early stage plasmaspheric refilling, *J. Geophys. Res.* **104** (1999) 10295.
51. Z. C. Dent, L. R. Mann, F. W. Menk, J. Goldstein *et al.*, *Geophys. Res. Lett.* **30** (2003) 1600.
52. J. C. Foster, P. J. Ericson, A. J. Coster and J. Goldstein, Ionospheric signatures of plasma tails, *Geophys. Res. Lett.* **29** (2002), 10.1029/2002GL015067.
53. J. L. Burch, M. S. B. Mitchell, D. G. Moore, T. E. Pollock *et al.*, *Science* **291** (2001) 619.
54. B. W. Reinisch, X. Huang, D. Haines *et al.*, First results from the Radio Plasma Imager on IMAGE, *Geophys. Res. Lett.* **28** (2001) 1167–1170.
55. B. W. Reinisch, X. Huang, P. Song *et al.*, *J. Geophys. Res.* **109** (2004) A01202, 10.1029/2003JA009948.
56. B. R. Sandel, R. A. King, W. T. Forrester *et al.*, Initial results from the IMAGE Extreme Ultraviolet Imager, *Geophys. Res. Lett.* **28** (2001) 1439–1442.
57. F. Darrouzet *et al.*, Density structures inside the Plasmasphere: Cluster observations, *Annales Geophysicae* **22** (2004) 2577–2585.

58. F. Darrouzet, D. L. Gallagher and J. Tu, Plasmaspheric density structures and dynamics: Properties obsewrved by Cluster and Image missions, *Space Science Review* **145** (2009a) 55–106.
59. B. W. Reinisch *et al.*, Augmented empirical models of plasmaspheric density and electric field using IMAGE and CLUSTER data, *Science Review* **145** (2009) 231–261.
60. V. Pierrard *et al.*, Recent progress in physics-based model of Plasmasphere, *Space Science Review* **145** (2009) 193–229.
61. F. Darrouzet, J. De Keyser and V. Pierrard (eds.), The Earth's Plasmasphere: CLUSTER and IMAGE — A Modern Perspective, p. 296, Springer, (2009b).

Advances in Geosciences
Vol. 21: Solar & Terrestrial Science (2008)
Ed. Marc Duldig

COMPLEX PLASMA PHYSICS AND THE EARTH'S DUSTY MESOSPHERE

S. V. VLADIMIROV[*,†] and B. A. KLUMOV[†,‡]

[*] *School of Physics, The University of Sydney, NSW 2006, Australia*
S.Vladimirov@physics.usyd.edu.au

[†] *Partner Group for Complex Plasmas, Joint Institute for High Temperatures RAS, 13/19 Izhorskaya St, Moscow 125412, Russia*

[‡] *Max-Planck-Institut für extraterrestrische Physik, 85748 Garching, Germany*
klumov@mpe.mpg.de

The middle atmosphere (mesosphere) of Earth is the site of numerous physical processes important to its evolution and dynamics especially in the region close to the mesopause at heights of 80–90 km. Some of the visible manifestations of these phenomena are clouds of icy "dust" particles called Polar Mesospheric Clouds (PMC), when viewed from space, and Noctilucent Clouds (NLC), when viewed by observers on Earth. Although most observations of NLC are in the northern hemisphere, they can also be observed in the southern hemisphere, with some differences, e.g. in the peak altitudes of NLC. Besides the NLC, dust leads to the polar mesospheric summer echoes (PMSE) appearing as a result of electromagnetic wave scattering from mesospheric electron inhomogeneities; the latter can result from charging of electrons onto subvisible aerosol dust particles. The presence of dust adds high level of complexity to plasma dynamics and collective processes. The electrostatic charge on dust grains is an important element of numerous dust-related plasma phenomena. The dust charging process is governed by the contributions of all electric currents to/from the grain surface, involving the plasma electron and ion currents, the photoemission and thermionic emission currents, etc. The dust grains convert a dusty plasma system into a highly dissipative open complex system where formation of various structures becomes energetically favourable.

1. Introduction

Particles of different sizes (dust), ranging from a few nanometers to tens of micrometers, have been often observed in nature as well as in various laboratory experiments.[1–5] Dust can be found in interstellar space, planetary atmospheres, ring structures, cometary tails, etc. For example, dust grains play an important role in the synthesis of molecular species in diffuse nebulae and directly participate in various astrophysical processes.[6] Planetary atmospheres, including the Earth's atmosphere, and planetary rings are yet another example of the complex plasma system in space environments.[2,7] Under terrestrial conditions, dust particles are important constituents of various colloidal suspensions, charged aerosols, high-pressure ionised gas systems, nucleation catalysts, and many others.[3,8] Dust particles with sizes in the nano-/micrometer range have been detected in low-temperature plasmas used for materials synthesis and processing applications, where fine dust powders appear in the mixtures of chemically active (reactive) gases.[3,4,9]

In various plasma environments, dust particles are usually charged.[4,5] Hannes Alfvèn was one of the very first scientists to suggest that electromagnetic forces acting on small charged dust particles played an important role in the formation of the entire Solar System.[10] In particular, he recognised planetary rings as an important laboratory for studying processes that not only act today, but also shaped the evolution of the early Solar System billions of years ago. There is a rich variety of phenomena observed in complex dusty plasmas complemented by the wide range of unsolved problems.[2,5,7,8] The outstanding problems are mostly related to the extremely high complexity of such plasmas often exhibiting unusual and unpredictable behaviour.[5,11,12]

The Earth's mesosphere, approximately situated at heights from 50 to 100 km as a part of the Earth's middle atmosphere,[13] is the site of a number of intriguing phenomena which are not well understood yet. It includes a rich variety of physical processes that are fundamentally important to its evolution and dynamics.[16–19] This is especially true for the region close to the mesopause appearing in summer at heights of 80–90 km[20–22] which is of interest here (in winter months, the mesopause in the polar regions is at altitudes above 90 km[14,15]). Some of the visible manifestations of these phenomena are related to mysterious clouds called Polar Mesospheric Clouds (PMC), when viewed from space, and Noctilucent Clouds (NLC), when viewed by observers on the Earth.[23,24]

While observations of the trails of meteors appear to be as old as mankind, the first reported observations of NLC are from 1885.[25] It is suggested that the occurrence of strong and easily visible NLC at that occasion was a consequence of the gigantic eruption of Krakatoa volcano in 1883, although it is also possible that their first observation is just due to more careful observations of the twilight sky.[26] However, it seems clear that NLC have increased in strength, occurrence rate and latitudinal coverage since they were first discovered.[19,23,24,27] This is probably mainly caused by an increased amount of water vapour in the mesosphere,[22,24] so that more water ice particles are formed during the NLC-season. Thus noctilucent clouds appear as an important indicator of global climate changes on the Earth.[28,29] Although most observations of NLC are in the northern hemisphere (NH),[21,30,31] they can also be observed in the southern hemisphere (SH), with some differences, e.g. in the peak altitudes of NLC:[32] the mean altitude of the SH-NLC distribution is approximately one kilometre higher than the mean altitude of the NH-NLC distribution.

2. Dust in the Earth's Mesosphere

There are several sources of solid dust particles in the mesosphere:[5] (a) ablation of meteorites; (b) volcanoes; (c) human activity; and (d) production of nanoscale dust agglomerates by cosmic rays. Since up to 100 tons of meteoritic material burns out each day and enters the Earth's atmosphere with large velocities, the ablation and recondensation of meteoritic material into smoke particles is one of the most important sources[33] that produces a persistent background of small and numerous dust particles.[34] Dust can cause global changes in the Earth's environment. Indeed, the presence of the dust in the atmosphere contributes to the radiative forcing of the present-day climate.[35,36] For example, it is suggested that deposition of dust at various periods, such as the last glacial maximum more than 20,000 years ago can be an order of magnitude higher than modern values over most of the world,[37,38,40] thus indicating a far higher long-term global atmospheric concentration of dust than is observed at present.

Besides the NLC, dust leads to the polar mesospheric summer echoes (PMSE):[41,42] strong radar echoes that were first observed in NH in the late 1970s[43] and are in the 50 MHz-1.3 GHz frequency range. PMSE appear as the result of electromagnetic wave scattering from mesospheric electron inhomogeneities; the latter can result from charging of electrons onto subvisible aerosol dust particles (~10 nm).[44] Note that NLC are formed primarily from larger (~50 nm) visible dust particles and are often observed slightly below the PMSE generation region altitude. Thus, polar mesospheric summer echoes appear as a crucial remote-sensing diagnostic for interpreting the creation, evolution and dynamics of noctilucent clouds.

Appropriate PMSE SH measurements were rare in the past, and limited to subpolar latitudes basically due to the lack of appropriate radars deployed in the southern polar region. The first SH PMSE were observed with the radar at Machu Picchu base on King George Island, Antarctica.[45] Later, a number of studies presented the morphology of daily and seasonal occurrence of SH-PMSE as well as comparison of PMSE NH and SH measurements.[46–50] The duration of the PMSE season at polar latitudes is highly correlated with the dynamical and thermal state of the mesopause region.[51–53] Modelling suggests that meridional winds directed towards the equator support the transport of PMSE particles from higher latitudes.[54–56] Thus PMSE are present at both hemispheres in summer as long as equator ward winds transport cold air from higher to lower latitudes.[55,57] It was concluded that the NH/SH similarities and differences of PMSE are most likely determined by the thermal structure of the mesosphere.[58] With indications that mesospheric temperatures are more sensitive to change than tropospheric temperatures,[28,59,60] observations of polar mesospheric summer echoes provide us with a powerful tool for monitoring global change.[61]

Figure 1 shows the parameters of the summer polar mesosphere that are important for the physics of NLC.[63] Conditions for the particle growth are satisfied at altitudes 80–90 km, where the local water vapors are supersaturated. Taking into account the presence of nanoparticles in the mesosphere, the most probable mechanism of the growth of particles is the heterogeneous nucleation on particles. Using data on the meteorite flux on the Earth that is ~100 ton/day (the mass m distribution of meteorites is close to the power law $\propto \mathrm{m}^{-s}$, where s is approximately in the range 2.3–2.5,[62] this gives the size distribution with the exponent ~ 5; centimeter-sized bodies precipitate primarily), the number density of nanoparticles in the mesosphere can be estimated as $n_{\mathrm{NLC}} \sim 10 - 10^3\,\mathrm{cm}^{-3}$. One of the key parameters is the illumination intensity of the mesosphere by solar radiation. The cumulative solar flux is defined

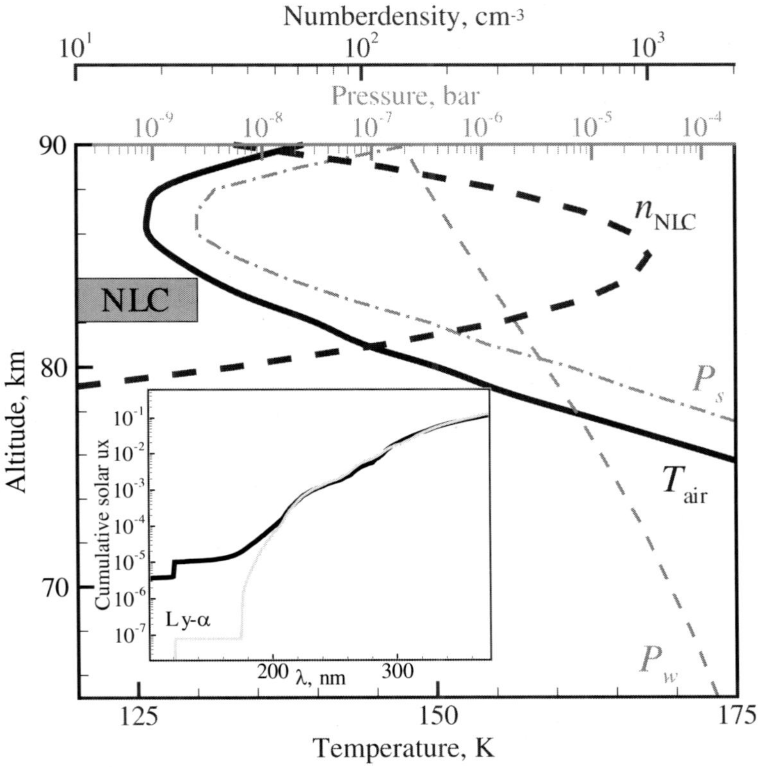

Fig. 1. Important (for the physics of NLC) parameters of the polar summer mesosphere: vertical profiles of the air temperature $T_{\rm air}$, the pressure of saturated water vapor over the ice surface P_s, the partial pressure of water vapor P_w, and the nanoparticle number density $n_{\rm NLC}$. It is clearly seen that the water vapor is supersaturated at altitudes of about 80 to 90 km, and, owing to this fact, nanoparticles may grow in the range of altitudes. The cumulative solar flux is defined Eq. (1); the solar zenith angle $\alpha = 85°$ was taken for the simulations. The inset shows calculated relative cumulative solar flux $\Phi_c(\lambda, H, \alpha)$ at two altitudes: $H = 100\,\rm km$ (black line) and $H = 82$ (grey line). Step-like behavior of Φ_c corresponds to Lyα photons ($\lambda \approx 121\,\rm nm$).

from relation

$$\Phi_c(\lambda, H, \alpha) = \int_0^\lambda F(\lambda, H, \alpha)d\lambda \Big/ \int_0^\infty F_0(\lambda)d\lambda, \qquad (1)$$

where $F_0(\lambda), F(\lambda, H, \alpha)$ are the extra-atmospheric solar flux and the solar flux at the altitude H, α is the solar zenith angle (the value $\alpha = 85°$ was taken for the simulations, see below Fig. 2). The inset in Fig. 1 shows the characteristic spectra of solar radiation as functions of the altitude for a

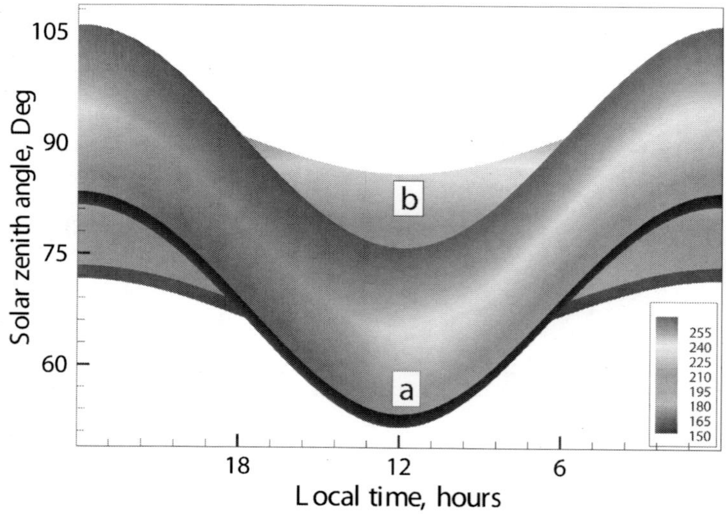

Fig. 2. The solar zenith angle α in the polar summer mesosphere (northern hemisphere) versus local time and day of the year (close to NLC season): latitude 75° (a) and 85° (b). The dependencies are greyscale-coded by the day of the year.

zenith angle of 85° that is characteristic of the summer polar mesosphere, Fig. 2. Note that short-wavelength radiation with wavelengths $\lambda \leq 170\,\text{nm}$ (i.e. with photon energies $h\nu \geq 7.3\,\text{eV}$) almost does not penetrate to the mesosphere, being strongly absorbed at altitudes $H \approx 100\,\text{km}$, except for the Lyα ($\lambda \approx 121\,\text{nm}$) atomic-hydrogen line, which is noticeably suppressed but still present at NLC altitudes.

3. Optical Properties and Heating of the Mesospheric Dust

Particles which absorb solar radiation are heated. The heating value depends on the reflective index and shape of the particle, the altitude, the solar zenith angle, etc. When the temperature of (water) ice particles exceeds $\simeq 150\,\text{K}$, they begin to evaporate (this refers to the mesopause pressures). This effect can be estimated from the balance equation for the temperature T_d of a dust particle (see, e.g.[64]). We write the balance equation as

$$P_{\text{sol}} + P_{\text{rad}} + P_{\text{col}} = 0, \tag{2}$$

where $P_{\rm sol}$ is the energy flux of solar radiation absorbed by the particle, $P_{\rm rad}$ stands for the thermal radiation of the microparticle, and $P_{\rm col}$ describes the dust particle cooling because of collisions with neutrals. Expressions for $P_{\rm sol}$ and $P_{\rm rad}$ are given by[63]

$$P_{\rm sol} = \pi a^2 hc \int \sigma_{\rm abs}(\lambda, a, m)(\Phi_\lambda^{\rm sol}(H, \alpha_{\rm za}) + \Phi_\lambda^{\rm sca}(H, \alpha_{\rm za}))d\lambda/\lambda \quad (3)$$

$$P_{\rm rad} = \pi a^2 hc \int \sigma_{\rm abs}(\lambda, a, m) B_\lambda(T_d) d\lambda/\lambda, \quad (4)$$

where $\sigma_{\rm abs}$ is the absorption efficiency (the cross-section for the radiation absorption by the dust particle of the radius a normalized to πa^2), $m = n + ik$ is the complex refractive index of the microparticle material, $\alpha_{\rm za}$ is the solar zenith angle, B_λ is the spectral flux of the thermal radiation from the particle, $\Phi^{\rm sol,sca}$ are the spectral fluxes of the direct (scattered by the atmosphere) solar radiation at the altitude H, c is the speed of light, and h is the Planck constant. The rate of the cooling of dust particles by neutrals is estimated by

$$P_{\rm col} \approx 4\pi a^2 \alpha_{\rm T} n_n v_{th} k_{\rm B}(T_d - T_n), \quad (5)$$

where $\alpha_{\rm T}$ is the probability of thermal accommodation (which is of order unity); n_n is the concentration of neutrals in the mesosphere ($n_n \sim 3 \times 10^{14}\,{\rm cm}^{-3}$ at NLC altitudes); v_{th} and T_n are the thermal velocity and the temperature of neutrals, respectively; and $k_{\rm B}$ is the Boltzmann constant. Numerical solution of the transport equation for solar radiation in the Earth's mesosphere shows that $\Phi_\lambda^{\rm sol}$ and $\Phi_\lambda^{\rm sca}$ can be of the same order at NLC altitudes. Note that contribution of the Earth's thermal radiation to the balance Eq. (2) is neglected since we are interested in small (sub-micron) particles for which the corresponding cross-section is small.[63]

Dependence of the absorption efficiency on the refractive index, size of a dust particle, and the radiation wavelength is complex. Figure 3 shows the absorption (scattering) efficiencies $\sigma_{\rm abs/sca}$ calculated for $\lambda = 121\,{\rm nm}$ radiation according to Mie theory as functions of a for spherical microparticles of various compositions.[63] We note that $\sigma_{\rm abs}$ for an ice-coated silicon particle differs slightly from $\sigma_{\rm abs}$ for an ice particle. The inset shows the complex refractive index as a function of the photon energy for several compounds. We point to a very small imaginary part of the refractive index k (which determines the absorption of radiation by a particle) in the optical

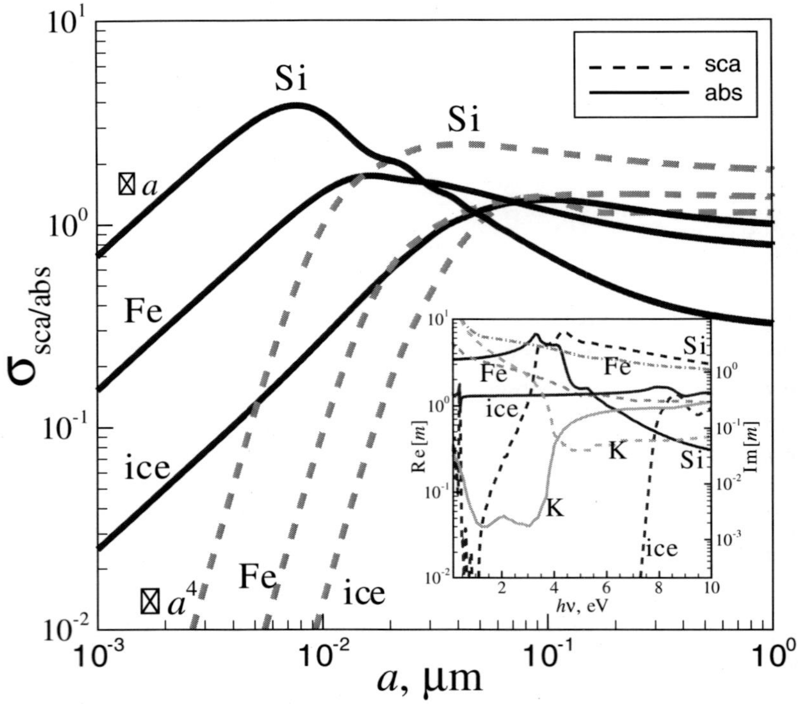

Fig. 3. Optical properties of different materials calculated using Mie theory. The efficiencies of scattering $\sigma_{\rm sca}$ and absorption $\sigma_{\rm abs}$ of 121-nm radiation (Lyα) versus the particle size a for various micro-particle materials: ice, iron, silicon, and water ice-coated silicon (at the corresponding radius ratio 1 : 2). The known scaling is shown for small a values ($\sigma_{\rm sca} \propto a^4$ and $\sigma_{\rm abs} \propto a$). The inset shows both the real (solid lines) and imaginary (dashed lines) parts of the refractive index m versus the photon energy $h\nu$ for several compounds that can be present in a dust particle.

range for ice particles. The values of k for the other compounds under consideration are relatively large.

Figure 4 shows $\sigma_{\rm abs}$ for microparticles of various compositions as a function of the photon energy $h\nu$ and particle size a.[63] The noticeable absorption of solar radiation by submicron potassium is observed for photons with energies $\sim 1 \div 3\,{\rm eV}$, which can easily penetrate to the mesosphere. At the same time, ice particles quite efficiently absorb short-wavelength ($h\nu \geq 8\,{\rm eV}$) and IR ($h\nu \leq 0.5\,{\rm eV}$) radiation.

The results can be used to estimate the heating and charging mesospheric dust particles. It is convenient to introduce the dimensionless

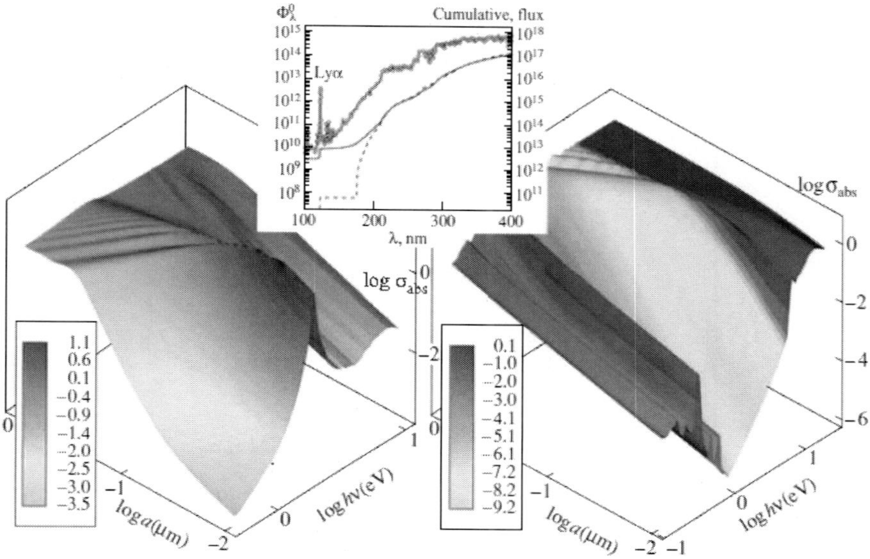

Fig. 4. Efficiency $\sigma_{\rm abs}$ of the absorption of radiation by a microparticle consisting of (left panel) potassium and (right panel) ice vs. the photon energy and particle size.[63] The inset shows the spectral fluxes of solar radiation: extra-atmospheric (black line) as well as at altitudes of 100 km (grey line) and 85 km (greay dashed line) (compare with inset of Fig. 1).

efficiency parameter $\eta(a,m)$ of the absorption of solar radiation energy by a particle as

$$\eta(a,m) = S_0^{-1} \int \sigma_{\rm abs} F_\lambda^0 d\lambda, \tag{6}$$

where F_λ^0 is the extra-atmospheric spectral energy flux (note that the main contribution to the heating of microparticles is due to radiation in the optical and near IR ranges that is almost not absorbed in the mesosphere) and $S_0 \approx 1366\,{\rm W/m^2}$ is the solar constant. For ice submicron-sized particles, η increases almost linearly with the particle radius: $\eta(a = 0.025\,\mu{\rm m}) \simeq 10^{-4}$ whereas $\eta(a = 0.8\,\mu\,{\rm m}) \simeq 3 \times 10^{-3}$. Using relation $\eta_{\rm ice} S_0 \approx P_{\rm col}$ (this is the absorption/cooling balance when the particle thermal emission is neglected) to estimate the heating value ΔT_d of such particles, we obtain $\Delta T_d \sim 3\,{\rm K}$; i.e. the heating of ice submicron particles in the mesosphere appears to be insufficient for their evaporation. For particles consisting of potassium, the situation changes radically:

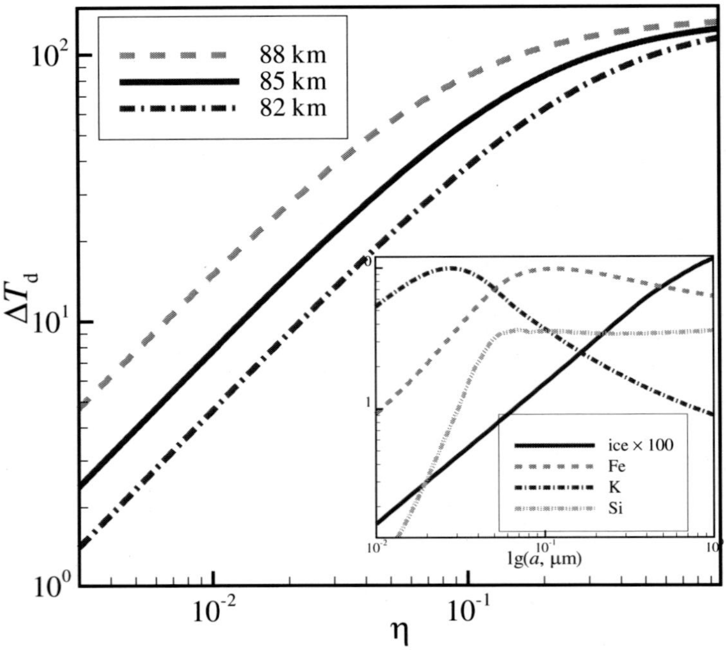

Fig. 5. Heating $\Delta T_d(H, \eta)$ of a mesospheric dust particle as a function of the efficiency parameter $\eta(a, m)$ at different altitudes H of mesosphere: H=85 km (solid line), 88 km (dashed line), 82 km (dash-dotted line). The inset shows dependence of $\eta(a, m)$ on the particle size a for the different materials (including water ice, this is multiplied by 100) that can be present in a dust particle.

in this case, $\eta_K \sim 0.1 \div 1$, the value of η_K reaches maximum for particles with $a \approx 30\,\text{nm}$ (η_{Fe} for iron microparticles has a maximum for $a \simeq 150\,\text{nm}$), and it is necessary to take into account the radiative cooling of such a particle. This circumstance determines the heating value $T_d \simeq 270\,\text{K}$.

Figure 5 shows the dependence of $\Delta T_d = T_d - T_n$ on the parameter η for a number of altitudes at which NLCs are located.[63] This dependence summarizes the above estimates. The inset shows the dependence $\eta(a)$ for some compounds. The obtained dependencies make it possible to determine how microparticles of various compositions are heated under the action of solar radiation in the mesosphere. The influence of the photoelectric effect on the charging of dust particles is discussed below in Sec. 4.

4. Recombination of the Mesosphere Plasma on Dust and Dust Charging

Plasma recombination on the dust particles' surfaces can significantly change ionization properties of the summer polar mesosphere. Such an effect of dust particles on the ionization balance can be estimated from the system of the continuity equations for the electron number density n_e, number density of simple n_i^s and cluster positive n_i^c ions, as well as for the dimensionless charge Z_d^a (the number of elementary charges on the particle) of the microparticle with the radius a, which are written in the local approximation:[67]

$$\frac{\partial n_e}{\partial t} = q_e - \alpha_s^{\rm rec} n_e n_i^s - \alpha_c^{\rm rec} n_e n_i^c + \pi\langle \nu^{\rm photo} n_d\rangle - \pi\langle \nu_e n_d\rangle, \tag{7}$$

$$\frac{\partial n_i^s}{\partial t} = q_e - \alpha_s^{\rm rec} n_e n_i^s - \beta_c n_i^s - \pi\langle \nu_s n_d\rangle, \tag{8}$$

$$\frac{\partial n_i^c}{\partial t} = \beta_c n_i^s - \alpha_c^{\rm rec} n_e n_i^c - \pi\langle \nu_c n_d\rangle, \tag{9}$$

$$\frac{\partial Z_d^a}{\partial t} = \nu^{\rm photo} + \nu_s + \nu_c - \nu_e, \tag{10}$$

Here, q_e is the ionization rate, $\alpha_{s/c}^{\rm rec}$ are the recombination coefficients for simple/cluster ions, respectively; the terms ν_j ($j = e, s, c$) describe, respectively, the losses of electrons as well as simple and cluster ions on dust particles; β_c is the rate of the conversion of simple ions to cluster ions; and $\nu^{\rm photo}$ describes the creation of photoelectrons when solar radiation is absorbed by a single particle with the radius a.

The expression determining $\nu^{\rm photo}$ is given by

$$\nu^{\rm photo} = \pi a^2 \int_0^{\lambda_W} \sigma_{\rm abs}(\lambda, a, m)\Phi_\lambda(H, \alpha_{\rm za}) Y(\lambda, m, a) d\lambda, \tag{11}$$

where λ_W is the maximum (threshold) radiation wavelength for the photoelectric effect. For example, λ_W for ice, iron, sodium, potassium, aluminum, and silicon corresponds to a photon energy of 8.7, 4.7, 2.4, 2.3, 4.1, and 4.85 eV, respectively. The photoelectron production probability $Y(\lambda, m, a)$ entering into Eq. (11) increases rapidly with the photon energy in the region which is above threshold ($\mid d\lambda/\lambda_W \mid \leq 1$) and is often estimated by the expression

$$Y(\lambda) = Y_\infty(1 - \lambda/\lambda^*) \tag{12}$$

(see, e.g.[68]) that interpolates the experimental data. The characteristic values are $Y_\infty \sim 1$, $\lambda^* \simeq \lambda_W$. It is worth mentioning that Y increases when the size of a microparticle decreases. This dependence can be explained by the fact that the photon mean free path l_ν in a substance is usually much longer than the electron mean free path l_e and thereby experiments with bulky samples provide underestimated values Y_b for the photoelectron yield, because some photoelectrons do not leave a sample. For $a \sim l_e$, the photoelectron yield can be higher than Y_b by a factor of 2.3.[69]

We note that solar radiation with wavelengths $\lambda \leq 170\,\mathrm{nm}$ (i.e. with photon energies $\geq 7.3\,\mathrm{eV}$) is almost fully absorbed at the mesosphere altitudes except for the Lyα ($\lambda \approx 121\,\mathrm{nm}$) atomic-hydrogen line. The damping of the Lyα line at NLC altitudes can be determined from Fig. 3. The ionization rate $\nu_{\mathrm{ice}}^{\mathrm{photo}}$ induced by the photoelectric effect on submicron ice particles can be estimated as

$$\nu_{\mathrm{ice}}^{\mathrm{photo}} \simeq \pi a^2 n_d \sigma_{\mathrm{abs}} \Phi_{\mathrm{Ly}\alpha} Y_{\mathrm{ice}}. \tag{13}$$

For $n_d \sim 10^2\,\mathrm{cm}^{-3}$ and $a \simeq 0.1\,\mu\mathrm{m}$, $\nu_{\mathrm{ice}}^{\mathrm{photo}} \sim 10 - 100\,\mathrm{cm}^{-3}\,\mathrm{s}^{-1}$, which is much higher than the equilibrium ionization rate q_e which is determined according to $q_e \simeq \alpha^{\mathrm{rec}} n_e^2 \sim 0.1 - 10\,\mathrm{cm}^{-3}\,\mathrm{s}^{-1}$. We emphasize that even nanosized ice particles can make certain contribution to the ionization of the mesosphere because $\sigma_{\mathrm{abs}} \propto a$ (see Fig. 2). If a dust particle consists of potassium, the role of photoelectrons becomes decisive beginning with moderate concentrations of submicron particles $n_d \sim 10\,\mathrm{cm}^{-3}$. Indeed, the cumulative flux of photons can be written as

$$\Phi(\lambda_W^{\mathrm{K}}) \sim 10^{17}\,\mathrm{cm}^{-2}\,\mathrm{s}^{-1}; \tag{14}$$

for such photons, $\sigma_{\mathrm{abs}} \sim 0.1 \div 1$ over a wide range of size a, and the photoemission rate induced by dust particles is very high. Indeed, we have

$$\nu_{\mathrm{K}}^{\mathrm{photo}} \simeq \pi \langle a^2 n_d(a) \sigma_{\mathrm{abs}}(a) \Phi(\lambda_W^{\mathrm{K}}) Y_{\mathrm{K}} \rangle \sim 10^3\,\mathrm{cm}^{-3}\,\mathrm{s}^{-1} \tag{15}$$

even for small values $Y_{\mathrm{K}} \sim 10^{-4}$. Such a high electron formation rate leads to a number of interesting photochemical effects. The electron number density in the cloud of such dust particles increases, whereas the concentration of ions (primarily cluster ions) decreases due to enhancement of the electron-ion recombination. Remarkably, such a behavior of the charged component under the NLC conditions is observed experimentally (see, e.g.[70]). Microparticles additionally acquire a considerable positive charge, which easily gives rise to PMSE-related radio reflections from the dusty layer. Figure 6 shows the impact of the dust particle number density n_d and the dust composition (water ice/iron) on the mesospheric plasma parameters.

5. The Dust Growth-Sedimentation Model

The kinetic equation describing the evolution of the microparticle distribution function $f_d(H, a, v, t)$ at the altitude H has the form

$$\frac{\partial f_d}{\partial t} + \frac{\alpha_{\rm w}}{4} m_{\rm w} v_{\rm w}^T \rho_d (n_{\rm w} - n_{\rm w}^s) \frac{\partial f_d}{\partial a} + v \frac{\partial f_d}{\partial H} + \left[g - \frac{\pi \rho c_s a^2 F_d (v + v_{\rm wind})}{m_d} \right] \frac{\partial f_d}{\partial v} = 0. \quad (16)$$

Here, ρ and ρ_d are the densities of air and microparticle material, respectively; m_d is the microparticle mass; $\alpha_{\rm w}$ is the probability of the accommodation of water molecules colliding with a dust particle ($\alpha_{\rm w} \sim 1$ for supersaturated water vapors); $v_{\rm w}^T$ is the thermal velocity of water molecules; c_s is the local speed of sound; $v_{\rm wind}$ and v are the vertical velocities of the wind and particle, respectively; F_d is the factor (of order unity) depending on the shape of a particle; and $n_{\rm w}$ and $n_{\rm w}^s$ are the concentrations of water vapor and saturated water vapors in the mesosphere, respectively. The second term in this kinetic equation describes the growth/evaporation of ice microparticles in dependence on the water vapor saturation, and the last term in Eq. (16) describes the sedimentation/rise of a dust particle interacting with neutrals (neutral drag). In the summer polar mesosphere, the average velocity component (directed against gravity) of neutral gas is $v_{\rm wind} \simeq 1 \div 10\,{\rm cm/s}$.

The solution of kinetic equation (16), together with the following equation describing the content of water vapors,

$$\frac{\partial n_{\rm w}}{\partial t} + \frac{\partial \Gamma_{\rm w}}{\partial h} = -P_{\rm w} - n_{\rm w} L_{\rm w} - \pi \alpha_{\rm w} v_{\rm w} n_{\rm w} \langle a^2 n_d \rangle, \quad (17)$$

where $\Gamma_{\rm w}$ is the vertical diffusion flow of water vapors (see, e.g.[65]) and $P_{\rm w}$ and $L_{\rm w}$ stand for photochemical sources and sinks of water vapors in the mesosphere (the last term on the right-hand side describes the loss of water

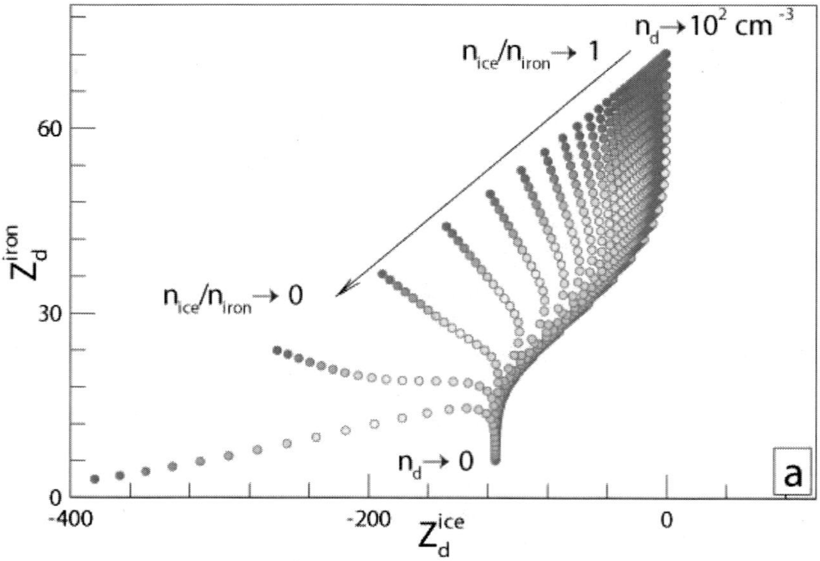

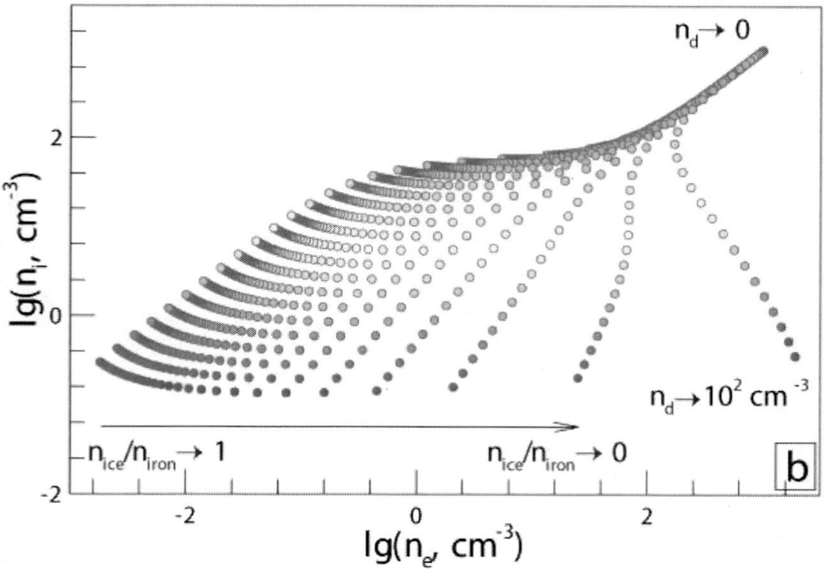

Fig. 6. Impact of the dust particle number density n_d and the dust composition (water ice/iron) on the mesospheric plasma parameters. Top panel (a) shows the variations of charges of both icy ($Z_d^{\rm ice}$) and iron ($Z_d^{\rm iron}$) sub-micrometer sized particle versus n_d at different ice/iron particles mixing ratio. Bottom panel (b) shows electron n_e and ion n_i densities at different dust number density n_d and different ice/iron particles mixing ratio.

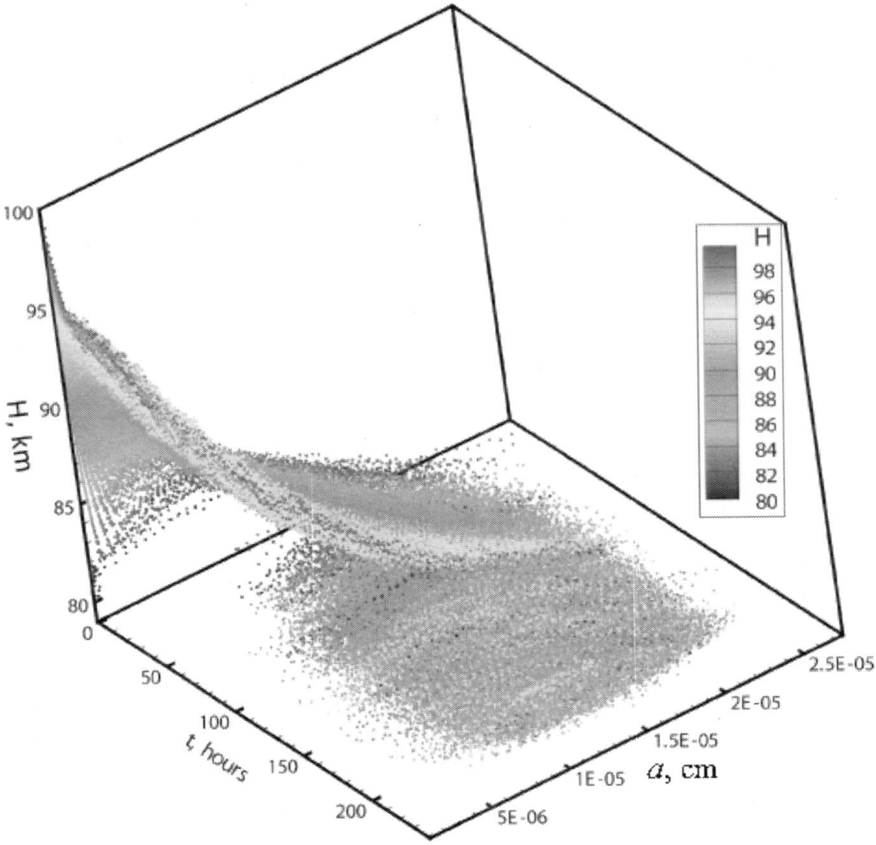

Fig. 7. Distribution of dust particles in the polar summer mesosphere versus altitude H, particle radius a and time t. Initially nanosized particles (with the bell-shape distribution and mean value of $a \simeq 5\,\text{nm}$) were distributed in between $90 \div 100\,\text{km}$ altitude range as shown in Fig. 1. Value of $v_{\text{wind}} \simeq 5\,\text{cm/s}$. For given parameters, formation of a steady-state dusty structure takes a few days.

molecules on growing dust particles), determines the characteristic size of particles in the summer polar mesosphere.

Figure 7 shows the time dependence of the distribution of dust particles with respect to the altitude and time for various particle number densities n_d and $v_{\text{wind}} \approx 5\,\text{cm/s}$. It can be seen that the particle growth leads to a decrease in the water concentration in the mesosphere, which in turn limits the further growth of microparticles. Thus, for typical NLC parameters, the effect of the mesosphere dehydration in the dust cloud limits the size of the particles in dependence on their concentration at a level of $\sim 0.05 \div 0.1\,\mu\text{m}$.

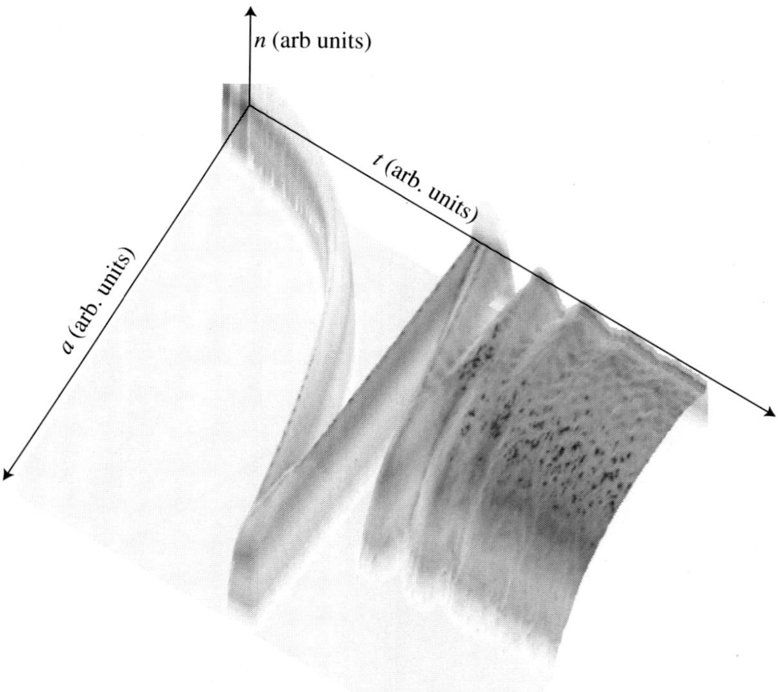

Fig. 8. Temporal (t) evolution of the size (a) distribution (n) of water ice particles in the polar summer mesosphere. The bimodal particle size distribution is formed at the steady-state stage.

Figure 8 demonstrates that bimodal particle size distributions are formed as a result of the temporal evolution.

6. Conclusion

Thus, the effect of the composition and optical properties of microparticles on the formation of dust layers in the mesosphere has been analyzed in detail. The parameters of the heating of dust particles and photoemission of electrons under the action of solar radiation have been determined. Certain photochemical consequences of the presence of dust in the upper atmosphere have been analyzed. In particular, it has been shown that microparticles may significantly reduce the concentration of water vapors in the upper atmosphere, and this decrease, in turn, limits the size of particles.

It has been shown that the composition of microparticles can determine the ionization features of the summer polar mesosphere.

Acknowledgments

This work was partially supported by the Australian Research Council, the Russian Foundation for Fundamental Research (project no. 07-02-01464a), and the Partner Group for Complex Plasmas Program, Max Planck Society (Germany).

References

1. L. Spitzer, *Physical Processes in the Interstellar Medium* (Wiley, New York, 1978).
2. C. K. Goertz, *Rev. Geophys.* **27** (1989) 271.
3. S. V. Vladimirov and K. Ostrikov, *Phys. Rep.* **393** (2004) 175.
4. S. V. Vladimirov *et al.*, *Physics and Applications of Complex Plasmas* (Imperial College, London, 2005).
5. V. N. Tsytovich *et al.*, *Elementary Physics of Complex Plasmas* (Springer, Berlin, 2007).
6. R. J. Gould and E. Salpeter, *Astrophys. J.* **138** (1963) 393.
7. F. Verheest, *Waves in Dusty Space Plasmas* (Kluwer, Dordrecht, 2000).
8. A. Bouchoule, Ed., *Dusty Plasmas. Physics, Chemistry and Technological Impacts in Plasma Processing* (Wiley, Chichester, 1999).
9. A. Garscadden *et al.*, *Plasma Sources Sci. Technol.* **3** (1994) 239.
10. H. Alfvén, *On the Origin of the Solar System* (Clarendon, Oxford, 1954).
11. H. M. Thomas and G. E. Morfill, *Nature* **379** (1996) 806.
12. M. Rubin-Zuzic *et al*, *Nature Physics* **2** (2006) 181.
13. D. G. Andrews, J. R. Horton, and C. B. Leovy, *Middle Atmosphere Dynamics* (Academic, London, 1987).
14. F.-J. Lübken *et al.*, *J. Geophys. Res.* **104** (1999) 9135.
15. J. Y. Xu *et al.*, *J. Geophys. Res.* **112** (2007) D09102.
16. T. Eidhammer and O. Havnes, *J. Geophys. Res. A* **106** 2001) 24831.
17. A. E. Hedin *et al.*, *J. Atmos. Terr. Phys.* **58** (1996) 1421.
18. R. G. Roble and E. C. Ridley, *Geophys. Res. Lett.* **21** (1994) 417.
19. G. E. Thomas, *Rev. Geophys.* **29** (1991) 553.
20. O. Havnes *et al.*, *J. Atmos. Terr. Phys.* **52** (1990) 637.
21. O. Havnes *et al.*, *J. Geophys. Res. A* **101** (1996) 10839.
22. O. Havnes *et al.*, *Phys. Scripta* **T89** (2001) 133; *ibid.* **T107** (2004) 70.
23. M. Gadsden and P. Parviainen, *Observing Noctilucent Clouds* (IAGA, 1995).
24. M. Gadsden and W. Schroder, *Noctilucent Clouds* (Springer, NY, 1989).
25. T. W. Backhouse, *Meteorol. Mag.* **20** (1885) 133.
26. G. E. Thomas *et al.*, *Nature* **338** (1989) 490.

27. U. von Zahn and W. Meyer, *J. Geophys. Res.* **94** (1989) 14647.
28. F.-J. Lübken *et al.*, *Adv. Space Res.* **40** (2007) 794.
29. W. A. Scales and G. Ganguli, *New J. Phys.* **6** (2004) 12.
30. O. Havnes *et al.*, *Geophys. Res. Lett.* **28** (2001) 1419.
31. G. von Cossart *et al.*, *Geophys. Res. Lett.* **26** (1999) 1513.
32. J. E. Wrotny and J. M. Russell III, *J. Atmos. Solar Terr. Phys.* **68** (2006) 1352.
33. J. J. Kornblum, *J. Geophys. Res.* **74** (1969) 1893.
34. D. M. Hunten *et al.*, *J. Atmos. Sci.* **37** (1980) 1342.
35. I. N. Sokolik and O. B. Toon, *Nature* **381** (1996) 681.
36. I. Tegen and I. Fung, *J Geophys. Res.* **99** (1994) 22897.
37. M. DeAngelis *et al.*, *J. Geophys. Res.* **102** (1997) 26681.
38. J. R. Petit, *Nature* **399** (1999) 429.
39. M. Rapp *et al.*, *J. Geophys. Res.* **107** (2002) D19.
40. L. G. Thompson and E. Mosley-Thompson, *Science* **212** (1981) 812.
41. J. Y. N. Cho and M. C. Kelley, *Rev. Geophys.* **31** (1993) 243.
42. J. Y. N. Cho and J. Rottger, *J. Geophys. Res.* **102** (1997) 2001.
43. P. Czechowsky *et al.*, *Geophys. Res. Lett.* **6** (1979) 459.
44. M. C. Kelley and J. C. Ulwick, *J. Geophys. Res.* **93** (1988) 7001.
45. R. F. Woodman *et al.*, *J. Geophys. Res.* **104** (1999) 22577.
46. M. M. Huaman and B. B. Balsley, *Geophys. Res. Lett.* **26** (1999) 1529.
47. S. Kirkwood *et al.*, *Geophys. Res. Lett.* **34** (2007) L15803.
48. R. Latteck *et al.*, *Geophys. Res. Lett.* **34** (2007) L14805.
49. R. Latteck *et al.*, *Ann. Geophys.* **26** (2008) 2795.
50. R. J. Morris *et al.*, *Geophys. Res. Lett.* **31** (2004) L16111.
51. S. Kirkwood *et al.*, *Geophys. Res. Lett.* **25** (1998) 1867.
52. R. J. Morris *et al.*, *Geophys. Res. Lett.* **34** (2007) L05805.
53. W. Singer *et al.*, *Adv. Space Res.* **31** (2003) 2055.
54. U. Berger and U. von Zahn, *J. Geophys. Res.* **112** (2007) D16204.
55. A. Dowdy *et al.*, *Geophys. Res. Lett.* **28** (2001) 1475.
56. A. J. Dowdy *et al.*, *J. Geophys. Res.* **112** (2007) D17104.
57. R. J. Morris *et al.*, *J. Atmos. Solar Terr. Phys.* **68** (2006) 418.
58. F.-J. Lübken and U. Berger, *J. Atmos. Solar Terr. Phys.* **69** (2007) 2292.
59. J. Perlwitz and N. Harnik, *J. Climate* **17** (2004) 4902.
60. S. Zhou *et al.*, *J. Climate* **15** (2002) 781.
61. G. E. Thomas, *Adv. Space Res.* **18** (1996) 149.
62. E. Grun *et al.*, *Icarus* **62** (1985) 244.
63. B. A. Klumov *et al.*, *JETP Lett.* **82** (2005) 632.
64. G. Grams and G. Fiocco, *J. Geophys. Res.* **82** (1977) 961.
65. B. J. Murray and J. M. C. Plane, *Atmos. Chem. Phys.* **5** (2005) 1027.
66. U. Berger and U. von Zahn, *J. Geophys. Res.* **107A** (2002) 1366.
67. B. A. Klumov *et al.*, *JETP Lett.* **72** (2000) 364.
68. B. T. Draine, *Astron. J. Suppl.* **36** (1978) 595.
69. W. D. Watson, *J. Opt. Soc. Am.* **63** (1973) 164.
70. M. Rapp and F.-J. Lübken, *Atmos. Chem. Phys.* **4** (2004) 2601.

Advances in Geosciences
Vol. 21: Solar & Terrestrial Science (2008)
Ed. Marc Duldig

DEVELOPMENT OF STIFF AND EXTENDIBLE ELECTROMAGNETIC SENSORS FOR SPACE MISSIONS

Y. KASABA and A. KUMAMOTO
Department of Geophysics, Tohoku University,
Aramaki-aza-Aoba 6-3, Aoba-ku, Sendai, Miyagi 981-0933, Japan
kasaba@pat.gp.tohoku.ac.jp

K. ISHISAKA
Department of Information System Engineering,
Toyama Prefectural University,
Imizu, Toyama 939-0398, Japan

H. KOJIMA
Research Institute of Sustainable Humanosphere (RISH),
Kyoto University, Uji, Kyoto 611-0011, Japan
kojima@rish.kyoto-u.ac.jp

K. HIGUCHI
Institute of Space and Astronautical Science (ISAS),
Japan Aerospace Exploration Agency (JAXA),
Sagamihara, Kanagawa 229-8510, Japan

A. WATANABE
Sakase Adtech, Co., Ltd., Sakai, Fukui 910-0363, Japan

K. WATANABE
Wel Research Co., Ltd., Ichikawa, Chiba 272-0021, Japan

We developed three types of stiff and extendible electromagnetic sensors in rigid monopole antenna, loop antenna, and Yagi-Uda antenna for future space missions. They are based on carbon fiber reinforced plastic (CFRP) technologies, in order to fulfill severe requirements, i.e. enough stiffness, light mass, compact storage, safe extension, and reasonable test efforts. One of them, rigid monopole antennas, coupled with an inflatable actuator system, was successfully used in the JAXA S-520-23 sounding rocket experiment in September 2007. Applications of those antennas are expected in space plasma missions including the SCOPE program, sounding rocket experiments, planetary radar remote sensing, and landing radio measurements.

1. Introduction

Electromagnetic sensors from dc to several 10 s MHz are one of key elements for various space missions. In space plasma observations, the detection of electric field and plasma waves provides a tool to identify global plasma dynamics and energetic processes in magnetospheres and ionospheres. For sounding rockets and landing missions, they can be used to study radio emissions and propagations in and from crusts, atmospheres, and ionospheres. They also enable to do remote soundings toward the subsurface of Moon, Mars, icy satellites, and asteroids.

For these objectives, long wire antenna can be used as a sensitive sensor. However, its extension requires a spinning platform. In order to implement these functions into wider missions, we have developed stiff and extendible antenna series, based on CFRP technologies under the collaboration of Japan Aerospace Exploration Agency (JAXA), Tohoku Univ., Toyama Pref. Univ., and Kyoto Univ. with the engineering supports from Sakase Adtech and Wel Research. They fulfill the severe requirements to the antenna system, i.e. light mass, enough stiffness, compact storage, safe extension, and reasonable test efforts. One of them, the rigid monopole antenna, was successfully used in the S-520-23 sounding rocket experiment in September 2007. In future, these antennas will be one of the core technologies for space plasma missions, sounding rocket experiments, planetary radar remote sensings, and landing radio measurements.

In this paper, we summarize the current development efforts of our stiff and extendible electromagnetic sensors. In Sec. 2, we describe the base concept, material, and mechanics. In Sec. 3–5, we introduce our three types of antennas: rigid monopole antenna, loop antenna, and Yagi-Uda antenna.

2. Base Concept

For the characteristics described in Sec. 1, required key technologies are (1) good antenna element and (2) simple extension mechanism.

For antenna element, material selections were based on light weight, high rigidity, softness for compact storage, straightness after extension, and small degradation after repeated extension and roll-ups. We selected tri-axially woven fabrics (TWF) carbon fiber reinforced plastic (CFRP). Figure 1 shows the structure (left) and the open section tubular member (right) made of TWF-CFRP. Its density is $0.7\,g/cm^3$, much smaller than that of other candidates such as BeCu, which we have used as antenna elements for sounding rocket experiments ($8.3\,g/cm^3$). In similar mechanical

Fig. 1. Tri-axially woven fabrics (TWF) carbon fiber reinforced plastic (CFRP).

strength, the mass can be about 1/4. It partially contributes to the reduction of weight and momentum of the extendible part.

Since CFRP is a conductive material, it can be used directly as antenna elements. We considered it's weaved structure shown in Fig. 1, because the conductivity of TWF-CFRP was assured only along its coilable fibers. These conductive fibers might create inductance and suffer antenna performance in the higher frequency range. In 2006, we conducted a field test of the radio detection capability of the monopole antennas made of Al, CFRP, and Al-coated CFRP, at Kyoto University. All antennas had a length of 2 m and detected broadcasting radio waves. Differences are only 1–2 dB in 0.6–1.3 MHz. We conducted another test in 100 MHz range by Yagi-Uda antenna elements, and confirmed good radio characteristics (see Sec. 5). We concluded that TWF-CFRP has enough potential for antenna applications up to 100 MHz range.

Another merit of CFRP is its length. By our current production system, the length of BeCu rod antennas used in sounding rocket experiments is 1 m in maximum. The current CFRP elements are longer, 2.5 m. The ones from the new production system that will be established at the end of 2009 will be 5 m, and fulfill the requirements of the spin-axial antenna for the next magnetospheric multi-spacecraft mission, SCOPE[1] (see Sec. 3).

For extension mechanism, we have three choices. The first is the self-extraction of the rolled CFRP element by its elastic force. This method is the simplest, using only wire cutter etc. The second is more traditional, and is driven by an electric motor. This type has long heritage for the extension of uni-STEM (Storable Tubular Extendible Member) and bi-STEM antenna

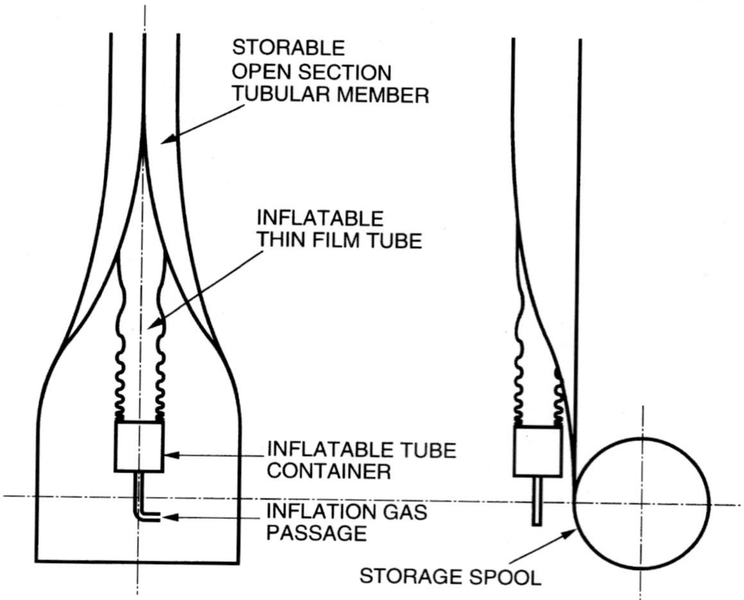

Fig. 2. Design concept of the inflatable actuation system for a monopole antenna.[2]

elements. This system is larger and more complex due to the motor and its drive unit.

The last one is the inflatable mechanism, using an inflatable thin film tube as an extension actuator. The design concept of "space inflatable structure"[2] adopted for a monopole antenna system is illustrated in Fig. 2. In this design, pressured N_2 gas is supplied into an inflatable tube made by thin soft film (ex. polyolefin). The tube pushes the rolled CFRP tubular member and extends it. After the extension, the CFRP member has enough rigidity to keep its straight structure even if the inflated inner tube contracts due to the loss of gas, and can act as a monopole antenna. In this extension system, motor and its drive unit (including heater unit, gear mechanism, power supply, etc.) can be replaced by a small inflatable actuation system with a simple gas supply unit.

3. Electric Monopole Antenna

As the most important and urgent application, we have developed the 5 m rigid-type monopole antenna for the magnetospheric multi-spacecraft mission, SCOPE (cross Scale COupling in Plasma universE),[1] which was officially proposed to JAXA in 2009. For this project, dipole antennas in 10 m tip-to-tip length extended along the spacecraft spin axis are required for plasma and radio wave measurements, in order to obtain 3-dimensional electric field wave vectors on all spacecraft. In our studies, the most feasible candidate which fulfills the severe mass and size restrictions is two stiff inflatable monopole antennas of 5 m length each. We started the development in 2005.

In order to test the concept, we developed a test dipole antenna, SPINAR (SPace INflataion Actuated Rod).[3] It was used in the sounding rocket experiment S-520-23 launched from JAXA Uchinoura Space Center at 19:20 JST on 2 September 2007. The aim of this experiment was to study the momentum transfer between neutral and plasma in the thermosphere. The SPINAR was used as the electric field dipole antenna for the Electric field and VLF/MF band Receiver (EVMR),[4] which measured plasma dynamics and density in the ionosphere at the altitude of 80–280 km.

The SPINAR system (Fig. 3) used the concept described in Fig. 2, i.e. to extend a rolled uni-STEM antenna element by an inflating gas tube with N_2 gas pressure. In this experiment, we aimed to test the actuating system, so the extension was set along the spin plane of the rocket in order to avoid the attitude turbulence associated with the extension along the rocket spin-axis. It means we did not need to use TWF-CFRP material to achieve enough stiffness. For ionospheric electric field measurements to meet the scientific objectives of this sounding rocket experiment, we set the antenna element non-conductive and add the electrode only at each top, in order to have longer effective length of the dipole antenna and reduce the capacity coupling with the rocket body. (The charging of non-conductive surface was not high in the mid-latitude ionosphere, several few V.)

Fig. 3. The SPINAR (left) and a picture taken after its extension in the flight (right).

Therefore, we used another non-conductive material, TWF quartz fiber reinforced plastic (QFRP), which is softer than TWF-CFRP. The length and diameter of the antenna elements were 2.4 m and 2 cm, respectively. Total mass of the system, which includes two monopole antennas and their extension control units, was 4.3 kg. The mass of each extendible part was 80 g.

To our knowledge, the SPINAR was the first successful use of an inflatable structure as a flight antenna. It extended and worked without any problems. The sounding rocket deployed its nosecone at +55 sec after firing. At 59.4 sec after launch (85 km altitude), the SPINAR started the extension by opening its pneumatic valve for N_2 gas, which pushed the inflatable tube at the pressure of 200 kPa. The full length was reached about 2 sec. later, i.e. with the extension rate of ~1.2 m/sec.

We now design the spin axis antenna for the SCOPE mission, using TWF-CFRP bi-STEM antenna elements with a length of 5 m and a diameter of 2.5 cm. In this design, the CFRP element itself has enough rigidity to endure the outer force turbulence even in the spinning spacecraft. The current design still has two potential problems. One is the fast extension rate by the inflatable mechanism, which might cause attitude instability by coupling with the spacecraft dynamics. For a solution, slower extension at lower N_2 pressure might be adapted. The other is the alignment of antenna element along the spacecraft spin axis. From a preliminary analysis for a spin rate of 20 rpm, the alignment required by the system design is less than 1 deg. Since our antenna does not have a stabilizer due to mass limitation, this requirement is marginally achievable for both the direction of extension and the straightness of the element. In order to avoid this problem, we may extend the bi-STEM elements by a motor unit. The base design will be established in 2009–2010.

We are also developing the Inflatable Tube Antenna (ITA), non-stiff antenna based on the inflatable system, as a medium-length antenna at lower cost for sounding rocket experiments. This antenna does not have CFRP elements, but the inflatable tube is made of poly-ethylene terephthalate (PET), which itself is conductive due to the copper coating. It can be extended to a length of 5 m in the spin plane, which can only be achieved by more expensive and complicated antennas like wire antenna system currently. Thanks to the refined design, the total mass will be less than 2 kg. Its design and base development was finished in the summer of 2009, and will be aboard the JAXA S-520-26 sounding rocket in the summer of 2010.

4. Magnetic Loop Antenna

In space plasma physics, the polarization and wave normal direction provide key information to identify the modes and origins of plasma turbulences. Such broad-band measurements have been made by loop antennas, from 0.1 to 1000 kHz. We developed a loop antenna system aboard the Akebono satellite (EXOS-D) launched in 1989.[5,6] The loop was square-shaped with an area of $0.36\,m^2$ (0.6×0.6 m) and the mass of $\sim$2 kg. The major part of its mass was due to antenna frames.

We have examined lighter loop antennas with CFRP technologies since 2007. Figure 4 shows a test loop antenna system developed in 2008. It has an area of $0.36\,m^2$, which is the same as that of the Akebono antenna. The TWF-CFRP tubes are used as antenna frames. Since the CFRP tube is conductive, it is also used as an electrostatic shield of the loop element. The antenna element is rectangular (0.6×0.6 m) open coil with 10 turns each. The weight of the loop antenna was 438 g (frame: 72 g, wire element: 135 g, joint parts: 231 g), 1/4 of the original Akebono design. As the next step, we will use CFRP joint parts. In that case, the mass will become half.

The model shown in Fig. 4 does not yet have a storage capability. The folding method of the loop antenna was examined in parallel. Figure 5 shows an example of the storage method investigated by a smaller model. As the next step, it will be tested by a model with realistic size.

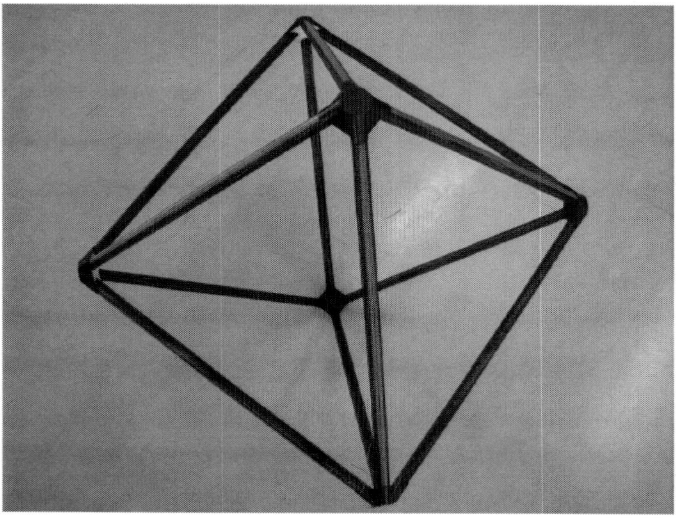

Fig. 4. Lightweight CFRP loop antenna.

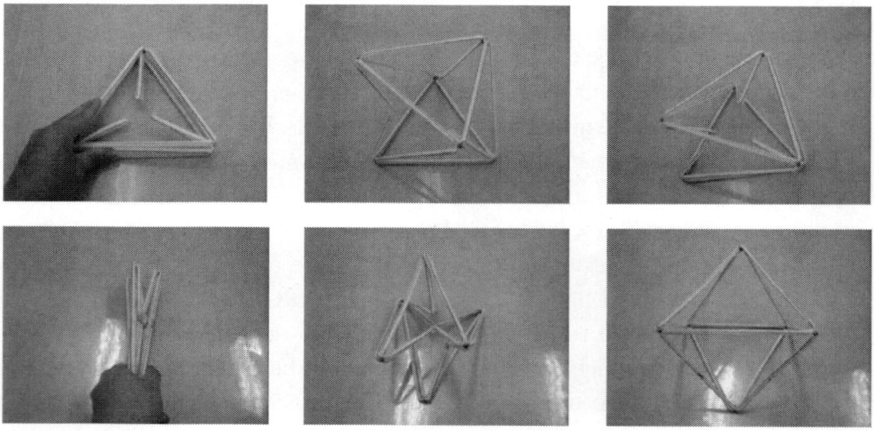

Fig. 5. A model of the folding method for lightweight loop antenna.

We expect to adopt the new loop antenna system to small-sized space missions for magnetospheric and ionospheric studies. It is also expected in landing missions, as a light sensor to detect radio waves from atmospheric discharges, subsurface radar echo, etc.

5. Yagi-Uda Type Antenna for Radar Sounding

It has been shown that spaceborne radar sounder is a powerful tool for exploring the surface, subsurface, and upper atmosphere of planets and satellites.

We have developed spaceborne sounder systems for remote sensing of the topside ionosphere from 1970's for sounding rocket experiments. We also implemented them in space missions Jikiken (EXOS-B: 1978–1985), Ohzora (EXOS-C: 1984–1988), and Akebono (EXOS-D: 1989-). The Nozomi spacecraft (Planet-B: 1998–2004) had the first radio sounder to Mars,[7,8] but it was unfortunately aborted by system troubles. Recently, we performed subsurface radar sounding of the moon by the Lunar Radar Sounder (LRS) aboard the Kaguya (SELENE) spacecraft.[9,10] Its observations revealed subsurface reflectors in a depth of several 100 m below the nearside maria, which were inferred to be old regolith covered by the basalt with a thickness of several 100 m.[11,12]

The subsurface radar sounding of Mars was also performed by the Mars Advanced Radar for Subsurface and Ionospheric Sounding (MARSIS) aboard Mars Express[13] and Shallow Radar (SHARAD) aboard the Mars

Reconnaissance Orbiter.[14] These sounders have made possible several new findings, such as buried craters in the northern lowlands, and the thickness of south Polar Layered Deposits.

Based on these successes, the expectations for the spaceborne radar sounding to other targets, such as Venus, icy moons of Jupiter, and asteroids are evolving.[15,16] The reduction of system mass and power while maintaining enough sensitivity is required.

The improvement of antenna gain is effective for the power reduction. In previous systems, dipole antenna was utilized for transmitting sounder pulses and receiving echoes. The dipole antenna has several merits, including its simple structure and shared observation with natural radio and plasma waves. However, its antenna gain is only 2.14 dBi, so that the radiation pattern is almost isotropic.

Yagi-Uda antenna, which is invented by Prof. Yagi and Dr. Uda at Tohoku University in 1926, has better characteristics. By its antenna gain of ~7 dBi, we can transmit sounder pulse and receive its echo with 5 dB higher gain. It allows us to reduce the transmitting power to 1/10. It has the additional merit of being less susceptible to external interferences. For example, the Kaguya/LRS was interfered by terrestrial broadcast radio waves and solar type III radio bursts. With Yagi-Uda type antenna, such noises from off-nadir directions can be suppressed.

Yagi-Uda antenna requires more antenna elements and more complex mechanism. In order to establish a light and simple design, we have developed a test model of Yagi-Uda antenna with TWF-CFRP elements since 2007. Assuming the application for subsurface sounding of icy moons of the Jupiter, operating frequency in the current design is set at 50 MHz, above the maximum frequency of strong Jovian decametric radiations. There are two ideas for its mechanism. One is the inflatable loop Yagi-Uda antenna with a size on the order of 3 m. A small test model was developed in 2007. Although its measured antenna characteristics are good, the problem is that it is not easy to keep the antenna shape without inner gas pressure. Another realistic scheme is to use the TWF-CFRP antenna elements attached to a traditional mast system. Each element is simply self-extracted from rolled elements (cf. Fig. 6). After the deployment, all antenna elements can be kept by their rigidity.

Figure 7 shows the half-sized model antenna developed in 2008. We tested the electric characteristics of this antenna. We confirmed that the sensitivity made by the CFRP elements is similar to that of antennas made with metal elements in the expected frequency range. We also tested

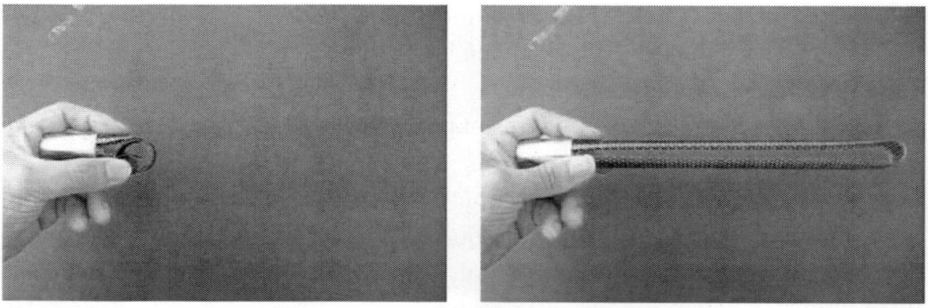

Fig. 6. A conceptual self-extension scheme of TWF-CFRP antenna elements.

Fig. 7. Evaluation model of spaceborne Yagi-Uda antenna system.

the low temperature characteristics of the TWF-CFRP material. The mechanical characteristics, like rigidity, softness, straightness, degradation after repeated extension and roll-ups, do not suffer any degradation even at 77 K. Further tests and quantitative analyses will be planned in the future.

6. Summary

In this paper, we summarized our development results of stiff and extendible electromagnetic sensors from dc to several 10 s of MHz for future space applications. We have studied three types based on CFRP technologies, which fulfill stringent design requirements, i.e. enough stiffness, light mass, compact storage, safe extension, and reasonable test efforts. One of these types, rigid monopole sensor with the inflatable extension system, was used in the JAXA S-520-23 sounding rocket experiment in September 2007, and became the world's first successful flight of the inflatable-type antenna. On the other hand, loop and Yagi-Uda type antennas have been tested by some engineering models. We established critical areas for the reduction of its sizing, which will be critical for future applications. They will be expected as one of the core elements for future projects, such as space electric field and plasma wave sensors for the SCOPE mission, compact radio sensors for landing missions, radar antennas for Lunar, Martian, Jovian, and asteroid missions which we are planning with JAXA and with scientists in our research fields world-wide.

Acknowledgments

The work described in this paper was supported by JAXA, in sounding rocket projects, the SCOPE project, and Fundamental Payload Technology Grants. The work in Tohoku University was supported by the Global COE (Global Education & Research Center for Earth & Planetary Dynamics) program.

References

1. C. F. Chyba, S. J. Ostro and B. C. Edwards, Radar detectability of a subsurface ocean on Europa, *Icarus*, **134**(2) (1998) 292–302, doi:10.1006/icar.1998.5961.
2. M. Fujimoto, Y. Tsuda, Y. Saito, I. Shinohara, T. Takashima, A. Matsuoka, H. Kojima and Y. Kasaba, The SCOPE Mission, *AIP Conf. Proc. "Future Perspectives of Space Plasma and Particle Instrumentation and International Collaborations"* **1144** (2009) 29–35.
3. K. Higuchi, K. Watanabe, A. Watanabe, H. Tsunoda and H. Yamakawa, Design and evaluation of an ultra-light extendible mast as an inflatable structure, *AIAA 2006-1809: 47th AIAA/ASME/ASCE/AHS/ASC Structures, Structural Dynamics, and Materials Conference, 7th AIAA Gossamer Spacecraft Forum*, Newport, Rhode Island, USA (2006).
4. K. Higuchi, Y. Ogi, K. Watanabe and A. Watanabe, Verification of practical use of an inflatable structure in space, *Trans. JSASS Space Tech. Japan.* **7**(26) (2009) Tc7–Tc11.

5. K. Ishisaka, Y. Kasaba, T. Okada, Y. Ashihara, T. Abe, T. Ono and S. Watanabe, Electric field measurement in the mid-latitude ionosphere by the S-520-23 sounding rocket experiment, *in preparation* (2009).
6. I. Kimura, K. Hashimoto, I. Nagano, T. Okada, M. Yamamoto, T. Yoshino, H. Matsumoto, M. Ejiri and K. Hayashi, VLF observations by the Akebono (EXOS-D) satellite, *J. Geomag. Geoelectr.* **42** (1990) 459–478.
7. J. C. Moore, Models of radar absorption in Europan ice, *Icarus*, **147**(1) (2000) 292–300, doi:10.1006/icar.2000.6425.
8. T. Okada, I. Nagano, K. Hashimoto, I. Kimura, H. Oya and A. Morioka, Design of a small loop antenna system for receiving waves in VLF and MF bands using a series-transformer network, *Trans. IEICE*, E70, (1987) 550–561.
9. T. Ono and H. Oya, Lunar Radar Sounder (LRS) experiment on-board the SELENE spacecraft, *Earth Planets Space* **52** (2000) 629–637.
10. T. Ono, A. Kumamoto, H. Nakagawa, Y. Yamaguchi, S. Oshigami, A. Yamaji, T. Kobayashi, Y. Kasahara and H. Oya, Lunar Radar Sounder observations of subsurface layers under the nearside maria of the Moon, *Science* **323** (2009) 909–912, doi:10.1126/science.1165988.
11. T. Ono, A. Kumamoto, Y. Yamaguchi, A. Yamaji, T. Kobayashi, Y. Kasahara and H. Oya, Instrumentation and observation target of the Lunar Radar Sounder (LRS) experiment on-board the SELENE spacecraft, *Earth Planets Space* **60** (2008) 321–332.
12. T. Ono, H. Oya, A. Morioka, A. Kumamoto, K. Kobayashi, T. Obara and T. Nakagawa, Plasma Waves and Sounder (PWS) experiment on-board the Planet-B Mars orbiter, *Earth Planets Space* **50** (1998) 213–222.
13. S. Oshigami, Y. Yamaguchi, A. Yamaji, T. Ono, A. Kumamoto, T. Kobayashi and H. Nakagawa, Distribution of the subsurface reflectors of the western nearside maria observed from Kaguya with Lunar Radar Sounder, *Geophys. Res. Lett.* in press, (2009) doi:10.1029/2009GL039835.
14. H. Oya and T. Ono, A new altimeter for Mars land shape observations utilizing the ionospheric sounder system on-board the Planet-B spacecraft, *Earth Planets Space* **50** (1998) 229–234.
15. G. Picardi, J. J. Plaut, D. Biccari, O. Bombaci, D. Calabrese, M. Cartacci, A. Cicchetti, S. M. Clifford, P. Edenhofer, W. M. Farrell, C. Federico, A. Frigeri, D. A. Gurnett, T. Hagfors, E. Heggy, A. Herique, R. L. Huff, A. B. Ivanov, W. T. K. Johnson, R. L. Jordan, D. L. Kirchner, W. Kofman, C. J. Leuschen, E. Nielsen, R. Orosei, E. Pettinelli, R. J. Phillips, D. Plettemeier, A. Safaeinili, R. Seu, E. R. Stofan, G. Vannaroni, T. R. Watters and E. Zampolini, Radar Soundings of the Subsurface of Mars, *Science*, **310** (2005) 5756, 1925–1928, doi:10.1126/science.1122165.
16. R. Seu, R. J. Phillips, G. Alberti, D. Biccari, F. Bonaventura, M. Bortone, D. Calabrese, B. A. Campbell, M. Cartacci, L. M. Carter, C. Catallo, A. Croce, R. Croci, M. Cutigni, A. D. Placido, S. Dinardo, C. Federico, E. Flamini, F. Fois, A. Frigeri, O. Fuga, E. Giacomoni, Y. Gim, M. Guelfi, J. W. Holt, W. Kofman, C. J. Leuschen, L. Marinangeli, P. Marras, A. Masdea, S. Mattei, R. Mecozzi, S. M. Milkovich, A. Morlupi, J. Mouginot,

R. Orosei, C. Papa, T. Paterno, P. P. del Marmo, E. Pettinelli, G. Pica, G. Picardi, J. J. Plaut, M. Provenziani, N. E. Putzig, F. Russo, A. Safaeinili, G. Salzillo, M. R. Santovito, S. E. Smrekar, B. Tattarletti and D. Vicari, Accumulation and erosion of Mars' south polar layered deposits, *Science* **317** (2007) 1715–1718, doi:10.1126/science.1144120.

Advances in Geosciences
Vol. 21: Solar & Terrestrial Science (2008)
Ed. Marc Duldig

MINIATURIZATION OF PLASMA WAVE RECEIVERS ONBOARD SCIENTIFIC SATELLITES AND APPLICATION TO THE SENSOR NETWORK SYSTEM FOR MONITORING ELECTROMAGNETIC ENVIRONMENTS IN SPACE

H. KOJIMA*, H. FUKUHARA and Y. MIZUOCHI
Research Institute for Sustainable Humanosphere,
Kyoto Univ., Kyoto 611–0011, Japan
**kojima@rish.kyoto-u.ac.jp*

S. YAGITANI
Graduate School of Natural Science & Technology,
Kanazawa Univ., Kanazawa 920–1192, Japan

H. IKEDA
Institute of Space and Astronautical Science, Japan
Aerospace Exploration Agency, Kanagawa 229–8510, Japan

Y. MIYAKE
Academic Center for Computing and Media Studies,
Kyoto Univ., Kyoto 606-8501, Japan

H. USUI
Graduate School of Engineering,
Kobe Univ., Hyogo 657-8501, Japan

H. IWAI
Faculty of Science and Engineering,
Doshisha Univ., Kyoto 610-0394, Japan

Y. TAKIZAWA
Faculty of Environmental and Urban Engineering,
Kansai Univ., Osaka 564-8680, Japan

Y. UEDA and H. YAMAKAWA
Research Institute for Sustainable Humanosphere,
Kyoto Univ., Kyoto 611-0011, Japan

Attempts have been made to achieve extreme miniaturization in plasma wave receivers for future space science missions. The plasma wave receivers onboard scientific satellites are classified into two types: waveform receivers and spectrum analyzers. The present paper describes one-chip waveform type receivers that are developed using analogue ASIC technology. We incorporate the necessary analogue components of the waveform receiver into a chip having dimensions of 3 mm × 3 mm and confirm the functionality and performance of the chip. The performance and quality of the chip are shown to be sufficient for use in science missions. Furthermore, we extend the application of the miniaturized plasma wave receivers to a sensor network system in space. The sensor network system consists of several small sensor nodes that are distributed randomly throughout the target area and monitor the spatial and time variation of electromagnetic environments in space. We then introduce the concept of this sensor network system and present results obtained during the development of the small sensor node for use in this sensor network in space.

1. Introduction

A plasma wave receiver onboard a scientific satellite is a sophisticated radio receiver that has the capability to observe electrostatic/electromagnetic waves excited in space plasmas with high sensitivity. Since space plasmas are essentially collisionless, their kinetic energies are exchanged and transferred via plasma waves. Therefore, the observation of plasma waves is essential to understanding the physical processes of phenomena that take place in space plasmas. This is why most space missions to investigate space plasma physics carry onboard plasma wave receivers.

Recently, the downsized satellites, such as satellites of the formation fights or small satellites, used for science missions require further reductions in the size and mass of the onboard instruments. The electronic circuits of plasma wave receivers are very complex, and the size and mass are larger than those of other instruments onboard scientific satellites. Therefore, the miniaturization of plasma wave receivers onboard scientific satellites is a crucial issue. A better way by which to realize the miniaturization of electronic circuits is digitization of the circuits. Digital technologies, such as the digital signal processor (DSP) and the field-programmable grid array (FPGA), enable part of analogue circuits of plasma wave receivers to be replaced by digital processing devices. Digital technologies have been adopted extensively in recent plasma wave receivers onboard sounding rockets and scientific satellites.[1,2] Hashimoto *et al.*[1] and Kasahara *et al.*[2] developed a digital spectrum analyzer based on the digital down-conversion method. Digital down-conversion is realized by a one-chip commercial device called the programmable down-converter (PDC). The sampled waveforms in the wide frequency range are digitized by an A/D converter and the data are input to the PDC. The PDC converts the digitized waveforms to low-frequency waveforms with narrow frequency bands, and the fast Fourier transform (FFT) spectra are calculated from the output of the PDC via an onboard DSP. Since the center frequency of the down-conversion is controllable by the outer signal, a kind of sweep frequency spectrum analyzer can be built using the PDC. The conventional sweep frequency analyzer requires several analogue band pass filters and local oscillators with phase locked loops (PLLs), as well as common analogue circuits, such as differential amplifiers and main amplifiers. Since the digital sweep frequency analyzer does not require analogue band pass filters or local oscillators, miniaturization of plasma wave receivers is possible.

The above digital sweep frequency analyzer provides a good example to the advantage of digitization. However, even after the above digitization, the common analogue parts of plasma wave receivers remain. The area of the electrical board of the plasma wave receiver occupied by the analogue parts is larger than that occupied in the other onboard instruments, because the plasma wave receivers must receive very weak signals picked up by sensors and amplify them to the input level of the A/D converter in a wide frequency range, typically from a few Hz to a few tens of MHz. Therefore, a breakthrough in the miniaturization of plasma wave receivers cannot be achieved without extreme miniaturization of the analog circuits.

In the present paper, we describe an attempt to develop a very small analogue part for the plasma wave receiver using analogue application specific integrated circuit (ASIC) technology. The analog ASIC enables several channels of plasma wave receivers to be installed in a small chip having dimensions of a few millimeters $\times$ a few millimeters. However, the design of the circuit is completely different from the design using discrete electronic parts. We newly designed a waveform receiver and implemented several channels of analogue components on a small chip. The results reveal the analogue ASIC is promising for use in the miniaturization of plasma wave receivers.

Recent breakthroughs in the miniaturization of plasma waves will enable the plasma wave receiver to be extended to other applications. We herein propose a sensor network system in space based on the use of miniaturized plasma wave receivers. The sensor network consists of several small sensor nodes that are distributed throughout the target area. This system simultaneously monitors plasma wave activities at multiple points.

Each small sensor node carries a very small plasma wave receiver with small sensors. The plasma wave receiver for the sensor node is developed based on analog ASIC technology. The present paper also describes the sensor network system in space as an application of small plasma wave receivers.

2. Plasma Wave Receiver

Plasma wave receivers onboard scientific satellites are mainly classified into two types: spectrum analyzers and waveform receivers. The spectrum analyzer provides the continuous spectrum of waves without the phase information and has the advantage of obtaining a spectrum with good frequency and time resolution continuously. On the other hand, the waveform receiver directly samples waveforms picked up by sensors. Although the observation of waveforms must be intermittent, due to the limitations of the telemetry capacity, the data contain the phase information of the waves. The significance of waveform observations was demonstrated by the discovery of electrostatic solitary waves as the waveform of the broadband electrostatic noise (BEN).[5] Waveform observations by the Geotail spacecraft revealed that, unlike the classical theory, which considers the BEN to be noise, the BEN is actually a series of isolated pulses.

The frequency coverage of plasma wave receivers is typically below 100 kHz for waveform receivers and below a few tens of MHz for spectrum receivers. Plasma wave receivers observe electric field and magnetic field components of plasma waves independently. As the sensors, we use long wire boom dipole antennas to measure the electric field components and search coil or loop-type antennas for magnetic field components. Figure 1 shows a block diagram of the typical plasma wave receivers, which are proposed for the scale coupling in plasma environment (SCOPE) mission.[4]

The high-frequency analyzer (HFA) is a spectrum receiver that operates up to a few tens of MHz based on the digital down conversion technique.[1,2] On the other hand, the waveform capture (WFC) instrument is a waveform receiver. The onboard frequency analyzer(OFA) is also shown in Fig. 1. This receiver is a kind of software receiver that provides FFT spectrum data calculated by the onboard processor from the waveform data observed by the WFC. Recently, low-frequency spectrum data have been generated from the waveform data using the onboard digital processing unit. Therefore, the analogue circuit of the OFA is common with the WFC. The electrical

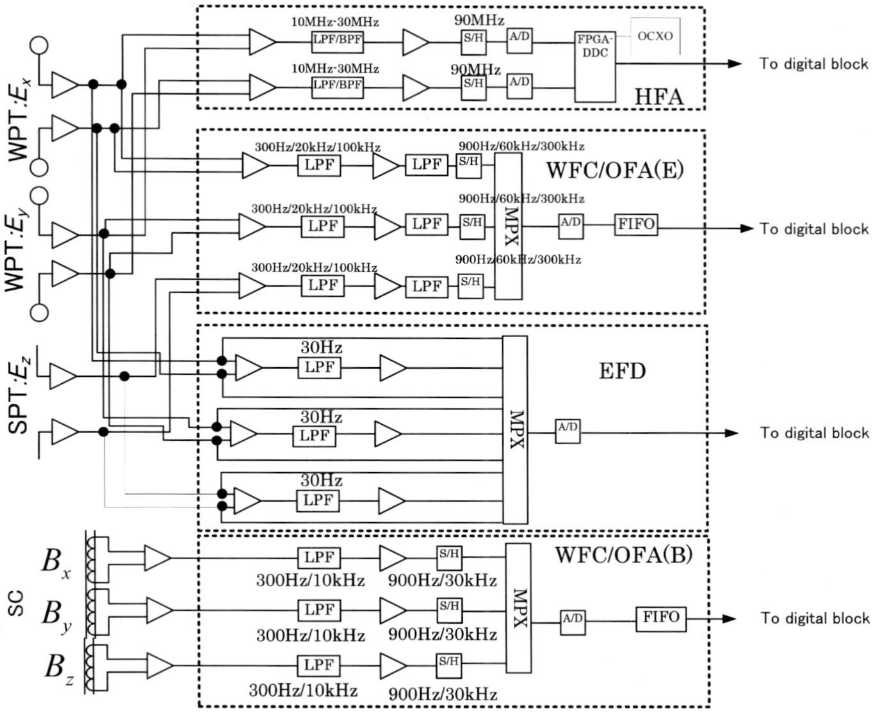

Fig. 1. Typical block diagram of the plasma wave receiver. The block diagram shown in this figure is the plasma wave instrument (PWI) proposed for the Japanese formation flight mission referred to as SCOPE.[4] The PWI consists of four main receivers: HFA, WFC for electric field components, WFC for magnetic field components, and EFD. The output of these receivers is connected to the digital circuits consisting of FPGA and mission data processor (MDP).

field detector (EFD) in Fig. 1 is the receiver for the DC and very slow perturbation of the electric field. Note that the EFD is beyond the scope of the present paper.

In the WFC block, signals picked up by sensors are fed to the first differential amplifier through each preamplifier. The observed signals are limited in the frequency band using the low pass filter (LPF) and are amplified by the main amplifier. The 2nd LPF is the anti-aliasing filter, which requires a very sharp reduction feature beyond the cutoff frequency. The sufficiently amplified signals are digitized by the A/D. The block diagram shown in Fig. 1 is the only the analogue part of the system. After digitization, the waveforms are processed by the onboard processor. We can obtain the frequency spectrum by FFT calculation.

In general, the most effective method for miniaturizing the electronic circuits is to use a digital technique. However, this is not always true in the case of the plasma wave receiver onboard satellites. The plasma wave receiver must detect very weak plasma wave signals with a good signal to noise (S/N) ratio.

This means that a low-noise amplifier with sufficient gain and filters for limiting the pass band is required in order to realize a good S/N ratio. Only analogue circuits can realize this feature in plasma wave receivers. Furthermore, to obtain the absolute intensities and vector orientation of plasma wave fields, a complete set of six receivers is necessary, i.e. three receivers for electric field components (E_x, E_y, and E_z) and three receivers for magnetic field components (B_x, B_y, and B_z). This also leads to the enlargement of the analogue circuits. Thus, in order to develop extremely small plasma wave receivers, it is necessary to miniaturize the analogue part of the plasma wave receiver.

3. Miniaturization of the Waveform Receiver

Figure 2 shows the simplified analogue part of the waveform receiver for one channel. We attempt to implement the analogue parts shown in Fig. 2 inside a silicon chip having dimensions of 3 mm × 3 mm. The design and layout of the circuits were conducted at Kyoto university, and the process itself was developed by the TSMC (Taiwan Semiconductor Manufacturing Company). In the present study, we use the TSMC 0.25 μm mixed-signal process.

The functions implemented inside the chip are: (i) a differential amplifier, (ii) a Gm-C type low pass filter (LPF), (iii) a main amplifier, and (iv) an anti-aliasing LPF. The target specifications of the developed chip are listed in Table 1. One significant difference in the waveform receiver implemented inside the ASIC, as compared to the typical waveform receiver, is that we cannot use inductance or capacitance with large values. Moreover, the resistance cannot be expected to have sufficient accuracy inside the

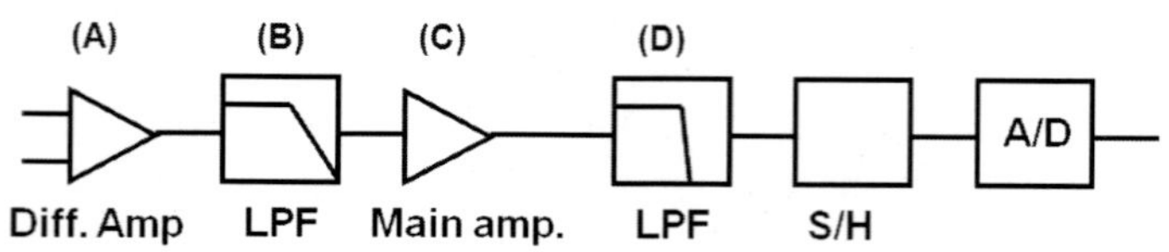

Fig. 2. Simple block diagram of the waveform receiver. (A) Differential amplifier, (B) 1st low pass filter, (C) main amplifier, and (D) 2nd low pass filter. The analogue circuits implemented in the ASIC do not include sample & hold (S/H) circuits or an A/D converter.

Table 1. Specifications of the designed one-chip waveform receiver.

Item	Target
Frequency range	<100 kHz
Components	3: electric field (E_x, E_y, E_z)
	3: magnetic field (B_x, B_y, B_z)
Dynamic range	>50 dB
Noise level at the input	<400 nV/$\sqrt{\text{Hz}}$ at 50 kHz
Crosstalk	< −50 dB

chip. These characteristics of the ASIC affect the design of filters, which are essential in plasma wave receivers. Therefore, the design of custom filters is necessary in the present study.

The role of the first low pass filter is to limit the frequency bandwidth to the target frequency range. Although the first low pass filter is not required to have the steep reduction feature above the cutoff frequency, the noise level should be low enough to receive the output signal of the differential amplifier. On the other hand, the second low pass filter should reduce the aliasing effects caused by the sampling by the A/D converter. Therefore, the second low pass filter is required to have steep reduction characteristics above the cutoff frequency. However, the noise level requirement not as strict, because the signals are already sufficiently amplified by the main amplifier to be input to the A/D. Based on the above conditions for each low pass filter, we use the Gm-C filter and switched-capacitor filter as the first and second LPFs, respectively. While the Gm-C filter has the advantage in its noise level, the switched-capacitor filter has one in the realization of the higher order filter.

Figure 3 shows the schematic of the typical 1st order Gm-C low pass filter we designed for the small waveform receiver. The Gm-C filter is constructed by the combination of the OTAs (Operation Transconductance Amplifier) and capacitances with small values (We do not use the inductance in the design). The OTA is a kind of operational amplifier that converts the voltage signals to current signals. It is a fundamental component in the analog ASIC design.[6]

The transfer function $H(s)$ and cutoff frequency f_c of the first-order Gm-C filter shown in Fig. 3 are as follows:

$$H(s) = -\frac{G_m}{sC + G_m} \tag{1}$$

$$f_c = \frac{1}{2\pi}\frac{G_m}{C}, \tag{2}$$

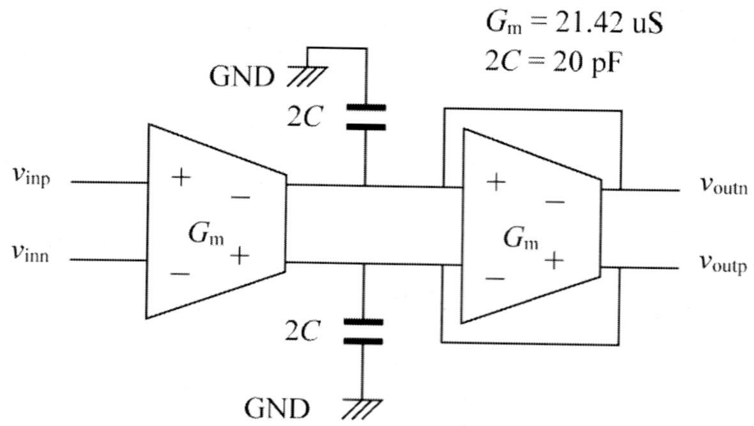

Fig. 3. Block diagram of the fully differential Gm-C low pass filter. The Gm-C filter is specialized for the use inside the analogue ASIC. The OTA is a kind of operational amplifier that converts voltage signals to current signals. The OTA is a fundamental element in the design of the analogue ASIC.

where G_m denotes the transconductance of the OTA. In the current Gm-C filter, f_c is designed to be 340 kHz by selecting $G_m = 21.42\,\mu\text{S}$ and $C = 10\,\text{pF}$.

Figure 4 shows an example of the layout inside the developed chip. In this version, which is shown in Fig. 4, various test components are implemented inside the chip having dimensions of 3 mm × 3 mm. The top area indicated by the white box designated by the red arrow corresponds to one channel of the waveform receiver, and contains the Gm-C LPF, the differential main amplifier with 20 dB gain, and the sixth-order switched capacitor filter (Chebyshev-type filter) with a cutoff frequency of 100 kHz. Figure 5 shows the total performance of the channel of the waveform receiver shown in Fig. 4. The solid lines, circles, and dotted line represent the measured data of the manufactured chip, the results of computer simulations by the electronic circuit simulator, and the theoretical values, respectively. The cutoff frequency is approximately equal to 100 kHz, and the gain of the passband is approximately 20 dB. The measured data agrees well with the simulation results, except in the frequency range above 500 kHz. This disagreement above 500 kHz arises from the noise level of the network analyzer that was used for the measurement.

The reduction feature above the cutoff frequency shows agreement with the present design. The reduction level at 200 kHz is greater than 60 dB, which means that if the sampling frequency of the waveform receiver is

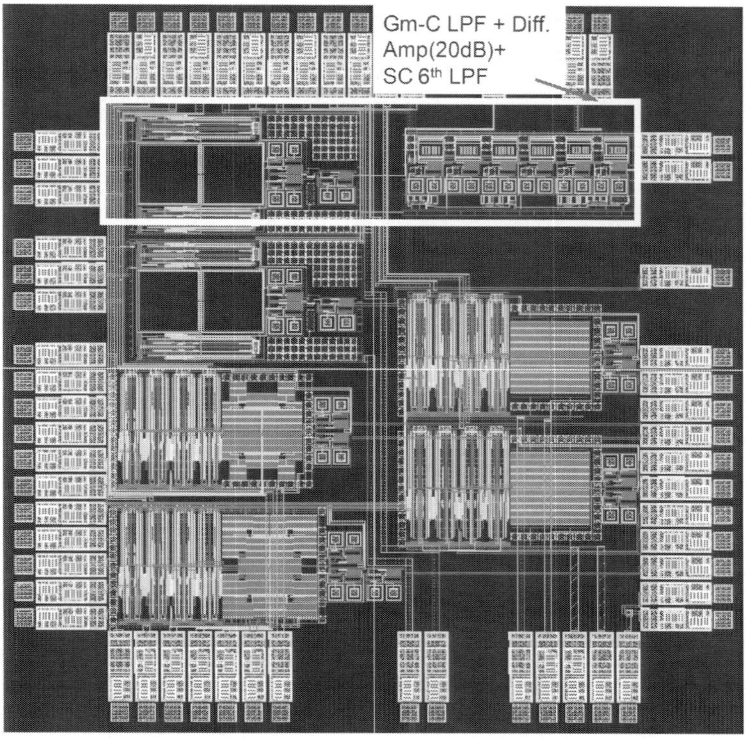

Fig. 4. Example of the layout inside the chip developed in the present study. This version includes various test components, as well as the waveform receiver component surrounded by the square.

400 kHz (two times oversampling), the dynamic range of the receiver can be expected to be 60 dB, which is equivalent to the 10-bit A/D converter. When we need greater dynamic range, the order of the switched capacitor filter should be increased or the sampling frequency of the A/D should be higher.

One important measure of performance in the plasma wave receiver is the crosstalk level. For instance, if the receiver does not have enough crosstalk reduction between an electric field channel and a magnetic field channel, pure electrostatic waves are mistaken as the electromagnetic waves. This leads to an incorrect physical understanding. The crosstalk itself takes place mainly due to the electromagnetic coupling between different channels through the floating capacitance. This gives cause for concern with respect to the possibility of increasing the crosstalk level

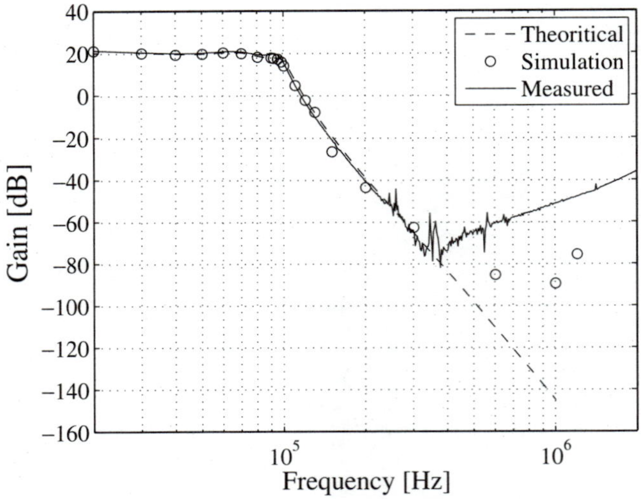

Fig. 5. Frequency response of the waveform receiver implemented inside the analogue ASIC. The upper cutoff frequency is 100 kHz, and the gain in the frequency pass band is 20 dB.

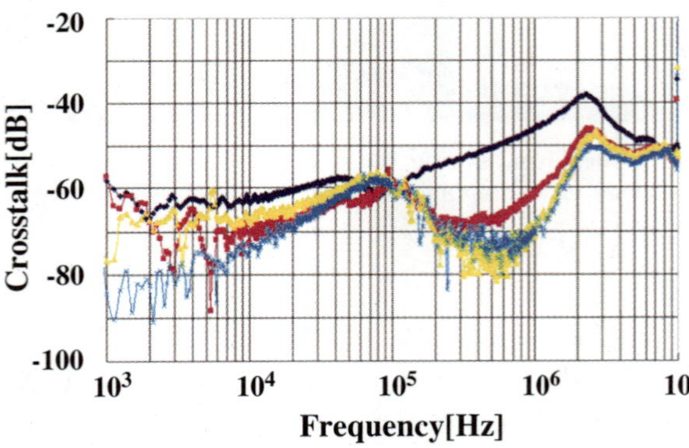

Fig. 6. Crosstalk level between different channels implemented inside the same ASIC. Since the different channels are located in close proximity to each other, the crosstalk between different channels should be considered. The measurement results show that the crosstalk level is maintained at greater than 50 dB in the frequency pass band.

of different channels implemented inside the analogue ASIC, because the relative distance of each channel is very small inside the chip. In order to reduce the crosstalk, we designed the layout inside the chip so that the ground line encloses each channel of the waveform receiver. Figure 6 shows the results of the crosstalk measurement in the present chip, in which several channels of the simple waveform receiver (main amplifier and switched capacitor LPF) are installed. In this simple waveform receiver, the upper cutoff frequency of the LPF is 100 kHz. Therefore, we focus on the frequency range below 100 kHz in Fig. 6. The different curves in Fig. 6 represent the crosstalk levels in the different combinations of the different channels of the implemented waveform receivers. The results show that the crosstalk level in the frequency range below 100 kHz is less than −55 dB. In the typical plasma wave receivers developed herein, the crosstalk level is less than approximately −60 dB to −50 dB. Therefore, the crosstalk level of the present small waveform receiver inside the analogue ASIC is of comparable order to that in the typical plasma wave receiver.

Another matter of concern with respect to the waveform receiver realized inside the analogue ASIC is its noise level. The intrinsic noise level of the MOS FET is generally higher than the junction type FET. Since the circuits inside the analogue ASIC are realized by various MOS FETs, the noise level at the input of the receiver should be considered in the design. In order to confirm the noise level of the present chip, we tested the version of the chip shown in Fig. 4, and found that its noise level at the input of the waveform receiver is approximately $200\,\mathrm{nV}/\sqrt{\mathrm{Hz}}$ at 50 kHz. Since the noise level of the preamplifier is approximately $400\,\mathrm{nV}/\sqrt{\mathrm{Hz}}$ at 50 kHz, the noise level of the waveform receiver is comparable to that of the output noise level of preamplifiers. This means that the noise level of the waveform receiver in the present study is small enough to detect weak signals having signal levels that are approximately equal to the output noise level of the preamplifier.

As stated above, the present attempt to develop a small waveform receiver using analogue ASIC technology was successful in the first step for the implementation of several channels. We found that the features such as the frequency response, crosstalk, and noise level are sufficient to observe plasma waves in space.

The exclusive area for one channel of the waveform receiver inside the chip of the current version is about equal to $1\,\mathrm{mm}^2$ in the current design. We need to add minor components to it in order to complete one channel waveform receiver. Furthermore, the I/O buffers and pads for the protection to electrostatic discharge (ESD) are also necessary. Therefore, the size $3\,\mathrm{mm} \times 3\,\mathrm{mm}$ of the bare chip is not enough to contain six channels of waveform receivers. However, even using larger size of the bare chip, the size will not exceed $5\,\mathrm{mm} \times 5\,\mathrm{mm}$.

4. Application to the Sensor Network System in Space

Since the limitation of single-point observations in space is easily understood, missions such as the Cluster or THEMIS satellites, which involve multiple satellites, have become the trend in recent scientific missions. However, even if multiple satellites observations can be expected, the number of the observation points is limited, because the number of the satellites is limited by cost and the mass budget of the launch vehicles, for example. Such multiple satellite missions are not particularly well suited to the observation of inhomogeneous phenomena localized in relatively small regions. One example of inhomogeneous phenomena in space plasmas is the interaction between space plasmas and artificial structures. Gurnett *et al.*[7] reported the existence of localized plasma wave turbulence and its inhomogeneity caused by the interaction between space plasmas and the space shuttle. The spatial distribution of such plasma wave turbulence is not easily identified by single-point observations or multiple-point observations from a limited number of satellites. However, knowledge about the turbulence around the artificial structures is crucial from a scientific point of view, as well as from the viewpoint of space development.

Based on the progress in the miniaturization of plasma wave receivers described in Sec. 3, we propose a new system for monitoring the electromagnetic environment in space. This system provides one solution to resolve the above problem concerning the observation of spatial inhomogeneous phenomena in space. We refer to this system as the Monitor System for space Electromagnetic Environments (MSEE). The MSEE is a sensor network system in space. Figure 7 shows a schematic illustration of the MSEE, which consists of palm-sized sensor nodes that are randomly distributed throughout the target area. Each sensor node carries a compact plasma wave receiver as well as other necessary components, such as communications and digital processing units. The observed data are transferred to the central station, such as a space station or a satellite/rocket, through the ad-hoc network system. The objective of the MSEE is to observe plasma wave activities in multiple points. The targets of the MSEE are the artificial disturbances caused by human activities in space as well as natural plasma waves.

Figure 8 shows a schematic illustration of the MSEE sensor node. The body is a cubic structure with sides of 7 cm. The body contains all of the necessary electrical components, such as a digital processing unit and a telemetry communication system, as well as a compact plasma wave

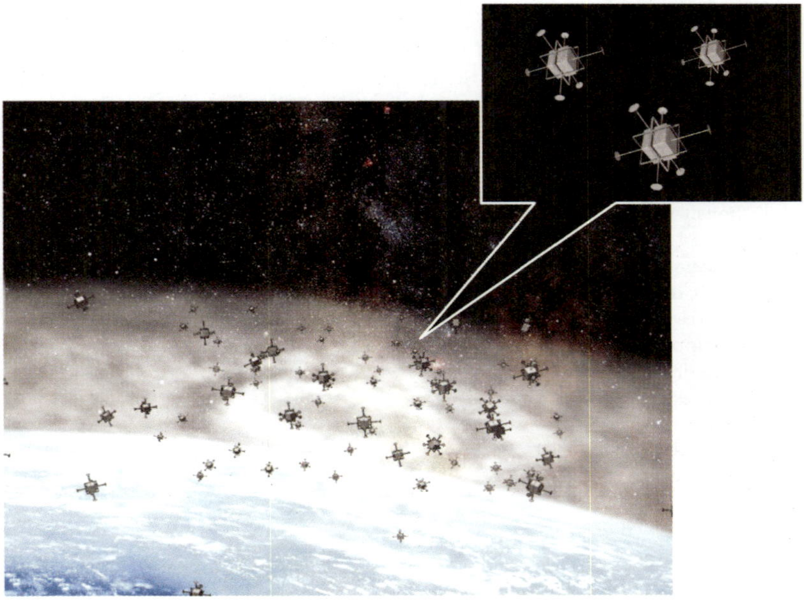

Fig. 7. Schematic illustration of the MSEE sensor network system. The system consists of several small sensor nodes. The sensor nodes are randomly distributed in the target area and simultaneously monitor plasma waves at multiple points.

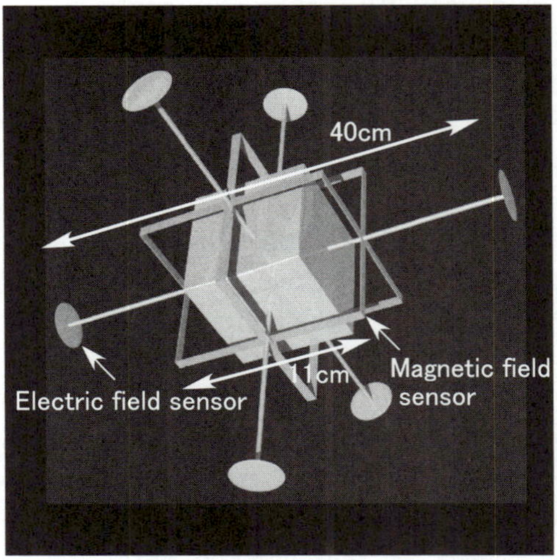

Fig. 8. Schematic illustration of the MSEE sensor node. All necessary electronics are installed inside the body of the sensor, which is in the form of a cube having sides of 7 cm. Tri-axes electric field sensors and tri-axes magnetic field sensors are attached outside the body.

Table 2. Specifications of the MSEE sensor node.

Sensors	
Dipole antenna (for the electric field)	
Length (radius)	40 cm (0.25 cm)
Sensitivity	$\sim 1\,\mu V/m/\sqrt{Hz}$ at 100 kHz
Loop antenna (for the magnetic field)	
Size	10 cm × 10 cm (15 turns)
Sensitivity	$\sim 0.1\,pT/\sqrt{Hz}$ at 100 kHz
Receiver	
Frequency	<100 kHz
Output data	Spectrum
Time resolution	a few seconds
Power supply	Dry battery (4.8 V)
Attitude detection	Magnetometer* and photo sensor*
Location estimation	GPS*
Life time	a few hours
Digital processor	PIC of Microchip Technology Inc.
Communication	Wireless LAN*

*Design is ongoing.

receiver. The body is made of aluminum plates, the surfaces of which are covered by permalloy. The plates suppress the noise radiating from the inner electronic circuits. The sensors of the electric field and the magnetic field are installed outside the body. While the electric field sensors are top-hat type dipole antennas with a tip-to-tip length of 0.4 m, the magnetic sensors are square loop antennas with 15 turns with edges of 10 cm. The sensors observe full sets of electromagnetic components (three for electric and three for magnetic field components). The target specifications of the MSEE sensor node are shown in Table 2. The concept of the MSEE sensor network is completely different from that of scientific satellites. The design of each sensor node is identical and the production costs are much cheaper than those of satellites. On the other hand, the lifetime of each sensor node after the distribution is approximately a few hours, and the nodes are completely disposal. However, we can repeatedly distribute the nodes in the target area. This system is similar to the radiosonde, which is used for monitoring meteorological phenomena in the earth's atmosphere.

Figure 9 shows a block diagram of the MSEE sensor node in the present design. The core part of the sensor node is the one-chip plasma wave receiver, which is realized by the analogue ASIC. We have not fixed the design of the ASIC for the MSEE sensor node. However, the prototypes of the necessary components such as filters and amplifiers have already been manufactured and tested. The signals picked up by six sensors are

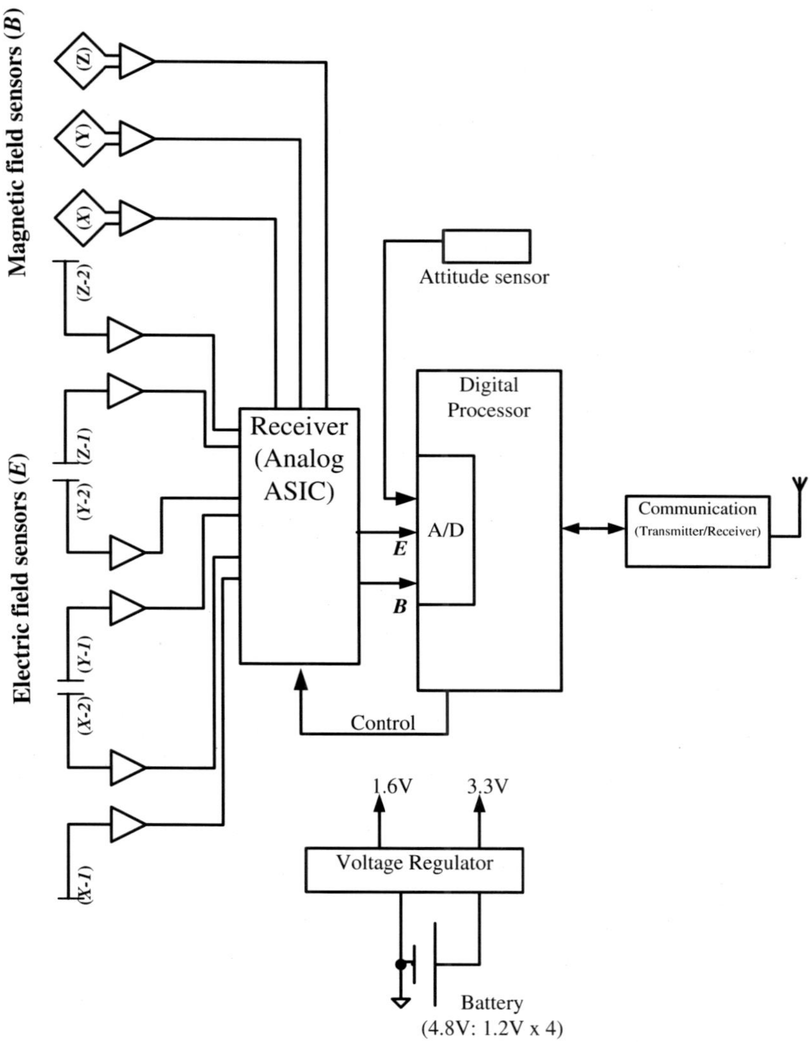

Fig. 9. Block diagram of the MSEE sensor node in the current design. The plasma wave receiver as the core is implemented inside the analogue ASIC. The power is supplied by a dry battery.

fed into the one-chip plasma wave receiver through their preamplifiers. The preamplifiers are required to have high input impedance and to have very low noise at the input. Therefore, they are not incorporated onto the analogue ASIC.

We designed the preamplifiers specifically for the MSEE sensor node. They should be compact and low-noise amplifiers with sufficient gain to compensate for the disadvantage of the sensitivities of compact electromagnetic sensors. Figure 10 shows the measured equivalent input noise of the preamplifier of the electric field sensor. The lowest noise level is approximately equal to $10\,\mathrm{nV}/\sqrt{\mathrm{Hz}}$ at 500 kHz. The electronic circuit is very simplified so that it can be installed inside the small body. However, its performance is sufficient for use in scientific missions.

The digital part of the MSEE sensor node performs any necessary digital processing, such as the frequency analysis, in addition to the control of the MSEE sensor node system. The design of the digital part is under way, and we use the PIC device of Microchip Technology Inc. Since the system of the MSEE sensor node does not require a powerful processor, the PIC with several channels of inner A/D converters is appropriate for this system.

Figure 11 shows a prototype MSEE sensor node. We installed three components of electric field sensors and three components of magnetic field sensors with their preamplifiers, as well as the power supply system. The

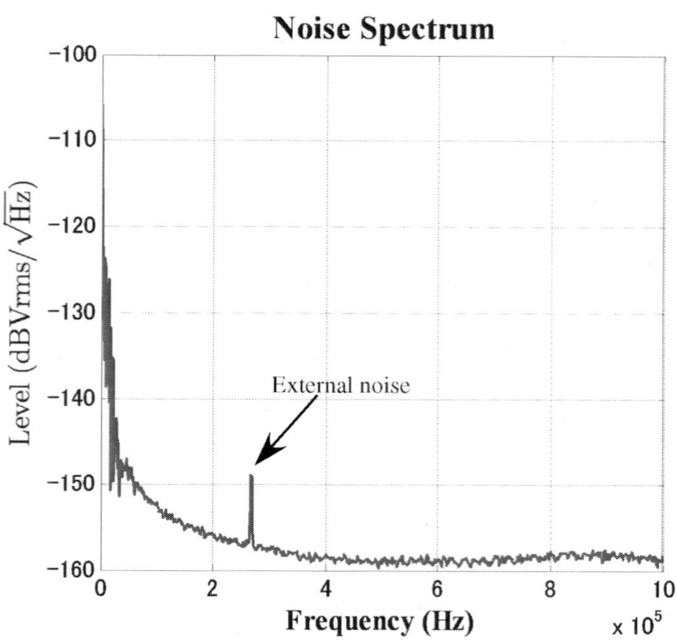

Fig. 10. Equivalent input noise of the preamplifier for the electric field component.

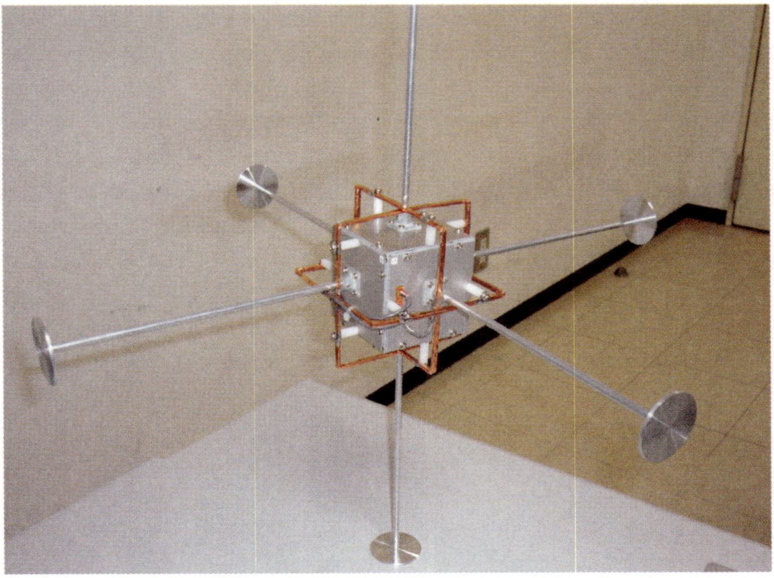

Fig. 11. Second prototype of the MSEE sensor node. The electric field and magnetic field sensors, along with their preamplifiers and a power supply system, are installed in this model.

equivalent input noise shown in Fig. 10 is measured in this prototype sensor node.

Whereas the length of the MSEE electric field sensor is 0.4 m tip-to-tip, the length of electric field sensors mounted on the Geotail spacecraft launched in 1992 is 100 m tip-to-tip.[5] The disadvantage of the MSEE electric field sensor, as compared to the electric field sensors used in scientific satellites, is its short antenna length. Therefore, it is important to evaluate the sensitivity of the MSEE electric field sensors.

In general, the antenna impedance of the short dipole antenna is capacitive in a vacuum. The capacitance of the dipole antenna is derived as follows:[8,9]

$$C_a = \frac{l}{120c \ln \dfrac{2l}{a}}, \tag{3}$$

where l, c, and a denote the length of one antenna element, the speed of light, and the radius of the antenna element, respectively. When we use $l = 0.2\,\mathrm{m}$ and $a = 0.0025\,\mathrm{mm}$, the antenna capacitance of the MSEE electric field sensor is calculated to be approximately 1 pF. Since the small disk is installed at the top of the MSEE electric field sensor, its capacitance is added to the capacitance of the dipole antenna. For reference, the antenna capacitance of the Geotail electric field sensor is 113 pF.[9] On the other hand, concerning the input capacitance at the input of the preamplifier, the MSEE has an advantage over scientific satellites, because the small input circuit leads to a small input capacitance. By measuring the prototype shown in Fig. 11, the input capacitance of the electric field preamplifiers was found to be approximately 5 pF. (The input capacitance of the Geotail electric field sensor preamplifier is approximately 40 pF.)

The antenna impedance, the input circuits of the preamplifier, and the input capacitance determine the transfer function of $G_p(f)$ as the pick up factor:

$$G_p(f) = \frac{V_i(f)}{V_a(f)} \tag{4}$$

where $V_i(f)$ and $V_a(f)$ denote the signal level at the input of the preamplifier and the signal level induced in the electric field sensor, respectively.

When we apply the above pickup factor to the following equation, we can obtain the sensitivity of the electric field sensors, as follows:

$$E_s(f) = \frac{N_i(f)}{|G_p(f)|} \cdot \frac{1}{l_{eff}}. \tag{5}$$

where $N_i(f)$ is the equivalent input noise level shown in Fig. 10, and l_{eff} is the antenna effective length, which is 0.2 m in the MSEE electric field sensor.

The antenna impedance in a vacuum described above is more complex in plasmas. In addition to the capacitance, the resistive effect occurs, depending on plasma densities and temperatures. Therefore, $G_p(f)$ varied depending on the target region in space plasmas. Figure 12 shows the summary of the sensitivities of electric field sensors in representative science missions (GEOTAIL and BepiColombo MMO) as well as the sensitivity of the MSEE compared with the typical intensities of plasma waves observed in the terrestrial magnetosphere. In Fig. 12, the antenna impedance of the MSEE is assumed to be that in the ionosphere. The sensitivity of the MSEE electric field sensor is worse than that of the scientific missions because of the shorter antenna length. Nevertheless, the MSEE electric field sensor has sufficient sensitivity to detect natural plasma wave phenomena. Since the plasma wave turbulence around the artificial structure, as shown by

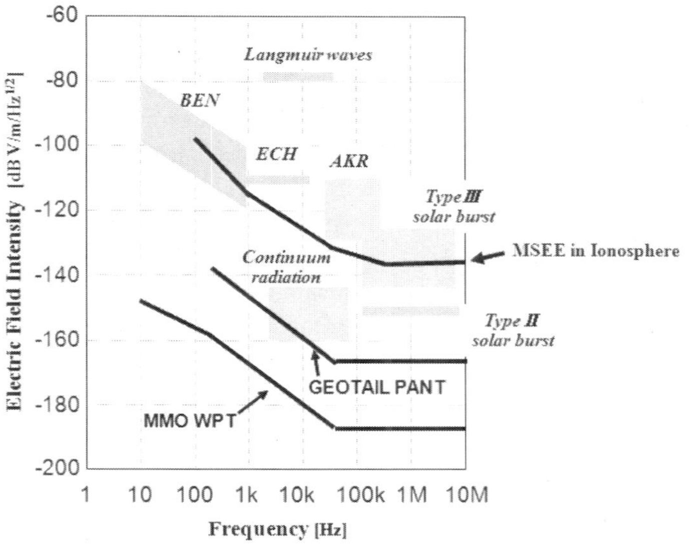

Fig. 12. Summary of the sensitivities of electric field sensors (MSEE, GEOTAIL and BepiColombo MMO) compared with the intensities of typical plasma waves observed in the terrestrial magnetosphere.

Gurnett *et al.*,[7] is expected to be more intense than that around natural phenomena, the MSEE electric field sensors are effective for observing both natural and artificial phenomena. The sensitivity of the MSEE magnetic field sensor is not discussed in the present paper. However, we have already confirmed that the MSEE magnetic field sensor has sufficient sensitivity to observe natural plasma waves in space.

The design of the MSEE sensor node has not yet been fixed. In particular, the location identification of each sensor node is very important. The sensor nodes are released and distributed randomly throughout the target area. The system is needed for identifying the location of each sensor node. The most convenient method of identifying the location of each sensor node is to use the GPS. There exist commercial small GPS receiver chips. Although such small GPS receivers can be used, the GPS is not always available in space. For instance, if the MSEE sensor nodes are distributed at a higher altitude than the GPS satellites, then the MSEE sensor node cannot identify its own location using the GPS.

The self-organizing map location (SOM) estimation[10] provides a method of location identification without using the GPS. The main feature of this method is to require only distance measurement among the neighborhood sensor nodes. The difference between the tentative location of a node and the measured distance is iterated in each sensor node. This means that the present SOM algorithm method does not require significant CPU resources for the location estimation of each sensor node. Since the SOM estimation was proposed for ad-hoc networks on the ground, it is basically two-dimensional. However, we have already succeeded in extending it to the three-dimensional algorithm and we have examined the efficiency of the three-dimensional SOM algorithm using computer simulations. The results indicate that the three-dimensional SOM is promising for the location estimation of each sensor node. However, an important issue that remains to be addressed is the development of a small distance measurement device. Therefore, the location estimation will depend on the GPS system until the SOM system can be completed.

5. Summary

The present paper describes our attempt to miniaturize plasma wave receivers for use on scientific satellites using analogue ASIC technology. Furthermore, we proposed a sensor network system for monitoring the status of the electromagnetic environment in space as an application of the small plasma wave receiver.

The miniaturized plasma wave receiver has the potential to expand its use in space science as well as space development. We demonstrated the feasibility of extreme miniaturization of plasma wave receivers using the analogue ASIC. The development of the analogue ASIC requires a great deal of skill and more time than the development of the digital chip, e.g. the FPGA. However, the miniaturization of analogue circuits is essential to achieving a breakthrough. We will continue our attempts to miniaturize the plasma wave receiver and develop a sensor network system for use in space. Our progress will be described in future publications.

Acknowledgments

The authors would like to thank Mr. Y. Sato of SE system company and Mr. H. Takano of Meiwa system company for their efforts in manufacturing the MSEE sensor node. We would also like to thank Ms. C. Noshi for providing the schematic illustration of the MSEE. This work is supported by VLSI Design and Education Center (VDEC), the University of Tokyo in collaboration with Cadence Design Systems, Inc. The development of the analog ASIC was supported by SiliConsortium Ltd. The present study was supported in part by a Grant-in-Aid for Scientific Research (A) 19204048.

References

1. K. Hashimoto, H. Iwai, Y. Ueda, H. Kojima and H. Matsumoto, Software wave receiver for the SS-520-2 rocket experiment, *IEEE Trans. On Geoscience and Remotesensing* **41** (2003) 2638–2647.
2. Y. Kasahara, Y. Goto and K. Hashimoto, Plasma wave observation using waveform capture in the Lunar Radar Sounder on board the SELENE spacecraft, *Earth Plasma Space* **60** (2008) 341–351.
3. H. Matsumoto, I. Nagano, R. R. Anderson, H. Kojima, K. Hashimoto, M. Tsutsui, T. Okada, I. Kimura, Y. Omura and M. Okada, Plasma Wave Observations with GEOTAIL Spacecraft, *J. Geomag. Geoelectr.* **46** (1994) 59–95.
4. SCOPE working group, Proposal of the SCOPE mission, 449, Japan Aerospace Exploration Agency, 2008 (in Japanese).
5. H. Matsumoto, H. Kojima, T. Miyatake, Y. Omura, M. Okada, I. Nagano and M. Tsutsui, Electrostatic Solitary Waves (ESW) in the Magnetotail:BEN Wave forms observed by GEOTAIL, *Geophys. Res. Lett.* **21** (1994) 2915–2918.
6. D. A. Johns and K. Martin, Analog integrated circuit design, John Wiley & Sons, Inc. (1997).
7. D. A. Gurnett, W. S. Kurth, J. T. Steinberg and S. D. Shawhan, Plasma wave turbulence around the shuttle: Results from the spacelab-2 flight, *Geophys. Res. Lett.* **15** (1988) 760–763.
8. R. S. Elliott, Antenna theory and design, 301–302, Prentice Hall, Englewood Cliffs, N. J. (1981).
9. M. Tsutsui, I. Nagano, H. Kojima, K. Hashimoto, H. Matsumoto, S. Yagitani and T. Okada, Measurements and analysis of antenna impedance aboard the Geotail spacecraft, *Radio Science* **32** (1997) 1101–1126.
10. Y. Takizawa, P. Davis, M. Kawai, H. Iwai, A, Yamaguchi and S. Obana, Self-organizing location estimation method using received signal strength, *IEICE Trans. Commun.* **E89-B** (2006) 2687–2695.